STRESS PROTEINS
IN BIOLOGY AND
MEDICINE

COLD SPRING HARBOR MONOGRAPH SERIES

The Lactose Operon
The Bacteriophage Lambda
The Molecular Biology of Tumour Viruses
Ribosomes
RNA Phages
RNA Polymerase
The Operon
The Single-Stranded DNA Phages
Transfer RNA:
- Structure, Properties, and Recognition
- Biological Aspects

Molecular Biology of Tumor Viruses, Second Edition:
- DNA Tumor Viruses
- RNA Tumor Viruses

The Molecular Biology of the Yeast Saccharomyces:
- Life Cycle and Inheritance
- Metabolism and Gene Expression

Mitochondrial Genes
Lambda II
Nucleases
Gene Function in Prokaryotes
Microbial Development
The Nematode *Caenorhabditis elegans*
Oncogenes and the Molecular Origins of Cancer
Stress Proteins in Biology and Medicine

STRESS PROTEINS IN BIOLOGY AND MEDICINE

Edited by

Richard I. Morimoto
Northwestern University

Alfred Tissières
University of Geneva

Costa Georgopoulos
University of Utah School of Medicine

COLD SPRING HARBOR LABORATORY PRESS
1990

STRESS PROTEINS IN BIOLOGY AND MEDICINE

Monograph 19

Printed in the United States of America
Book design by Emily Harste

Library of Congress Cataloging-in-Publication Data

Stress proteins in biology and medicine / edited by
Richard I. Morimoto, Alfred Tissières, Costa Georgopoulos.
p. cm. — (Cold Spring Harbor monograph series; 19)
Includes bibliographical references.
ISBN 0-87969-337-1
1. Heat shock proteins. I. Morimoto, Richard I., 1952–
II. Tissières, Alfred. III. Georgopoulos, Costa. IV. Cold Spring Harbor
Laboratory Press. V. Series.
QP552.H43S78 1990 89-23926
616.98--dc20 CIP

All Cold Spring Harbor Laboratory Press publications may be ordered directly from Cold Spring Harbor Laboratory, Box 100, Cold Spring Harbor, New York 11724. (Phone: 1-800-843-4388.) In New York State (516) 367-8325.

Contents

Preface

The heat shock response was originally described as a phenomenon of inducible gene expression in *Drosophila* and has rapidly become a widely studied adaptive response to a diverse array of environmental conditions that cause physiological stress. Initial studies on the heat shock proteins concentrated almost entirely on the cloning and characterization of *Drosophila* heat shock gene expression. The convergence of studies utilizing higher eukaryotes and prokaryotes revealed that the heat shock or stress response was ubiquitous and, furthermore, that the genes encoding these highly conserved heat shock proteins were closely related to the level of nucleotide sequences. More recently, the field of stress protein research has grown and diversified. Studies on stress proteins now encompass questions in molecular biology of gene expression on the complexity of signals that control the transcriptional regulation of stress genes, in cell biology on the role of stress proteins as chaperones that regulate aspects of protein folding and transport, in immunobiology and infectious diseases on the involvement of stress proteins in the immune response during tissue damage and infectious disease, and in pathophysiology and medicine on the role of the stress response in human disease. The role of the stress response in human disease is unclear, yet there is every expectation that aberrant expression of stress proteins during diseases including hypertrophy, fever, inflammation, infection with pathogens, ischemic recovery, tissue trauma, hyperthermia, and certain forms of neoplasia will have useful predictive and diagnostic value. This monograph represents a selective representation of some themes in the field of heat shock and stress-induced proteins in biology and medicine.

We are grateful to the staff of Cold Spring Harbor Laboratory Press: Nancy Ford, Managing Director of Publications; Patricia Barker, technical editor; and Mary Cozza, editorial assistant.

R.I. Morimoto

A. Tissières

C. Georgopoulos

STRESS PROTEINS IN BIOLOGY AND MEDICINE

1

The Stress Response, Function of the Proteins, and Perspectives

Richard I. Morimoto

Department of Biochemistry, Molecular Biology and Cell Biology
Northwestern University
Evanston, Illinois 60208

Alfred Tissières

Department of Molecular Biology
University of Geneva
1211 Geneva 4, Switzerland

Costa Georgopoulos

Department of Cellular, Viral and Molecular Biology
University of Utah School of Medicine
Salt Lake City, Utah 84132

This introductory chapter is divided into three sections. In the first section, we introduce the nonspecialist to the biology of the stress response through a brief history of the early studies on the stress or heat shock

Stress Proteins in Biology and Medicine.

response. In the second section, we highlight aspects of the stress response, including the latest developments in the field, with emphasis on the function of the stress proteins. Finally, in the third section, an attempt is made to place the stress response in the perspective of human disease. Throughout this book, the terms heat shock and the stress response will be used interchangeably.

I. INTRODUCTORY REMARKS

A. The Heat Shock or Stress Response

Living organisms respond at the cellular level to unfavorable conditions such as heat shock, or other stressful situations of many different origins, by the rapid, vigorous, and transient acceleration in the rate of expression of a small number of specific genes (heat shock genes). Consequently, the products of these genes (commonly referred to as heat shock proteins [hsps] or the stress proteins), which are also present under normal conditions but in lesser amounts, increase and accumulate in cells to reach, in some instances, fairly large concentrations. The activation of heat shock gene expression by stress takes place in all cells of organisms with only a few exceptions, such as during early embryonic development (Dura 1981; Wittig et al. 1983; Banerji et al. 1984, 1987). A great deal of circumstantial evidence supports the belief that the function of these proteins is to protect cells from the ill effects of stress, allowing them to recover and survive (see Chapter 4).

Besides the activation of heat shock genes, the expression of most other genes is inhibited as a result of stress. Thus, heat shock or stress leads to a perturbation of normal gene expression, which, if prolonged, can have drastic consequences. For example, in cultured *Drosophila* cells, the overall rate of transcription of cells in culture, and specifically the rate of transcription of the ribosomal genes, is sharply reduced by a heat shock at 37°C, which is 12°C above the temperature of cultivation (Arrigo 1980; J. Vazquez, pers. comm.). Gene expression is also altered in other ways, such as by RNA processing (Yost and Lindquist 1986), mRNA stability and translation (Lindquist 1980), and transcriptional termination (J. Vazquez, pers. comm.; see Chapter 16). Moreover, other cellular events are perturbed. Mitosis and DNA synthesis are arrested (Kramhoft and Zeuthen 1971; Zeuthen 1971; A.-P. Arrigo, pers. comm.), and several morphological changes take place. In nuclei, filaments containing actin are affected: The nucleoli, including the granular ribonucleoprotein components, are transiently damaged, and the assembly and export of ribosomes from these organelles are blocked (Pelham 1984; Welch and Suhan 1985). The Golgi complex is disrupted and vimentin-containing filaments around the nucleus collapse and aggregate (Welch and Suhan 1985).

B. The Inducers of the Response

Besides heat shock, the induction of the stress response can be achieved by the release from anoxia, addition of 2,4-dinitrophenol (an uncoupler of oxidative phosphorylation), or addition of sodium azide (a poison of cellular respiration). Ritossa (1962, 1964), who made these early observations working with *Drosophila*, asked whether the respiratory chain might be the target of the stress response. Although a direct connection with this system cannot so far be excluded, crucial evidence is still lacking. Other inducers of the response include sodium arsenite, ethanol, sulfhydryl reagents, hydrogen peroxide, transition series metals, amino acid analogs, and even viral infection (for reviews, see Nover 1984; Craig 1985; Neidhardt et al. 1985; Lindquist and Craig 1988; Chapters 10 and 14). The effectiveness of some of these agents often varies, depending on the organism. Evidence exists that abnormal proteins in *Escherichia coli* as well as in eukaryotes can trigger the activation of the heat shock genes (Goff and Goldberg 1985; Ananthan et al. 1986; Parsell and Sauer 1989). In parallel, an increase in the rate of protein degradation has been noted as a result of stress (Parag et al. 1987), although the correlation with increased proteolysis appears to be dependent on the specific form of stress (Westwood and Steinhardt 1989). Whether a common target of the response exists, and if so, what its nature is, remain so far open questions.

C. The Heat Shock Proteins

The major heat shock proteins have been classified into four protein families (Table 1): the large-molecular-weight heat shock proteins of 83–90 kD; the hsp70 family, ranging from ~66 to 78 kD; the hsp60 family, present in bacteria, mitochondria, and chloroplasts, and which have been termed "chaperonins" (Hemmingsen et al. 1988); and the small heat shock proteins, a diverse group of proteins from ~15 to >30 kD, with a single form found in yeast and human cells and approximately 30 different forms found in higher plants. In addition, some larger heat shock proteins of ~100–110 kD with properties that differ from those of the other families have been characterized in mammalian cells (Subjeck et al. 1983; Shyy et al. 1986; Chapter 10).

It has been clear for some years that, in the absence of stress, the heat shock proteins are present at a lower level in cells, suggesting that they

Table 1 Conserved stress proteins with known function

Family	Alternate designations		Function	Reference
	eukaryotes	prokaryotes		
hsp90	hsp83, 87	HtpG(C62.5)	Bind to specific polypeptides, and either silence their function (e.g., glucocorticoid receptor), and/or escort them to their proper cellular compartment (e.g., $pp60^{src}$)	This chapter; Chapters 9 and 10
hsp70(DnaK)	hsp68, 72, 73; clathrin uncoating ATPase; BiP; grp75, 78, 80; hsc70; KAR2; SSA1, 2, 3, 4; SSB1, 2; SSC1; SSD1	DnaK	Dissociate some protein aggregates; maintain some polypeptides in unfolded state, thus facilitating their translocation across membranes, and/or accelerating proper folding and oligomerization; bind to specific polypeptides (e.g., p53)	This chapter; Chapters 8–14
hsp60(GroEL)	hsp58; RuBisCO-subunit binding protein; MIF-4	GroEL(MopB)	Maintain some polypeptides in unfolded state, thus facilitating their translocation across membranes, and/or accelerating proper folding and assembly (by preventing misfolding?)	This chapter; Chapters 8–10

DnaJ	SEC63 (NPL)[a], SIS1[b] (K. Arndt, pers. comm.)	DnaJ	Binds to specific polypepties (e.g., RepA protein of phage P1; S. Wickner, pers. comm.); accelerates the release of some of the polypeptides bound to DnaK (?); interacts with DnaK and GrpE	This chapter; Chapters 8 and 9
GrpE	antigenically related protein is present (D. Ang, pers. comm.)	GrpE	Interacts with DnaK and DnaJ; accelerates the release of some of the polypeptides bound to DnaK (?)	This chapter; Chapters 8 and 9
Small hsps[c]	hsp17.5, 22, 23 26; GENE 3	HtrC(?)[d]	Eukaryotic function is largely unknown; homology to α-crystallin family; HtrC may be necessary for septum formation in prokaryotes (?)[d]	Arrigo et al. (1988); de Jong et al. (1988); Lindquist and Craig (1988)
Ubiquitin	UBI1, 2, 3, 4	none	Conjugates to and "tags" unfolded polypeptides destined to be destroyed	Hershko (1988); Lindquist and Craig (1988)

The references for this table can be found in this introductory chapter or in the appropriate chapters or references indicated.

[a]The homology to DnaJ is confined only to a 70-amino-acid stretch (Sadler et al. 1990); region that mediates interaction with hsp70 proteins (?).

[b]The homology to DnaJ is extensive and is found almost throughout the protein (K. Arndt, pers. comm.). SIS1 was isolated as multicopy, extragenic suppressor of mutations in SIT4, a gene encoding a putative cell-cycle-dependent phosphatase (Arndt et al. 1989).

[c]A diverse family of proteins of 15–30 kD, with tendency to aggregate.

[d]The *htrC* gene is under σ^{32}-transcriptional regulation. The HtrC protein is needed for growth only at high temperatures and shares homology with the small hsps at its carboxyl end (S. Raina and C. Georgopoulos, unpubl.).

are needed at all times to serve basic, indispensable cellular functions (Chapters 8–14). In the case of the hsp70 gene family, the expression of which is almost negligible in *Drosophila* cells without stress, homolog genes, called cognates, exist. These are expressed constitutively (Palter et al. 1986) and appear to have functions similar to hsp70 genes. Other heat shock genes are regulated in the course of development (Chapter 15) or the growth state of the cell (Chapter 14).

For many years, the search for the functions of the heat shock proteins seemed fruitless, and interesting hypotheses were difficult to test. This situation changed recently when it was realized that hsp60, hsp70, and hsp90 are all involved in subtle interactions with other proteins to perform particular tasks. These recent developments are discussed in the second section of the Introduction and in several chapters in this volume. It is conceivable that the small heat shock proteins are also engaged in similar protein-protein interactions, although functional evidence is still lacking. Curiously, the small heat shock proteins are related to α-crystallin (Ingolia and Craig 1982) and share with this polypeptide the property of interacting with each other to form large aggregates (Arrigo et al. 1988).

D. The Heat Shock or Stress Response Is Universal

The heat shock or stress response has been observed in all organisms examined from archaebacteria to eubacteria, yeasts, plants, invertebrates, and vertebrates, including humans. Although there are differences among the various organisms, for instance in the molecular weights and the number of heat shock proteins, it is remarkable how similar the response has remained in very distant species throughout evolution. Comparison of the sequences of the respective heat shock genes from *E. coli*, plants, yeasts, *Drosophila*, and man, to name only a few, and the use of antibodies to test similarities of structure between heat shock proteins from a number of distant species, have indicated that the heat shock proteins are among the most highly conserved proteins in nature (Bardwell and Craig 1984, 1987; Hunt and Morimoto 1985; Chapters 9, 10, 13, and 14).

E. Thermotolerance and Phenocopy Induction

When organisms are exposed to sufficiently severe heat shock or stress conditions, the majority die. If, however, prior to this lethal heat shock they undergo a mild heat treatment, a considerable proportion of them survive. For example, most or all third-instar *Drosophila* larvae die when exposed to a temperature of approximately 40°C for 20 minutes. However, if the larvae are pretreated at 33–35°C for 30 minutes, and then exposed to the lethal temperature, as many as 50% of the animals survive and develop normally (Mitchell et al. 1979). Observations of this type have been made with a wide variety of organisms and cells in culture, including whole mice, *Drosophila* at all stages of development (except early embryos which are very sensitive to heat), soybean seedlings, slime molds, yeast, and bacterial cells. Experiments with *Drosophila* have shown that the optimum temperature for the pretreatment (33–35°C) was such that heat shock proteins were induced while overall rates of RNA and protein synthesis were only slightly affected (Mitchell et al. 1979). The pretreatment with heat can be replaced by other inducers of heat shock protein synthesis, such as ethanol or arsenite, with similar effects. Since thermotolerance induction appears to closely correlate with heat shock protein synthesis, particularly with hsp70 (Li and Laszlo 1985), it is likely that the function of at least one heat shock protein is implicated in the production of thermotolerance (Chapter 4).

Phenocopies are nonheritable, stress-induced abnormalities of development that closely resemble developmental defects in homeotic mutants. They were first described in *Drosophila* by Goldschmidt (1935) and were extensively studied in that organism ever since. Specific phenocopies result from stress applied at specific stages of embryonic development or metamorphosis. For instance, Gloor (1947) described bithorax phenocopies resulting from treatments of 4-hour-old embryos with heat shock or ether, a study that was later extended by Capdevila and Garcia-Bellido (1974). Many different chemicals and toxic agents have been used to induce phenocopy formation. The phenomenon is not limited to insects and has been the object of extensive studies with mammalian embryos (Pleet et al. 1981; Webster et al. 1985). As a mild heat pretreatment protects the organisms from death caused by a more severe heat shock immediately thereafter, so does a mild heat pretreatment afford specific protection from phenocopy induction when applied immediately preceding the more severe heat shock treatment. The effective heat pretreatment is such that the synthesis of heat shock proteins is stimulated without drastically altering normal protein synthesis. Heat shock proteins resulting from heat pretreatment are therefore implicated in the protection from phenocopy induction.

F. Initial Studies of the Heat Shock Response

Here we describe briefly the initial part of the work and the chronological developments. We will touch upon the more recent developments later in this article.

The earliest work on the effect of heat on gene expression in *Drosophila* was probably that of Goldschmidt (1935), who obtained developmental defects he called phenocopies because they resembled mutant defects. This work was later continued by others, e.g., Gloor (1947), who used both heat and ether; Hadorn (1955); Milkman (1966); and Mitchell (1966). The results can be interpreted as the effect of stress on the expression of specific genes at crucial times during development, leading to phenocopy formation (Mitchell and Peterson 1982).

Ritossa (1962, 1963, 1964), a geneticist then working in Naples, reported that when *Drosophila (D. busckii, D. melanogaster)* larvae raised at 25°C, were exposed at 30–32°C for about 30 minutes, several new puffs appeared on the giant salivary gland chromosomes. Following a 30-minute heat shock, the new puffs disappeared after about 1 hour. If the larvae were maintained at 30–32°C, the new puffs regressed after 3 hours. It was also reported that, in the case of prolonged heat shock treatment, the puffs active prior to the heat shock treatment regressed and, in some experiments, disappeared altogether. These early observations had demonstrated, first, the transient induction of the heat shock puffs and second, the inhibition of the puffs already present at the permissive temperature, before heat shock. In addition, it was noted that the heat-shock-induced puffing pattern was observed in salivary gland chromosomes during different stages of larval development, and in the midgut and the hindgut as well, indicating that it was not limited to salivary glands and hence was not tissue-specific. The heat shock puffs could be induced by treating excised salivary glands at the higher temperature. The puffing pattern seen after temperature elevation could also be induced by treating the isolated salivary glands with 2,4-dinitrophenol, sodium salicylate, sodium azide, or dicumarol, or by release from anoxia. Experiments with radioactive precursors showed that the heat-induced puffs were the sites of intense RNA transcription of active, induced genes. Thus, several important features of the heat shock response were already apparent from Ritossa's experiments.

In the late 1960s, Berendes et al. (1965) with *D. hydei* and Ashburner (1970) with *D. melanogaster* made a thorough study of the heat-shock-induced puffs in these two strains. They tested a number of different inducers and found that the heat shock puff induction was very rapid, occurring within minutes after temperature increase.

Could the presumed protein products, directed from mRNAs made at heat shock puff sites from the induced genes, be seen on protein polyacrylamide gels? This was the question Mitchell and Tissières asked at Caltech in 1973. The results were striking: Heat shock led to the appearance of approximately seven new protein species, whereas many proteins, made constitutively before the stress was applied, regressed and, in some experiments, disappeared altogether. This was reminiscent of the appearance of seven heat shock puffs on the polytene chromosomes (Ashburner [1970] described nine heat shock puffs in *D. melanogaster*) and the parallel disappearance of most puffs active before heat shock. These results were obtained with all the different tissues examined, including salivary glands, brain, Malpighian tubes, and wing imaginal discs at several stages during development (Tissières et al. 1974).

From which heat shock puff does the mRNA for a particular protein band originate? The heat shock messages had to be characterized, and the answer would come from the newly devised technique of in situ hybridization of RNA to the polytene chromosomes (Gall and Pardue 1971). Two reports appeared in 1975 making use of this technique and *Drosophila* tissue culture cells, which are convenient for RNA work and high specific-activity labeling. Lindquist-McKenzie et al. (1975) found that temperature elevation caused the rapid disappearance of preexisting polysomes, followed by the buildup of new larger polysomes. The newly made RNA in these polysomes consisted mainly of poly(A) RNA of a size sufficient to code for hsp70. This RNA hybridized in situ to chromosomal DNA at the sites of the two largest heat shock puffs, band sites 87A and 87C. Spradling et al. (1975) used a different strategy: They hybridized in situ to polytene chromosomes poly(A) RNA extracted from control cells kept at 25°C. The labeled control cell RNA hybridized to about 50 chromosomal bands, whereas poly(A) RNA from heat-shocked cells hybridized to seven sites not seen with mRNA from control cells. These seven sites corresponded with those of the heat shock puffs.

Further work showed that purified fractions of heat shock mRNA hybridized in situ to specific heat shock puffs. Moreover, when these purified heat shock mRNA fractions were added to in vitro systems for protein synthesis, they were translated into specific heat shock proteins (Lindquist-McKenzie and Meselson 1977; Spradling et al. 1977; Mirault et al. 1978). Taken together, these experiments clearly indicated that heat shock mRNAs, made at heat shock puff sites, were translated into heat shock proteins.

For many years heat shock research was solely done with *Drosophila*, although there was the question of how universal the phenomenon would turn out to be. At the end of the 1970s, however, a typical heat shock response was reported in chicken embryo fibroblasts (Kelley and Schlesinger 1978) and in *E. coli* (Lemaux et al. 1978; Yamamori et al. 1978). Similar observations were made with yeast (McAlister and Finkelstein 1980), plants (Barnett et al. 1980), and other organisms, soon providing evidence that the response would turn out to be universal.

Taking advantage of the partially purified mRNA preparations for heat shock proteins, the heat shock genes from *Drosophila* were among the first eukaryotic genes to be cloned and to have their organization studied in detail (Livak et al. 1978; Schedl et al. 1978; Craig et al. 1979; Moran et al. 1979; Craig and McCarthy 1981; Voellmy et al. 1981). The nucleotide sequences of these genes soon became available (Holmgren et al. 1981; Karch et al. 1981), demonstrating in particular the presence of homologous regions with dyad symmetry upstream of the heat shock genes, which might serve for their heat-induced activation (Chapter 17).

II. LATEST DEVELOPMENTS AND INTERPRETATIONS

A. Heat Shock Factor(s)

Although the heat shock response is universal, its quality and magnitude vary widely in nature. In general, eukaryotes possess at least two copies of most heat shock genes, one under heat shock regulation and one under constitutive control, e.g., yeast possesses at least nine genes of the hsp70 family (Chapter 13); a similar number of hsp70 genes are present in the human genome (Chapter 14). However, prokaryotes possess only a single copy of each heat shock gene. This fact, coupled with the essential function of most heat shock genes, necessitates the constitutive expression, under all conditions, of most prokaryotic heat shock genes.

Prokaryotic control. The major heat shock genes of *E. coli* have been shown to be under the control of the *rpoH* (*htpR*) gene, which codes for the σ^{32} factor (Chapters 8 and 9). The expression of the heat shock genes in *E. coli* has been shown to correlate well with the intracellular σ^{32} levels (Straus et al. 1987). Interestingly, a shift to high temperature affects the level of σ^{32} in various ways, namely, *rpoH* mRNA is transiently stabilized, *rpoH* gene transcription is stimulated, and the half-life of σ^{32} is transiently increased (Straus et al. 1987). Because of the extremely short half-life of the σ^{32} polypeptide (as short as 45 sec; Straus et al. 1987; Tilly et al. 1989), the intracellular level of σ^{32} immediately reflects variations in the parameters that control σ^{32} production and stability (Chapter 8).

In a recent study, Zhou et al. (1988) have shown that the *rpoH* (σ^{32}) gene of *E. coli* can be deleted as long as the bacteria are maintained at temperatures below 20°C. In the absence of σ^{32}, none of the three *dnaK* gene promoters are utilized, and very little, if any, DNAK antigen is present (Zhou et al. 1988). Other heat shock genes normally under σ^{32}-regulation, such as *groE* and *grpE*, are expressed, in the absence of σ^{32}, from $E\sigma^{70}$-dependent promoters that overlap the $E\sigma^{32}$-dependent ones, and whose activity is normally silenced in wild-type bacteria (Zhou et al. 1988; D. Ang, pers. comm.). When Kusukawa and Yura (1988) examined the nature of extragenic suppressors that allow *rpoH*-deleted *E. coli* to grow to temperatures up to 39°C, they found that the suppressor mutations result in the overexpression of the *groE* operon. Additional suppressor mutations that allow bacterial growth at temperatures up to 42°C were shown to be due to overexpression of the *dnaK* operon. Thus, it appears that the GroE proteins play a major role in *E. coli* growth at all temperatures, supported by the fact that neither the *groES* nor the *groEL* gene can be deleted from *E. coli* (Fayet et al. 1989). The role of the GroE and DnaK proteins in *E. coli* physiology is discussed in a later section.

Recently, three interesting twists in *rpoH* gene expression have been found. The first is that *rpoH* is expressed from at least four different promoters (Erickson et al. 1987; Fujita and Ishihama 1987; Erickson and Gross 1989; Wang and Kaguni 1989a; Chapter 8). The second twist is that two of the promoters are negatively regulated by the DnaA protein (required for initiation of *E. coli* DNA replication as well as global regulation [Georgopoulos 1989]) (Wang and Kaguni 1989b), a third promoter is under CAP-cAMP catabolite control (Chapter 8), and a fourth is under the control of a newly discovered σ factor, σ^{24} (σ^{E}) (Erickson and Gross 1989; Wang and Kaguni 1989a). Although the gene coding for σ^{24} has not been identified yet, it is clear that σ^{24}-promoted transcription is accelerated at high temperatures, including 50°C. Thus, it appears that there are at least two heat shock systems in *E. coli*, the "classic" one regulated by σ^{32}, and the newly discovered one, under σ^{24} regulation. At 50°C, heat shock genes are transcribed exclusively by either $E\sigma^{32}$ or $E\sigma^{24}$, and the σ^{32} levels are continuously replenished by $E\sigma^{24}$-promoted transcription of the *rpoH* gene. A newly discovered heat shock gene, *htrA* (*degP*) (whose product has been shown to be essential for *E. coli* viability only at high temperatures and to be under heat shock regulation [Lipinska et al. 1988a,b; Strauch et al. 1989]), recently has been shown to be under the exclusive transcriptional control of $E\sigma^{24}$ (Erickson and Gross 1989). Thus, the heat shock response of prokaryotes appears to be more complex than previously thought, and the two heat shock regulons appear to be interconnected. The third twist is that *rpoH* mRNA translation appears to be under negative repression control as well (Chapter 8). Signals that affect the activity of the putative repressor element would result in altered intracellular levels of σ^{32}.

B. Heat Shock Element

1. Prokaryotes

As mentioned above, there are two heat shock regulation systems in *E. coli*, one under $E\sigma^{32}$ and the other under $E\sigma^{24}$ regulation. The two RNA polymerase holoenzymes recognize distinct promoter sequences (Chapter

8). One of the four *rpoH* (σ^{32}) gene promoters is under the control of Eσ^{24}, allowing continuous σ^{32} production at 50°C (Chapter 8).

2. *Eukaryotes*

All eukaryotic genes that are transcriptionally induced following exposure to heat shock contain a sequence motif located in the 5′-flanking region referred to as the heat shock element (HSE). The HSE corresponds to inverted multimers of the sequence NGAAN (Chapter 17) and has been shown to bind with the heat shock transcription factor (HSF) obtained from a number of different organisms (Chapter 18). Eukaryotic heat shock genes are also transcriptionally regulated during a number of non-stress conditions including serum stimulation, hormonal regulation, viral infection, and development (Chapters 13–15). In general, the multiple forms of transcriptional regulation are mediated by the presence of additional non-HSE promoter elements.

C. Signals Controlling the Heat Shock Response

1. *Prokaryotes*

In *E. coli*, the magnitude of the Eσ^{32}-promoted heat shock response appears to be proportional to the intracellular levels of the short-lived σ^{32} polypeptide. This "simple" relationship appears to hold at high temperatures (Straus et al. 1987). At low temperatures, additional parameters such as σ^{32} modification or conformational state may also play a role, since σ^{32} overproduction does not result in a strictly proportional expression of the heat shock genes (Straus et al. 1989). Very little is known about the regulation of the newly discovered σ^{24} factor, since the corresponding gene has not been cloned and large quantities of purified σ^{24} are not yet available (for details, see Chapter 8).

Transcription of the *rpoH* gene has been shown to be under temperature regulation (Tilly et al. 1986; Erickson et al. 1987; Erickson and Gross 1989). Hence, an increase in temperature, coupled with the fact that the *rpoH* mRNA is transiently stabilized at high temperature, allows the production of high levels of σ^{32}, leading to higher heat shock gene expression. Since σ^{32} is extremely unstable, any "signals" that influence either the level or expression of proteinase(s) degrading σ^{32} will directly affect the intracellular concentration of σ^{32}, and ultimately heat shock gene expression. In all likelihood, more than one proteinase system is active in σ^{32} degradation. The half-life of σ^{32} has been shown to be stabilized during bacteriophage λ infection. It has been shown that the phage λcIII gene product is by itself both necessary and sufficient to stabilize σ^{32}, leading to high levels of heat shock gene expression (Bahl et al. 1987). Although it is known that the λcIII protein interferes with the ability of the HflA and HflB proteinase systems to degrade the λcII protein, mutations in the *hflA* and *hflB* genes do not show substantially higher levels of heat shock gene expression. It could be that the λcIII protein acts either as a generalized proteinase inhibitor, or itself is a substrate for such proteinases. Its overexpression could "saturate" or "inactivate" the appropriate proteinase(s) leading to an increase in σ^{32} half-life, and hence increased hs gene expression.

Mutations in the *dnaK*, *dnaJ*, or *grpE* heat shock genes also result in overexpression of the heat shock response, demonstrating that these heat shock proteins themselves negatively modulate the heat shock response (Tilly et al. 1983, 1989; Sell 1987; D. Straus and C. Gross, unpubl.; Chapters 8 and 9). How do the heat shock proteins autoregulate their own synthesis? Since mutations in *dnaK*, *dnaJ*, or *grpE* result in lower overall proteinase levels in *E. coli* (Keller and Simon 1988; Straus et al. 1988), the stabilization of σ^{32} in these backgrounds could be simply due to this phenomenon. Alternatively, because these three heat shock proteins may be involved in binding and stabilizing "unfolded" polypeptides (see below), Straus et al. (1988) and Gross et al. (Chapter 8) have suggested that they aid proteinase action by simply maintaining polypeptides in an "unfolded" state, thus making them more accessible to proteinases. Georgopoulos et al. (Chapter 9) have proposed that the DnaK, DnaJ, and GrpE proteins help release σ^{32} from the $E\sigma^{32}$ holoenzyme, and that "free" σ^{32} is more susceptible to proteinase action. This proposal is based on the known ability of these three heat shock proteins to release the λP replication protein from the λ DNA initiation complex assembled at *ori*λ (see below). Regardless of the exact mechanism by which *dnaK*, *dnaJ*, and *grpE* modulate σ^{32} stability, the fact that they do so provides us with an excellent system for studying the autoregulation of the heat shock response. In eukaryotes, there is also evidence that the intracellular levels of the hsp70 protein negatively modulate heat shock gene expression (DiDomenico et al. 1982).

2. *Eukaryotes*

Proper levels of the various heat shock proteins are ensured through a multitude of mechanisms, including gene duplication, constitutive and inducible expression through the use of multiple *cis*-acting promoter elements, and modulation of heat shock protein levels by posttranslational effects such as message stability and translational control (Chapter 16).

Common to all heat shock genes, regardless of species, is transcriptional activation following heat shock and other forms of physiological stress. However, the mechanism by which this is accomplished differs among organisms. In yeasts, HSF is constitutively bound in vivo to the HSE, whereas in *Drosophila*, HSF binding is inducible (Chapter 18). The transcriptional activity of the constitutively bound yeast HSF increases upon heat shock and is associated with extensive phosphorylation; the high-affinity binding requires trimerization (Sorger and Nelson 1989). In *Drosophila*, the HSF-HSE complex forms following heat shock, and the increase in transcription rate is thought to occur by the release of preinitiated and partially elongated transcripts bound to an engaged RNA polymerase II molecule (Gilmour and Lis 1986; Rougvie and Lis 1988). Human cells contain two HSE-specific DNA-binding activities; one found predominantly in extracts from unstressed human cells, the other in extracts from stressed cells (Kingston et al. 1987; Sorger et al. 1987; Mosser et al. 1988). The relationship of these two DNA-binding activities has not been firmly established; however, the level of the stress-induced activity increases during stress at which time the constitutive form disappears. Upon return to 37°C, the level of the stress-induced HSE-binding complex diminishes. In vitro studies on human HSF (Larson et al. 1988) have also provided support for phosphorylation in transcriptional activation.

An important question that remains to be solved is the mechanism by which the heat shock transcriptional apparatus detects heat shock or exposure to chemical inducers of the response. The simplest model suggests that HSF is the sensor or "thermometer." Alternatively, a number of molecules (proteins) could serve as sensors and induce a stress-dependent activation of HSF. A related question is the mechanism by which the DNA-binding activity is sequestered. This could be accomplished by maintaining HSF in a complex. One such molecule with which HSF could associate would be hsp70. Disruption of such an HSF:hsp70 complex, for example during heat shock, would release hsp70 to associate with unfolded domains of substrate proteins and HSF for binding to the HSE, thus resulting in heat shock gene transcriptional activation. In this model, the trigger for the response would be the appearance of denatured proteins that would compete with the HSF for association with hsp70. Such a model could also provide an explanation for the autoregulatory aspect of the heat shock response. As the level of stress-induced hsp70 increases for association with unfolded proteins, the HSE remains transcriptionally active. However, during recovery from stress, the pool of free hsp70 increases and results in the HSF:hsp70 complex and attenuation of the heat shock response. Similar models have been proposed for complexes composed of hsp90 and steroid receptors as a mechanism to maintain the DNA-binding domain in an inaccessible state (for review, see Beato 1989).

D. Function of the Heat Shock Proteins

Although the heat shock proteins appear to possess many diverse functions, the major heat shock proteins fall into four discrete groups. Not only have these proteins been highly conserved in nature at the amino acid sequence level, but they also appear to carry out similar functions, such as binding to other polypeptides and allowing them to either reach their correct intracellular destination in the cell, transporting other proteins across membranes, or preventing other proteins from misfolding. Ultimately, all three of these functions could be based on the ability of these proteins to recognize "unstructured" protein regions present in other polypeptides (for review, see Pelham 1989; Rothman 1989). In some instances, both the prokaryotic and eukaryotic analogs recognize and bind the same polypeptide with equally high avidity, e.g., the eukaryotic hsc70 and the prokaryotic DnaK homologs bind very tightly to the tumor-promoting protein p53 and release it only in the presence of hydrolyzable ATP (Clarke et al. 1988). Table 1 briefly summarizes the members of these conserved protein classes and their known properties. (For details, see Chapters 8–14.)

1. Prokaryotes

The identity and properties of the known $E\sigma^{32}$-dependent hs proteins of *E. coli* are summarized in part in Table 1 and in detail in Table 1 of Chapter 9 by Georgopoulos et al. It can be seen that these proteins possess diverse functions ranging from the major *E. coli* σ factor, σ^{70} (the *rpoD* gene product), to a cryptic form of lysyl-tRNA-synthetase (the *lysU* gene product). Here, we highlight those properties of the *E. coli* heat shock proteins that are shared by their eukaryotic counterparts.

a. DNAK, DNAJ, and GRPE. The *dnaK*, *dnaJ*, and *grpE* genes (see below) were originally discovered because mutations in any of them blocked phage λ DNA replication in vivo (for review, see Friedman et al. 1984; Chapter 9). Subsequent studies established that mutations in these genes exerted global, pleiotropic effects on host metabolism, including defects in DNA and RNA synthesis, proteolysis, cell division, and overproduction of heat shock proteins (for details, see Chapters 8 and 9).

The DnaK protein possesses an autophosphorylation activity and a weak 5′-nucleotidase activity, cleaving the 5′-phosphate groups from both ribose- and deoxyribose nucleotides (Zylicz et al. 1983; Bochner et al. 1986). Both genetic and biochemical evidence exists demonstrating protein-protein interactions between the DnaK protein and the λO and λP DNA replication proteins. In addition, DnaK has been shown to interact with the DnaJ and GrpE hs proteins (for review, see Chapter 9). The DnaK protein has also been shown to bind preferentially to the unfolded form of bovine pancreatic trypsin inhibitor as opposed to the native protein (Johnson 1988).

The exact role that the DnaK, DnaJ, and GrpE proteins play in λ DNA replication is examined in detail in Chapter 9, Zylicz et al. (1989), and Alfano and McMacken (1989). Briefly, these three heat shock proteins, through specific protein-protein interactions and with the aid of ATP hydrolysis, disassemble a very stable *ori*λ DNA-λO-λP-DnaB complex, thus liberating the DnaB helicase from its tight complex with λP and allowing it to perform its vital function in λ DNA replication. Thus, these three heat shock proteins appear to play the role of a "molecular crowbar" in λ DNA replication. Such a molecular role can be easily extended to normal *E. coli* physiology, leading to disassembly of improperly formed protein complexes and aggregates, essentially as first proposed by Hightower (1980) and Pelham (1986). Such improperly folded structures or aggregates may be especially abundant at high temperatures or under other conditions that can also lead to protein denaturation. This may be the reason why cells in which the *rpoH* (σ^{32}) gene has been deleted, and which possess very low levels of dnaK protein, can only grow at 42°C in the presence of excess DnaK protein (Kusukawa and Yura 1988; see Chapter 8).

Recent in vivo genetic experiments implicate the DnaK protein as playing a role in the transport of a *lamB-lacZ* mutant fusion protein across the *E. coli* inner membrane (G. Phillips and T. Silhavy, pers. comm.). This result suggests that in vivo DnaK protein may be capable of binding to and maintaining the nascent form of some polypeptides, thus allowing them to either be exported or folded properly. The apparently disparate abilities of the DnaK protein, namely, to both disassemble protein complexes and to bind to nascent polypeptide forms, are probably due to its ability to recognize and bind to some structural feature of "improperly folded" or "unstructured" polypeptides. A polypeptide bound to DnaK can be released by utilizing the energy of ATP hydrolysis (as in the case of the complex of DnaK with λP or GrpE) or be exchanged with the membrane transport apparatus (as in the case with the *lamB-lacZ* fusion). The ability of the *E. coli* DnaK protein to facilitate polypeptide transport across membranes also has been demonstrated in the heterologous yeast system with preproalpha factor (Waters et al.

1989), a function usually carried out by some members of the yeast hsc70 family (Chirico et al. 1988; Deshaies et al. 1988). As mentioned above, both the eukaryotic hsc70 and the prokaryotic DnaK proteins are also capable of recognizing and binding to the p53 eukaryotic protein. The ability of the eukaryotic hsc70 protein to disassemble clathrin triskelions (Braell et al. 1984) is reminiscent of DnaK's ability to disassemble the large protein complex assembled at *ori*λ. Thus, both the eukaryotic and prokaryotic hsp70 families of proteins appear to play similar roles in cell physiology.

The discovery of the existence of DnaJ and GrpE analogs in eukaryotes (see Table 1) argues also for a conservation in their function across evolution. We envision the following role for the molecular mode of action of the *E. coli* DnaJ and GrpE proteins. This model is based on the known properties of these heat shock proteins and the recent demonstration that these three proteins form an assayable complex; i.e., M. Zylicz's laboratory (pers. comm.) has recently shown that DnaK's weak ATPase activity is stimulated at least fivefold only in the presence of both the GrpE and DnaJ proteins. The mutant GrpE280 protein, unable to form a stable complex with DnaK (Ang 1988), does not functionally substitute for wild-type GrpE in this assay (M. Zylicz, pers. comm.). The proposed model is reminiscent of the "cogwheel" action of the GroES protein shown in Figure 1. It is proposed that DnaK (and the hsp70 family members) can bind to a variety of nascent polypeptide chains. Although the nature of the "polypeptide structure" recognized is not known, it can be both hydrophilic and hydrophobic in nature. In addition, the avidity of binding can unpredictably vary among various polypeptide substrates, as demonstrated in the assay for BiP protein (a member of the hsp70 family; Table 1) used by Flynn et al. (1989). Some of the polypeptides bound to DnaK (hsp70) protein may dissociate without ATP hydrolysis, especially if the binding is weak; others may require ATP hydrolysis for their release; still others may require the action of the GrpE/DnaJ-like proteins for their release (a scheme exactly analogous to the GroEL, GroES mode of action shown in Fig. 1; see below). The cogwheel release function of the GrpE/DnaJ-like proteins could either be due to an actual displacement of the bound polypeptide from DnaK (hsp70), or simply an indirect consequence of the observed increase in DnaK's ATPase activity in their presence. This model is not only economical, but also helps explain why mutations in the *dnaK*, *dnaJ*, or *grpE* genes exhibit an equivalent plethora of bacterial phenotypes.

In the eukaryotic system, there appears to be at least another protein factor needed for maximal hsp70 protein function, as demonstrated in the

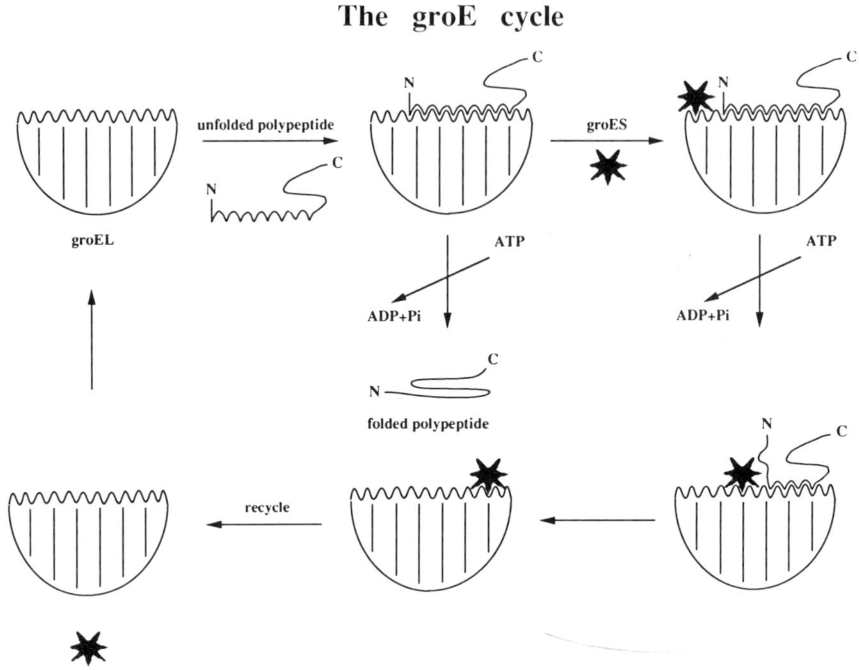

Figure 1 Hypothetical scheme depicting the mode of action of the GroEL (hsp60) and GroES heat shock proteins. The GroEL protein binds an "unfolded" polypeptide or a nascent polypeptide as it emerges from the ribosome. The bound polypeptide may or may not be released, depending primarily on the strength of binding. Some polypeptides may be "exchanged" with the cell's export machinery situated in a membrane. Other polypeptides may be bound tightly, so that energy in the form of ATP hydrolysis may be absolutely required for their release. The ATP is hydrolyzed by GroEL, and it may lead to a change in conformational state, resulting in the release of the bound polypeptide. Still other polypeptides may be so tightly bound to GroEL that the depicted, hypothetical cogwheel function of the GroES protein may be required in addition to ATP hydrolysis. An exactly analogous model for the mode of action of the DnaK (hsp70) protein can be proposed in binding and releasing "denatured," nascent polypeptides. The DnaJ and GrpE hs proteins, known to interact intimately with DnaK (see text), may play a role analogous to that depicted for GroES, i.e., they facilitate the release of some of the polypeptides avidly bound to DnaK (see text for details).

studies by Chirico et al. (1988) and Deshaies et al. (1988). Perhaps this NEM-sensitive factor(s) is coded by a gene homologous to either *grpE* and/or *dnaJ*.

b. GROE. The *groES* and *groEL* genes of *E. coli* were originally discovered because mutations in them resulted in a block to bacteriophage λ growth (for review, see Friedman et al. 1984; Chapter 9). Subsequently, it was shown that mutations in the *groE* genes blocked various phage assembly processes (either head or tail) and bacterial growth at high temperatures as well (see Chapter 9).

Both GroES and GroEL proteins possess a sevenfold symmetric structure, the GroES protein composed of seven 97-amino-acid monomers (Chandrasekhar et al. 1986; R. Hendrix and T. Zeigelhoffer, pers. comm.) and GroEL composed of 14 subunits of 548-amino-acid monomers (Hendrix et al. 1979; Hohn et al. 1979). The GroEL protein possesses a weak ATPase activity, modulated in vitro by the GroES protein (Chandrasekhar et al. 1986). A great deal of genetic and biochemical evidence exists demonstrating that the GroES and GroEL proteins interact both in vivo (Tilly and Georgopoulos 1982) and in vitro (Chandrasekhar et al. 1986). This result helps explain why mutations in either gene result in mostly equivalent phenotypes in bacterial and phage growth.

Figure 1 proposes a detailed molecular model that incorporates all known properties of the GroES and GroEL proteins (C. Georgopoulos and D. Ang, in prep.). The GroEL protein recognizes and binds to "unstructured" peptide regions exposed in nascent or misfolded proteins. Although the nature of the unstructured peptide features recognized by GroEL protein is not known, it will not be surprising if both hydrophilic and hydrophobic regions are recognized, as in the case with the hsp70 family of proteins mentioned before (Flynn et al. 1989). The bound polypeptide may or may not be spontaneously released, depending on the quality and strength of its interaction with GroEL. The model envisions that some of the stably bound polypeptides may be released following ATP hydrolysis, whereas others simultaneously need assistance by the GroES protein. The GroES protein is proposed to displace tightly bound polypeptides through the cogwheel action depicted in Figure 1. The released polypeptide may assume a different conformation, leading to intra- or even intermolecular folding. In the absence of GroES, the release of some of the GroEL-bound polypeptide may occur too slowly, or even not at all, essentially preventing the recycling of the GroEL protein. In this instance, mutations in either the *groES* or *groEL* genes will lead to the same phenotype, i.e., unavailability of active and free GroEL polypeptides. In the absence of sufficient wild-type GroEL levels or mutant GroEL protein, the nascent polypeptide chain may misfold, due to the presence of reactive peptide sequences, e.g., extremely hydrophilic or hydrophobic sequences, which lead to premature binding to other sections of the polypeptide.

What is the evidence in support of such a model? The first pioneering observation in this field was made by Bochkareva et al. (1988), who demonstrated that newly synthesized pre-β-lactamase and chloramphenicol transacetylase polypeptides transiently associate with the GroEL protein in crude extracts and that this interaction is disrupted by hydrolyzable ATP. The purified GroEL protein of *E. coli* has been shown to bind to and stabilize the unfolded form of denatured pro-ompA and pre-phoE polypeptides, thus facilitating their transport through the *E. coli* inner membrane (Lecker et al. 1989). One decatetrameric molecule of GroEL has been shown to bind to one molecule of pre-β-lactamase and to completely prevent it from folding in vitro. The presence of hydrolyzable ATP leads to the release of pre-β-lactamase, and the rate of release is slightly accelerated by the presence of GroES. Surprisingly, the GroEL protein can induce "unfolding" of pre-β-lactamase, folded previously in vitro, suggesting that GroEL protein may have limited unfoldase activity (A. Laminet et al., in prep.). Kusukawa et al. (1989) have recently presented data demonstrating that the GroES and GroEL proteins participate in pre-β-lactamase folding in vivo.

In a recent study, Van Dyk et al. (1989) demonstrated that overproduction of both the GroES and GroEL proteins of *E. coli* leads to the in vivo suppression of mutations in genes encoding diverse proteins. This work extends the prior demonstration by Fayet et al. (1986) and Jenkins et al. (1986) that overexpression of GroES and GroEL leads to the allele-specific suppression of mutations in the *dnaA* gene of *E. coli*. Thus, it appears that the GroES and GroEL pair of proteins is intimately involved with proper folding and/or oligomeric assembly of many proteins in *E. coli*. Overproduction of the GroE proteins also has been shown to facili tate the export of a *lamB-lacZ* mutant protein in vivo (G. Phillips and T. Silhavy, pers. comm.).

Recently, Goloubinoff et al. (1989) have succeeded in reconstituting active, dimeric, prokaryotic ribulose bisphos-phate carboxylase (RuBisCO) from unfolded polypeptides. The unfolded RuBisCO subunit was shown to bind to the GroEL decatetramer. The presence of both ATP and GroES was absolutely required for the release of the bound RuBisCO subunit in a folded form. Thus, the facilitated folding of pre-β-lactamase by GroEL follows the first part of the pathway shown in Figure 1 (i.e., the release of the bound polypeptide is mostly independent of GroES), whereas that of RuBisCO follows the second part of the pathway (i.e., an absolute requirement of GroES for the bound polypeptide). Goloubinoff et al. (1989) purified the GroEL homologs from plant chloroplasts (RuBisCO-subunit binding protein) and yeast mitochondria (hsp60) and asked whether they could also assist the assembly of purified prokaryotic RuBisCO. They showed that the chloroplast and mitochondrial eukaryotic homologs were 25% and 10% as effective,

wide variety of environmental agents including xenobiotics, aromatic hydrocarbons, and photochemical air pollutants. Free radicals are known to have a broad range of cellular targets including lipid oxidation, DNA damage and effects on amino acid modification, cross-linking and general effects on protein conformation and activity. Prokaryotic cells also have an adaptive response to the toxic side effects of molecular oxygen, for example, H_2O_2 induces a set of genes that overlaps with the family of heat shock proteins (Lee et al. 1983; Christman et al. 1985; Morgan et al. 1986).

One tissue that is likely to be exposed to a wide range of suboptimal physiological conditions is blood. Macrophages metabolize oxygen to generate highly reactive intermediates that cause cell damage; for example, H_2O_2 oxidizes hemoglobin to methemoglobin in erythrocytes. It may be relevant to this issue that the genes for the constitutive P72 and HSP70 are expressed during erythroid differentiation and the corresponding proteins accumulate to high levels in the mature erythrocyte (Morimoto and Fodor 1984; Banerji et al. 1987; Theodorakis et al. 1989). Many chemical inducers of the heat shock response, including hydrogen peroxide, arsenite, cadmium, and ethanol, are oxidizing agents. The deleterious effects of metals, either as oxidants or sulfhydryl reagents, may be important to diseases involving metal metabolism, such as hemochromacytosis, Menke's and Wilson's diseases that affect metal uptake and secretion.

Activation of the stress response through oxidant damage or by fever and inflammation may have negative effects on the growth stimulatory role of mononuclear cells. Monocytes are unusually sensitive to the effects of elevated temperature; activation of heat shock protein synthesis occurs at 39.5°C, whereas other mononuclear cells such as T and B cells require prolonged incubation at temperatures above 41°C before equivalent levels of stress protein synthesis are detected (C. Richman et al., unpubl.). Induction of the stress response in human monocytes by heat shock or heavy metals resulted in decreased synthesis of IL-1β (Schmidt and Abdulla 1988). IL-1 belongs to a family of related polypeptides that are critical in the febrile and inflammatory response. The negative regulation of IL-1β biosynthesis following heat shock may represent a form of autoregulation for endogenous pyrogen activity in the febrile response. Related to the observation that heat shock might have anti-inflammatory effects is the original observation by Ritossa (1963) and more recently by Zimarino and Wu (1987) that the *Drosophila* heat shock response can be induced by salicylates. Recently, Ohno et al. (1988) and Santoro et al. (1989) demonstrated that prostaglandin treatment of mammalian cells will also induce members of the HSP70 family.

Thus, by a variety of mechanisms, activation of the stress response may have a direct effect on the inflammatory response and, therefore, affect cell growth.

The evidence that oxidant injury induces the stress response has led to a widely held hypothesis that cellular events common to senescent cells, such as oxidant injury and other cumulative forms of cellular damage, might result in a chronic stress response. This hypothesis predicts that older cells with reduced proliferative potential will constitutively express stress proteins at an abnormally high level and consequently mount a less vigorous stress response if exposed to a subsequent stress challenge. In one study, the accumulation of damaged proteins in aged *Drosophila* may have been responsible for the elevated expression of an unusually large number of heat shock proteins (Fleming et al. 1988). Other studies performed with cultured mammalian cells have yielded different conclusions in that the heat shock response as measured by a variety of criteria was less vigorous in aged cells (Liu et al. 1989).

Through a variety of pathways, the chronic activation and maintenance of a stress response could have damaging effects. Such an example is found in humans who have a homozygous deficiency of α_1-antitrypsin and develop chronic liver disease. An amino acid substitution in α_1-antitrypsin could affect protein folding and transport resulting in abnormally low plasma levels of functionally active α_1-antitrypsin. This genetic deficiency causes the elevated and constitutive expression of hsp90, hsp70, and ubiquitin (Perlmutter et al. 1989). The activation of the 70-kD stress proteins could therefore be in response to the accumulation of abnormal proteins. Likewise, the activation of ubiquitin could be due to a general stress response or a specific response to increased levels of components of the proteolytic machinery.

The relationship between proteolysis and the stress response, although somewhat unclear, is based in part on the evidence that ubiquitin (Bond and Schlessinger 1985) and the *lon* proteinase in *E. coli* are heat shock proteins (Goff and Goldberg 1985) and in part on ubiquitin's role in protein degradation (Hershko 1988). It is not unreasonable to consider that a biochemical mechanism that is capable of recognizing newly synthesized proteins and responds to denatured substrates may be linked to protein degradative pathways (Chapter 8). Direct evidence to support the model that unfolded proteins signal the induction of the heat shock response in *E. coli* has been provided from experiments using unfolded variants of λ repressor (Parsell and Sauer 1989). These results are consistent with experiments in which the injection of denatured proteins into eukaryotic cells stimulated the expression of coinjected heat shock genes (Ananthan et al. 1986).

B. Immunobiology and the Stress Response

During infection with bacterial and parasitic pathogens, high levels of antibodies specific to the infectious agent stress proteins appear (Chapter 7). The major antigens of *Mycobacteria* that act as potent T-cell immunogens include, for example, the 65-kD protein related to the bacterial GroEL protein and the 70–71-kD antigen, which is a DnaK homolog (Young et al. 1988). These observations have led to the proposal to use these highly immunodominant stress proteins as candidates for vaccines. However, one consideration is that chronic stimulation with antigens that have closely related homologs in the host may have detrimental consequences and lead to autoimmune disease. Stress proteins are major immunogens for both $CD4^+$ and $CD8^+$ T cells and B cells (Ottenhoff et al. 1988). Recently, this form of antigenic recognition has been extended to $\gamma\delta$ T cells that do not express $CD4^+$ or $CD8^+$ molecules on the cell surface (Haregewoin et al. 1989). Perhaps it is not surprising that stress proteins are potent antigens, given their abundance and that infection often leads to elevated synthesis of stress proteins. The implication of these observations is not certain; one possibility is that the immune system has evolved multiple classes of T cells that recognize both pathogen-infected and damaged cells.

Additional support for involvement of stress proteins in the immune response comes from chromosomal mapping studies that place *HSP70* genes to human chromosomes 6, 14, and 21 (Harrison et al. 1987). Of particular interest, Sargent et al. (1989) mapped the human *HSP70* gene to the class III region of HLA, between the TNF-α and -β genes and the complement genes. Similar results have been obtained for the rat *HSP70* gene. Therefore, another potential link between stress proteins and the response to infection and neoplasia may be suggested by proximity of the human *HSP70* gene with the tumor necrosis factor (TNF) α and β genes. TNF, a member of the cytokine family, elicits a broad range of reactions in inflammation and immunity and in response to cell injury. Both TNF and IL-1 are involved in hypothalamic thermoregulation in the febrile response. The interplay between evoking an inflammatory response, yet maintaining a balance during infection and injury, could involve the stress response as a potent negative regulatory mechanism.

C. Stress Proteins and Neoplasia

Given the vast array of conditions that result in modulation of stress gene expression, we end with a discussion to consider whether there is any evidence to support a hypothesis that links stress proteins and neoplasia. Although there is little direct experimental evidence, the aberrant expression of stress proteins is likely to have deleterious effects on a variety of cellular processes with long-term consequences for the control of cell growth. The activation of heat shock genes during infection with DNA and RNA viruses occurs through the actions of specific oncogenes. At the transcriptional level, common promoter elements are required for both growth-regulated and adenovirus-E1A-induced activation (Williams et al. 1989; Chapter 14). At the level of protein interactions, members of the hsp70 family (including DnaK) are associated with the product of the recessive cellular oncogene, p53 (Pinhasi-Kimhi et al. 1986; Clarke et al. 1988; Finlay et al. 1988). In SV40-infected cells, p53 also associates with T antigen; therefore, the possibility exists that the normal p53:hsp70 complex represents normal cellular complexes that are disrupted by the presence of the SV40 T antigen.

Comparison of the levels of hsp70 protein synthesis or mRNA levels in normal or transformed mammalian tissue culture cells reveals that transformed cells often express higher constitutive levels of stress proteins than do primary cells (R. Morimoto, unpubl.). However, it is unclear whether this correlation represents a chronic stress response to aberrant cellular proteins expressed in viral or chemically transformed cells. The elevated synthesis of stress proteins in transformed cells could simply reflect the increased fraction of growing cells during neoplasia and the growth-regulated expression of stress genes. As discussed previously, there is no direct evidence to indicate whether the disregulation or altered function of stress proteins during conditions of uncontrolled cell growth leads to specific disease states.

An interesting story has emerged with a class of structurally dissimilar hypolipidemic compounds that induce a pleiotropic response consisting of peroxisome proliferation in hepatic parenchymal cells, hepatomegaly, and hepatocarcinogenesis (for review, see Rao and Reddy 1987). These compounds include the phthalate ester plasticizers, which are widespread among the human population. In general, these compounds are not mutagenic; therefore, it has been proposed that hepatocarcinogenesis might result from metabolic disturbances related to the proliferation of peroxisomes (Reddy and Rao 1986). The peroxisome proliferators, although structurally diverse, have predictable effects on liver cell function, such that Reddy and Rao (1986) proposed that these agents might exert their effect though a receptor-ligand mechanism. Lalwani et al. (1987) tested this hypothesis using three related compounds, nafenopin, clofibric acid, and ciprofibrate, as affinity ligands and eluting agents to identify specific binding proteins. With each ligand, the same 70,000-dalton protein was affinity-purified; the microsequence analysis of tryptic peptides revealed a sequence that shared complete identity with the constitutively expressed rat p72 and nearly complete identity with human HSP70 (K. Alvares and J. Reddy, unpubl.). Furthermore, the protein purified from rat liver by chromatography on a ciprofibric acid affinity column binds to an ATP affinity column, a feature common to members of the hsp70 family, and cross-reacts with antisera raised against members of the hsp70 family. As might be expected, purified constitutive p72 and hsp70 from chicken and human cells or purified p72 from rabbit reticulocytes also binds with high affinity to the ciprofibric acid affinity column. The issue of structural diversity of compounds that bind to members of the hsp70 family was further supported by the observation that clofibric acid will elute p72 and hsp70 bound to an ATP affinity column and that ATP will elute the same proteins bound to a clofibric acid column (K. Alvares et al., unpubl.). With this information, it seems reasonable to suggest that the hypolipidemic compounds could exert their effects by interfering with the normal function of the 70K heat shock proteins, an immediate consequence of which is a perturbation of organellar biogenesis, as represented by peroxisome proliferation, and a long-term consequence of which is neoplasia. A related possibility that may have broad-reaching implications is that binding of these xenobiotics to the hsp70 proteins could provide a mechanism of transport into the nucleus.

A direct relationship between elevated stress protein expression and neoplasia has been suggested from studies on estrogen-induced proteins in breast cancer cells. The expression of a 24-kD protein was shown to correlate with steroid hormone receptors in cell lines, organs, and tumors (Edwards et al. 1981; Ciocca et al. 1983; Adams and McGuire 1985). Of particular interest, the 24-kD estrogen-induced protein was found to be expressed at high levels in carcinoma of the breast, suggesting the possibility for its use as a tumor marker. Another study has suggested that the levels of hsp27 correlate with endometrial adenocarcinomas (Ciocca et al. 1985; Puy et al. 1989). Upon cloning and partial nucleotide sequence analysis of the gene encoding the 24-kD protein, comparisons with the nucleotide sequence data base revealed complete identity with the small heat shock protein hsp27 (Fuqua et al. 1989). Curiously, the regulation of human HSP27 by estrogens and heat shock is reminiscent of the ecdysteroid and heat shock activation of the *Drosophila* HSP20 genes (Chapter 15). Monitoring the expression or levels of stress proteins may offer a diagnostic tool for many disease states including cancer.

REFERENCES

Adams, D.J. and W.L. McGuire. 1985. Quantitative enzyme-linked immunosorbent assay for the estrogen-regulated M_r 24,000 protein in human breast tumors: Correlation with estrogen and progesterone receptors. *Cancer Res.* **45:** 2445–2449.

Alfano, C. and R. McMacken. 1989. Heat shock protein-mediated disassembly of nucleoprotein structures is required for the initiation of bacteriophage DNA replication. *J. Biol. Chem.* **264:** 10709–10718.

Ananthan, J., A.L. Goldberg, and R. Voellmy. 1986. Abnormal proteins serve as eukaryotic stress signals and trigger the activation of heat shock genes. *Science* **232:** 522–524.

Ang, D. 1988. "The role of the *Escherichia coli* heat shock protein, grpE, in *Escherichia coli* growth and λ DNA replication." Ph.D. thesis, University of Utah, Salt Lake City.

Arndt, K.T., C.A. Styles, and G.R. Fink. 1989. A suppressor of a HIS4 transcriptional defect encodes a protein with homology to the catalytic subunit of protein phosphatases. *Cell* **56:** 527–537.

Arrigo, A.-P. 1980. Investigation of the function of the heat shock proteins in *Drosophila melanogaster* tissue culture cells. *Mol. Gen. Genet.* **178:** 517–524.

Arrigo, A.-P., J. Suhan, and W.J. Welch. 1988. Dynamic changes in the structure and intracellular locale of the mammalian low-molecular-weight heat shock protein. *Mol. Cell. Biol.* **8:** 5059–5071.

Ashburner, M. 1970. Pattern of puffing activity in the salivary gland chromosomes of *Drosophila.* V. Response to environmental treatments. *Chromosoma* **31:** 356–376.

Bahl, H., H. Echols, D.B. Straus, D. Court, R. Crowl, and C.P. Georgopoulos. 1987. Induction of the heat shock response of *E. coli* through stabilization of σ^{32} by the phage λ cIII protein. *Genes Dev.* **1:** 57–64.

Banerji, S.S., K. Laing, and R.I. Morimoto. 1987. Erythroid lineage-specific expression and inducibility of the major heat shock protein HSP70 during avian embryogenesis. *Genes Dev.* **1:** 946–953.

Banerji, S.S., N.G. Theodorakis, and R. Morimoto. 1984. Heat shock induced translational control of HSP70 and globin synthesis in chicken reticulocytes. *Mol. Cell. Biol.* **4:** 2437–2448.

Bardwell, J.C.A. and E.A. Craig. 1984. Major heat shock gene of *Drosophila* and the *Escherichia coli* heat inducible *dnak* gene are homologous. *Proc. Natl. Acad. Sci.* **81:** 848–852.

———. 1987. Eukaryotic M_r 83,000 heat shock protein has a homologue in *Escherichia coli. Proc. Natl. Acad. Sci.* **84:** 5177–5181.

Barnett, T., M. Altschuler, C.N. McDaniel, and J.P. Mascarenhas. 1980. Heat shock induced proteins in plant cells. *Dev. Genet.* **1:** 331–340.

Beato, M. 1989. Gene regulation by steroid hormones. *Cell* **56:** 335–344.

Berendes, H.D., F.M.A. van Breugel, and T.K.H. Holt. 1965. Experimental puffs in salivary gland chromosomes of *D. hydei. Chromosoma* **16:** 35–46.

Bochkareva, E.S., N.M. Lisson, and A.S. Girshovich. 1988. Transient association of newly synthesized unfolded proteins with the heat-shock GroEL protein. *Nature* **336:** 254–257.

Bochner, B.R., M. Zylicz, and C. Georgopoulos. 1986. *Escherichia coli* dnaK protein possesses a 5′-nucleotidase activity that is inhibited by AppppA. *J. Bacteriol.* **168:** 931–935.

Bond, U. and M.J. Schlesinger. 1985. Ubiquitin is a heat shock protein in chicken embryo fibroblasts. *Mol. Cell. Biol.* **5:** 949–956.

Braell, W.A., D.M. Schlossman, S.L. Schmid, and J. Rothman. 1984. Dissociation of clathrin coats coupled to hydrolysis of ATP: Role of an uncoating ATPase. *J. Cell Biol.* **99:** 734–741.

Buttyan, R., Z. Zakeri, R. Lockshin, and D. Wolgemuth. 1988. Cascade induction of c-*fos*, c-*myc*, and heat shock 70k transcripts during regression of the rat ventral prostrate gland. *Mol. Endocrinol.* **2:** 650–657.

Capdevila, M.D. and A. Garcia-Bellido. 1974. Development and genetic analysis of bithorax phenocopies in *Drosophila. Nature* **250:** 500–502.

Chandrasekhar, G.N., K. Tilly, C. Woolford, R. Hendrix, and C. Georgopoulos. 1986. Purification and properties of the groES morphogenetic protein of *Escherichia coli. J. Biol. Chem.* **261:** 12414–12419.
Chirico, W.J., M.G. Waters, and G. Blobel. 1988. 70 K heat shock related proteins stimulate protein translocation into microsomes. *Nature* **332:** 805–810.
Christman, M.F., R.W. Morgan, F.S. Jacobson, and B.N. Ames. 1985. Positive control of a regulon for defenses against oxidative stress and some heat shock proteins in *Salmonella typhimurium. Cell* **41:** 753–762.
Ciocca, D.R., D.J. Adams, D.P. Edwards, R.J. Bjercke, and W.L. McGuire. 1983. Distribution of an estrogen-induced protein with a molecular weight of 24,000 in normal and malignant human tissues and cells. *Cancer Res..* **43:** 1204–1210.
Ciocca, D.R., L.A. Puy, D.P. Edwards, D.J. Adams, and W.L. McGuire. 1985. The presence of an estrogen-regulated protein detected by monoclonal antibody in abnormal human endometrium. *J. Clin. Endocrinol. Metabol.* **60:** 137–143.
Clarke, C.F., K. Cheng, A.B. Frey, R. Stein, P.W. Hinds, and A.J. Levine. 1988. Purification of complexes of nuclear oncogene p53 with rat and *Escherichia coli* heat shock protein: In vitro dissociation of hsc70 and dnaK from murine p53 by ATP. *Mol. Cell. Biol.* **8:** 1206–1215.
Craig, E.A. 1985. The heat shock response. *CRC Crit. Rev. Biochem.* **18:** 239–280.
Craig, E.A. and B.J. McCarthy. 1981. Four *Drosophila* heat shock genes at 67B: Characterization of recombinant plasmids. *Nucleic Acids Res.* **8:** 4441–4457.
Craig, E.A., B.J. McCarthy, and S.C. Wadsworth. 1979. Sequence organization of two recombinant plasmids containing genes for the major heat shock induced protein of *Drosophila melanogaster. Cell* **16:** 575–583.
Currie, R.W. 1987. Effects of ischemia and perfusion temperature on the synthesis of stress-induced (heat shock) proteins in isolated and perfused rat hearts. *J. Mol. Cell. Cardiol.* **19:** 795–808.
Currie, R.W. and F.P. White. 1981. Trauma-induced protein in rat tissues: A physiological role for a "heat shock" protein? *Science* **214:** 72–73.
de Jong, W.W., W. Hendriks, J.W.M. Mulders, and H. Bloemendal. 1989. Evolution of eye lens crystallins: The stress connection. *Trends Biochem. Sci.* **14:** 365–368.
Deshaies, R.J., B.D. Koch, M. Werner-Washburne, E.A. Craig, and R. Schekman. 1988. A subfamily of stress proteins facilitates translocation of secretory and mitochondrial precursor polypeptides. *Nature* **332:** 800–805.
DiDomenico, B.J., G.E. Bugaisky, and S. Lindquist. 1982. The heat shock response is self-regulated at both the transcriptional and post-transcriptional level. *Cell* **31:** 593–603.
Dillman, W.H., H.B. Mehta, A. Barrieux, B.D. Guth, W.E. Neeley, and J. Ross, Jr. 1986. Ischemia of the dog heart induces the appearance of a cardiac mRNA coding for a protein with migration characteristics similar to heat-shock/stress protein 71. *Circ. Res.* **59:** 110–114.
Dura, J.M. 1981. Hsp synthesis is induced only after treatment at blastoderm and later stages in development. *Mol. Gen. Genet.* **184:** 73–79.
Edwards, D.P., D.J. Adams, and W.L. McGuire. 1981. Estradiol stimulates synthesis of a major intracellular protein in a human breast cancer cell line (MCF-7). *Breast Cancer Res. Treat.* **1:** 209–223.
Ellis, J. 1987. Proteins as molecular chaperones. *Nature* **328:** 387–379.
Erickson, J.W. and C.A. Gross. 1989. Identification of the σE subunit of *Escherichia coli* RNA polymerase: A second alternate σ factor involved in high-temperature gene expression. *Genes Dev.* **3:** 1462–1471.
Erickson, J.W., V. Vaughn, W.A. Walter, F.C. Neidhardt, and C.A. Gross. 1987. Regulation of the promoters and transcripts of *rpoH*, the *Escherichia coli* heat shock regulatory gene. *Genes Dev.* **1:** 419–432.
Fayet, O., J.-M. Louarn, and C. Georgopoulos. 1986. Suppression of the *Escherichia coli*

*dnaA*46 mutation by amplification of the *groES* and *groEL* genes. *Mol. Gen. Genet.* **202:** 435–445.

Fayet, O., T. Ziegelhoffer, and C. Georgopoulos. 1989. The *groES* and *groEL* heat shock genes of *Escherichia coli* are essential for bacterial growth at all temperatures. *J. Bacteriol.* **171:** 1379–1385.

Findlay, C.A., P.W. Hinds, T.H. Tan, D. Eliyahu, M. Oren, and A.J. Levine. 1988. Activating mutations for transformation by p53 produce a gene product that forms a hsc70-p53 complex with an altered half-life. *Mol. Cell. Biol.* **8:** 531–539.

Fleming, J.E., J.K. Walton, R. Dubitsky, and K.G. Bensch. 1988. Aging results in an unusual expression of *Drosophila* heat shock proteins. *Proc. Natl. Acad. Sci.* **85:** 4099–4103.

Flynn, G.C., T.G. Chappell, and J.E. Rothman. 1989. Peptide binding and release by proteins implicated as catalysts of protein assembly. *Science* **245:** 385–390.

Friedman, D.E., E.R. Olson, C. Georgopoulos, K. Tilly, I. Herskowitz, and F. Banuett. 1984. Interactions of bacteriophage and host macromolecules in the growth of bacteriophage λ. *Microbiol. Rev.* **48:** 299–325.

Fujita, N. and A. Ishihama. 1987. Heat-shock induction of RNA polymerase σ^{32} synthesis in *Escherichia coli:* Transcriptional control and a multiple promoter system. *Mol. Gen. Genet.* **210:** 10–15.

Fuqua, S.A.W., M. Blum-Salingaros, and W.L. McGuire. 1989. Induction of the estrogen-regulated "24k" protein by heat shock. *Cancer Res.* **49:** 4126–4129.

Gall, J.G. and M.L. Pardue. 1971. Nucleic acid hybridization in cytological preparations. *Methods Enzymol.* **21:** 470–480.

Georgopoulos, C.P. 1989. The *Escherichia coli* dnaA initiation protein: A protein for all seasons. *Trends Genet.* **5:** 319–321.

Gilmour, D.S. and J.T. Lis. 1986. RNA polymerase II interacts with the promoter region of the noninduced hsp70 gene in *Drosophila melanogaster* cells. *Mol. Gen. Genet.* **6:** 3984–3989.

Gloor, H. 1947. Phanokopie-Versuche mit Aether an *Drosophila. Rev. Suisse Zool.* **54:** 637–712.

Goff, S.A. and A.L. Goldberg. 1985. Production of abnormal protein in *E. coli* stimulates transcription of *lon* and other heat shock genes. *Cell* **41:** 587–595.

Goldschmidt, R. 1935. Gen und Ausseneigenschaft. 1. (Untersuchung an *Drosophila*). *Z. Indukt. Abstammungs Vererbungsl.* **69:** 38–131.

Goloubinoff, P., J.T. Christeller, A.A. Gatenby, and G.H. Lorimer. 1989. Reconstitution of active dimeric ribulosebiphosphate carboxylase from an unfolded state depends on two chareronin proteins and magesium ATP. *Nature* **342:** 844–889.

Hadorn, E. 1955. Letalfaktoren, pp. 233–253. Georg Thieme Verlag, Stuttgart.

Haregewoin, A., G. Soman, A.C. Hom, and R.W. Finberg. 1989. Human γδ T cells respond to mycobacterial heat-shock protein. *Nature* **340:** 309–310.

Harrison, G.S., H.A. Drabkin, F.-T. Kao, I.M. Hartz, E.H.Y. Chu, B.J. Wu, and R.I. Morimoto. 1987. Chromosomal location of human genes encoding major heat shock protein hsp70. *Somatic Cell Mol. Genet.* **13:** 119–130.

Hemmingsen, S.M., C. Woolford, S.M. van der Vies, K. Tilly, D.T. Dennis, C.P. Georgopoulos, R.W. Hendrix, and R.J. Ellis. 1988. Homologous plant and bacterial proteins chaperone oligomeric protein assembly. *Nature* **33**: 330–334.

Hendrix, R.W. 1979. Purification and properties of GroE, a host protein involved in bacteriophage assembly. *J. Mol. Biol.* **129:** 375–392.

Hershko, A. 1988. Ubiquitin-mediated protein degradation. *J. Biol. Chem.* **263:** 15237–15240.

Hightower, L.W. 1980. Cultured animal cells exposed in amino acid analogues or puromycin rapidly synthesize several polypeptides. *J. Cell. Physiol.* **102:** 407–427.

Hohn, T., B. Hohn, A. Engel, M. Wurtz, and P.R. Smith. 1979. Isolation and characterization of the host protein GroE involved in bacteriophage λ assembly. *J. Mol. Biol.* **129:** 359–373.

Holmgren, R., V. Corces, R. Morimoto, R. Blackman, and M. Meselson. 1981. Sequence homologies in the 5′ regions of four *Drosophila* heat shock genes. *Proc. Natl. Acad. Sci.* **78:** 3775–3778.

Hunt, C.R. and R.I. Morimoto. 1985. Conserved features of eukaryotic hsp70 genes revealed by comparison with the nucleotide sequence of human hsp70. *Proc. Natl. Acad. Sci.* **82:** 6455–6459.

Ingolia, T.D. and E.A. Craig. 1982. Four small *Drosophila* heat shock proteins are related to each other and to mammalian α-crystallin. *Proc. Natl. Acad. Sci.* **79:** 2360–2364.

Jenkins, A.J., J.B. Marsh, I.R. Oliver, and M. Master. 1986. A DNA fragment containing the groE genes can suppress mutations in the *Escherichia coli* dnaA gene. *Mol. Gen. Genet.* **202:** 446–454.

Johnson, C. 1988. "The *Escherichia coli* dnaK protein: Aspects of its functions during heat stress." Ph.D. thesis, University of Utah, Salt Lake City.

Karch, F., I. Torok, and A. Tissières. 1981. Extensive regions of homology in front of the two hsp70 heat shock variant genes in *Drosophila melanogaster*. *J. Mol. Biol.* **148:** 219–230.

Keller, J.A. and L.D. Simon. 1988. Divergent effects of a dnaK mutation on abnormal protein degradation in *Escherichia coli*. *Mol. Microbiol.* **2:** 31–41.

Kelley, P.M. and M.J. Schlesinger. 1978. The effect of amino acid analogues and heat shock on gene expression in chicken embryo fibroblasts. *Cell* **15:** 1277–1286.

Kingston, R.E., T.J. Schuetz, and Z. Larin. 1987. Heat-inducible human factor that binds to a human hsp70 promoter. *Mol. Cell. Biol.* **7:** 1530–1534.

Kramhoft, B. and E. Zeuthen. 1971. Synchronization of cell division in the fission yeast, *Schizosaccharomyces pombe,* using heat shocks. *C.R. Trav. Lab. Carlsberg* **38:** 351–368.

Kubo, T., C.A. Towle, H.J. Mankin, and B.V. Treadwell. 1985. Stress-induced proteins in chondrocytes from patients with osteoarthritis. *Arthritis Rheum.* **28:** 1140–1145.

Kusukawa, N. and T. Yura. 1988. Heat shock protein GroE of *Escherichia coli:* Key protective roles against thermal stress. *Genes Dev.* **2:** 874–882.

Kusukawa, N., T. Yura, C. Ueguchi, Y. Akiyama, and K. Ito. 1989. Effects of mutations in heat-shock genes *groES* and *groEL* on protein export of *Escherichia coli*. *EMBO J.* **8:** 3517–3521.

Lalwani, N.D., K. Alvares, M.K. Reddy, M.N. Reddy, I. Parikh, and J.K. Reddy. 1987. Peroxisome proliferator-binding protein: Identification and partial characterization of nafenopin-, clofibric acid-, and ciprofibrate-binding proteins from rat liver. *Proc. Natl. Acad. Sci.* **84:** 5242–5246.

Larson, J.S., T.J. Schuetz, and R.E. Kingston. 1988. Activation in vitro of sequence specific DNA binding by a human regulatory factor. *Nature* **335:** 372–375.

Lecker, S., R. Lill, T. Ziegelhoffer, C. Georgopoulos, P.J. Bassford, Jr., C.A. Kumamoto, and W. Wickner. 1989. Three pure chaperone proteins of *Escherichia coli*, secB, trigger factor and groEL, form soluble complexes with precursor proteins in vitro. *EMBO J.* **8:** 2703–2709.

Lee, P.C., B.R. Bochner, and B.N. Ames. 1983. AppppA, heat shock stress, and cell oxidation. *Proc. Natl. Acad. Sci.* **80:** 7496–7500.

Lemaux, P.G., S.L. Herendeen, P.L. Bloch, and F.C. Neidhardt. 1978. Transient rates of synthesis of individual polypeptides in *E. coli* following temperature shifts. *Cell* **13:** 427–434.

Li, G.C. and A. Laszlo. 1985. Thermotolerance in mammalian cells: A possible role for the heat shock proteins. In *Changes in eukaryotic gene expression in response to environmental stress* (ed. B.G. Atkinson and D.B. Walden), pp. 349–371. Academic Press, Orlando.

Lindquist, S. 1980. Translational efficiency of heat induced messages in *Drosophila melanogaster* cells. *J. Mol. Biol.* **137:** 151–158.

Lindquist, S. and E.A. Craig. 1988. The heat-shock proteins. *Annu. Rev. Genet.* **22:** 631–677.

Lindquist-McKenzie, S.L. and M. Meselson. 1977. Translation in vitro of *Drosophila* heat shock messages. *J. Mol. Biol.* **117:** 279–283.

Lindquist-McKenzie, S.L., S. Henikoff, and M. Meselson. 1975. Localization of RNA from heat-induced polysomes at puff sites in *Drosophila melanogaster*. *Proc. Natl. Acad. Sci.* **72:** 1117–1121.

Lipinska, B., S. Sharma, and C. Georgopoulos. 1988a. Sequence analysis and transcriptional regulation of the *htrA* gene of *Escherichia coli:* A σ^{32}-independent mechanism of heat-inducible transcription. *Nucleic Acids Res.* **16:** 10053–10067.

Lipinska, B., O. Fayet, L. Baird, and C. Georgopoulos. 1988b. Identification, characterization and mapping of the *Escherichia coli htrA* gene, whose product is essential for bacterial viability only at elevated temperatures. *J. Bacteriol.* **171:** 1574–1584.

Liu, A.Y., Z. Lin, H.S. Choi, F. Sorhage, and B. Li. 1989. Attenuated induction of heat shock gene expression in aging diploid fibroblasts. *J. Biol. Chem.* **264:** 12037–12045.

Livak, K.J., R. Freund, M. Schweber, P.C. Wensink, and M. Meselson. 1978. Sequence organization and transcription at two heat shock loci in *Drosophila*. *Proc. Natl. Acad. Sci.* **75:** 5613–5617.

McAlister, L. and D.B. Finkelstein. 1980. Heat shock proteins and thermal resistance in yeast. *Biochem Biophys. Res. Commun.* **93:** 819–824.

Mehta, H.B., B.K. Popovich, and W.H. Dillman. 1988. Ischemia induces changes in the level of mRNAs coding for stress protein 71 and creatin kinase M. *Circ. Res.* **63:** 512–517.

Milkman, R. 1966. Analysis of some temperature effects on *Drosophila* pupae. *Biol. Bull.* **131:** 331–345.

Mirault, M.-E., M. Goldschmidt-Clermont, L. Moran, A.-P. Arrigo, and A. Tissières. 1978. The effect of heat shock on gene expression in *Drosophila melanogaster*. *Cold Spring Harbor Symp. Quant. Biol.* **42:** 819–827.

Mitchell, H.K. 1966. Phenol oxidases and *Drosophila* development. *J. Insect Physiol.* **18:** 755–765.

Mitchell, H.K. and N.S. Petersen. 1982. Developmental abnormalities in *Drosophila* induced by heat shock. *Dev. Genet.* **3:** 91–102.

Mitchell, H.K., G. Moller, N.S. Petersen, and L. Lipps-Sarmiento. 1979. Specific protection from phenocopy induction by heat shock. *Dev. Genet.* **1:** 181–192.

Moran, L., M.-E. Mirault, A. Tissières, J. Lis, P. Schedl, S. Artavanis-Tsakonas, and W.J. Gehring. 1979. Physical map of two *Drosophila melanogaster* DNA segments containing sequences coding for the 70,000 dalton heat shock protein. *Cell* **17:** 1–8.

Morgan, R.W., M.F. Christman, F.S. Jacobson, G. Storz, and B.N. Ames. 1986. Hydrogen peroxide-inducible proteins in *Salmonella typhimurium* overlap with heat shock and other stress proteins. *Proc. Natl. Acad. Sci.* **83:** 8059–8063.

Morimoto, R.I. and E. Fodor. 1984. Cell specific expression of heat shock proteins in chicken reticulocytes and lymphocytes. *J. Cell Biol.* **99:** 1316–1323.

Mosser, D.D., N.G. Theodorakis, and R.I. Morimoto. 1988. Coordinate changes in heat shock element binding activity and hsp70 gene transcription rates in human cells. *Mol. Cell. Biol.* **8:** 4736–4744.

Neidhardt, F.C., R.A. VanBogelen, and V. Vaughn. 1985. The genetics and regulation of heat shock proteins. *Annu. Rev. Genet.* **18:** 295–329.
Nover, L., ed. 1984. *Heat shock response of eukaryotic cells*, pp. 7–10. Springer-Verlag, Berlin.
Nowak, T.S., Jr. 1985. Synthesis of a stress protein following transient ischemia in the gerbil. *J. Neurochem.* **45:** 1635–1641.
Ohno, K., M. Fukushima, and S. Narumiya. 1988. Induction of 68,000-dalton heat shock proteins by cyclopentenone prostaglandins. *J. Biol. Chem.* **263:** 19764–19770.
Ottenhoff, T.H., A.B. Kale, J.D.A. Van Embden, J.E.R. Thole, and R. Kiessling. 1988. The recombinant 65-kd heat shock protein of *Mycobacterium bovis* bacillus calmette-guerin/*M. tuberculosis* is a target molecule for CD4+ cytotoxic T lymphocytes that lyse human monocytes. *J. Exp. Med.* **168:** 1947–1952.
Ostermann, J., A.L. Horwich, W. Neupert, and F.-U. Hartl. 1989. Protein folding in mitochondria requires complex formation with hsp60 and ATP-hydrolysis. *Nature* **341:** 125–130.
Palter, K.B., M. Watanabe, L. Stinson, A.P. Mahowald, and E.A. Craig. 1986. Expression and localization of *Drosophila melanogaster* hsp70 cognate proteins. *Mol. Cell. Biol.* **6:** 1187–1203.
Parag, H.A., B. Raboy, and R.G. Kulka. 1987. Effect of heat shock on protein degradation in mammalian cells: Involvement of the ubiquitin system. *EMBO J.* **6:** 55–61.
Parsell, D.A. and R.T. Sauer. 1989. Induction of a heat shock-like response by unfolded protein in *Escherichia coli:* Dependence on protein level not protein degradation. *Genes Dev.* **3:** 1226–1232.
Pelham, H.B.R. 1984. Hsp70 accelerates the recovery of nucleolar morphology after heat shock. *EMBO J.* **3:** 3095–3100.
———. 1986. Speculations on the functions of the major heat shock and glucose-regulated proteins. *Cell* **46:** 959–961.
———. 1989. Heat shock and the sorting of luminal ER proteins. *EMBO J.* **8:** 3171–3176.
Perlmutter, D.H., M.J. Schlesinger, J.A. Pierce, P.I. Punsal, and A.L. Schwartz. 1989. Synthesis of stress proteins is increased in individuals with homozygous PiZZ a1-antitrypsin deficiency and liver disease. *J. Clin. Invest.* **84:** 1555–1561.
Pinhasi-Kimhi, O., D. Michalovitz, A. Ben-Zeev, and M. Oren. 1986. Specific interactions between the p53 cellular tumour antigen and major heat shock proteins. *Nature* **320:** 182–184.
Pleet, H., J.M. Graham, Jr., and D.W. Smith. 1981. Central nervous system and facial defects associated with maternal hyperthermia at four to 14 weeks gestation. *Pediatrics* **67:** 785–789.
Puy, L.A., G. Lo Castro, J.E. Olcese, H.O. Lofti, H.R. Brandi, and D.R. Ciocca. 1989. Analysis of a 24-kilodalton (kd) protein in the human uterine cervix during abnormal growth. *Cancer* **64:** 1067–1073.
Rao, M.S. and J.K. Reddy. 1987. Peroxisome proliferation and carcinogenesis. *Carcinogenesis* **8:** 631–636.
Reddy, J.K. and M.S. Rao. 1986. Peroxisome proliferators and cancer: Mechanisms and implications. *Trends Pharmacol. Sci.* **7:** 438–443.
Ritossa, F.M. 1962. A new puffing pattern induced by a temperature shock and DNP in *Drosophila. Experientia* **18:** 571–573.
———. 1963. New puffs induced by temperature shock. DNP and salicylate in salivary chromosomes of *Drosophila melanogaster. Drosophila Inf. Service* **37:** 122–123.
———. 1964. Specific loci in polytene chromosomes of *Drosophila. Exp. Cell. Res.* **35:** 601–607.
Rothman, J.E. 1989. Polypeptide chain binding proteins: Catalysts of protein folding and related processes in cells. *Cell* **59:** 591–601.
Rougvie, A.E. and J.T. Lis. 1988. The RNA polymerase II molecule at the 5′ end of the

uninduced hsp70 gene in *D. melanogaster* is transcriptionally engaged. *Cell* **54:** 795–804.

Sadler, I., A. Chiang, T. Kurihara, J. Rothblatt, J. Way, and P. Silver. 1989. A yeast gene important for protein assembly into the endoplasmic reticulum and the nucleus has homology to dnaJ and *Escherichia coli* heat shock protein. *J. Cell Biol.* **109:** 2665–2675.

Santoro, M.G., E. Garaci, and C. Amica. 1989. Prostaglandins with antiproliferative activity induce the synthesis of a heat shock protein in human cells. *Proc. Natl. Acad. Sci.* **86:** 8407–8411.

Sargent, C.A., I. Dunhan, J. Trowsdale, and R.D. Campbell. 1989. Human major histocompatibility complex contains genes for the major heat shock protein HSP70. *Proc. Natl. Acad. Sci.* **86:** 1968–1972.

Sciandra, J.J., J.P. Subjeck, and C.S. Hughes. 1984. Induction of glucose-regulated proteins during anaerobic exposure and of heat shock proteins after reoxygenation. *Proc. Natl. Acad. Sci.* **81:** 4843–4847.

Schedl, P., S. Artavanis-Tsakonas, R. Steward, W.J. Gehring, M.-E. Mirault, M. Goldschmidt-Clermont, L. Moran, and A. Tissières. 1978. Two hybrid plasmids with *Drosophila melanogaster* DNA sequences complementary to mRNA coding for the major heat shock protein. *Cell* **14:** 921–929.

Schmidt, J.A. and E. Abdulla. 1988. Down-regulation of IL-1β biosynthesis by inducers of the heat shock response. *J. Immunol.* **141:** 2027–2034.

Sell, S. 1987. "Studies of bacterial genes whose products are involved in heat shock regulation and bacteriophage lambda DNA replication." Ph.D. thesis, University of Utah, Salt Lake City.

Shyy, T.T., J.R. Subjeck, R. Heinaman, and G. Anderson. 1986. Effect of growth state and heat shock on nucleolar localization of the 110,000-Da heat shock protein in mouse embryo fibroblasts. *Cancer Res.* **46:** 4738–4745.

Sorger, P.K. and H.C.M. Nelson. 1989. Trimerization of a yeast transcriptional activator via a coiled-coil motif. *Cell* **59:** 807–813.

Sorger, P.K., M.J. Lewis, and H.R.B. Pelham. 1987. Heat shock factor is regulated differently in yeast and HeLa cells. *Nature* **329:** 81–84.

Spradling, A., M.L. Pardue, and S. Penman. 1977. Messenger RNA in heat-shocked *Drosophila* cells. *J. Mol. Biol.* **109:** 559–587.

Spradling, A., S. Penman, and M.L. Pardue. 1975. Analysis of *Drosophila* mRNA by in situ hybridization: Sequences transcribed in normal and heat shock cultured cell. *Cell* **4:** 395–404.

Strauch, K.L., K. Johnson, and J. Beckwith. 1989. Characterization of *degP*, a gene required for proteolysis in the cell envelope and essential for growth of *Escherichia coli* at high temperature. *J. Bacteriol.* **171:** 2689–2696.

Straus, D.B., W.A. Walter, and C.A. Gross. 1987. The heat shock response of *E. coli* is regulated by changes in the concentration of σ^{32}. *Nature* **329:** 348–351.

———. 1988. *Escherichia coli* heat shock gene mutants are defective in proteolysis. *Genes Dev.* **2:** 1851–1858.

———. 1989. The activity of σ^{32} is reduced under conditions of excess heat shock protein production in *Escherichia coli*. *Genes Dev.* **3:** 2003–2010.

Subjeck, J.R., T. Shyy, J. Shenand, and R. Johnson. 1983. Association between the mammalian 110,000 dalton heat shock protein and nucleoli. *J. Cell Biol.* **97:** 1389–1395.

Theodorakis, N.G., D.J. Zand, P.T. Kotzbauer, G.T. Williams, and R.I. Morimoto. 1989. Hemin induced transcriptional activation of the hsp70 gene during erythroid maturation in K562 cells is due to a heat shock factor mediated stress response. *Mol. Cell. Biol.* **9:** 3166–3173.

Tilly, K. and C. Georgopoulos. 1982. Evidence that the two *Escherichia coli* groE morphogenetic gene products interact in vivo. *J. Bacteriol.* **149:** 1082–1088.

Tilly, K., J. Spence, and C. Georgopoulos. 1989. Modulation of the stability of *Escherichia coli* heat shock regulatory factor σ^{32}. *J. Bacteriol.* **171:** 1585–1589.

Tilly, K., J. Erickson, S. Sharma, and C. Georgopoulos. 1986. Heat shock regulatory gene *rpoH* mRNA level increases after heat shock in *Escherichia coli*. *J. Bacteriol.* **168:** 1155–1158.

Tilly, K., N. McKittrick, M. Zylicz, and C. Georgopoulos. 1983. The dnaK protein modulates the heat-shock response of *Escherichia coli*. *Cell* **34:** 641–646.

Tissières, A., H.K. Mitchell, and V.M. Tracy. 1974. Protein synthesis in salivary glands of *Drosophila melanogaster:* Relation to chromosome puffs. *J. Mol. Biol.* **84:** 389–398.

Van Dyk, T.K., A.A. Gatenby, and R.A. LaRossa. 1989. Demonstration by genetic suppression of interaction of GroE products with many proteins. *Nature* **342:** 451–453.

Voellmy, R., M. Goldschmidt-Clermont, R. Southgate, A. Tissières, R. Levis, and W. Gehring. 1981. A DNA segment isolated from chromosomal site 67B in *Drosophila melanogaster* contains four closely linked heat shock genes. *Cell* **23:** 261–270.

Wang, Q. and J.M. Kaguni. 1989a. A novel σ factor is involved in expression of the *rpoH* gene of *Escherichia coli*. *J. Bacteriol.* **171:** 4248–4253.

———. 1989b. dnaA protein regulates transcription of the *rpoH* gene of *Escherichia coli*. *J. Biol. Chem.* **264:** 7338–7344.

Waters, M.G., W.J. Chirico, R. Henriquez, and G. Blobel. 1989. Purification of yeast stress proteins based on their ability to facilitate secretory protein translocation. *UCLA Symp. Mol. Cell. Biol.* **96:** 163–174.

Webster, W.S., M.A. Gernain, and M.J. Edwards. 1985. The induction of microphthalmia, encephalocele, and other heat defects following hyperthermia during the gastrulation process in the rat. *Teratology* **31:** 73–82.

Welch, W.J. and J.P. Suhan. 1985. Morphological study of the mammalian stress response: Characterization of changes in cytoplasmic organelles, cytoskeleton, and nucleoli, and appearence of intracellular actin filaments in rat fibroblasts after heat shock treatment. *J. Cell. Biol.* **101:** 1198–1211.

Westwood, J.T. and R.A. Steinhardt. 1989. Effects of heat and other inducers of the stress response on protein degradation in Chinese hamster and *Drosophila* cells. *J. Cell. Physiol.* **139:** 196–209.

Williams, G.T., T.K. McClanahan, and R.I. Morimoto. 1989. E1A-transactivation of the human hsp70 promoter is mediated through the basal transcription complex. *Mol. Cell. Biol.* **9:** 2574–2587.

Wittig, S., S. Hensse, C. Keitel, C. Elsner, and B. Wittig. 1983. Heat shock gene expression is regulated during teratocarcinoma cell differentiation and early embryonic development. *Dev. Biol.* **96:** 507–514.

Yamamori, T., K. Ito, Y. Nakamura, and T. Yura. 1978. Transient regulation of protein synthesis in *Escherichia coli* upon shift-up of growth temperature. *J. Bacteriol.* **134:** 1133–1140.

Yost, H.J. and S. Lindquist. 1986. RNA splicing is interrupted by heat shock and is rescued by heat shock protein synthesis. *Cell* **45:** 185–193.

Young, D., R. Lathigra, R. Hendrix, D. Sweetser, and R.A. Young. 1988. Stress proteins are immune targets in leprosy and tuberculosis. *Proc. Natl. Acad. Sci.* **85:** 4267–4270.

Zeuthen, E. 1971. Synchrony in *Tetrahymena* by heat shocks spaced a normal cell generation appart. *Exp. Cell. Res.* **68:** 49–60.

Zhou, Y., N. Kusukawa, J.W. Erickson, C.A. Gross, and T. Yura. 1988. Isolation and characterization of *Escherichia coli* mutants that lack the heat shock sigma factor (σ^{32}). *J. Bacteriol.* **170:** 3640–3649.

Zimarino, V. and C. Wu. 1987. Induction of sequence specific binding of *Drosophila* heat shock activator protein without protein synthesis. *Nature* **327:** 727–730.

Zylicz, M., D. Ang, K. Liberek, and C. Georgopoulos. 1989. Initiation of λ DNA replication with purified host- and bacteriophage-encoded proteins: The role of the dnaK, dnaJ and grpE heat shock proteins. *EMBO J.* **8:** 1601–1608.

Zylicz, M., J.H. LeBowitz, R. McMacken, and C. Georgopoulos. 1983. The dnaK protein of *Escherichia coli* possesses an ATPase and autophosphorylating activity and is essential in an in vitro DNA replication system. *Proc. Natl. Acad. Sci.* **80:** 6431–6435.

2

Physiological Adjustments to Fluctuating Thermal Environments: An Ecological and Evolutionary Perspective

Raymond B. Huey
Department of Zoology NJ-15
University of Washington
Seattle, Washington 98195

Albert F. Bennett
School of Biological Sciences
University of California
Irvine, California 92717

I. INTRODUCTION

The physical environment of organisms is not constant but changes through time and space. Major environmental changes cause stress, damage, or even death to organisms, but even minor changes may hamper the physiological capacities of organisms to grow, reproduce, or interact socially. Not surprisingly, organisms have evolved a variety of adjustments (e.g., behavior, acclimation, heat stress response) that help buffer the physiological impact of environmental change. These adjustments, which can have a profound effect on evolutionary fitness, can be studied from diverse perspectives. For example, what molecular and physiological bases underlie a given adjustment? What ecological and physiological circumstances favor its evolution and use?

In this paper, we adopt an ecological and evolutionary framework that has been developed to study general adjustments that organisms use to

Stress Proteins in Biology and Medicine.

buffer natural environmental variation (Levins 1968; Slobodkin and Rapoport 1974; Hochachka and Somero 1984). We apply this framework in an analysis of adjustments specifically associated with heat stress. We begin with a brief discussion of the general effects of temperature on organisms. Then, for each class of adjustment to stress, we evaluate the ecological circumstances favoring its use, describe some organismal examples from the comparative literature, and also evaluate its relevance to studies of induced aspects of the heat stress response. Our primary goal is to encourage an expanded dialogue between biologists studying the heat stress response and those interested in ecological and evolutionary physiology. As we will argue, both groups have much to contribute to each other.

That organisms are sometimes exposed to potentially stressful conditions is well illustrated with canyon lizards (*Sceloporus merriami*) living in Big Bend National Park in southern Texas (Grant and Dunham 1988). Big Bend can be very hot in midsummer, and canyon lizards die if their body temperature reaches 40–42°C. Throughout most of the day, a lizard that emerged from the relative coolness of its crevice would quickly overheat and die. Indeed, equilibrium body temperatures of these lizards exceed their upper lethal temperatures on more than 90% of the available ground surface (Fig. 1). For canyon lizards in summer, heat stress is a daily risk.

Heat stress can even occur in seemingly cold environments. For example, humans usually think of alpine areas as being cold. However, on cloudless days at 4000 meters in the Rocky Mountains, radiation loads can be intense: *Colias* butterflies suffer heat stress if the cooling winds die for more than a few minutes (Kingsolver and Watt 1983)! Heat stress can be a problem even for organisms not restricted to deserts or hot springs. Stress depends not only on local meteorological conditions, but also on the behavior, morphology, and physiology of a particular organism (Bartholomew 1958, 1964).

II. THERMAL EFFECTS AND ORGANISMAL RESPONSES

Body temperature has a profound effect on biological processes, particularly on rates of those processes. Even changes on the order of only 1°C may cause the rates of biological reactions to change by nearly 10% (Hochachka and Somero 1984; Prosser 1986; Cossins and Bowler 1987). Larger fluctuations can disrupt metabolic pathways and may lead to rapid death. Regulation of temperature or adaptation to temperature change is therefore crucial to all organisms.

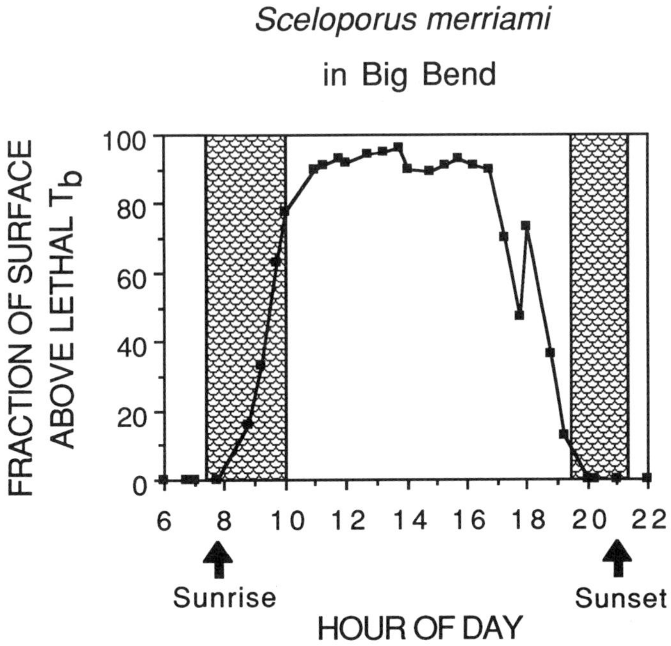

Figure 1 On a clear summer day in Big Bend National Park, Texas, equilibrium body temperatures for the canyon lizard (*Sceloporus merriami*) would exceed the lizard's lethal temperature on most of the ground surface. Lizards restrict their activity on the ground surface to early and late in the day ("scaly" regions). (Original data courtesy of B.W. Grant and A.E. Dunham.)

The range of body temperatures within which an ectothermic organism can live is often called its "tolerance zone." Within that zone, an organism's physiological capacities increase with body temperature up to a maximal value (the "optimal" body temperature) and then rapidly decrease as the organism approaches the highest temperatures it can survive (Fig. 2). The thermal sensitivity of many physiological processes shows this general pattern (Fig. 3).

Organisms cope with the profound effects of temperature on physiological capacities in two general ways. One solution involves regulation of body temperature (i.e., thermoregulation), which enables an organism to gain some control over its physiological capacities as well as to avoid extreme temperatures. The complementary solution involves the reorganization of an organism's thermal sensitivity: This solution includes both short-term (e.g., acclimation or acclimatization) and long-term (evolutionary adaptation) processes.

Many animals go to great lengths to regulate body temperatures. Mammals and birds ("endotherms"), for example, expend major portions

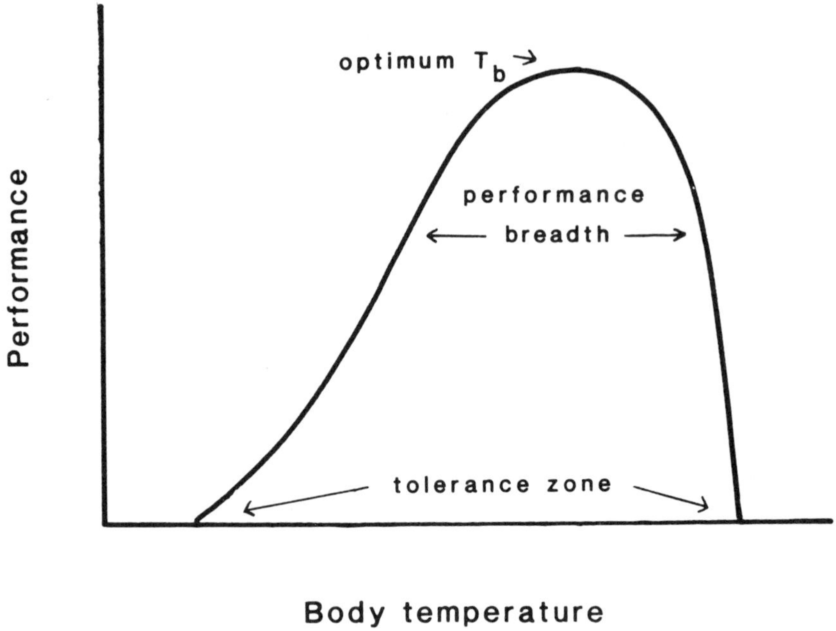

Figure 2 Effect of body temperature on the physiological performance capacity of a hypothetical ectotherm. (Modified, with permission, from Huey and Stevenson 1979.)

of their total energy budgets to stabilize body temperature metabolically (Cossins and Bowler 1987). Many arthropods, fish, and reptiles ("ectotherms") behaviorally regulate body temperature (Cowles and Bogert 1944; Casey 1981) at levels near optimal for physiological functioning (Fig. 3).

Under certain circumstances, the thermoregulatory abilities—metabolic or behavioral—of organisms are inadequate or inappropriate, but animals can still buffer environmental fluctuations by adjusting the thermal sensitivity of their physiological capacities. Comparative physiologists (Prosser 1986) often distinguish between adjustments that affect survival at extreme temperatures ("resistance" adaptations) and those that affect physiological capacities at non-extreme body temperatures ("capacity" adaptations). Included among these capacity and resistance adaptations are acclimation responses, developmental effects, and evolutionary adaptations.

III. TIME SCALES AND SEVERITY OF FLUCTUATIONS

To understand the types of adjustments to environmental fluctuations, one must consider the time scales both of the fluctuation and of the ad-

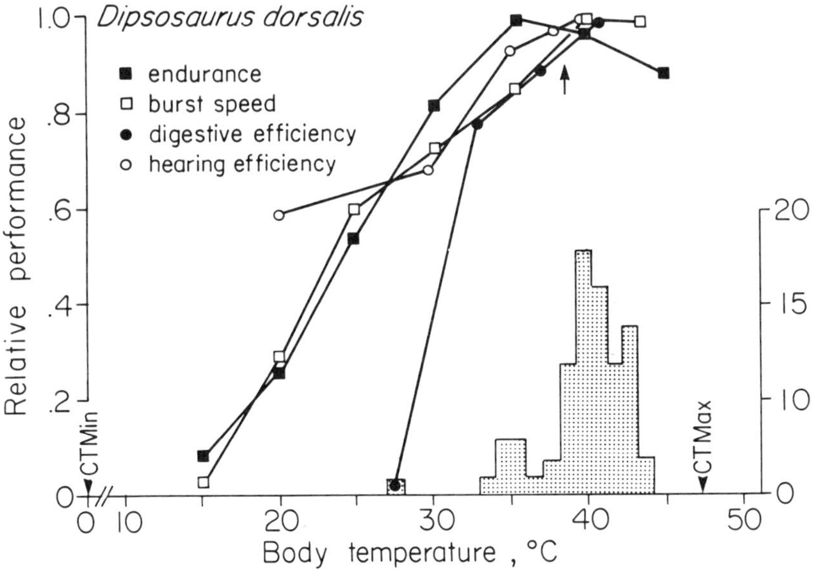

Figure 3 Sensitivity of several physiological functions to body temperature in a lizard, the desert iguana *Dipsosaurus dorsalis*. The body temperatures of lizards in nature (histogram) or in a laboratory thermal gradient (up arrow) enable the lizards to perform at or near maximal physiological capacities. Critical thermal limits (loss of righting response) are also shown (down arrow). Original sources of data in Huey (1982). (Modified, with permission, from Huey 1982.)

justment (Levins 1968; Slobodkin and Rapoport 1974; Hochachka and Somero 1984). An adjustment that takes a week to activate is clearly inappropriate if the stress itself is likely to last only for an hour or two (Levins 1968; Hochachka and Somero 1984). Thus, the response should be matched in time with the environmental fluctuation. Moreover, the magnitude of the response should be matched to the potential severity of the stress: Mobilizing a major physiological response may prove maladaptive if the stress is relatively minor (Palumbi 1984).

Fortunately for organisms, both the magnitude and duration of environmental fluctuations are sometimes predictable. For example, organisms can use environmental cues (e.g., photoperiod) to anticipate general environmental shifts associated with daily and seasonal cycles in temperature.

The various adjustments animals use to deal with the environmental fluctuations have various lag times (Hochachka and Somero 1984; Prosser 1986): Some are activated instantaneously, but others take days or even weeks. To some extent, variation in response time of different physiological adjustments is probably adaptive. For example, fast-acting responses can be activated quickly to cope with brief, but still dangerous perturbations (Palumbi 1984). However, major physiological adjustments such as acclimatization, which may be metabolically expensive, are activated only by sustained environmental fluctuations, such as those associated with weather fronts or especially with seasons. Thus, differences in response times among adjustments effectively enable an organism to deal with variation in the likely persistence and seriousness of a given environmental fluctuation.

IV. RESPONSES TO SHORT-TERM FLUCTUATIONS

Short-term thermal fluctuations, those on the order of a minute or a few hours, may not even be noticed by large endotherms or even ectotherms (e.g., Komodo dragons, Galapagos tortoises), whose body size is an effective thermal buffer (Spotila et al. 1973; Stevenson 1985). However, short-term fluctuations can be very stressful to small organisms (e.g., copepods, fruitflies, butterflies, and lizards), because their body temperatures equilibrate quickly with environmental temperatures. Adjustments that deal effectively with short-term fluctuations must respond quickly, otherwise the fluctuation might have done its damage by the time the compensatory adjustment is activated. Several behavioral and physiological adjustments meet this criterion of a fast response time (Table 1).

Simple behavioral adjustments can be effective in helping animals avoid short-term stress. For example, *Sceloporus merriami* in Big Bend avoid stress during the summer by restricting above-ground activity to very early and very late in the day, the only times when much of the ground surface area is relatively cool (Fig. 1). During the heat of the day, the lizards retreat into crevices and caves (Grant and Dunham 1988). Ground squirrels (*Xerus inauris*) use their fluffy tails as a parasol on hot days in the Kalahari desert of Africa but must still shuttle to cool burrows at midday (Bennett et al. 1984). Behavior is often the most effective way of avoiding stress in a dangerously hot locality (Bartholomew 1958, 1964; Stevenson 1986).

Animals may supplement behavioral adjustments with fast-acting ones involving physiology and morphology. The first four of these responses (Table 1) can quickly change rates of heating and cooling or can change the equilibrium temperature of an animal (Turner and Tracy 1983; Stevenson 1986), but they do not directly alter the way that temperature influences physiological performance. In contrast, the last three adjustments do modify the effect of temperature on physiology. Because these last two adjustments are of special relevance to the theme of this book, we discuss them in the next section.

Table 1 Behavioral and physiological adjustments to short-term temperature fluctuations

Behavioral adjustments	regulate time of activity—avoid times of extreme conditions
	shuttling—regulate time in sun versus shade
	postural changes—change surface area exposed to heat sources or sinks
Physiological and morphological adjustments	circulatory adjustments—vasoconstriction vasodilation
	evaporative cooling—panting
	color change—alter reflectivity
	thermogenesis
	tolerance of hypothermia and of hyperthermia
	immediate compensation
	heat hardening
	heat shock response

For elaboration on behavioral adjustments, see Stevenson (1986); on physiological adjustments, see Bartholomew (1982) and Cossins and Bowler (1987).

V. HEAT HARDENING AND THE HEAT STRESS RESPONSE

"Heat hardening," a transient increase in the heat tolerance of an organism, is stimulated by briefly preexposing the animal to a near-lethal temperature (Hutchison and Maness 1979). This is an organismal-level phenomenon, and its mechanistic bases are unclear. The "heat stress response," which can also be triggered by brief exposure to high temperature (or by other environmental stimuli, Lindquist 1986), is also associated with an increase in heat tolerance. However, the heat stress response is a cellular-level phenomenon that specifically involves the induced synthesis of some proteins and often the repression of synthesis of others (Ashburner and Bonner 1979). It appears to have repair and protective functions (Li and Werb 1982; Pelham 1986; Schlesinger 1986; Landry et al. 1987; Riabowol et al. 1988; Chapters 1, 4, 10, 12, 14). Both heat hardening and the heat stress response have very fast response times, sometimes as fast as 15–30 minutes (Burns et al. 1986; Easton et al. 1987); both are transient phenomena. Moreover, both phenomena are activated by exposure to high body temperature (although the heat stress response is activated by many factors). In effect, these induced responses appear to reflect last-ditch, go-for-broke responses to short but dangerous environmental conditions. However, some salamanders in nature voluntarily expose themselves to temperatures high enough to induce hsp70 mRNA (Easton et al. 1988), suggesting that the adaptive bases of the heat stress response may be complex.

Studies of heat hardening seem to have developed largely independently from those of the heat stress response. Attention to heat hardening has come largely from organismal biologists (Hutchison and Maness 1979). An example of heat hardening in a salamander is shown in Figure 4. Easton and his co-workers (1987) have shown that the temperature necessary to induce heat spasms can be increased transiently by briefly exposing a salamander to a temperature 2°C below that which normally induces spasms. Interestingly, heat hardening can also be induced by exposing the salamander for 1 hour to temperatures as low as 5°C below the spasm temperature (P.S. Rutledge, pers. comm.). It is tempting to speculate that the heat shock response is the primary physiological mechanism behind heat hardening. However, several studies show that the acquisition of organismal heat tolerance can be uncoupled from hsp synthesis (Burns et al. 1986; Easton et al. 1987; Hightower and Renfro 1988). Thus, heat hardening and the heat stress response may sometimes be mechanistically unrelated.

To an evolutionary physiologist, two aspects of the heat stress response are striking. First, the omnipresence and remarkable conservatism of many heat stress proteins (at least the hsp70 family, Bienz 1984; Lindquist 1986) implies either (1) that stress is a common and universal experience of organisms or (2) that the proteins also play a fundamental role in normal physiological responses (Chapters 10, 12, 14). Otherwise, protein sequences would have long ago diverged by genetic drift. In any

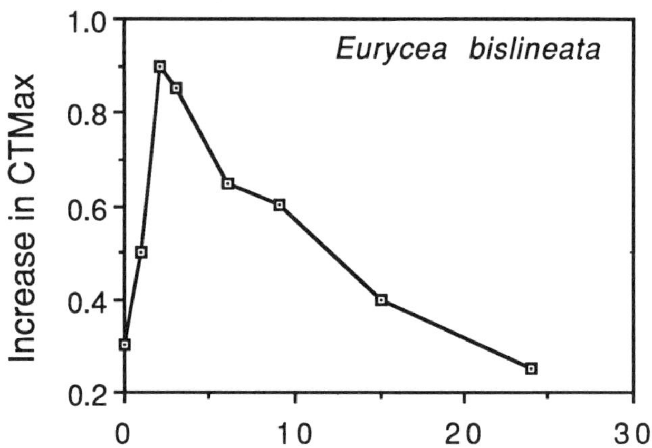

Figure 4 Heat hardening (i.e., increase in critical thermal maximum, CTM, measured as onset of spasms) in the salamander *Eurycea bislineata* following a 1-hr exposure to a temperature 2°C below an individual's (previously established) CTM. (Modified, with permission, from Easton et al. 1987.)

case, the conservatism in protein sequence strongly implies the response arose very early in evolution. Although the evolutionary importance of stress has long been recognized (Haldane 1932; Kingsolver and Watt 1983; Parsons 1983, 1987), the omnipresence of the heat stress response encourages organismal and evolutionary physiologists to pay closer attention to the role of stress in physiological adaptation. Second, the observation that some heat stress proteins are not routinely expressed is interesting, given the possible thermoprotective function they serve (Chapters 1, 4, 12, 14). A priori, one might indeed expect that organisms would maintain heat stress proteins as a "prophylactic" against stress. The nonconstitutive nature of some heat stress proteins might reflect either the excessive metabolic costs of maintaining the proteins (see Levins 1968, p. 11) or perhaps some detrimental effects on normal physiological functioning.

Nevertheless, heat shock proteins are present in some animals even in the apparent absence of a heat stress. For example, some heat stress proteins are seemingly constitutive in *Fundulus heteroclitus* (M. Koban, pers. comm.), a fish living in salt marshes and estuaries where temperatures frequently become very hot. Koban and his co-workers speculate that these proteins may provide the fish with "anticipatory" protection against sudden heat shock.

VI. RESPONSES TO SEASONAL FLUCTUATIONS

Some environmental fluctuations take place over a week or so (a weather front) or months (seasonal changes). For fluctuations of such length, the types of short-term adjustments discussed in the previous section would be inappropriate or perhaps inadequate. Accordingly, organisms can supplement these short-term adjustments with others that reversibly alter the animal's sensitivity to temperature. Such physiological reorganizations are called acclimation (in controlled laboratory environments) or acclimatization (in nature) responses.

The thermal sensitivities of many physiological processes can be radically shifted during acclimation, thereby compensating for the effects of temperature. For example, during chronic exposure to low temperature, many ectotherms increase metabolic rates, in effect thereby compensating for the depressing effects of cold on metabolism (Precht et al. 1973; Prosser 1986). Thus, when measured at common body temperatures, processing rates of cold-acclimated individuals are often greater than those of warm-acclimated ones. (Not all acclimation responses are "compensatory," and the presumed significance of these alternative responses is discussed in Tsuji [1988].) Acclimation is normally reversible if the animal is returned to its original body temperature, but only after a period of reacclimation. The mechanisms of capacity acclimation are highly varied, including synthesis of different isozymes, differential utilization of metabolic pathways, altered rates of protein synthesis and degradation, and changes in membrane composition (Hochachka and Somero 1984; Prosser 1986; Cossins and Bowler 1987).

Resistance adaptations (e.g., critical thermal limits) also respond to acclimation, perhaps even more frequently than do capacity adaptations such as metabolic rate. A particularly striking example is seen in seasonal shifts in critical temperatures (Fig. 5) of the crayfish (*Orconectes rusticus*), studied by Layne et al. (1987). Critical temperatures in crayfish in nature change dramatically, roughly paralleling seasonal changes in stream and probably body temperatures (Fig. 5).

Does the heat stress response show similar seasonal shifts? If so, the threshold temperature for protein induction (or the magnitude of the response) should be lower in winter than in summer. Although such seasonal shifts have not yet been examined for natural populations, several studies have compared the heat stress response in laboratory experiments where acclimation temperature was varied. For example, the threshold temperature for the heat stress response of hepatocytes of channel catfish (*Ictalurus punctatus*) is seemingly unaffected by acclimation temperature (however, the temperature steps were 2.5°C, which may obscure a subtle effect of acclimation on threshold temperatures), but the temperature for maximum synthesis is higher for warm-acclimated fish (Koban et al.

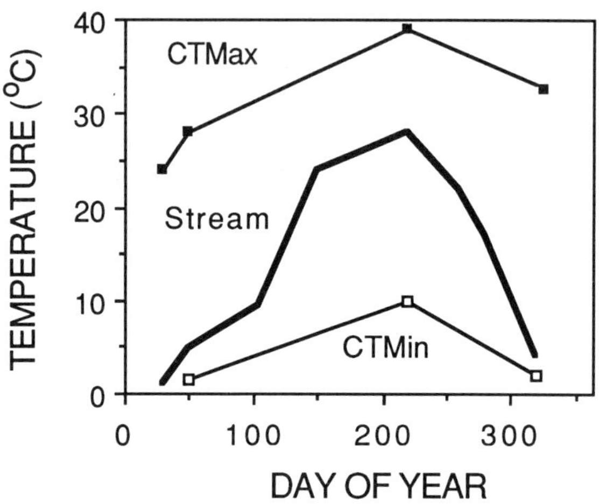

Figure 5 Critical thermal limits of the crayfish *Orconectes rusticus* and water temperatures at different seasons. Note the parallel shifts in critical limits and in water temperatures. (Redrawn, with permission, from Layne et al. 1987.)

1987). In unpublished studies, Koban and co-workers (pers. comm.) find that the heat stress response of the killifish (*Fundulus heteroclitus*) is also sensitive to acclimation temperature, but the patterns are complex. Interestingly, Bradley et al. (1988) found that copepods raised at low temperature (4°C or 15°C) induced somewhat different proteins following heat shock than did copepods raised at higher temperatures (20°C). These studies imply that some aspects of the dynamics of the heat stress response are affected by the prior thermal history of an organism. If so, acclimation protocols might prove to be a useful experimental tool for molecular biologists interested in manipulating the heat stress response.

Further studies of the effects of acclimation on the heat stress response of whole organisms would clearly be of interest, especially of temperate-zone organisms, which often experience dramatic seasonal shifts in body temperature (Fig. 5; Tsuji 1988). However, a caveat is warranted: Experiments designed to stimulate acclimation may need to vary photoperiod, not just temperature (Tsuji 1988). As Levins (1968, p. 13) noted, the type of environmental information (e.g., photoperiodic cues) an animal uses to anticipate environmental fluctuations may be different from the factor (e.g., temperature) for which the organism's physiology is actually compensating.

Although most acclimation responses are reversible, one special class—sometimes called "developmental switching"—results in phenotypic modifications that are fixed throughout life (Maynard Smith 1958; Levins 1968). Thus, the physiological or morphological phenotype of an organism is determined by the conditions experienced during development. For example, butterflies raised at low ambient temperatures have darker wings than those raised at higher ambient temperatures, and this switch is presumably adaptive by enhancing the effectiveness of basking at low ambient temperatures (Watt 1969). Similarly, exposure of newly emerged *Drosophila* females to high temperature (30.5°C for 5–12 days) has a profound effect on life span at 20°C, possibly by causing a permanent reduction in metabolic rate (Maynard Smith 1958).

Developmental switching is rarely studied in thermal biology, and we know of no studies examining whether the temperature threshold or the magnitude for the heat stress response depends on developmental experiences. In theory (Levins 1968), developmental switching should be restricted to species for which the projected environmental conditions last at least as long as the life span of the organism. Otherwise, the organism would have an inappropriate phenotype for part or much of its life. Accordingly, developmental switching should be found in animals (e.g., many microorganisms, insects) that go through several nonoverlapping generations during the course of a year. However, developmental switching could be favored in long-lived organisms that are distributed along sharp thermal gradients, such as up a mountain.

VII. RESPONSES TO LONG-TERM FLUCTUATIONS

As environmental temperatures change or as organisms colonize areas with different thermal conditions, selection may favor the evolution—not just acclimation—of altered thermal sensitivity of physiological processes (Levins 1968; Slobodkin and Rapoport 1974; Parsons 1983; Bradley and Ketzner 1982; Huey and Kingsolver 1989). Specifically, selection should favor evolutionary changes in thermal sensitivity that improve the match between the body temperatures experienced by the animals and those optimizing physiological and ecological performance (Huey and Bennett 1987; Lynch and Gabriel 1987). For example, because a lineage invading a hot desert will experience higher body temperatures, selection should favor the evolution of higher optimal temperatures and critical thermal limits. Do such evolutionary changes occur? We examine both interspecific and intraspecific patterns.

Most of the data germane to this topic are based on interspecific studies comparing species living in areas with different thermal regimes, and these data frequently document differences in both capacity and resistance adaptations (Hochachka and Somero 1984; Prosser 1986). Among different species of fish, for example, those from polar regions have substantially higher metabolic rates than do those from temperate waters when measured at common body temperatures; likewise, metabolic rates of temperate fish exceed those of tropical fish, a situation reminiscent of warm and cold temperature acclimation (Brett and Groves 1979). Resistance adaptations also vary dramatically with latitude. Specifically, upper and lower lethal (or critical) temperatures are typically high in the tropics and lower in temperate and especially polar zones (e.g., fish, Fig. 6; frogs [John-Alder et al. 1988], lizards [van Berkum 1988]). Remarkably, arctic fish die of heat stress at temperatures that would cause tropical fish to die of cold stress (Fig. 6). Resistance adaptations often vary with habitat (Prosser 1986). For example, bacteria show exceptional variation in heat resistance, and "thermophilic" bacteria in particular are heat tolerant (Henle 1987).

Have induction temperatures for the heat stress response similarly evolved to match environmental thermal regimes? Unfortunately, few data are available for heat stress responses of whole organisms, which may be different from those of cell cultures. Nevertheless, insects (e.g., desert locusts vs. *Drosophila*, Whyard et al. 1986) or fish (killifish or

Talapia vs. rainbow trout; Kothary and Candido 1982; Chen et al. 1988; M. Koban et al., pers. comm.) living in warm environments typically synthesize heat stress proteins at higher temperatures than do species living in cooler environments. Interestingly, initial studies of the Antarctic fish *Notothenia rossii,* which can tolerate temperatures no warmer than 12°C, suggest that hsp70 gene (incubated spleen) may be maximally expressed at 8°C (Maresca et al. 1988). Overall, these limited comparative data suggest that the evolution of the heat stress response may well be involved with the evolution of habitat selection and of geographic distribution (Parsons 1987; Spotila et al. 1989).

The interpretation of available comparative data on the heat stress response is, however, weakened by two considerations. First, most studies have looked at responses of cultured cells rather than of whole organisms: Organismal responses are more relevant both ecologically and evolutionarily (Bartholomew 1958, 1964; Huey and Stevenson 1979). Second, most studies involve comparisons of distantly related taxa rather than of closely related ones. Although desert locusts and *Drosophila* (or killifish, arctic fish, and rainbow trout) differ in their heat stress responses, these species have long had separate evolutionary histories. Consequently, any observed differences in their heat stress responses might merely reflect those separate histories and thus need not have anything to do with adaptation to specific thermal environments

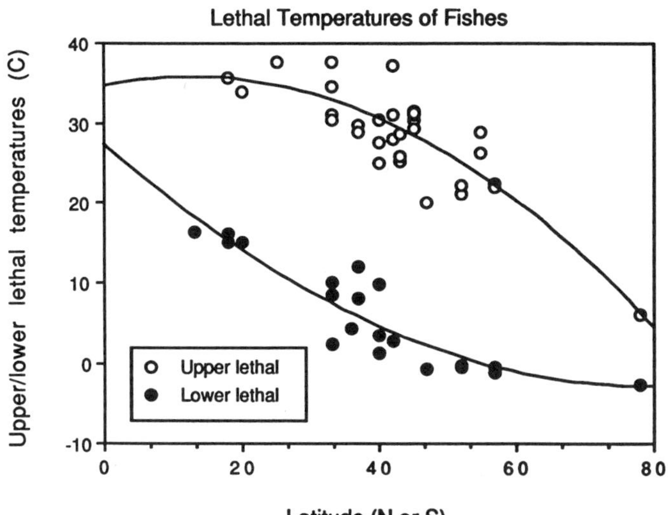

Figure 6 Upper and lower lethal temperatures for fishes (post-embryonic states only) from different latitudes. Curves fitted by polynomial regression. (Redrawn, with permission, from Brett 1970.)

that modern species currently occupy (Hochachka and Somero 1984; Huey and Bennett 1986; Huey 1987). This is the classic ambiguity plaguing any comparison of "apples and oranges."

One solution to the apples and oranges problem inolves studying geographic variation within rather than between species (Powers 1987). Very often heat tolerance of a population is directly related to the ambient temperature of the environment in which it lives (Hertz et al. 1983; Hochachka and Somero 1984; Prosser 1986; but see Coyne et al. 1983). However, geographic variation in the heat stress response has thus far been examined only in killifish, *Fundulus heteroclitus* (M. Koban, pers. comm.). Distributed along the eastern coast of the United States, this fish is subjected to an extraordinarily sharp thermal gradient (1°C per degree of latitude, Powers 1987). Different populations have different threshold temperatures for heat stress proteins, but the actual divergence between *Fundulus* from Maine and Georgia is minor. Curiously, northern populations initiate synthesis at a temperature slightly higher than do southern populations (M. Koban, pers. comm.).

An alternative solution to historical complications in comparative studies involves studying closely related species and then interpreting observed patterns with explicit reference to the phylogenetic relations and history of those species (Lauder 1981; Huey and Bennett 1986, 1987; Huey 1987). We recently developed an evolutionary approach for studying interspecific patterns of capacity and resistance adaptations, including the analysis of heat tolerance, with a method that uses phylogenetic information (Huey and Bennett 1987). We studied the thermal dependence of sprint speed and the critical thermal limits (temperatures at which the righting response is lost) of several species of scincid lizards from Australia that live either in temperate coastal regions or in hot interior deserts. These lizards differ strikingly in both optimal temperature for sprinting and in their critical thermal maxima, the upper temperature at which righting response is lost. Using a minimum-evolution or "parsimony" algorithm, we estimated the probable preferred temperature, critical thermal limits, and habitat associations of the hypothetical ancestors of this group (Fig. 7). This method allowed us to examine the probable *changes* in optimal temperatures and in heat tolerance that occurred when a given lineage invaded a new habitat.

A coastal distribution is thought to be ancestral in these Australian skinks (Greer 1980). However, two genera *(Eremiascincus* and *Ctenotus)* that we studied appear to have invaded the hot interior deserts of Australia (Fig. 7). Paralleling this change in habitat are increases in the critical thermal maxima (Fig. 7) and optimal sprint temperature (not shown) of both genera, presumably reflecting selection for higher heat tolerance in

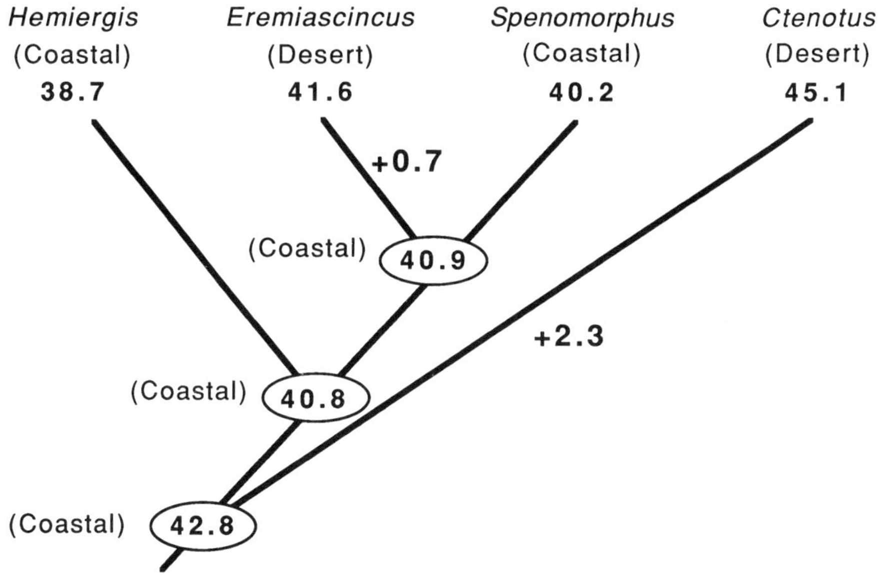

Figure 7 A partial phylogeny of some Australian lizards (Scincidae), showing critical thermal maxima (loss of righting response) and current habitat associations (desert vs. coastal) for the different genera and for their presumed ancestors. (Intrageneric variation in critical thermal maxima is minor in these lizards.) The values of hypothetical ancestors are estimated using an algorithm that minimizes the evolutionary change in the tree (see text). Note that the invasion of the Australian desert by two genera *(Eremiascincus, Ctenotus)* is associated with an apparent increased heat tolerance (critical thermal maximum). From data in Huey and Bennett (1987).

deserts. Whether the heat stress response has also evolved in the Australian skinks is currently unknown.

Drosophila would be an ideal genus for a phylogenetic analysis of the heat stress response: Good phylogenies are available for many species groups, species occur in diverse thermal habitats, species can differ dramatically in heat tolerance (Levins 1969; David et al. 1983; Parsons 1983), and the techniques for studying the heat stress response of *Drosophila* are well developed. Preliminary evidence (Table 2; S. Lindquist, pers. comm.) suggests that larvae of two desert species *(D. mojavensis* and *D. arizonensis)* have a higher temperature for maximal hsp70 synthesis than do larvae of *D. melanogaster*.

Phylogenetic techniques may also provide molecular biologists with a new comparative tool for testing functional hypotheses generated by conventional molecular techniques. Given both a phylogeny and protein sequences for a series of related species, one can use parsimony algorithms

Table 2 Heat stress responses of two desert species (*mojavensis* and *arizonensis*) of *Drosophila* are shifted to high temperature (°C) relative to those of *D. melanogaster*

		Temperature at	
Species	Habitat	maximum hsp70 synthesis	shutdown of all protein synthesis
melanogaster	widespread	35–37°C	37°C
mojavensis	desert	39°C	41°C
arizonensis	desert	41°C	43°C

Preliminary data, courtesy of S. Lindquist (pers. comm.). Larvae were raised at 25°C and acutely exposed to a graded series of temperatures.

to reconstruct ancestral protein sequences and then attempt to associate specific evolutionary changes in protein sequence with specific evolutionary changes in function. Indeed, this general approach has been used to trace the evolution of hemoglobin (Goodman et al. 1975; Goodman 1976), and it could well be used with the heat stress response.

Phylogenetic reconstructions are not without limitations. Historical reconstructions vary somewhat depending on the parsimony algorithm that is used, and parsimony itself can sometimes produce misleading inferences concerning history (Felsenstein 1983). Moreover, historical reconstructions are not amenable to experimental validation (Huey and Bennett 1986). Nevertheless, we argue that phylogenetic and historical considerations must be incorporated into comparative analyses. Otherwise, one is forced to accept the false assumption that organismal phenotypes are adapted only to contemporary conditions and thus bear no vestiges of history (Huey 1987).

VIII. SELECTION EXPERIMENTS

Evolutionary biologists sometimes use selection experiments to explore the dynamics of evolutionary change in a given trait. Selection experiments include two general classes (Rose and Graves 1989): (1) selective breeding of individuals with specified phenotypes ("artificial selection," often used to improve crop yields), or (2) raising outbred lines for several generations under specified environmental conditions ("natural selection" in the laboratory). Both types of studies provide information on the evolutionary lability of traits as well as of the degree to which their evolution might be constrained by genetic correlations with other traits (Falconer 1981; Arnold 1987). Selection experiments hold considerable potential as a tool for comparative or molecular physiologists (Hill 1980; Service 1987; Huey and Kingsolver 1989; Rose and Graves 1989).

Selection experiments have now been conducted on heat and cold resistance in flagellates (Dallinger 1887), wasps (White et al. 1970), *Drosophila* (David et al. 1983), and fish (Shah 1985). In all cases, heat and cold tolerance was altered by selection (for review, see Huey and Kingsolver 1989).

Natural and artificial selection have also been used to explore changes in heat resistance as well as in the heat stress response of *Drosophila melanogaster*. For example, Alahiotis and Stephanou (1982; see also Alahiotis 1983; Stephanou and Alahiotis 1983; Kilias and Alahiotis 1985) maintained two lines at different temperatures (14°C or 26°C), thereby subjecting them to natural selection for growth and reproduction at low versus high temperature. After 7 years, the high-temperature line had higher resistance to heat stress and also had a higher heat stress response than did the low-temperature line. Both differences seem to be maternally inherited (Stephanou and Alahiotis 1983). Interestingly, both heat sensitivity and the heat stress response apparently evolved even though the flies were kept at temperatures (14°C, 26°C) well below the supposed temperature threshold for the induction of heat stress proteins in *Drosophila*. Perhaps the heat stress response is active during chronic exposure to temperatures *well below* those required to trigger an acute response. In other words, perhaps induction of the response is sensitive not only to temperature, but also to the duration of exposure to temperature (Lindquist 1986; Easton et al. 1987; Rutledge et al. 1987a). If so, the tendency of most studies to focus exclusively on acute heat *shock* experiments may have inadvertently masked an inherent flexibility of the heat stress response.

Flies were also selectively bred for resistance or sensitivity to acute heat shock (40°C for 25 min; Alahiotis and Stephanou 1982; Stephanou and Alahiotis 1983; Kilias and Alahiotis 1985). Only 6.5% of flies from the sensitive line survived the heat shock, whereas 76.2% from the resistant line survived. Moreover, the capacity for synthesis of hsp70 following heat shock was seemingly higher for the resistant line than for the sensitive line (Stephanou et al. 1983).

Further selection studies on the heat stress response are certainly warranted. Of particular interest would be a comparison of the evolution of heat stress responses in lines subject to artificial selection for tolerance versus intolerance to high temperature (e.g., White et al. 1970; Morrison and Milkman 1978; Stephanou et al. 1983). Individual variation (Bennett 1987) in heat hardening and heat resistance is well known in diverse organisms (Bradley and Ketzner 1982; Hosgood and Parsons 1968; Rutledge et al. 1987b), suggesting that the heat stress response has the genetic potential to evolve (Bennett 1987; see also Stephanou et al. 1983; Rutledge et al. 1987b). Because of the conservatism in some hsp sequences, evolutionary change might be most conspicuous at the control level. If so, selection experiments might prove a useful tool for molecular biologists interested in control mechanisms.

IX. FUTURE STUDIES ON THE EVOLUTION AND ECOLOGY OF HEAT STRESS RESPONSES

The heat stress response is one of many ways (Table 1) that organisms use to buffer environmental stresses. On the basis of its kinetics and physiological side effects, the heat stress response appears to be a "last-resort" response to brief but severe perturbations. It may convey some thermoprotection and repair, but at some probable (though undefined) cost to the organism.

To an evolutionary biologist, the heat stress response offers a largely unexplored system for comparative study. Organismal biologists have only recently begun to exploit the opportunities provided by the heat stress response and by the technology of molecular biology (Alahiotis 1983; Parsons 1987; Spotila et al. 1989). Obvious topics in need of further work are acclimation sensitivity, ontogenetic changes in heat resistance and in the heat stress response (Lindquist 1986), and patterns of interspecific and intraspecific divergence (e.g., Table 2). These comparative studies may be of interest to molecular biologists because divergence in responses among related organisms may provide useful clues as to the regulation and functions of the response itself.

To a physiological ecologist, heat stress proteins may also serve as a useful bioassay of how frequently organisms are stressed in nature (Saunders 1988; Spotila et al. 1989) and perhaps even of the type or severity of the stress (Neidhardt et al. 1984; Rutledge et al. 1987a)—issues significant to debates in population biology and in physiological evolution (Haldane 1932; Kingsolver and Watt 1983; Parsons 1987; Spotila et al. 1989). The omnipotence and apparent conservatism of the heat stress response could suggest that organisms routinely experience stress in nature (see above). However, by determining the proportion of individuals expressing heat stress proteins as a function of geography or season or habitat, one could quantify the actual frequency of stress. Work along these lines has recently begun (Easton et al. 1988).

X. CONCLUDING REMARKS

The heat stress response encompasses an area of biology where organismal and molecular biologists can learn much from each other. Molecular techniques (e.g., hsp assays, genetic engineering) can help evolutionary physiologists measure the frequency of stress in nature as well as document the patterns of the evolution of the response. Conversely, organismal and evolutionary techniques (e.g., acclimation manipulations, phylogenetic analysis, selection experiments) provide molecular biologists with novel sources of "experimental" variation, which in turn may lead to new insights as to the mechanics and regulation of the heat stress response.

An expanded dialogue between molecular and evolutionary biologists will almost certainly be mutually beneficial. We suspect that organismal and evolutionary biologists will increasingly become involved in heat stress research now that the physiological and protective functions (if not the physiological costs) of the response at the *organismal* level are being clarified.

ACKNOWLEDGMENTS

We thank R. Morimoto and W. Welch for the opportunity to contribute to this volume. We are especially indebted to D. Easton, B. Grant, A. Dunham, M. Koban, S. Lindquist, D. Powers, P. Rutledge, J. Spotila, and B. Wu for graciously sharing unpublished data, for constructive advice, or for discussion. We also thank B. Bradley, D. Easton, J. Herron, R. Morimoto, P. Rutledge, J. Spotila, and M. Koban for useful comments on the manuscript. The preparation of this chapter was supported by National Science Foundation grants BSR-8414855 and BSR-8718063 to R.B.H. and by National Science Foundation grants DCB-8502218 and BSR-8600066 to A.F.B.

REFERENCES

Alahiotis, S.N. 1983. Heat shock proteins. A new view on the temperature compensation. *Comp. Biochem. Physiol.* **75B:** 379–387.

Alahiotis, S.N. and G. Stephanou. 1982. Temperature adaptation of *Drosophila* populations. The heat shock proteins system. *Comp. Biochem. Physiol.* **73B:** 529–533.

Arnold, S.J. 1987. Genetic correlation and the evolution of physiology. In *New directions in ecological physiology* (ed. M.E. Feder et al.), pp. 189–212. Cambridge University Press, United Kingdom.

Ashburner, M. and J.J. Bonner. 1979. The induction of gene activity in *Drosophila* by heat shock. *Cell* **17:** 241–245.

Bartholomew, G.A. 1958. The role of physiology in the distribution of terrestrial vertebrates. In *Zoogeography* (ed. C.L. Hubbs), pp. 81–95. American Association for the Advancement of Science, Washington, D.C.

———. 1964. The roles of physiology and behaviour in the maintenance of homeostatis in the desert environment. *Symp. Soc. Exp. Biol.* **18:** 7–29.

———. 1982. Physiological control of body temperature. In *Biology of the Reptilia* (ed. C. Gans and F.H. Pough), vol. 12 (Physiology C), pp. 167–211. Academic Press, New York.

Bennett, A.F. 1987. Interindividual variability: An underutilized resource. In *New directions in ecological physiology* (ed. M.E. Feder et al.), pp. 147–166. Cambridge University Press, United Kingdom.

Bennett, A.F., R.B. Huey, H. John-Alder, and K.A. Nagy. 1984. The parasol tail and thermoregulatory behavior of the Cape ground squirrel *Xerus inauris*. *Physiol. Zool.* **57:** 57–62.

Bienz, M. 1984. Developmental control of the heat shock response in *Xenopus*. *Proc. Natl. Acad. Sci.* **81:** 3138–3142.

Bradley, B.P. and P.A. Ketzner. 1982. Genetic and non-genetic variability in temperature tolerance of the copepod *Eurytemora affinis* in live temperature regimes. *Biol. Bull.* **162:** 233–245.

Bradley, B.P., R. Hakimzadeh, and J.S. Vincent. 1988. Rapid responses to stress in *Eurytemora affinis*. *Hydrobiologica* **167/168:** 197–200.

Brett, J.R. 1970. Temperature. Animals. Fishes. In *Marine ecology* (ed. O. Kinne), vol. 1, pp. 515–560. John Wiley, New York.

Brett, J.R. and T.D.D. Groves. 1979. Physiological energetics. In *Fish physiology* (ed. W.A. Hoar et al.), vol. 7, pp. 279–352. Academic Press, New York.

Burns, C.P., B.J. Lambert, B.N. Haugstad, and M.M. Guffy. 1986. Influence of rate of heating on thermosensitivity of 1-1210 leukemia membrane lipids and 70000-molecular-weight heat shock protein. *Cancer Res.* **46:** 1882–1887.

Casey, T. M. 1981. Behavioral mechanisms of thermoregulation. In *Insect thermoregulation* (ed. B. Heinrich), pp. 79–114. John Wiley, New York.

Chen, J.D., F.H. Yew, and G.C. Lid. 1988. Thermal adaptation and heat shock responses of *Tilapia* ovary cells. *J. Cell. Physiol.* **134:** 189–199.

Cossins, A.R. and K. Bowler. 1987. *Temperature biology of animals*. Chapman & Hall, London.

Cowles, R.B. and C.M. Bogert. 1944. A preliminary study of the thermal requirements of desert reptiles. *Bull. Am. Mus. Nat. Hist.* **83:** 265–296.

Coyne, J.A., J. Bundgaard, and T. Prout. 1983. Geographic variation of tolerance to environmental stress in *Drosophila pseudoobscura*. *Am. Nat.* **122:** 474–488.

Dallinger, W.H. 1887. Transactions of the Society. V. The President's Address. *J. R. Micros. Soc.*, pp. 185–199.

David, J.R., R. Allemand, J. Van Herrewege, and Y. Cohet. 1983. Ecophysiology: Abiotic factors. In *The genetics and biology of* Drosophila (ed. M. Ashburner et al.), pp. 105–170. Academic Press, New York.

Easton, D.P., P.S. Rutledge, and J.R. Spotila. 1987. Heat shock protein induction and induced thermal tolerance are independent in adult salamanders. *J. Exp. Zool.* **241:** 263–267.

Easton, D.P., J.C. Near, P.S. Rutledge, D.P. Dickinson, and J.R. Spotila. 1988. Induction and accumulation of hsp70 mRNA in adult salamanders in response to different heat stresses, including naturally occurring thermal stress conditions in the field. *J. Cell. Biochem.* (suppl.) **12D:** 289. (Abstr.)

Falconer, D.S. 1981. *Introduction to quantitative genetics,* 2nd ed. Longman, New York.

Felsenstein, J. 1983. Parsimony in systematics: Biological and statistical issues. *Annu. Rev. Ecol. Syst.* **14:** 313–333.

Goodman, M. 1976. Protein sequences in phylogeny. In *Molecular evolution* (ed. F. Ayala), pp. 141–159. Sinauer Associates, Sunderland, Massachusetts.

Goodman, M., G.W. Moore, and G. Matsuda. 1975. Darwinian evolution in the genealogy of haemoglobin. *Nature* **253:** 603–608.

Grant, B.W. and A.E. Dunham. 1988. Thermally imposed time constraints on the activity of the desert lizard *Sceloporus merriami*. *Ecology* **69:** 167–176.

Greer, A.E. 1980. Critical thermal maximum temperatures in Australian scincid lizards: Their ecological and evolutionary significance. *Aust. J. Zool.* **28:** 91–102.

Haldane, J.B.S. 1932. *The causes of evolution*. Harper Brothers, New York.

Henle, K.J., ed. 1987. *Thermotolerance and thermophily*. CRC Press, Boca Raton, Florida.

Hertz, P.E., R.B. Huey, and E. Nevo. 1983. Homage to Santa Anita: Thermal sensitivity of sprint speed in agamid lizards. *Evolution* **37:** 1075–1084.

Hightower, L.E. and J.L. Renfro. 1988. Recent applications of fish cell culture to biomedical research. *J. Exp. Zool.* **248:** 290–302.

Hill, W. G. 1980. Design of quantitative genetic selection experiments. In *Selection experiments in laboratory and domestic animals* (ed. A. Robertson), pp. 1–13. Commonwealth Agricultural Bureaux, Edinburgh, Scotland.

Hochachka, P.W. and G.N. Somero. 1984. *Biochemical adaptation*. Princeton University Press, New Jersey.

Hosgood, S.M.W. and P.A. Parsons. 1968. Polymorphisms in natural populations of *Drosophila* to withstand temperature shocks. *Experientia* **24:** 727–728.

Huey, R.B. 1982. Temperature, physiology, and the ecology of reptiles. In *Biology of the Reptilia* (ed. C. Gans and F.H. Pough), vol. 12 (Physiology C), pp. 25–91. Academic Press, New York.

———. 1987. Phylogeny, history, and the comparative method. In *New directions in ecological physiology* (ed. M.E. Feder et al.), pp. 76–101. Cambridge University Press, United Kingdom.

Huey, R.B. and A.F. Bennett. 1986. A comparative approach to field and laboratory studies in evolutionary biology. In *Predator-prey relationships: Perspectives and approaches from the study of lower vertebrates* (ed. M.E. Feder and G.V. Lauder), pp. 82–98. University of Chicago Press, Illinois.

———. 1987. Phylogenetic studies of coadaptation: Preferred temperatures versus optimal performance temperatures of lizards. *Evolution* **41:** 1098–1115.

Huey, R.B. and J.G. Kingsolver. 1989. Evolution of thermal sensitivity of ectotherm performance. *Trends Ecol. Evol.* **4:** 131–135.

Huey, R.B. and R.D. Stevenson. 1979. Integrating thermal physiology and ecology of ectotherms: A discussion of approaches. *Am. Zool.* **19:** 357–366.

Hutchison, V.H. and J.D. Maness. 1979. The role of behavior in temperature acclimation and tolerance in ectotherms. *Am. Zool.* **19:** 367–384.

John-Alder, H.B., P.J. Morin, and S. Lawlor. 1988. Thermal physiology, phenology, and distribution of tree frogs. *Am. Nat.* **132:** 506–520.

Kilias, G. and S.N. Alahiotis. 1985. Indirect thermal selection in *Drosophila melanogaster* and adaptive consequences. *Theor. Appl. Genet.* **69:** 645–650.

Kingsolver, J.G. and W.B. Watt. 1983. Thermoregulatory strategies in *Colias* butterflies: Thermal stress and the limits to adaptation in temporally varying environments. *Am. Nat.* **121:** 32–55.

Koban, M., G. Graham, and C.L. Prosser. 1987. Induction of heat-shock protein synthesis in teleost hepatocytes: Effects of acclimation temperature. *Physiol. Zool.* **60:** 290–296.

Kothary, R.K. and E.P.M. Candido. 1982. Induction of a novel set of polypeptides by heat shock or sodium arsenite in cultured cells of rainbow trout, *Salmo gairdneri*. *Can. J. Biochem.* **60:** 347–355.

Landry, J., S. Lamarche, and P. Chretien. 1987. Heat shock proteins: A lead to the understanding of cell thermotolerance. In *Thermotolerance* (ed. K.J. Henle), vol. I, pp. 146–172. CRC Press, Boca Raton, Florida.

Lauder, G.V. 1981. Form and function: Structural analysis in evolutionary morphology. *Paleobiology* **7:** 430–442.

Layne, J.R., Jr., D.L. Claussen, and M.L. Manis. 1987. Effects of acclimation temperature, season, and time of day on the critical thermal maxima and minima of the crayfish *Orconectes rusticus*. *J. Therm. Biol.* **12:** 183–187.

Levins, R. 1968. *Evolution in changing environments.* Princeton University Press, New Jersey.

———. 1969. Thermal acclimation and heat resistance in *Drosophila* species. *Am. Nat.* **103:** 483–499.

Li, G.C. and Z. Werb. 1982. Correlation between synthesis of heat shock proteins and development of thermal tolerance in Chinese hamster fibroblasts. *Proc. Natl. Acad. Sci.* **79:** 3218–3222.

Lindquist, S. 1986. The heat-shock response. *Annu. Rev. Biochem.* **55:** 1151–1191.

Lynch, M. and W. Gabriel. 1987. Environmental tolerance. *Am. Nat.* **129:** 283–303.

Maresca, B., E. Patriarca, C. Goldenberg, and M. Sacco. 1988. Heat shock and cold adaptation in Antarctic fishes: A molecular approach. *Comp. Biochem. Physiol.* **90B:** 623–629.

Maynard Smith, J. 1958. Prolongation of the life of *Drosophila subobscura* by a brief exposure of adults to a high temperature. *Nature* **181:** 496–497.

Morrison, W. and R. Milkman. 1978. Modification of heat resistance in *Drosophila* by selection. *Nature* **273:** 49–50.

Neidhardt, F.C., R.A. Van Bogelen, and V. Vaughn. 1984. The genetics and regulation of heat-shock proteins. *Annu. Rev. Genet.* **18:** 295–329.

Palumbi, S.R. 1984. Tactics of acclimation: Morphological changes of sponges in an unpredictable environment. *Science* **225:** 1478–1480.

Parsons, P.A. 1983. *Evolutionary biology of colonizing species.* Cambridge University Press, United Kingdom.

———. 1987. Evolutionary rates under environmental stress. *Evol. Biol.* **21:** 311–347.

Pelham, H.R.B. 1986. Speculations on the functions of the major heat shock and glucose-regulated proteins. *Cell* **46:** 959–961.

Powers, D.A. 1987. A multidisciplinary approach to the study of genetic variation within species. In *New directions in ecological physiology* (ed. M.E. Feder et al.), pp. 102–130. Cambridge University Press, United Kingdom.

Precht, H., J. Cristophersen, H. Hensel, and W. Larcher. 1973. *Temperature and life.* Springer-Verlag, New York.

Prosser, C.L. 1986. *Adaptational biology: Molecules to organisms.* John Wiley, New York.

Riabowol, K.T., L.A. Mizzen, and W.J. Welch. 1988. Heat shock is lethal to fibroblasts microinjected with antibodies against hsp 70. *Science* **242:** 433–436.

Rose, M.R. and J.L. Graves, Jr. 1989. What evolutionary biology can do for gerontology. *J. Gerontol.* **44:** B27–B29.

Rutledge, P.S., D.P. Easton, and J.R. Spotila. 1987a. Heat shock proteins from the lungless salamanders *Eurycea bislineata* and *Desmognathus ochrophaeus*. *Comp. Biochem. Physiol.* **87B:** 13–18.

Rutledge, P.S., J.R. Spotila, and D.P. Easton. 1987b. Heat hardening in response to two types of heat shock in the lungless salamanders *Eurycea bislineata* and *Desmognathus ochrophaeus*. *J. Therm. Biol.* **12:** 235–241.

Saunders, B.M. 1988. The role of the stress proteins response in physiological adaptation of marine molluscs. *Mar. Environ. Res.* **24:** 207–210.

Schlesinger, M.J. 1986. Heat shock proteins: The search for functions. *J. Cell Biol.* **103:** 321–325.

Service, P.M. 1987. Physiological mechanisms of increased stress resistance in *Drosophila melanogaster* selected for postponed senescence. *Physiol. Zool.* **60:** 321–326.

Shah, M.S. 1985. Thermal effects on allozyme polymorphism in the guppy *(Poecilia reticulata,* Peters). *Bangladesh J. Fish.* **8:** 35–51.

Slobodkin, L.B. and A. Rapoport. 1974. An optimal strategy of evolution. *Quart. Rev. Biol.* **49:** 181–200.

Spotila, J.R., P.W. Lommen, G.S. Bakken, and D.M. Gates. 1973. A mathematical model for body temperatures of large reptiles: Implications for dinosaur ecology. *Am. Nat.* **107:** 391–404.

Spotila, J.R., E.A. Standora, D.P. Easton, and P.S. Rutledge. 1989. Bioenergetics, behavior, and resource partitioning in stressed habitats: Biophysical and molecular approaches. *Physiol. Zool.* **62:** 253–285.

Stephanou, G. and S.N. Alahiotis. 1983. Non-mendelian inheritance of "heat-sensitivity" in *Drosophila melanogaster*. *Genetics* **103:** 93–107.

Stephanou, G., S.N. Alahiotis, C. Christodoulou, and V.J. Marmaras. 1983. Adaptation of *Drosophila* to temperature: Heat-shock proteins and survival in *Drosophila melanogaster*. *Dev. Genet.* **3:** 299–308.

Stevenson, R.D. 1985. Body size and limits to the daily range of body temperature in terrestrial ectotherms. *Am. Nat.* **125:** 102–117.

———. 1986. The relative importance of behavioral and physiological adjustments controlling body temperature in terrestrial ectotherms. *Am. Nat.* **126:** 362–386.

Tsuji, J.S. 1988. Thermal acclimation of metabolism in *Sceloporus* lizards from different latitudes. *Physiol. Zool.* **61:** 241–253.

Turner, J.S. and C.R. Tracy. 1983. Blood flow to appendages and the control of heat exchange in American alligators. *Physiol. Zool.* **56:** 195–200.

van Berkum, F.H. 1988. Latitudinal patterns of the thermal sensitivity of sprint speed in lizards. *Am. Nat.* **132:** 327–343.

Watt, W.B. 1969. Adaptive significance of pigment polymorphism in *Colias* butterflies. II. Thermoregulation and photoperiodically controlled melanin variation in *Colias eurytheme*. *Proc. Natl. Acad. Sci.* **63:** 767–774.

White, E.B., P. DeBach, and M.J. Garber. 1970. Artificial selection for genetic adaptation to temperature extremes in *Aphytis lingnanensis compere* (Hymenoptera: Aphelinidae) *Hilgardia* **40:** 161–192.

Whyard, S., G.R. Wyatt, and V.K. Walker. 1986. The heat shock response in *Locusta migratoria*. *J. Comp. Physiol. B* **156:** 813–817.

3
The Febrile Response

Matthew J. Kluger
Department of Physiology
The University of Michigan Medical School
Ann Arbor, Michigan 48109

I. FEVER VERSUS HYPERTHERMIA

Although fever has often been defined as any elevation in body temperature, this definition would result in considerable confusion between true fever and hyperthermia. In fever, as a result of the elevated thermoregulatory "set-point," a person's heat production increases, while at the same time heat loss decreases. Snell and Atkins (1968) defined four categories of body temperature based on the set-point theory of temperature regulation:

1. Normothermia—where set-point and actual body temperature coincide. This condition exists most of the time.
2. Hypothermia—where set-point may or may not be normal, but actual body temperature is below this set-point. Hypothermia can occur as a response to a drug, disease, or exposure to the cold.
3. Hyperthermia—where set-point may or may not be normal, but actual body temperature is higher than this set-point. Hyperthermia can occur as a response to a drug, disease, or exposure to a warm environment.
4. Fever—where set-point is raised and deep body temperature may or may not be raised to the same level.

Studies with human subjects (e.g., Cooper et al. 1964; Cabanac and Massonnet 1974) and with experimental animals (e.g., Cabanac et al. 1970; Sharp and Hammel 1972; Vaughn et al. 1974; Satinoff et al. 1976) have shown that the thermoregulatory set-point is indeed raised during fever.

Stress Proteins in Biology and Medicine.

II. MEDIATORS OF FEVER

For many years, fever in response to endotoxin or purified lipopolysaccharide (LPS) or other "exogenous pyrogen," was assumed to be caused by the release into the blood of a small protein known as "endogenous pyrogen" (EP) from activated macrophages (for reviews, see Atkins and Bodel 1974; Kluger 1979; Atkins 1982). A variety of studies led to the conclusion that circulating EP produced fever by acting on the central nervous system. For example, King and Wood (1958) found that an injection of EP into the carotid arteries of rabbits resulted in fevers of shorter latencies and greater magnitude than those resulting from injections into veins. Cooper et al. (1967), Jackson (1967), and others showed that injection of minute quantities of EP into the anterior hypothalamus produced fever. As a result of an inability to quantitatively measure the concentrations of EP in plasma or other body fluids, the question of whether fever in response to injection is actually attributable to EP was never addressed.

Within the past few years, it has often been stated that interleukin-1 (IL-1 α and β) is equivalent to EP (see, e.g., Dinarello 1984; Oppenheim 1986). IL-1 α and β are peptides produced by a variety of phagocytic cells that are thought to influence virtually every aspect of the stereotyped "acute-phase response" to infection and inflammation (Dinarello 1984; Kluger et al. 1985; Oppenheim 1986). These cytokines are thought to be responsible for stimulating T cells, for lowering the plasma concentrations of iron and zinc and raising that of copper, for increasing slow-wave sleep, for causing many of the aches and pains associated with infection, for suppressing food appetite, for stimulating the production of many acute-phase proteins, and for other actions (see Table 1). However, it is now apparent that other cytokines such as tumor necrosis factor (TNF) (Dinarello et al. 1986; Kettelhut and Goldberg 1988), interferon α (Dinarello et al. 1984; Feldberg and Scott 1984), interferon γ (Morimoto et al. 1987), and interferon β_2 (now known as interleukin-6 or IL-6; Nijsten et al. 1987) are also capable of producing many of the above acute-phase responses, including causing fever, when injected into experimental animals. On the basis of historical development of our understanding of the pathogenesis of fever, the assumption has generally been made that intravenous or intraperitoneal injection of LPS results in an elevation in the circulating concentration of IL-1 (or these other mediators). It is further assumed that IL-1 then crosses areas in the blood-brain barrier that are leaky and, in or near the anterior hypothalamus, causes an elevation in the thermoregulatory set-point, resulting in both physiologic and behavioral responses that cause body temperature to rise.

The fever that occurs following injection of these cytokines is thought to be mediated primarily by the release of prostaglandins in or near the region of the hypothalamus (Bernheim 1986; Stitt 1986). Antipyretic drugs, such as aspirin, are thought to exert their temperature-lowering action by blocking the synthesis of prostaglandins, thus returning the thermoregulatory set-point to its afebrile level.

There is no direct quantitative evidence that IL-1, TNF, or any other mediator currently assumed to be an endogenous pyrogen actually accounts for the fevers (or other acute-phase responses) induced by injection of LPS or other exogenous pyrogen. Evidence purporting to show this relationship suffers from the same serious problem as the earlier work with EP; that is, in virtually all of the studies used to determine the effects of a mediator, it is not known whether the concentrations of the mediator used in the experiments achieved plasma concentrations similar to those that occur in vivo. For example, Tocco-Bradley et al. (1986) have shown that in order to induce a fever in the laboratory rat, they had to infuse intravenously 7500 units of recombinant murine IL-1 per hour. Even were this IL-1 being cleared with a half-life of 2 minutes, as suggested by Kampschmidt and Jones (1985), the plasma concentration of IL-1 would probably be considerably greater than 10 units of IL-1 per milliliter, a concentration much higher than that measured by anyone to date. Kettelhut and Goldberg injected rats with 10^5 units/g of TNF (~10^7 units of TNF) to produce large fevers that last about 7 hours. In ongoing studies conducted in our laboratory, we have found that the plasma concentration of TNF in rats injected with a dose of LPS that produced a fever similar to that reported by Kettelhut and Goldberg (1988) resulted in variable increases in the plasma concentration of TNF from ~10 units/ml to about 600 units/ml; probably far less than would be found following intraperitoneal (i.p.) injection of 10^7 units of TNF (Long et al. 1989a).

Another putative EP is IL-6. At the International Workshop on Monokines and Nonlymphocytic Cytokines, Billiau and van Damme (1987) have shown that IL-1, TNF, and other mediators stimulate the production and release of IL-6. Nijsten et al. (1987) have shown that body temperature of patients with severe burns correlates with plasma concentration of IL-6, and concluded that "IL-6 plays a causal role in the acute phase response." However, they provided no data to document that the concentration of IL-6 found in these patients actually is capable of inducing the acute-phase responses seen. Helle et al. (1988) have shown that purified IL-6 induces fever in rabbits. However, as in the studies using IL-1 or TNF, it is not known whether the amount injected that caused fever (0.5–5 μg/kg) achieved plasma concentrations within the

Table 1 Reported actions of IL-1 and TNF

Target	Action
T lymphocytes	co-mitogenesis chemotaxis
B lymphocytes	co-mitogenesis Ia antigen expression chemotaxis
Natural killer cells	increased cytotoxicity induction of IL-2 receptors
Mononuclear phagocytes	differentiation of precursors induction of cytotoxic, antimicrobal, and reactive oxygen intermediates secretory capacity secretion of IL-1 secretion of TNF production of thromboxane chemotaxis
Neutrophils	inhibition of precursors degranulation, lysosomal enzyme release secretion of reactive oxygen intermediates production of thromboxane enhancement of phagocytosis, ADCC adhesion to endothelium surface display of C3bi receptors chemotaxis
Eosinophils	toxicity to schistosomes adhesion to endothelium
Endothelium	enhanced proliferation induction of procoagulant activity decreased synthesis of plasminogen activator decreased activation of protein C increased synthesis of plasminogen activator inhibitor increased platelet-activating factor increased PGE_2 and prostacyclin secretion of Von Willebrand factor secretion of IL-1 secretion of G/M-CSF, G-CSF, M-CSF induction of HLA-A, B antigens, p100.120 antigen induction of ELAM-1, ICAM-1

Table 1 (Continued.)

Target	Action
Endothelium (cont'd.)	adhesion of neutrophils, monocytes, lymphocytes, natural killer cells
	secretion of platelet-derived growth factor
Fibroblasts	increased proliferation
Synoviocytes	synthesis of PGE_2, collagenase
	secretion of interferons
	secretion of proteinase inhibitor
	secretion of hyaluronidase
	secretion of CSF-G/M
	secretion of IL-1
	expression of HLA-A, B antigens
	antiviral activity
Chondrocytes	deceased synthesis of proteoglycans
	increased proteoglycanase, collagenase, PGE_2
Osteoclasts	increased proliferation and bone resorption
Osteoblasts	decreased bone synthesis
Adipocytes	decreased synthesis of lipoprotein lipase
Hepatocytes	induction of serum amyloid A and P
	complement factor B haptoglobin
	α_2 macroglobin
	inhibition of albumin synthesis
Astroglia	increased proliferation
Mesangial cells	increased proliferation
	neutral proteinase secretion
Pancreatic islets	cytotoxicity
Tumor cells	cytotoxicity
	antiviral activity
In vivo, loci of action not certain	hemorrhagic necrosis of some tumors
	increased plasma ACTH, corticosteroids
	increased turnover of plasma amino acids
	decreased plasma iron, zinc
	decreased plasma glucose
	accumulation of neutrophils
	epidermal hyperplasia, fibrosis
	decreased lethality from radiation
	decreased food intake
	fever
	increased slow-wave sleep

Based largely on Nathan (1987), as modified by Kluger and Liu (1990).

normal physiologic or pathophysiologic range. In studies recently carried out in my laboratory, we have found that IL-6 injected intracerebroventricularly into rats produced dose-dependent fevers that were blockable with indomethacin (LeMay et al. 1989). When we injected rats with fever-inducing doses of LPS (2.5 μg/kg and 10 μg/kg), plasma and cerebrospinal fluid IL-6 activity increased. During the rising and plateau phases of fever, there was a positive correlation of 0.84 ($p < 0.005$) between the magnitude of the rises in plasma IL-6 activity and the temperature increases with the two doses of LPS. These data are consistent with the hypothesis that circulating IL-6 is a mediator of LPS fever. However, this relationship between plasma IL-6 and fever following LPS must be compared to that following exogenous administration of IL-6, and attempts must be made to block the LPS-induced temperature rise with antibodies to IL-6.

Ansel et al. (1987) have found that both fever and increased serum IL-1 activity occurred in rabbits as a result of ultraviolet (UV) irradiation. These authors suggested that the chills and fever that often accompany sunburn may be attributable to IL-1. However, the temporal relationship between fever and plasma IL-1 activity was poor. In response to UV irradiation, rabbits developed a fever that peaked at 5 hours, fell rapidly toward control values, and was close to baseline sometime between 7 and 24 hours; in contrast, the peak in plasma IL-1 activity occurred at 8 hours and was still at this elevated level 24 hours later. As noted by the authors of this study, "the post-UV fever curve was not identical to the post-UV serum IL-1 activities, with the rabbit core body temperature decreasing even as the serum IL-1 activity was increasing." They suggested that the dissociation between fever and IL-1 activity may be a regulatory compensatory mechanism that prevented the harmful catabolic activities associated with prolonged fevers. Another obvious possibility is that circulating IL-1 is not the physiological endogenous pyrogen in this situation.

One of the major problems preventing quantitation in this field has been lack of sensitive assays that would enable investigators to measure the plasma or extracellular concentrations of mediators such as IL-1. None of the published immunoassays for IL-1 (e.g., Kenney et al. 1987) is sensitive enough to measure the plasma concentrations of IL-1 that occur during infection. Moreover, although at the Workshop on Monokines and Other Nonlymphocytic Cytokines several individuals did report on radioimmunoassays sensitive enough to measure the plasma concentrations of IL-1 in human beings, there is a poor correlation between this immunoactivity and bioactivity. A recent report shows that plasma concentrations of TNF increase, as measured by immunoassay, following an injection of modest amounts (4 ng/kg) of LPS in human subjects (Michie et al. 1988). However, there also appears to be a poor relationship between immunological reactivity and bioactivity for TNF (Petersen and Moller 1988), and so the relationship between bioactive circulating TNF and fever is not clear.

To summarize, it has not been shown that the administration of any endogenous pyrogen in doses adequate to produce the plasma concentration of that endogenous pyrogen induced by LPS or other exogenous pyrogen elicits a fever equal to that elicited by the exogenous pyrogen.

A second approach to testing a causal role of any mediator in the induction of fever in response to exogenous pyrogen is to use specific antibodies to block the action of the mediator. In an experiment in my laboratory, we found that i.p. injection of rats with antiserum against IL-1 α did not block LPS fever (Long et al. 1989b). The dose of antiserum injected i.p. was sufficient to block fever caused by subsequent intramuscular injection of IL-1. These data do not support the hypothesis that circulating IL-1 α is involved in LPS fever in the rat. Experiments are currently in progress to assess whether injection of antisera against IL-1 β would influence LPS fever.

To attempt to determine whether TNF is involved in LPS-induced fever, we have injected rats i.p. with antisera to TNF or with control sera (Long et al. 1989b). After 2 hours, each rat was injected with a fever-inducing dose of LPS (10 μg/kg). The rats injected with antisera to TNF had a significantly attenuated initial rise in body temperature; however, the second peak of the "bimodal" fever was significantly larger. These data suggest to us that TNF may be involved in the initiation or first phase of fever. Endogenously produced TNF might then act (either directly or indirectly) as an endogenous antipyretic limiting the magnitude of fever; the removal or inactivation of this TNF by antisera to TNF removes this antipyretic action, thus resulting in the enhanced fever.

It is possible that a circulating endogenous pyrogen is not the critical variable, but that elevated brain levels of EP are essential for fever development. Within the past few years, it has become apparent that cells within the central nervous system are capable of producing IL-1 (Fontana et al. 1984; Guilian and Lachman 1985; Nieto-Sampedro and Berman 1987). In the study by Fontana et al. (1984), mice were injected with 50 μg LPS i.p., and IL-1 was assayed for and found in the brain of these mice. These observations have raised the possibility that in response to injection of LPS or other mediators of inflammation, it is the IL-1 produced locally by the brain (rather than the systemic plasma concentration of IL-1) that may be important in the development of fever. Unfortunately, no dose response was reported in the study by Fontana et al. (1984).

Assuming the mice weighed 25 g, the dose of LPS they used would be equivalent to 2000 μg/kg. This dose, in terms of fever induction, is an enormous one; it is unclear whether a dose of LPS that would produce a modest fever (e.g. half-maximal) would result in a significant rise in brain concentration of IL-1. In a more recent study, Coceani et al. (1988) found that injection of a fever-inducing dose of endotoxin did not result in any increase in IL-1 in the cerebrospinal fluid of the cat.

To our knowledge, there are no published studies indicating that TNF is produced by brain tissue. Grau et al. (1987) have shown that TNF is likely to be responsible for the pathogenesis of cerebral malaria in a murine model. This TNF, however, is presumably derived from adherence of macrophages to cerebral blood vessels, rather than being brain-derived TNF. At the Workshop on Monokines and Other Non-lymphocytic Cytokines, A. Billiau (pers. comm.) mentioned that IL-6 has been found in cerebrospinal fluid of infected patients. G. Wong (pers. comm.) (from Genentech) then noted that glioblastoma cell lines have a high constitutive level of IL-6 mRNA. These data raise the possibility that IL-6 produced in the central nervous system may play a role in fever.

In summary, although many cytokines can induce fever when injected or infused into experimental animals, it is still unclear which, if any, of the putative endogenous pyrogens actually play a role in infection-induced fever.

III. IS STRESS HYPERTHERMIA A FEVER?

Exposure to novel environments or handling produces rises in body temperature of as much as 2°C in rats (Gollnick and Ianuzzo 1968; Briese and De Quijada 1970; Goldstein and Lowery 1975; Blasig et al. 1978; Stewart and Eikelboom 1979; Briese and Cabanac 1980; York and Regan 1982; Singer et al. 1986; Kluger et al. 1987, 1989; Long et al. 1989b). Rises in body temperature of people in response to psychological stress have also been reported (Renbourn 1960; Hasan and White 1979). Several studies have shown that part of this stress-induced hyperthermia in rats can be blocked with naloxone, suggesting that endogenous opioids are responsible for a portion of this rise in body temperature. The role of other mediators remains to be determined.

In studies conducted in our own laboratory using biotelemetry, we have found that exposing rats for 30 minutes to the novel environment of an open field (Singer et al. 1986), or simply switching the cages of rats (Kluger et al. 1989), results in a rise in body temperature of similar magnitude to that found during endotoxin or LPS fever. Most of the rise in body temperature can be blocked by pretreatment of the rats with an antipyretic dose of sodium salicylate or of indomethacin (a more potent prostaglandin blocker). A similar reduction in body temperature was found following intracerebroventricular injection of sodium salicylate (Kluger et al. 1987). Since hyperthermias attributable to endogenous opioids are not blocked by antipyretic drugs such as sodium salicylate (for review, see Clark 1981), these data suggest to us that at least part of the stress-induced rise in body temperature may be a true fever, mediated by prostaglandins, and ultimately by the acute release of IL-1 α and β, TNF, IL-6, or other putative endogenous pyrogens.

IV. IS FEVER BENEFICIAL?

With few exceptions, both endothermic and ectothermic vertebrates (as well as invertebrates) develop fevers in response to injections of endotoxin or other substances pyrogenic to mammals (Table 2). There have been only a few reports of ectothermic vertebrates that have failed to develop fever: the lizard *Cordylus cataphractus* (Laburn et al. 1981) to rabbit endogenous pyrogen and to heat-killed bacteria, the leopard tortoise *Geochelone pardalis* to LPS (Zurovsky et al. 1987), the teleost fish *Lepomis gibbosus* to endotoxin and to prostaglandin E_1 (Marx et al. 1984), and the snail *Limnaea auricularia* to a variety of pyrogens (Cabanac and Rossetti 1987).

Negative results should always be viewed with caution, particularly in the area of the biology of fever. A fever might not develop in an animal for any number of reasons. One is that the pyrogen used is not the appropriate stimulus for that species. For example, a dose of LPS that causes a large fever in the rabbit (e.g., 5 ng/kg) does not cause any detectable fever in the rat (our unpublished observations). A bacterium that might cause sickness and perhaps even death in one species might be "seen" as an innocuous stimulus in another and, therefore, not trigger an elevation in the thermoregulatory set-point. The dose of pyrogen might be either too low or too high. For example, in the rabbit, a dose-dependent fever will develop in response to injections of LPS; however, at high doses, perhaps because of the induction of cardiovascular shock, body temperature often falls. Another reason to be particularly cautious when interpreting negative results in the area of fever research is that both behavioral and physiologic responses are often used to raise body temperature. A "stressed"' animal, for several reasons, might not be able to develop a fever. One possible reason for this could be related to the suppressive effects of glucocorticoids on the release of endogenous pyrogens or on the production of prostaglandins. Another reason might be related to the phenomenon of stress-induced hyperthermia described

Table 2 Febrile responses of ectothermic vertebrates and invertebrates

	Activator of fever	References
Reptiles		
Dipsosaurus dorsalis	bacteria, IL-1	1, 2
Iguana iguana	bacteria	3
Crotaphytus collaris	bacteria	4
Terrepene carolina	bacteria	5
Chrysemys picta	bacteria	5
Amphibians		
Hyla cinerea	bacteria	6
Rana pipiens	bacteria	7
Rana catesbeiana	bacteria	7
Rana esculenta	bacteria, PGE_1, IL-1?	8
Necturus maculosus	PGE_1	9
Fishes		
Micropterus salmoides	bacteria	10
Lepomis macrochirus	endotoxin, bacteria	10, 11
Carassius auratus	endotoxin, bacteria	12, 13
Invertebrates (Arthropods)		
Cambarus bartoni (crayfish)	bacteria	14
Gromphadorhina portentosa (cockroach)	endotoxin, bacteria	15
Gryllus bimaculatus (cricket)	*Rickettesiella grylli*	16
Melanoplus sanguinipes (grasshopper)	*Nosema acridophagus*	17
Homarus americanus (lobster)	PGE_1	18
Penaeus duorarum (shrimp)	PGE_1	18
Limulus polyphemus (horseshoe crab)	PGE_1	18
Buthus occitanus (scorpion)	PGE_1	19
Androctonus australis (scorpion)	PGE_1	19
Onymacris plana (tenebrionid beetle)	endotoxin	20

Reference key. [1]Vaughn et al. (1974); [2]Bernheim and Kluger (1977); [3]Kluger (1978); [4]Firth et al. (1980); [5]Monagas and Gatten (1983); [6]Kluger (1977); [7]Casterlin and Reynolds (1977); [8]Myhre et al. (1977); [9]Hutchison and Erskine (1981); [10]Reynolds et al. (1976); [11]Reynolds et al. (1978a); [12]Reynolds et al. (1976); [13]Reynolds et al. (1978b); [14]Casterlin and Reynolds (1980); [15]Bronstein and Conner (1984); [16]Louis et al. (1986); [17]Boorstein and Ewald (1987); [18]Casterlin and Reynolds (1979); [19]Cabanac and Guelte (1980); [20]McClain et al. (1988).

above. The stressed animal may already have an elevated temperature, thus obscuring any fever-inducing effect of a pyrogen.

The results of studies of fever in "lower" animals strongly suggest that fever has an ancient phylogenetic history. Fever probably evolved hundreds of millions of years ago, as a means of enhancing host defense responses to infection. The rationale for speculating about the role of fever in disease, based on its evolutionary history, is as follows. It is improbable that fever would have evolved and been retained throughout the vertebrates and invertebrates without being of benefit to the host. There is a large metabolic cost associated with elevating and then maintaining body temperature even 1 or 2°C above normal. Most biochemical reactions have a Q_{10} of approximately 2 to 3, and this means that for every 1°C rise in body temperature, metabolic rate increases about 10% or more.

Occasionally the argument is made that some host responses to infection are harmful. There are few physiological (including immunological) responses that are, predominantly, maladaptive. Although certain inflammatory responses are thought to be injurious (such as inflammation in knee joints or allergic reactions to pollen), it is generally agreed that the inflammatory response is essential to survival in a hostile world. Without the movement of phagocytes into a damaged area to destroy or inactivate the pathogens and the subsequent activation of repair mechanisms, minor breaks in epithelial linings could result in massive infection and potentially death.

Other than the "evolutionary" argument developed above, do hard data exist in favor of the idea that fever is beneficial? Within the past decade, numerous studies have demonstrated that small elevations in body temperature, similar to those observed during fever, result in an enhancement of the immune response. Four examples will be briefly discussed. These include increased mobility and activity of white blood cells, stimulation of interferon production and function, activation of T lymphocytes, and the effect of hypoferremia on the growth of pathogens.

Once a potential pathogen breaks through the protective skin or epithelial barriers lining the respiratory or digestive system, the next line of defense is probably the activation of the polymorphonuclear leukocyte or neutrophil. These cells rapidly migrate to the site of infection and then phagocytize the foreign substances. Phagocytosis results in a burst of activity leading to the production of many antibacterial substances, including hydrogen peroxide, superoxide anion, lysozyme, and lactoferrin (Davis and Gallin 1981). Studies from several laboratories have shown that febrile temperatures result in more rapid neutrophil migration (Nahas et al. 1971; Bernheim et al. 1978) and secretion of antibacterial chemicals (van Oss et al. 1980; Johansen et al. 1983).

The interferons are a family of proteins that exert potent antiviral and antitumor, as well as antibacterial, effects. These actions of interferon are enhanced at febrile temperatures (Heron and Berg 1978; Yerushalmi et al. 1982). In addition, the in vivo production of interferon in the rhesus monkey is increased at febrile temperatures (Downing et al. 1984, 1987). Not only do the production and activity of interferon increase at febrile temperatures, but interferon itself appears to be pyrogenic (Dinarello et al. 1984; Feldberg and Scott 1984; Morimoto et al. 1987).

As mentioned above, IL-1 is thought to exert many effects on host defense mechanisms. Perhaps the most widely studied is its effect on the activation of a group of white blood cells responsible for "cell-mediated" immunity, the T lymphocyte. The "activated" T lymphocyte undergoes proliferation, thus enabling the cell to exert its antiviral and antitumor actions. This T-cell proliferation is facilitated by fever (Roberts and Steigbidel 1977; Gupta and Agarwal 1980; Duff and Durum 1982; Hanson et al. 1983; Jampel et al. 1983; Hanson and Murphy 1985; Lederman et al. 1987).

Another effect of IL-1 (and presumably IL-6) is to reduce the plasma iron concentration. Several studies have shown that this hypoferremia serves to reduce the growth rate of many species of bacteria (for review, see Weinberg 1978). Although the reduction in plasma iron concentration is independent of the presence of fever (Tocco et al. 1983), there appears to be a synergism between fever and hypoferremia in reducing the growth of bacteria (Garibaldi 1972; Kochan 1977; Grieger and Kluger 1978; Kluger and Rothenburg 1979), since many species of bacteria are less able to produce iron-chelating proteins at febrile temperatures and, therefore, are unable to obtain adequate iron for growth.

In addition to the above studies that have focused on the effects of fever on specific immune functions, there have been several investigations involving the effects of fever on mortality and morbidity during bacterial and viral infections. In general, these studies have shown that moderate fevers have a beneficial effect on the outcome of infections. For example, lizards or goldfish infected with bacteria have higher survival rates when febrile (Kluger et al. 1975; Covert and Reynolds 1977). Newborn mammals infected with a variety of viruses also have higher survival rates when febrile (Carmichael et al. 1969; Haahr and Mogensen 1977). Suppression of fever using antipyretic drugs results in increased influenza virus in ferrets (Husseini et al. 1982) and increased mortality rate in bacterially infected rabbits (Vaughn et al. 1980). Clinical studies have also shown a correlation between fever and decreased morbidity and mortality rate during a variety of infections (Weinstein et al. 1978; Hoefs et al. 1980; Yerushalmi and Lwoff 1980).

It is important to emphasize that although fever probably evolved as an adaptive host-defense response to infection, not all fevers must be beneficial. In terms of evolution, a trait merely needs to have survival value in order to have evolved and been retained. This simply means, as alluded to above, that on a statistical basis, fever is beneficial. However, in individual cases, fever may be maladaptive. For example, in people with heart conditions or wasting due to cachexia, fever might pose an unmanageable stress. In addition, high fevers during pregnancy might result in an increased incidence of birth defects.

Finally, it is not known whether the temperatures observed during most fevers trigger a heat shock response. If a high fever induces the synthesis of heat shock proteins, these may protect the *cell* against the potentially harmful effects of elevated temperature. At the same time, the elevated temperature may rev up host defenses and protect the organism.

REFERENCES

Ansel, J.C., T.A. Luger, and I. Green. 1987. Fever and increased serum IL-1 activity as a systemic manifestation of acute phototoxicity in New Zealand white rabbits. *J. Invest. Dermatol.* **89:** 32–37.

Atkins, E. 1982. Fever: Its history, cause and function. *Yale J. Biol. Med.* **55:** 283–289.

Atkins, E. and E. Bodel. 1974. Fever. In *The inflammatory process*, 2nd edition, (ed. B.W. Zweibach et al.), vol. 3, pp. 467–511. Academic Press, New York.

Bernheim, H.A. 1986. Is prostaglandin E_2 involved in the pathogenesis of fever: Effects of interleukin-1 on the release of prostaglandins. *Yale. J. Biol. Med.* **59:** 151–158.

Bernheim, H.A. and M.J. Kluger. 1977. Endogenous pyrogen-like substance produced by reptiles. *J. Physiol.* **267:** 659–666.

Bernheim, H.A, B.T. Bodel, P.W. Askenase, and E. Atkins. 1978. Effects of fever on host defence mechanisms after infection in the lizard *Dipsosaurus dorsalis*. *Br. J. Exp. Pathol.* **59:** 76–84.

Billiau, A. and J. van Damme. 1987. The role of interferon B_1 (IFN-B_1) and the 26kDa protein (INF-B_2, HPGF, BSF-2) as mediators of the antiviral effect of interleukin-1 (IL-1) and tumor necrosis factor (TNF_a). *J. Leukocyte Biol.* **42:** 553–554.

Blasig, J., V. Hollt, U. Bauerle, and A. Herz. 1978. Involvement of endorphins in emotional hyperthermia of rats. *Life Sci.* **23:** 2525–2532.

Boorstein, S.M. and P.W. Ewald. 1987. Costs and benefits of behavioral fever in *Melanoplus sanguinipes* infected by *Nosema acridophagus*. *Physiol. Zool.* **60:** 586–595.

Briese, E. and M. Cabanac. 1980. Emotional fever and salicylate. In *The Satellite Symposium on Thermal Physiology*, Abstract no. 9. pp. 161–163. Pecs, Hungary.

Briese, E. and M.G. De Quijada. 1970. Colonic temperature of rats during handling. *Acta Physiol. Latinoam.* **20:** 97–102.

Bronstein, S.M. and W.E. Conner. 1984. Endotoxin-induced behavioural fever in the Madagascar cockroach, *Gromphadorhina portentosa*. *J. Insect Physiol.* **30:** 327–330.

Cabanac, M. and L.L. Guelte. 1980. Temperature regulation and prostaglandin E_1 fever in scorpions. *J. Physiol.* **303:** 365–370.

Cabanac, M. and B. Massonnet. 1974. Temperature regulation during fever: Change of set point or change of gain? A tentative answer from a behavioral study in man. *J. Physiol.* **238:** 561–568.

Cabanac, M. and Y. Rossetti. 1987. Fever in snails, reflection on a negative result. *Comp. Biochem. Physiol.* **87A:** 1017–1020.

Cabanac, M., R. Duclauz, and A. Gillet. 1970. Thermoregulation comportementale chez le chien: Effets de la fievre et al thyroxine. *Physiol. Behav.* **5:** 697–704.

Carmichael, L.E., F.D. Barnes, and D.H. Percy. 1969. Temperature as a factor in resistance of young puppies. *J. Infect. Dis.* **120:** 669–678.

Casterlin, M.E. and W.W. Reynolds. 1977. Behavioral fever in anuran amphibian larvae. *Life Sci.* **20:** 593–596.

———. 1979. Fever induced in marine arthropods by prostaglandin E_1. *Life Sci.* **25:** 1601–1604.

———. 1980. Fever and antipyresis in the crayfish *Cambarus bartoni*. *J. Physiol.* **303:** 417–421.

Clark, W.G. 1981. Effects of opioid peptides on thermoregulation. *Fed. Proc.* **40:** 2754–2759.

Coceani, F., J. Lees, and C.A. Dinarello. 1988. Occurrence of interleukin-1 in cerebrospinal fluid of the conscious cat. *Brain Res.* **446:** 245–250.

Cooper, K.E., W.I. Cranston, and A.J. Honour. 1967. Observations on the site and mode of action of pyrogens in the rabbit brain. *J. Physiol.* **191:** 325–337.

Cooper, K.E., W.I. Cranston, and E.S. Snell. 1964. Temperature regulation during fever in man. *Clin. Sci.* **27:** 345–356.

Covert, J.B. and W.W. Reynolds. 1977. Survival value of fever in fish. *Nature* **267:** 43–45.

Davis, J.M. and J.I. Gallin. 1981. The neutrophil. In *Cellular functions in immunity and inflammation* (ed. J.J. Oppenheim et al.), pp. 77–102. Elsevier/North Holland, New York.

Dinarello, C.A. 1984. Interleukin-1. *Rev. Infect. Dis.* **6:** 51–95.

Dinarello, C.A., H.A. Bernheim, G.W. Duff, H.V. Le, T.L. Nagabhushan, N.C. Hamilton, and F. Coceani. 1984. Mechanisms of fever induced by recombinant human interferon. *J. Clin. Invest.* **74:** 906–913.

Dinarello, C.A., J.G. Cannon, S.M. Wolff, H.A. Bernheim, B. Beutler, A. Cerami, I.S. Figari, M.A. Palladino, Jr., and J.V. O'Connor. 1986. Tumor necrosis factor (cachectin) is an endogenous pyrogen and induces production of interleukin 1. *J. Exp. Med.* **163:** 1433–1450.

Downing, J.F., M.W. Taylor, K.-M. Wie, and R.S. Elizondo. 1987. In vivo hyperthermia enhances plasma antiviral activity and stimulates peripheral lymphocytes for increased synthesis of interferon gamma. **7:** 185–193.

Downing, J.F., H. Martinez-Valdez, T. Long, R. Elizondo, and M.W. Taylor. 1984. In vivo induction of IFN-gamma at febrile temperatures. In *The biology of the interferon system* (ed. H. Kirchner and H. Schellekens), pp. 429–432. Elsevier Science Publishers BV, Amsterdam.

Duff, G.W. and S.K. Durum. 1982. Fever and immunoregulation: Hyperthermia, interleukins 1 and 2, and T-cell proliferation. *Yale J. Biol. Med.* **55:** 437–442.

Feldberg, W. and G.M. Scott. 1984. Interferon fever in cats. *J. Physiol.* **355:** 399–28P.

Firth, B.T., C.L. Ralph, and T.J. Boardman. 1980. Independent effects of the pineal and a bacterial pyrogen in behavioural thermoregulation in lizards. *Nature* **285:** 399–400.

Fontana, A., E. Weber, and J.-M. Dayer. 1984. Synthesis of interleukin 1/endogenous pyrogen in the brain of endotoxin-treated mice: A step in fever induction? *J. Immunol.* **133:** 1696–1698.

Garibaldi, J.A. 1972. Influence of temperature on the biosynthesis of iron transport compounds by *Salmonella typhimurium*. *J. Bacteriol.* **110:** 262–265.

Goldstein, A. and P. Lowery 1975. Effect of the opiate antagonist naloxone on body temperature in rats. *Life Sci.* **17:** 927–932.

Gollnick, P.D. and C.D. Ianuzzo. 1968. Colonic temperature response of rats during exercise. *J. App. Physiol.* **24:** 747–750.

Grau, G.E., L.F. Fajardo, P.-F. Piguet, B. Allet, P.-H. Lambert, and P. Vassalli. 1987. Tumor necrosis factor (cachectin) as an essential mediator in murine cerebral malaria. *Science* **237:** 1210–1212.

Grieger, T.S. and M.J. Kluger. 1978. Fever and survival: The role of serum iron. *J. Physiol.* **279:** 187–196.

Guilian, D. and L.B. Lachman. 1985. Interleukin-1 stimulation of astroglial proliferation after brain injury. *Science* **228:** 497–499.

Gupta, S. and S. Agarwal. 1980. Effect of hyperthermia on *in vitro* proliferative response of lymphocyte subpopulations. *Fed. Proc.* **39:** 916.

Haahr, S. and S. Mogensen. 1977. Function of fever. *Lancet* **II:** 613.

Hanson, D.F. and P.A. Murphy. 1985. Temperature sensitivity of interleukin-dependent murine T cell proliferation: Q_2 mapping of the responses of peanut agglutinin-negative thymocytes. *J. Immunol.* **135:** 3011–3020.

Hanson, D.F., P.A. Murphy, R. Silican, and H.S. Shin. 1983. The effect of temperature on the activation of thymocytes by interleukins I and II. *J. Immunol.* **130:** 216–221.

Hasan, M.K., and A.C. White. 1979. Psychogenic fever: Entity or nonentity. *Postgrad. Med.* **66:** 152–154.

Helle, M., J.P.J. Brakenhoff, E.R. de Groot, and L.A. Aarden. 1988. Interleukin 6 is involved in interleukin 1-induced activities. *Eur. J. Immunol.* **18:** 957–959.

Heron, I. and K. Berg. 1978. The actions of interferon are potentiated at elevated temperature. *Nature* **274:** 508–510.

Hoefs, J., F.L. Sapico, H.N. Canawati, and J.Z. Montgomerie. 1980. The relationship of white blood cell (WBC) and pyrogenic response to survival in spontaneous bacterial peritonitis (SBP). *Gastroenterology* **78:** 1308.

Husseini, R.H., C. Sweet, M.H. Collie, and H. Smith. 1982. Elevation of nasal viral levels by suppression of fever in ferrets infected with influenza viruses of differing virulence. *J. Infect. Dis.* **145:** 520–524.

Hutchison, V.H. and D.J. Erskine. 1981. Thermal selection and prostaglandin E_1 fever in the salamander *Necturus maculosus. Herpetologica* **37:** 195–198.

Jackson, D.L. 1967. A hypothalamic region responsive to localized injection of pyrogens. *J. Neurophysiol.* **30:** 586–602.

Jampel, H.D., G.W. Duff, R.K. Gershon, E.A. Atkins, and S.K. Durum. 1983. Fever immunoregulation: III. Hyperthermia augments the primary in vitro humoral immune response. *J. Exp. Med.* **157:** 1229–1238.

Johansen, K.S., E.M. Berger, and J.E. Repine. 1983. Effect of temperature on polymorphonuclear leukocyte function. *Acta Pathol. Microbiol. Immunol. Scand. Sect. C Immunol.* **91:** 355–359.

Kampschmidt, R.F. and T. Jones. 1985. Rate of clearance of interleukin-1 from the blood of normal and nephrectomized rats. *Proc. Soc. Exp. Biol. Med.* **180:** 170–173.

Kenney, J.S., M.P. Masada, E.M. Eugui, B.M. Delustro, M.A. Mulkins, and A.C. Allison. 1987. Monoclonal antibodies to human recombinant interleukin 1 (IL-1) β: Quantitation of IL 1 β and inhibition of biological activity. *J. Immunol.* **138:** 4236–4242.

Kettelhut, I.C. and A.L. Goldberg. 1988. Tumor necrosis factor can induce fever in rats without activating protein breakdown in muscle or lipolysis in adipose tissue. *J. Clin. Invest.* **81:** 1384–1389.

King, M.K. and W.B. Wood, Jr. 1958. Studies on the pathogenesis of fever. IV. The site of action of leucocytic and circulating endogenous pyrogen. *J. Exp. Med.* **107:** 291–303.

Kluger, M.J. 1977. Fever in the frog *Hyla cinerea. J. Therm. Biol.* **2:** 79–81.

———. 1978. The evolution and adaptive value of fever. *Am. Sci.* **66:** 38–43.

———. 1979. *Fever: Its biology, evolution, and function.* Princeton University Press, New Jersey.

Kluger, M.J., and D.Y. Liu. 1990. The role of interleukin-1 and tumor necrosis factor in ischemic injury. In *Clinical ischemic syndromes: Mechanisms and consequences of tissue injury* (ed. G.B. Zelenock et al.), pp. 287–302. Mosby Press, St. Louis.

Kluger, M.J. and B.A. Rothenburg. 1979. Fever and reduced iron: Their interaction as a host defense response to bacterial infection. *Science* **188:** 166–168.

Kluger, M.J., J.J. Oppenheim, and M.C. Powanda, eds.. 1985. *Progress in leukocyte biology. Proceedings of International Symposium held in Ann Arbor, Michigan, June 4-6 1985*. Alan R. Liss, New York.

Kluger, M.J., B. O'Reilly, and A.J. Vander. 1989. Polymyxin attenuates stress induced hyperthermia. In *Thermoregulation: Research and clinical applications. 7th International Symposium of Thermoregulation, Odense, Denmark 1988* (ed. P. Lomax and E. Schonbaum), pp. 171–175. Karger, Basel.

Kluger, M.J., D.H. Ringler, and M.R. Anver. 1975. Fever and survival. *Science* **188:** 166–168.

Kluger, M.J., B.O. O'Reilly, T.R. Shope, and A.J. Vander. 1987. Further evidence that stress hyperthermia is a fever. *Physiol. Behav.* **39:** 763–766.

Kochan, I. 1977. Role of siderophores in nutritional immunity and bacterial parasitism. In *Microorganisms and minerals.* (ed. E.D. Weinberg), pp. 251–288. Marcel Dekker, New York.

Laburn, H.P., D. Mitchell, E. Kenedi, and G.N. Louw. 1981. Pyrogens fail to produce fever in cordylid lizard. *Am. J. Physiol.* **241:** R198–R202.

Lederman, H.M., C.R. Brill and P.A. Murphy. 1987. Interleukin 1-driven secretion of interleukin 2 is highly temperature-dependent. *J. Immunol.* **138:** 3808–3811.

LeMay, L.G., A.J. Vander, and M.J. Kluger. 1989. The role of IL-6 in fever in the rat. *Am. J. Physiol.* (in press).

Long, N.C., M.J. Kluger, and A.J. Vander. 1989a. Antiserum against mouse IL-1 does not block stress hyperthermia or LPS fever in the rat. In *Thermoregulation: Research and clinical applications. 7th International Symposium of Pharmacology of Thermoregulation, Odense, Denmark 1988* (ed. P. Lomax and E. Schonbaum), pp. 78–84. Karger, Basel.

Long, N.C., S.L. Kunkel, A.J. Vander, and M.J. Kluger. 1989b. Antiserum against TNF enhances LPS fever in the rat. *Am. J. Physiol.* (in press).

Louis, C., M. Jourdan, and M. Cabanac. 1986. Behavioral fever and therapy in a rickettsia-infected Orthoptera. *Am. J. Physiol.* **250:** R991–R995.

Marx, J., R. Hilbig, and H. Rahmann. 1984. Endotoxin and prostaglandin E_1 fail to induce fever in a teleost fish. *Comp. Biochem. Physiol.* **77A:** 483–487.

McClain, E., P. Magnuson, and S.J. Warner. 1988. Behavioral fever in a Namib desert tenebrionid beetle, *Onymacris plana. J. Insect Physiol.* **34:** 279–284.

Michie, H.R., K.R. Manogue, D.R. Spriggs, A. Revhaug, S. O'Dwyer, C.A. Dinarello, A. Cerami, S.M. Wolff, and D.W. Wilmore. 1988. Detection of circulating tumor necrosis factor after endotoxin administration. *N. Engl. J. Med.* **318:** 1481–1486.

Monagas, W.R. and R.E. Gatten, Jr. 1983. Behavioural fever in the turtles *Terrapene carolina* and *Chrysemys picta. J. Therm. Biol.* **8:** 285–288.

Morimoto, A., N. Murakami, M. Takada, S. Teshirogi, and T. Watanabe. 1987. Fever and the acute phase response induced in rabbits by human recombinant interferon-γ. *J. Physiol.* **391:** 209–218.

Myhre, K.M., M. Cabanac, and G. Myhre. 1977. Fever and behavioural temperature regulation in the frog *Rana esculenta*. *Acta Physiol. Scand.* **101:** 219–229.

Nahas, G.G., M.L. Tannieres, and J.F. Lennon. 1971. Direct measurement of leukocyte motility: Effects of pH and temperature. *Proc. Soc. Exp. Biol. Med.* **138:** 350–352.

Nathan, C.F. 1987. Secretory products of macrophages. *J. Clin. Invest.* **79:** 319-326.

Nieto-Sampedro, M. and M.A Berman. 1987. Interleukin-1-like activity in rat brain: Sources, targets, and effect of injury. *J. Neurosc. Res.* **17:** 214–219.

Nijsten, M.W.N., E.R. De Grrot, H.J. Ten Duis, H.J. Klasen, C.E. Hack, and L.A. Aarden. 1987. Serum levels of interleukin-6 and acute phase response. *Lancet* **II:** 921.

Oppenheim, J.J. 1986. Interleukins and interferons and inflammation. In *Current concepts*. The Upjohn Company, Kalamazoo, Michigan.

Petersen, C.M., and B.K. Moller. 1988. Immunological reactivity and bioactivity of tumour necrosis factor. *Lancet* **I:** 934–935.

Renbourn, E.T. 1960. Body temperature and pulse rate in boys and young men prior to sporting contests. A study in emotional hyperthermia: With a review of the literature. *J. Psychosom. Res.* **4:** 149–175.

Reynolds, W.W., M.E. Casterlin, and J.B. Covert. 1976. Behavioural fever in teleost fishes. *Nature* **259:** 41–42.

———. 1978a. Febrile responses of bluegill *(Lepomis macrochirus)* to bacterial pyrogens. *J. Therm. Biol.* **3:** 129–130.

Reynolds, W.W., J.B. Covert, and M.E. Casterlin. 1978b. Febrile responses of goldfish *Carassius auratus* to *Aeromonas hydrophila* and to *Escherichia coli* endotoxin. *J. Fish Dis.* **1:** 271–273.

Roberts, N.J., Jr. and R.T. Steigbigel. 1977. Hyperthermia and human leukocyte functions: Effects on response of lymphocytes to mitogen and antigen and bactericidal capacity of monocytes and neutrophils. *Infect. Immunol.* **18:** 673–679.

Satinoff, E., G.N. McEwen, Jr., and B.A. Williams. 1976. Behavioral fever in newborn rabbits. *Science* **193:** 1139–1140.

Sharp, F.R. and H.T. Hammel. 1972. Effects of fever on salivation response in the resting and exercising dog. *Am. J. Physiol.* **223:** 77–82.

Singer, R., C.T. Harker, A.J. Vander, and M.J. Kluger. 1986. Hyperthermia induced by open-field stress is blocked by salicylate. *Physiol. Behav.* **6:** 1179–1182.

Snell, E.S. and E. Atkins. 1968. The mechanisms of fever. In *The biological basis of medicine* (ed. E.E. Bittar and N. Bittar), pp. 397–419. Academic Press, New York.

Stewart, J. and R. Eikelboom. 1979. Stress masks the hypothermic effect of naloxone in rats. *Life Sci.* **25:** 1165–1172.

Stitt, J.T. 1986. Prostaglandin E as the neural mediator of the febrile response. *Yale J. Biol. Med.* **59:** 137–149.

Tocco, R.J., L.L. Kahn, M.J. Kluger, and A.J. Vander. 1983. Relationship of trace metals to fever during infection: Are prostaglandins involved? *Am. J. Physiol.* **244:** R368–R373.

Tocco-Bradley, R.T., L.L. Moldawer, C.T. Jones, B. Gerson, G.L. Blackburn, and S.R. Bistrian. 1986. The biological activity in vivo of recombinant murine interleukin in the rat. *Proc. Soc. Exp. Biol. Med.* **182:** 263–271.

van Oss, C.J., D.R. Absolom, L.L. Moore, B.H. Park, and J.R. Humbert. 1980. Effect of temperature on chemotaxis, phagocytic engulfment, digestion and O_2 consumption of human polymorphonuclear leukocytes. *J. Reticuloendothelial. Soc.* **27:** 561–565.

Vaughn, L.K., H.A. Bernheim, and M.J. Kluger. 1974. Fever in the lizard *Dipsosaurus dorsalis*. *Nature* **252:** 473–474.

Vaughn, L.K., W.L. Veale, and K.E. Cooper. 1980. Antipyresis: Its effect on mortality rate of bacterially infected rabbits. *Brain Res. Bull.* **5:** 69–73.

Weinberg, E.D. 1978. Iron and infection. *Microbiol. Rev.* **42:** 45–66.

Weinstein, M.P., P.B. Iannini, C.W. Stratton, and T.C. Eickhoff. 1978. Spontaneous bacterial peritonitis. A review of 28 cases with emphasis on improved survival and factors influencing prognosis. *Am. J. Med.* **64:** 592–598.

Yerushalmi, A. and A. Lwoff. 1980. Traitement du coryza et des rhinites persistantes allergizues par la thermotherapie. *C.R. Acad. Sci. Ser. D* **291:** 857–959.

Yerushalmi, A., M. Tovey, and I. Gresser. 1982. Antitumor effect of combined interferon and hyperthermia in mice. *Proc. Soc. Exp. Biol. Med.* **169:** 413–415.

York, J.L. and S.G. Regan. 1982. Conditioned and unconditioned influences on body temperature and ethanol hypothermia in laboratory rats. *Pharmacol. Biochem. Behav.* **17:** 119–124.

Zurovsky, Y., D. Mitchell, and H. Laburn. 1987. Pyrogens fail to produce fever in the leopard tortoise, *Geochelone pardalis*. *Comp. Biochem. Physiol.* **87A:** 467–469.

4

Thermotolerance, Thermoresistance, and Thermosensitization

George M. Hahn
Department of Radiation Oncology
Stanford University School of Medicine
Stanford, California 94305

Gloria C. Li
Department of Radiation Oncology
Radiation Oncology Research Laboratory
University of California, San Francisco, California 94143

I. INTRODUCTION

When organisms or cells in culture are exposed to temperatures that are outside their normal growth range, the response to that change is modulated by many environmental factors. These include parameters such as availability of nutrients, including glucose and oxygen, and the pH of the old and the new milieu, to name two of the more important ones. Our concern in this chapter is, however, with an intrinsic determinant that also influences responses to temperature changes, namely, the thermal history of the organism. The importance of thermal history has been known for a considerable time, particularly to plant pathologists (Alexandrov 1977) and, for that matter, to farmers and gardeners. These know from experience that if a plant is to be moved from a cold to a warm environment, the probability of its survival is vastly enhanced if the move is not made in one step, but gradually. In other words, if the plant is given a chance to adapt to small changes in temperature, it acquires additional thermal resistance. This acquired characteristic increases the temperature range of survival. In the last few years, it has been shown that the same considerations apply not only to plants, but also to bacteria, mammalian cells, and even complex organisms such as rodents. Humans survive fevers; these reach temperatures that kill unadapted cells in vitro. Human cells in vivo must, therefore, also develop temporary heat resistance. The molecular basis of this adaptive process has, until recently, been totally unknown. We now have begun to realize that the expression of the heat-tolerant phenotype is part of a complex series of responses, primarily on a cellular level. The organism initiates these responses after (or during) presentation of thermal and other stresses. Acquired, transient heat resistance has been termed thermotolerance. It has been studied most extensively in mammalian cells, largely because of the upsurge in interest in the use of heat as an adjunctive technique for the treatment of some cancers in humans (Hahn 1982; Field and Franconi 1987; Paliwal et al. 1988). The clinical use of hyperthermia usually involves multiple exposures of tumors to elevated temperatures. Clearly, it is of considerable importance to know whether a past treatment affects the response of cells to subsequent heat exposures. Therefore, the kinetics of induction and decay of thermotolerance are of great interest. Furthermore, not only is it important to know whether the heat response of cells is modified by their thermal history, but it is at least equally important to determine whether the radiation (or drug) sensitivity of cells surviving hyperthermic exposures has been altered. Thus, many studies have examined the response of thermotolerant cells to drugs or X-irradiation (Dewey and Holahan 1987; Hahn 1987).

There is another reason for studying the thermotolerant or, indeed, the heat-resistant cell. Many investigators have attempted to determine why mammalian (and other) cells die when exposed to mildly elevated temperatures. Medical applications have also provided the impetus for performing these studies that attempt to correlate morphological or biochemical changes with cell survival. Heat, however, modifies many organelles and affects cellular functions. Therefore, correlations are easily established between temperature exposure and damage to various cellular subsystems. However, so many of these correlations have been found that these studies have not provided convincing data on what constitutes the critical lesion (or if indeed one such target exists). For this reason, other investigators have attempted to determine what organelles or functions are protected by thermotolerance. The idea here is that if only a few specific organelles or functions are protected, it is reasonable to assume that among these must be the lesion responsible for cell death.

A variant of this approach has been to examine the chemosensitivity of thermotolerant and genetically resistant cells. If, for individual agents or classes of agents, consistent cross-correlations can be found between thermotolerance and drug tolerance, it can be argued that the mode of action of cell killing by these drugs has a specific relationship to cell killing by heat. Since the mode of cytotoxicity for most drugs is well known, inferences regarding how heat kills cells can then reasonably be drawn.

II. HEAT RESISTANCE AND THERMOTOLERANCE: ARE THEY THE SAME?

Mammalian cells vary appreciably in their intrinsic resistance to heat. Differences in the ability of cells to survive elevated temperatures occur in cells of widely differing origins, as well as in cells derived from a common parent line. In a study performed by Raaphorst et al. (1979), it was found that cells derived from different species varied in their heat sensitivities. These investigators examined several cellular parameters to see if any of these correlated with heat response. The only cross-correlation found by them was with the normal body temperature of the animal that was the source of the particular cell line. Cells derived from animals whose normal body temperature was high were heat-resistant; cells derived from animals with low body temperatures were heat-sensitive. Among the most heat-resistant cells are those derived from birds; during flight the normal temperatures of avians may rise to 43°C. For most mammals, 35°C–40°C constitutes the normal range. This finding suggested a reasonable hypothesis, namely, that permanent heat resistance is simply a genetic reflection of a constitutive level of whatever is the molecular basis for thermotolerance.

In at least one cell system, this hypothesis, despite its attractiveness, does not appear to provide the correct answer. We have selected a series of heat-resistant strains from a cell line derived from a murine radiation-induced fibrosarcoma (RIF-1) (Hahn and van Kersen 1988). These heat-resistant mutants (TR) were derived by repeated cycles of heat exposures followed after each exposure by regrowth of the surviving cells at 37°C. After seven cycles, each consisting of a 45°C heat exposure lasting 45 minutes followed by 6–10 days at 37°C, clones derived from single cells were selected and examined for a variety of properties. These are listed in Table 1 (Hahn and van Kersen 1988). When the heat resistance of the most resistant strains was compared to the resistance of cells from the parent line that had been made maximally thermotolerant, survival curves were indeed quite similar, consistent with the hypothesis stated earlier. If, however, the TR cells were preheated (30 min at 45°C) and 8 hours later subjected to graded exposures at 46°C, these cells were found

Table 1 Comparison of several responses of RIF-1 cells and the heat-resistant TR strains

	RIF-1	RIF-TR-1	RIF-TR-3	RIF-TR-4	RIF-TR-5	RIF-TR-6	RIF-TR-8	RIF-TR-10	RIF-TR-20
Plating efficiency in vitro (%)	79	75	79	79	100	100	72	84	72
Cells surviving 90 min at 45°C in vitro (%)	10^{-5}	1	4	76	0.6	2	0.5	0.2	2.8
Doubling time of cells in vitro (hours)	9.6	14.4	12	14.4	9.0	n.m.	14.4	16.8	19.2
Plating efficiency in vivo-in vitro (%)	48	43	6	57	19	82	6	38	51
Cells surviving 60 min at 45°C in vivo (%)	10^{-5}	4	n.m.	10	32	4[a]	n.m.	40	25
Tumor doubling time (days)	3	>10	>10	3	2.5	n.m.	4.5	3	8
Relative antigenicity[b]	1	0.3	n.m.	0.8	4.0	n.m.	n.m.	0.3	1.5

n.m.: Not measured.

[a]After exposure to 45°C for 45 min.

[b]Relative antigenicity is a measure of the ability of pre-irradiated cells to inhibit tumor formation by live cells in syngeneic hosts. There is no statistical difference between any of the values shown. (Data from Hahn and van Kersen 1987.)

to have taken on considerable additional thermotolerance. The TR strains made additionally thermotolerant were far more resistant to heating than thermotolerant wild-type cells. This is illustrated in Figure 1. If the degree of thermotolerance is defined as the change in slope of the respective survival curves, then the heat-resistant strains took on as much additional thermotolerance as did the parent line. Thermotolerance decayed when cells were returned to 37°C in both the wild and TR strains at rates that were not appreciably different. These sets of experiments suggest that, at least in the RIF-TR system, genetic heat resistance and phenotypic thermotolerance are different phenomena.

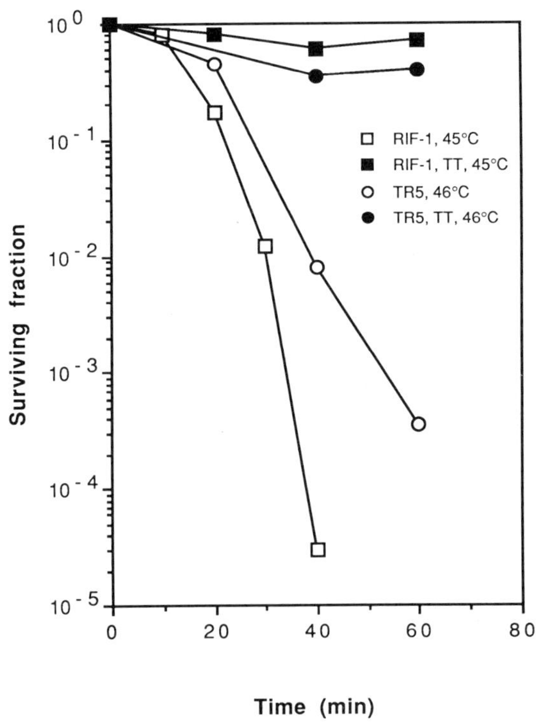

Figure 1 Comparison of thermotolerance induced in RIF-1 cells and in a heat-resistant mutant. Cells of the radiation-induced fibrosarcoma (RIF-1) and cells from a heat-resistant mutant (RIF-TR5) isolated from the RIF-1 cells were given isosurvival exposures (RIF-1: 15 min at 45°C; RIF-TR5: 20 min at 46°C). The response of cells that had not been made thermotolerant is shown for comparison. Induction is obvious for both the RIF-1 cells and the TR mutants. Furthermore, if thermotolerance is quantified by taking the ratios of the slopes of the appropriate survival curves, the amount of tolerance induced is similar in the two strains. Thus, the data indicate that, at least in these cells, thermotolerance and thermoresistance are not coupled. (Data from Hahn and van Kersen 1988.)

The TR strains raise an interesting point. When we made an estimate of the frequency of a TR cell occurring in a population of RIF-1 cells, this frequency was calculated to be roughly 1 in 10^{11}. At the initial heating, however, only about 10^7 cells were at risk. This makes it very unlikely that the initial population contained any of the very heat-resistant cells. Heat in mammalian cells is known to be minimally, if at all, mutagenic. If selection is unlikely, and induced mutagenesis is also unlikely, how then did TR cells arise? One possibility, which is highly unlikely, is that of heat-directed mutagenesis specifically for heat resistance. Another, perhaps more likely, explanation is that among cells showing partial heat resistance, the probability of mutations leading to even more increased resistance remains more or less constant until a final plateau is reached.

III. INDUCTION OF THERMOTOLERANCE BY HEAT

The earliest studies on mammalian cells that demonstrated that heat induces a transient state of heat resistance were those of Henle and Leeper (1976) and Gerner and Schneider (1975). For example, Henle and Leeper found that thermotolerance can be induced in Chinese hamster cells by a short (5–15 min) 45°C heat treatment followed by incubation at 37°C. Resistance was demonstrated by showing that the survival curve to a second heating of 45°C differed sharply from that of unheated cells in the direction of greater heat resistance. Examination of the kinetics of induction and decay of tolerance showed that immediately after heating, the cells were maximally sensitive; an interval of about 8 hours at 37°C was required before maximum thermotolerance manifested itself. Resistance persisted for at least 48 hours. Macromolecular synthesis was also inhibited following the initial heat exposure; protein synthesis resumed in parallel to the development of tolerance. RNA and DNA synthesis was slower to recover. Below about 43°C, tolerance can also be induced *during* continuous heating: At these lower temperatures, protein synthesis continues. An Arrhenius plot of the induction of thermotolerance is shown in Figure 2B; there, it is compared to the Arrhenius plot for cell killing (Fig. 2A) and for induction of synthesis of one of the heat shock proteins (hsp70) (Fig. 2C). The slopes of the straight lines can be interpreted as being a measure of the activation energy for the rate-limiting reactions involved. The value ~120 kcal/mole suggests protein unfolding or other protein denaturation. The steep change at ~43°C is indicative of a highly cooperative process, such as a phase change, either in the "critical target" or in its immediate surroundings. The similarity in the three graphs (in Fig. 2A above 43°C) strongly suggests that cell killing by

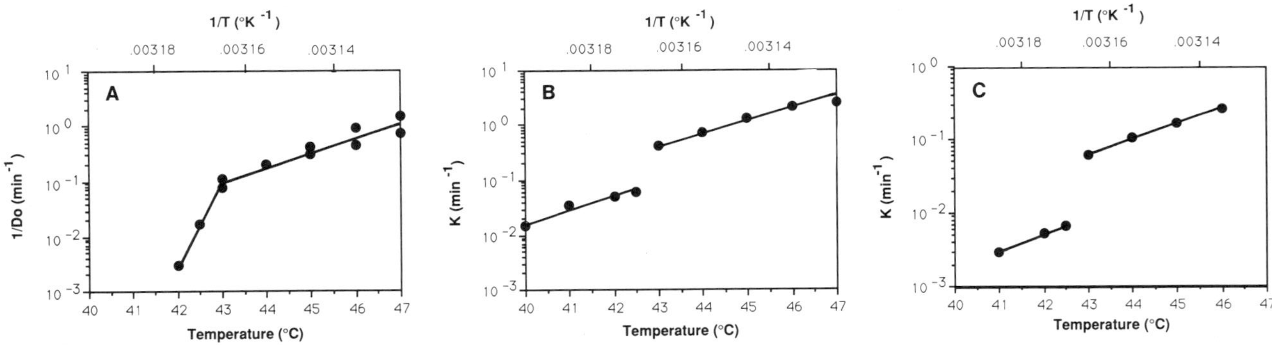

Figure 2 Arrhenius plots for heat-killing, induction of thermotolerance, and induction of heat shock proteins in Chinese hamster cells (HA-1). (*A*) Plot of the inverse of the log of the slope of the survival curves obtained at various temperatures against the inverse of the absolute temperature. (*B*) Rate at which thermotolerance is induced as a function of the duration of the triggering. This rate is plotted vs. the inverse of the absolute temperature of the appropriate treatment. (*C*) Similar to *B*, except that the time to induce a maximum amount of hsp70 was used to determine the ordinate. Thus, *C* indicates rate of induction of hsp70 synthesis. The last two Arrhenius plots are statistically indistinguishable, suggesting that the induction of either thermotolerance and synthesis of hsps involve the same or at least very similar mechanisms. The sharp break in all three Arrhenius plots suggests a cooperative process, perhaps involving a phase change either in proteins or in lipids. The slope of each line is a measure of the activation energy of the process; this is ~120 kcal/mole in *A* and *B*, consistent with the suggestion that protein denaturation is involved. (Panel *A*: data from panel A, Figure 4. Panel *B*: data from Li 1982. Panel *C*: G.C. Li, unpubl.)

heat, induction of tolerance, and induction of the heat shock response all have their origin in this target.

IV. PHENOMENOLOGICAL MODEL OF THERMOTOLERANCE

Partially on the basis of the data discussed in the previous paragraph, we proposed a phenomenological model of thermotolerance (Li and Hahn 1980). Our basic hypothesis was that the acquisition of resistance involves at least two distinct steps: an initial signal ("trigger"), and the subsequent processing of the initial step ("development"). Survival is further modified by the slow disappearance of thermotolerance ("decay") in the absence of continuous stress. At temperatures below about 43°C, the three processes go on simultaneously. At higher temperatures, development is blocked and triggering is the only mechanism involved. At higher temperatures, development and (very likely) decay do not take place. Only if the cells are returned to their normal growth temperature, i. e., 37°C, can development occur.

A diagram of the model is shown in Figure. 3. The rate constant, k_1, governs the conversion by heat of untreated, sensitive cells to cells in an activated form (triggered cells). These cells are, however, still heat sensitive. Conversion to resistant cells occurs only under conditions where the highly temperature-sensitive rate constant k_2 takes on a nonzero value. This permissive range is at temperatures below 43°C for rodent cells, but below 44°C for most human cells (Hahn et al. 1989). In the temperature range where k_2 is at or near its maximum value, simulation studies show

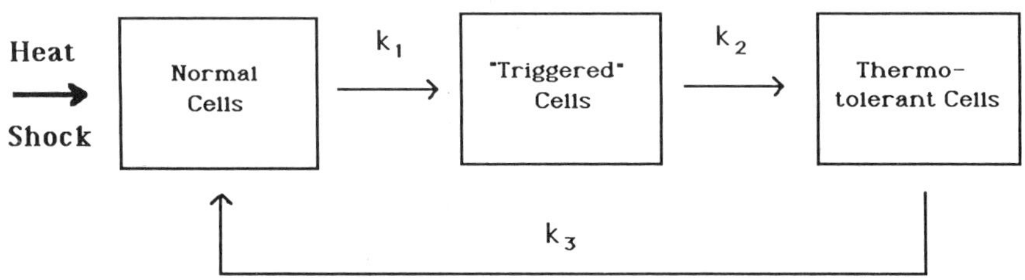

Figure 3 Model of thermotolerance. Thermotolerance develops in two steps: First, the triggering event converts normal cells to the triggered state with a rate constant k_1. Second, below about 42.5°C, these triggered cells are almost immediately converted to thermotolerant cells ($k_2 \gg k_1$). Above 42.5°C, $k_2 \to 0$; the triggered cells remain sensitive, only upon transfer to 37°C can they become converted to thermotolerant cells. Finally, thermotolerant cells all reconvert to their sensitive state at a slow rate governed by k_3. (After Li and Hahn 1980.)

that $k_2 \gg k_1 \gg k_3$ (G.C. Li and G.M. Hahn, unpubl.). Thus, k_1 can only be measured indirectly, since all triggered cells are almost immediately converted to resistant cells. The decay constant, k_3, is also temperature sensitive; its sensitivity varies with temperature in a manner similar to that for many enzymatic processes. For example, below about 25°C, $k_3 \approx 0$, i.e., thermotolerance decays very slowly. This suggests that decay results from catabolysis of one or more intracellular constituents. At 37°C, k_3 is at or near its maximum, although at higher temperatures (but below 43°C) the decay constant is difficult to measure because trigger, development, and decay are going on simultaneously (Li and Hahn 1980).

Although a phenomenological model such as that described above has its obvious limitations, nevertheless, it has proved useful in predicting or explaining various aspects of thermosensitivity. For example, it was shown by Henle and Leeper (1976) that mammalian cells exposed to a short heat shock at temperatures above 43°C (for rodent cells) are then sensitized to exposure at temperatures below 43°C ("step-down heating"). Experimental results are shown in Figure 4 for Chinese hamster cells. To explain what at first glance appears to be a paradoxical result, we need to examine the explanation for heat resistance below 43°C. The explanation provided by the model for thermoresistance (in previously unheated cells) below 43°C is that k_1, k_2, and k_3 all have nonzero values during exposure of the cells at these mildly elevated temperatures, so that thermotolerance is continually activated and decays. This continuous activation requires protein synthesis. The heat shock at the higher temperature, however, temporarily stops protein synthesis and thereby reduces k_2 to zero; until the rate constant recovers to control values, the cells remain sensitive.

To test this assumption, and in view of the importance of protein synthesis, we examined the ability of heat-shocked cells to synthesize these macromolecules after heat shock. Protein synthesis was completely inhibited by a 10-minute exposure of CHO cells to 45°C. When these cells were then placed at 37°C, protein synthesis resumed within 1–2 hours. If, however, cells were placed at 40°C after the initial treatment, protein synthesis remained essentially undetectable for several hours (Hahn and Shiu 1985). Protein synthesis appears to be required for tolerance development, as stated before. Heat-shocked cells at 40°C did not rapidly recover the ability to synthesize proteins; k_2 remained zero and the cells remained sensitive. Furthermore, the model predicts that once cells have been made thermotolerant, they are no longer subject to increased rate of killing by step-down heating; this too has been verified (Hahn and Shiu 1985). The model is, however, obviously oversimplified. Each of its components almost surely represents a series of coupled

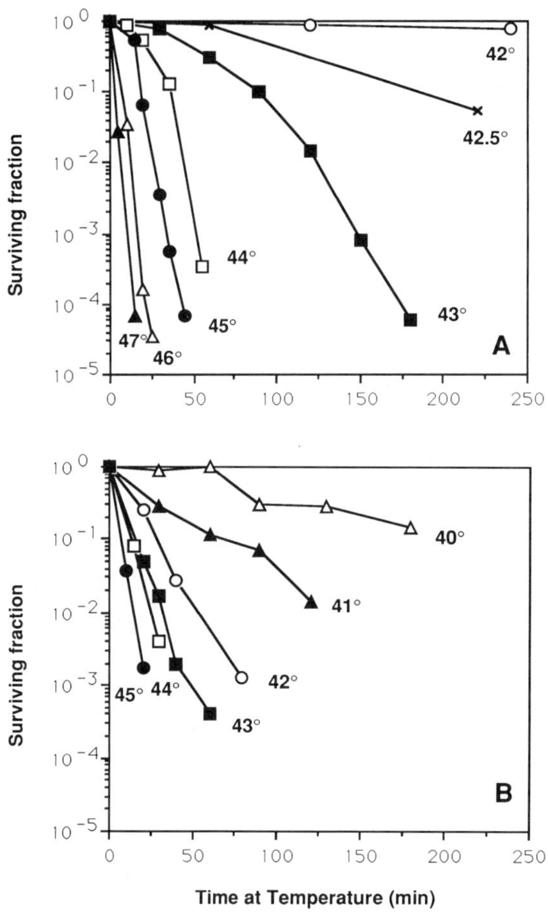

Figure 4 Survival kinetics of Chinese hamster (HA-1) cells: effect of "step-down" heating. (*A*) Survival of HA-1 cells exposed to graded doses of heat. (*B*) Same cells preexposed to 45°C for 20 min, and then treated as in *A*. Note the appreciable sensitization of cells by the preexposure, particularly at the lower temperatures (40°C–42°C). The results suggest that the initial heat shock prevents development of thermotolerance during subsequent heating at these temperatures.

biochemical reactions. In this context, k_1, k_2, and k_3 must be interpreted to be rate constants of the reaction that is rate-limiting under the specific circumstances of the individual experiment and for the particular cell line. It is important to be aware that for different cell lines, and even for the same cell line under different experimental conditions, different critical reactions may become rate-limiting.

V. INDUCTION OF THERMOTOLERANCE BY AGENTS OTHER THAN HEAT

In addition to heat itself, many agents (primarily chemicals, but also reoxygenation after hypoxia, glucose starvation, and some other stresses) can induce a state of thermotolerance. This state persists for some time after exposure of the cells; i.e., the cells remain thermotolerant, even in the absence of the inducer. This distinguishes inducers from thermoprotectors such as glycerol or heavy water; these latter substances need to be present during heating in order to exert their heat-sparing effects. Inducers include a variety of sulfhydryl compounds (e. g., sodium arsenite, carbonyl cyanide *m*-chlorophenyl hydrazone, diamide), alcohols (ethanol and other particularly longer-chain aliphatic alcohols), and other nonpolar solvents such as dimethylsulfoxide. Henle, in his review (1987), lists many of these. Induction of thermotolerance by drugs can also be divided into triggering and development stages, at least for those agents where this question has been tested. For example, if cells are exposed to pentanol at 24°C, no thermotolerance is seen if cells are tested immediately after drug exposure (G.M. Hahn, unpubl.). If, however, the alcohol is washed out and the cells are returned to 37°C, tolerance develops after an initial lag (Fig. 5). At 24°C, no protein synthesis is measurable. At 37°C, and in the absence of pentanol, protein synthesis resumes, but only after a lag that mirrors the lag in the development of thermotolerance. These results suggest that for pentanol, triggering is an event that requires no macromolecular synthesis or other, enzyme-mediated reactions; development, however, does require these reactions.

VI. MOLECULAR MECHANISMS OF TRIGGERING

The ability of so many diverse agents to induce tolerance is puzzling. In mammalian cells at least, the response is not a nonspecific stress response. For example, X-irradiation and other modalities that damage DNA specifically do not induce the heat shock response (Anderson et al. 1988). (In yeast, the situation appears to be different; both X- and UV-irradiation are reported to be inducers of the stress response.) However, there is one common property among heat and the drugs mentioned. All cause protein denaturation or other protein modification. This realization caused Hightower (1980) to suggest that the cells regard as "foreign" damaged proteins and that these are responsible for inducing the signals that lead to the heat shock response and presumably to thermotolerance. Several experiments have now demonstrated that indeed foreign (or nonfunctional) proteins introduced into cells by a variety of mechanisms induce hsps (Ananthan et al. 1986), but only one such study has measured the concurrent induction of thermotolerance. Lee and Hahn (1988) ex-

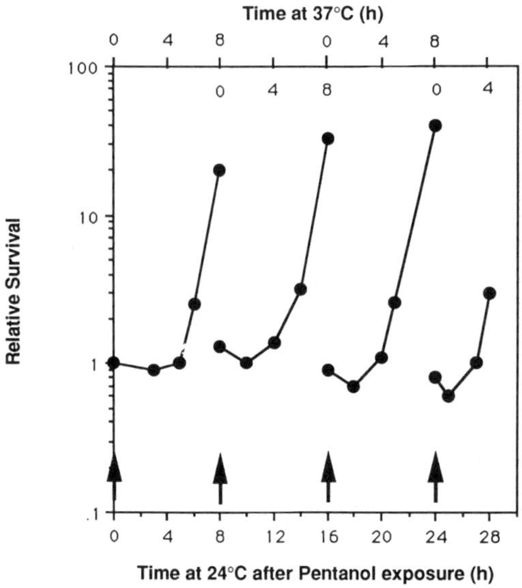

Figure 5 Inhibition of pentanol-induced thermotolerance at 24°C. Chinese hamster cells were exposed to 9×10^{-2} M pentanol in medium without serum for 1 hr at 24°C. The drug-containing medium was then removed and the cells were rinsed; fresh serum-containing medium at 24°C was then added and the cells were returned to the 24°C environment. Every 8 hr, some of the cells were transferred to an incubator at 37°C and their heat resistance was tested at various times thereafter by an exposure of 45°C for 30 min. The pattern of survival in cells removed from the 24°C environment immediately after the medium change was very similar to that in cells removed up to 24 hr later, indicating that no tolerance developed during the 24°C interval. In all three groups of cells, thermotolerance developed within about 3–5 hr following transfer to 37°C. These cells were, therefore, in a triggered state but could not develop tolerance until transferred to 37°C. The arrows indicate when cells were transferred from 24°C to 37°C (G.M. Hahn, unpubl.)

posed CHO cells to diamide and to dithio-bis (succinimidyl propionate). Both these drugs cross-link proteins, the former via formation of S–S bonds; the latter, an alkylating agent, via bifunctional binding to amino groups. In the study cited, both drugs induced thermotolerance (and also hsps). In both cases, induction of tolerance occurred only at doses that led to detectable levels of a high-molecular-weight protein complex. This complex formed rapidly and disappeared somewhat less rapidly when the cross-linking drugs were removed. The time needed for induction of maximum thermotolerance coincided approximately with the time required to "repair" or catabolize the high-molecular-weight complex. These results suggest strongly that the complex was the trigger and that only during its continued presence did the cells continue to develop thermotolerance.

Assuming that this hypothesis is correct (i.e., that nonfunctional, foreign proteins are the trigger for induction of thermotolerance), it is reasonable to ask whether all proteins are equally likely to become the initiators of tolerance. Although at this time no definite answer can be provided, several pieces of evidence suggest that proteins embedded in lipids are of primary concern. The first was the finding that the concentration of aliphatic alcohols required for the induction of thermotolerance was related to their water-lipid partitioning coefficient (Hahn et al. 1985). The higher this coefficient, i.e., the greater the partitioning of the alcohol into lipids, the lower the required concentration in the aqueous medium. Indeed, although the latter concentration varied by a factor of about 10^2 (Table 2), the membrane concentration required to induce tolerance remained sensibly constant, clearly suggesting that membrane rather than cytosolic proteins were primarily at risk. Of equal interest is the finding that the thermodynamics of trigger involve a very sharp temperature transition, near 43°C for CHO cells (Fig. 2B). The half-width of this transition is on the order of 0.5° or less, far too narrow to be a reflection of phase changes in proteins in aqueous solutions. For these, as denaturation proceeds, volume changes occur that are accompanied by "hydrophobic hydration"; it is this phenomenon that is thought to be responsible for the dispersion over several degrees of phase transitions of proteins in water. For proteins in lipid bilayers, however, the volume changes would be expected to be much smaller, and there is no direct analog to hydration. Therefore, the data of Li and Hahn (1980) very likely reflect a phase transition of one or more membrane proteins. V.S. Basrur and G.M. Hahn (unpubl.) have recently demonstrated that the polyene antibiotic amphotericin B is capable of initiating thermo-

Table 2 Induction of thermotolerance by aliphatic alcohols

Alcohol	Molar dose required (in medium) for induction of thermotolerance	Membrane/water partition coefficient	Molar dose required (in membrane) for induction of tolerance
1-Propanol	7×10^{-1}	0.46	3.2×10^{-1}
1-Pentanol	5×10^{-2}	3.6	1.6×10^{-1}
1-Hexanol	9×10^{-3}	13.0	1.2×10^{-1}
1-Octanol	1.6×10^{-3}	151.8	2.4×10^{-1}

Note that although the alcohol concentration in medium differed by more than 100-fold, the concentration in membranes was almost constant, suggesting that the alcohol in the membrane is responsible for the induction of tolerance. (Data from Hahn et al. 1985.)

tolerance. This drug is restricted to lipid regions within the cell; there its only known action involves binding to cholesterol, and it is not inconceivable that cholesterol/amphotericin B complexes are regarded as foreign proteins by the cell. Finally, Calderwood et al. (1988) showed that heat induces all the known second-messenger cascades. These, of course, have their initiating signals in the plasma membrane.

These data are all consistent with the view that protein modifications, primarily in membranes, are the events that constitute the first step in induction of tolerance. Development may or may not follow, depending on factors in the environment of the cells. A caveat, however: Even if these somewhat speculative conclusions regarding triggering mechanisms are correct, they in no way imply that thermotolerance could not also be induced by other very different mechanisms. It must be recognized that heat and most of the other inducers discussed affect the cell in many different ways. It is not only not inconceivable, but quite likely that more than one intracellular lesion can activate events that lead to development of thermotolerance.

An unusual induction of thermotolerance results from exposing cells to glucocorticoids. Fisher et al. (1986) exposed Chinese hamster cells for 24 hours to a variety of steroids (testosterone, progesterone, hydrocortisone, dexamethasone, and ecdysterone). Of these, only the glucocorticoids, hydrocortisone and dexamethasone, induced thermotolerance. Tolerance was induced by hydrocortisone at doses as low as 10^{-8} M, i.e., at physiological concentrations. Increasing dosage did not increase the amount of protection as measured by cell survival. The rate of induction was slow; thermotolerance reached its maximum only about 24 hours after the cells had been exposed to the steroid. After removal of hydrocortisone, thermotolerance persisted for at least 3 days. Competition studies between hydrocortisone and progesterone, a steroid that competes for binding glucocorticoid receptors, showed that binding of hydrocortisone to its receptor was necessary for the induction of thermotolerance. Presumably, triggering in this case did not involve foreign proteins, unless the receptor/hormone complex is so considered by the mechanism that the cell employs to identify such proteins.

VII. SYNTHESIS OF HSPS: THE MOLECULAR MECHANISM FOR THE DEVELOPMENT OF THERMOTOLERANCE?

We have already emphasized that protein synthesis appears to be a requirement to permit the development of thermotolerance. Although there are studies in the literature suggesting that a limited amount of tolerance can develop even in the absence of protein synthesis, these are usually open to alternate interpretation. Furthermore, it has been shown in many experiments that during the period where the development of thermotolerance goes on at its maximum rate, preferential synthesis of heat shock proteins occurs. This has led to the frequently voiced suggestion that one or more of the hsps are responsible for protecting the cell against heat and other stresses. This argument is strengthened by the fact that not only heat itself, but also almost all the other inducers of thermotolerance that have been examined are also inducers of one or more of the hsps. Several studies have pointed out that a quantitative relationship can be established between the concentration of some of the hsps and levels of thermotolerance (Landry et al. 1987), as determined by survival studies. The most detailed of these studies are those carried out by Li and her associates (Li and Werb 1982). These studies show that not only after heating, but also after initial exposure of cells to ethanol or to sodium arsenite, hsps are synthesized. Li suggests that, in particular, intracellular levels of hsp70 might be used as an indicator of retained thermotolerance. Thermotolerance and induction of hsps occur not only in tissue culture cells, but also in tumors and in normal tissues (Rhee and Song 1987). For these reasons, it has been suggested that quantification of levels of hsp70 could be used as a clinical tool to monitor the kinetics of induction and decay of thermotolerance both in tumors and in normal tissue during hyperthermia treatments. Recently, data from Cretien and Landry (1988) suggest that hsp28 may be an equally good candidate.

Do the hsps actually protect cells from thermal damage? Despite the large amount of data that suggests a direct connection between hsps and thermotolerance, the answer to this question still remains controversial. Several isolated experiments suggest that exceptions exist where thermotolerance develops in the absence of detectable hsp production. The induction of thermotolerance by glucocorticoids has already been described; the steroids, however, did not appear to induce hsps. B. Fisher et al. (unpubl.) induced a limited amount of thermotolerance in a murine lymphoma line and verified the finding of Mehdi and Hahn (1989) that this cell line was incapable of preferentially synthesizing heat shock proteins. Carper et al. (1987) have reviewed the relationship between hsp synthesis and thermotolerance. Their somewhat controversial conclusion is that the synthesis of hsps is neither necessary nor sufficient for the development of thermotolerance.

One possible explanation for the controversy may be technical. Almost all the studies are based on quantification of protein concentrations as measured on densitometric tracings of autoradiograms from one-dimensional polyacrylamide gels. It is likely that this methodology is not sufficiently sensitive to detect small changes in the levels of individual proteins (R.L. Anderson, unpubl.) and that these undetected increases are sufficient to induce measurable levels of thermotolerance.

Two states of thermotolerance have been postulated by Boon-Niermeijer et al. (1986) and Laszlo (1988). Only the latter work is concerned with mammalian cells. The data in both studies are based on induction of a low level of thermotolerance that decays rapidly and does not appear to require hsp synthesis, and a more intense level of thermotolerance that decays more slowly and does seem to require hsp induction. Again, the findings may involve technical and also interpretational difficulties. At present, all the available data are highly suggestive that the synthesis of hsps leads to development of thermotolerance, but definite proof is lacking. However, the inverse is likely not to be true; i.e., a modest amount of thermotolerance may be induced in the absence of hsp synthesis. Even among investigators who believe that hsps are responsible for thermotolerance, two schools of thought exist. One group of investigators suggests that the presence of increased amounts of hsps prevents denaturation of other proteins and associated cellular damage (Minton et al. 1982). Another group believes that the hsps have a specific repair function and that an increased amount of hsps allows increased amounts of repair (Pelham 1986). Experiments to distinguish between these two hypotheses are difficult to design because the critical lesion that leads to heat-induced cell death is as yet unknown. The repair hypothesis needs to be examined with some skepticism: cell biologists have reported repair phenomena for many years; with the one auspicious exception of repair of DNA, all these reports have proven to be artifacts. There is no evidence that DNA repair in mammalian cells is inducible by heat.

VIII. DECAY OF THERMOTOLERANCE

After tolerance reaches its maximum, and in the absence of additional triggering, resistance begins to decay immediately. This immediate disappearance is not apparent from many experimental data; if the test dose is not sufficiently severe, the results appear to show that resistance remains constant for a period that in mammalian cells may exceed several days before decay starts (Fig. 6A). If the experimental conditions remain constant, the rate of decay is essentially independent of the triggering dose (Fig. 6B). It is, however, dependent on the environmental temperature (Li and Hahn 1987), as shown in Figure 6C, and on the rate of proliferation of the cells involved (Gerweck and Epstein 1985), although this is not a universal finding (Armour et al. 1985).

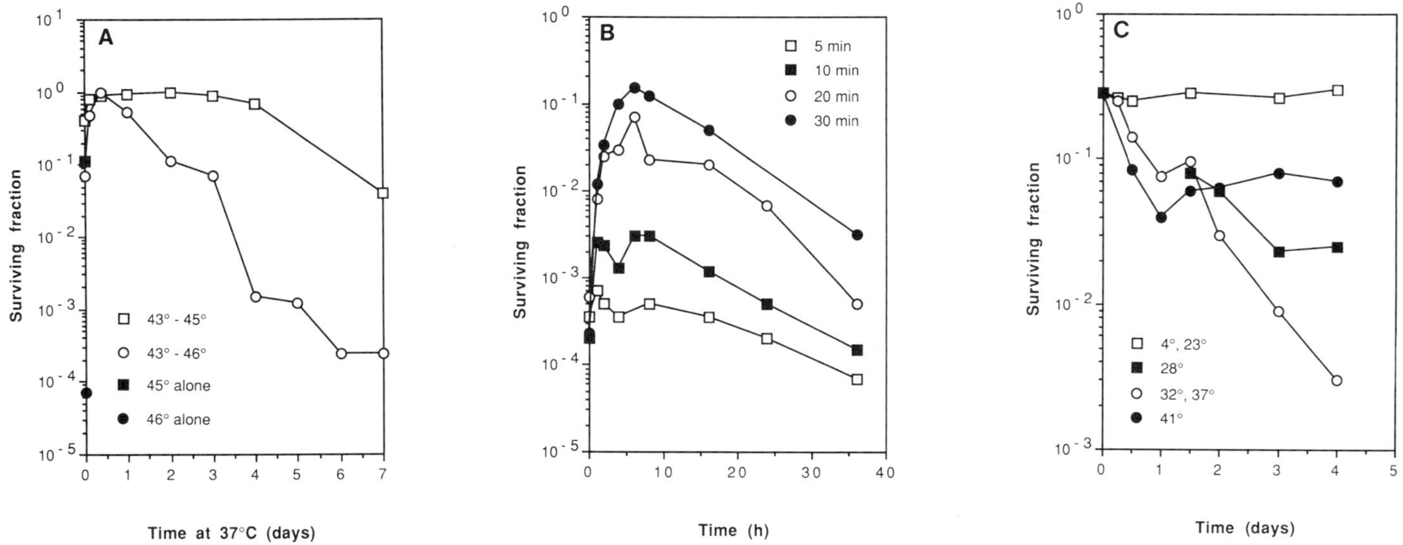

Figure 6 Dependence of decay of thermotolerance on test dose, induction dose, and temperature. Once tolerance is induced, it remains constant for a time, then decays. *A* shows that the plateau at which tolerance remains constant may reflect only the measurement dose. HT1088 cells from a human fibrosarcoma were exposed to 20 min at 45°C and then reincubated at 37°C. Periodically, the cells were challenged with an exposure of 45 min at either 45°C or 46°C. Cell survival remained high for the 45°C tested cells between 8 hr and 80 hr; but if the test dose was 46°C, tolerance started decaying immediately after reaching its peak. *B* shows that only the duration of the plateau, not the subsequent rate of decay, is a function of the intensity of the triggering dose. Chinese hamster cells (HA-1) were exposed to 43°C for 5, 10, and 20 min, and the cells were then returned to 37°C. Periodically, some of the cells were challenged by exposure to 45°C for 45 min, and survival was assayed. Slopes of the decay curves are similar for the three treatments. In *C*, HA-1 cells were exposed for 10 min to 45°C, and thermotolerance was permitted to develop at 37°C for 12 hr. Cells were then transferred to incubators at the indicated temperature and challenged with a uniform test (45°C, 45 min). Thermotolerance did not measurably decay during the 100 hr of the experiment; most rapid decay occurred at 37°C. (Data from Li and Hahn 1987.)

The temperature dependence of the rate of decay is of some interest. In Chinese hamster ovary cells (HA-1) at temperatures below 23°C, no measurable decay was seen over a 100-hour interval; this suggests that decay involves a catabolic process. At 37°C, in nonproliferating or minimally proliferating density-inhibited hamster cell cultures, resistance decayed exponentially with a half time of about 30–50 hours (in human cells, decay may be appreciably slower [Hahn et al. 1989]) and was similar to the turnover rate of several proteins including actin and the 70-kD heat shock protein, suggesting that catabolysis of a protein is involved in decay. In rapidly proliferating cells, Gerweck and Epstein (1985) found a decay time of perhaps 15 hours, not substantially different from the cells' doubling time. They suggest that only two factors are involved in determining the decay: "dilution," i.e., the progressive reduction of the concentration of a protective material (or of protected molecules) due to cell division, and catabolism or other degradation. As the data of Armour et al. (1985) show, however, things very likely are not that simple. The intracellular concentration of the putative protective material depends not only on dilution and turnover, but also on its synthesis, and this may be under a mixture of transcriptional and translational controls. These, in turn, may depend on the level of the protective substance itself or even on that of an unrelated substance. For example, Lindquist (1986) has shown that a complicated control exists that governs hsp levels in *Drosophila* cells. Such controls may influence synthesis of protective materials during decay of tolerance. Hence, degradation and dilution of an as yet undefined substance are probably not the only determinants that influence the rate of decay of tolerance.

The rate at which cellular resistance disappears is of considerable importance during a course of multiple hyperthermia treatments. Because treatment temperatures are limited to 43°C–45°C by practical considerations and by possible damage to normal tissue, the temporal behavior of thermosensitivity after an initial heat treatment mimics that of the 45°C kinetics curve of Figure 2A. In other words, tolerance remains high for several days; retreatment during that period would be expected to have very little effect on cell survival. Therefore, tolerance and its decay need to be taken carefully into account when scheduling a course of fractionated hyperthermia.

IX. CHARACTERIZATION OF THE THERMOTOLERANT CELL

Thermotolerant cells respond differently to a variety of treatments than do cells that were not previously heated or exposed to other stresses. In a review, Henle (1987) lists 15 such responses. In addition, the thermotolerant cell is resistant to some (although not all) anticancer drugs. This phenomenon is both temperature- and pH-sensitive (Hahn 1987). Table 3 lists several agents that affect cells differently after induction of thermotolerance. Also shown in that table are several agents whose cytotoxicity is not affected by the thermotolerant state.

Some things of interest should be noted. The only drugs that show cross-tolerance with heat at 37°C are Adriamycin, actinomycin D, and amphotericin B. The cross-tolerance with at least two of these reflects membrane properties. Less Adriamycin gets into the thermotolerant cells than into the unstressed cell. The reduction in cytoplasmic Adriamycin levels reflects at least partially the reduced cytotoxicity of the agent against thermotolerant cells (Rice and Hahn 1987). The mode of action of amphotericin-B-induced cytotoxicity has already been discussed. As far as is known, this agent does not penetrate into the cytoplasm and is confined to membrane structures. Its cytotoxicity involves interaction with cholesterol. Since cholesterol, in mammalian cells, is found almost

Table 3 Effect of thermotolerance on drug response

		Exposure at	
Drug	Mode of Action	37°C	43°C
Adriamycin	intercalates DNA damages plasma membrane (lipid peroxidation?)	?	+
Amphotericin B	interacts with cholesterol to form ionophores in plasma membranes	++	++
Arsenites	cell "poison" (as arsenate) binds to sulfhydryl groups	?	++
Ethyl alcohols	denatures proteins fluidizes membranes	++	++
Bifunctional alkylating agents (e.g., BCNU)	cross-link DNA	–	–
Bleomycin	induces single-strand breaks in DNA	–	+
Cisplatin	cross-links DNA	–	–
VP-16	inhibits topoisomerase II	–	–

(–) No protection; (?) equivocal results; (+) some protection; (++) strong protection.

exclusively in the plasma membrane, cross-resistance of amphotericin B and heat in thermotolerant cells suggests specific membrane modifications in the thermotolerant cells. The finding that protection at 43°C is extended to agents other than the three antibiotics just mentioned suggests a different molecular basis for protection at the higher temperature. The pH dependence is of interest because at least portions of most tumors are at a pH well below that of normal tissues (pH 7.4).

X. SUMMARY

Thermotolerance is a phenomenon of considerable interest to biologists and to clinicians who use hyperthermia in the treatment of cancer. It is the most readily studied example of a wider class of stress responses. The biochemical nature of the heat-induced increased resistance to heat is not completely understood; several lines of evidence suggest involvement of heat shock proteins. If these are involved, whether they act as "repair" molecules or somehow provide protection by their presence during heating, is not known. Protein synthesis does appear to be a requirement for the development of tolerance.

In mammalian cells, thermotolerance is induced during the exposure of cells at 41°C–42.5°C; at these temperatures, protein synthesis is only partially inhibited. Heat shocks at higher temperatures do inhibit protein synthesis: The cells that survive need an interval at 37°C before protein synthesis resumes and thermotolerance develops. Thermoresistance, a genetic trait of some cell strains, does not appear to be simply a case of retained thermotolerance, since tolerance can be induced readily even in cells from thermoresistant strains. Step-down heating, i.e., sensitization to exposures of 42.5°C or lower by a previous heat shock at higher temperatures, can be explained on the basis of inhibition of protein synthesis by the high temperature and lack of recovery at 41°C–42.5°C.

Once thermotolerance has been induced, it decays slowly compared to the time it takes for development. In proliferating human fibrosarcoma cells, tolerance is still detectable 150 hours after its induction. Therefore, fractionated hyperthermia for cancer treatment must be scheduled with some care in order to minimize the protective effect of thermotolerance.

Not only do thermotolerant cells survive heat shock in greater numbers than do cells not previously heated, but many organelles and functions are also protected. Among the former are nucleoli and stress fibers; among the latter, perhaps the most important is protein synthesis. Thermotolerant cells are more resistant to some drugs, including several anticancer agents such as Adriamycin and actinomycin D. These cells are also less likely to be sensitized by heat to X-irradiation or to the cytotoxic action of drugs. All these characteristics of thermotolerant cells suggest either multiple protective mechanisms or a nonspecific stabilization of heat-sensitive cellular functions.

ACKNOWLEDGMENTS

Most of the work described here was supported by grants CA-19386 and CA-44665 from the National Institutes of Health.

REFERENCES

Alexandrov, V.Y. 1977. *Cells, molecules and temperature: Conformational flexibility of macromolecules and ecological adaptation.* Springer-Verlag, Berlin.

Ananthan, J., A.L. Goldberg, and R. Voellmy. 1986. Abnormal proteins serve as eukaryotic stress signals and trigger the activation of heat shock genes. *Science* **232:** 522–524.

Anderson, R.L., E. Shiu, G.A. Fisher, and G.M. Hahn. 1988. DNA damage does not appear to be a trigger for thermotolerance in mammalian cells. *Int. J. Radiat. Biol.* **54:** 285–298.

Armour E.P., G.C. Li, and G.M. Hahn. 1985. Effects of proliferation on the decay of thermotolerance in Chinese hamster cells. *Radiat. Res.* **103:** 351–362.

Boon-Niermeijer, E.K., M. Tuyl, and H. van der Scheur. 1986. Evidence for two states of thermotolerance. *Int. J. Hyperthermia* **1:** 93–105.

Calderwood, S.K., M.A. Stevenson, and G.M. Hahn. 1988. Effects of heat on cell calcium and inositol lipid metabolism. *Radiat. Res.* **113:** 414–425.

Carper, S.W., J.J. Duffy, and E.W. Gerner. 1987. Heat shock proteins in thermotolerance and other cellular processes. *Cancer Res.* **47:** 5249–5255.

Cretien, P. and J. Landry. 1988. Enhanced constitutive expression of the 27-kDa heat shock proteins in heat resistant variants from Chinese hamster cells. *J. Cell. Physiol.* **137:** 157–166.

Dewey, W.C. and P.K. Holahan. 1987. Thermotolerance as a modifier of radiation toxicity. In *Thermotolerance: Thermotolerance and thermophily* (ed. K. Henle), vol. 1, pp. 113–125. CRC Press, Boca Raton, Florida.

Field, S.B. and C. Franconi, eds. 1987. *Physics and technology of hyperthermia.* Martinus Nijhoff, The Hague.

Fisher, G.A., R.L. Anderson, and G.M. Hahn. 1986. Glucocorticoid-induced heat resistance in mammalian cells. *J. Cell. Physiol.* **128:** 127–132.

Gerner, E.W. and M.J. Schneider. 1975. Induced thermal resistance in Hela cells. *Nature* **256:** 500–502.

Gerweck, L.E. and L.F. Epstein. 1985. Cell proliferation, protein turnover, and the decay of thermotolerance in CHO cells. *Int. J. Radiat. Oncol. Biol. Phys.* **8:** 1935–1941.

Hahn, G.M. 1982. *Hyperthermia and cancer.* Plenum Press, New York.

———. 1987. Thermotolerance and drugs. In *Thermotolerance: Thermotolerance and thermophily* (ed. K. Henle), vol. 1., pp. 97–112. CRC Press, Boca Raton, Florida.

Hahn, G.M. and E.C. Shiu. 1985. Protein synthesis, thermotolerance and step down heating. *Int. J. Radiat. Oncol. Biol. Phys.* **11:** 159–164.

Hahn, G.M. and I. van Kersen. 1988. Isolation and initial characterization of thermoresistant RIF tumor cell strains. *Cancer Res.* **48:** 1803–1807.

Hahn, G.M., S.C. Ning, M. Elizaga, D.S. Kapp, and R.L. Anderson. 1989. A comparison of thermal responses of human and rodent cells. *Int. J. Radiat. Biol.* (in press).

Hahn, G.M., E.C. Shiu, B. West, L. Goldstein, and G.C. Li. 1985. Mechanistic implications of the induction of thermotolerance in Chinese hamster cells by organic solvents. *Cancer Res.* **45:** 4138–4143.

Henle, K.J. 1987. Thermotolerance in cultured mammalian cells. In *Thermotolerance: Thermotolerance and thermophily* (ed. K. Henle), vol. 1, pp. 13–71. CRC Press, Boca Raton, Florida.

Henle, K.J. and D.B. Leeper. 1976. Interaction of hyperthermia and radiation in CHO cells: Recovery kinetics. *Radiat. Res.* **66:** 505–518.

Hightower, L.E. 1980. Cultured animal cells exposed to amino acid analogues or puromycin rapidly synthesize several polypeptides. *J. Cell. Physiol.* **102:** 407–427.

Landry, J., S. Lamarche, and P. Chretien. 1987. Heat shock proteins: A lead to the understanding of cell thermotolerance. In *Thermotolerance: Thermotolerance and thermophily* (ed. K. Henle), vol. 1, pp. 145–173. CRC Press, Boca Raton, Florida.

Laszlo, A. 1988. Evidence for two states of thermotolerance in mammalian cells. *Int. J. Hyperthermia* **4:** 513–526.

Lee, K.J. and G.M. Hahn. 1988. Abnormal proteins as the trigger for the induction of stress responses: Heat, diamide and sodium arsenite. *J. Cell. Physiol.* **136:** 411–420.

Li, G.C. and G.M. Hahn. 1980. A proposed operational model of thermotolerance based on effects of nutrients and the initial treatment temperature. *Cancer Res.* **40:** 4501–4508.

———. 1987. Influence of temperature on the development and decay of thermotolerance and heat shock proteins. *Radiat. Res.* **112:** 517–524.

Li, G.C. and Z. Werb. 1982. Correlation between synthesis of heat shock proteins and development of thermotolerance in Chinese hamster fibroblasts. *Proc. Natl. Acad. Sci.* **79:** 3218–3222.

Li, G.C., G.A. Fisher, and G.M. Hahn. 1982. Modification of the thermal response by D_2O: II. Thermotolerance and the specific inhibition of development. *Radiat. Res.* **92:** 541–551.

Lindquist, S.L. 1986. The heat shock response. *Annu. Rev. Biochem.* **55:** 1151–1191.

Mehdi, S.Q. and G.M. Hahn. 1989. Effects of hyperthermic temperatures and the synthesis of heat-shock proteins on the lateral diffusion of H-$2K^k$. *Int. J. Hyperthermia* (in press).

Minton, K.W., P. Karmin, G.M. Hahn, and A.P. Minton. 1982. Non-specific stabilization of stress-susceptible proteins by stress-resistant proteins: A model for the biological role of heat shock proteins. *Proc. Natl. Acad. Sci.* **79:** 7107–7111.

Paliwal, B.R., F. Hetzel, and M.W. Dewhirst, eds. 1988. *Biological, physical and clinical aspects of hyperthermia.* AAPM Monograph No. 16, American Institute of Physics, New York.

Pelham, H.R. 1986. Speculation on the functions of the major heat shock and glucose-regulated proteins. *Cell* **46:** 959–961.

Raaphorst, G., S. Romano, J. Mitchell, J. Bedford, and W.C. Dewey. 1979. Intrinsic differences in heat and/or X-ray sensitivity of seven mammalian cell lines cultured and treated under identical conditions. *Cancer Res.* **39:** 396–401.

Rhee, J.G. and C.W. Song. 1987. Thermotolerance of organized tissues and tumors. In *Thermotolerance: Thermotolerance and thermophily* (ed. K. Henle), vol. 1, pp. 73–95. CRC Press, Boca Raton, Florida.

Rice, G.C. and G.M. Hahn. 1987. Modulation of Adriamycin transport by hyperthermia as measured by fluorescence-activated cell sorting. *Cancer Chemother. Pharmacol.* **20:** 183–187.

5

Hyperthermia in Cancer Therapy: Model Systems

Mark W. Dewhirst
Division of Radiation Oncology
Duke University Medical Center
Durham, North Carolina 27710

I. INTRODUCTION

Over the past two decades, a number of institutions have gained extensive clinical experience in the use of hyperthermia, particularly in conjunction with radiation therapy, in the treatment of human malignancies. Results have pointed consistently to the potential therapeutic benefit to be gained from the use of this combination therapy as compared with radiotherapy alone. In 1984, Overgaard reviewed the world's literature and found a total of 213 published studies with nearly 11,000 patients having been treated (Overgaard 1985). Of particular interest were patients in whom two or more tumors of the same type were treated. In several studies of this type, involving several hundred patients, it was shown that the complete response rate for local hyperthermia combined with radiotherapy was roughly double that achieved with radiotherapy alone, i.e., approximately 60% versus 30%. More recently, randomized Phase III studies from pet dogs with oral squamous cell carcinomas (Gillette et al. 1987) and human patients with recurrent head and neck lesions (Valdagni et al. 1988) have shown that the same dose of radiotherapy plus heat yields a larger percentage of complete responses and long-term tumor controls than that achieved with radiotherapy alone. These results have been obtained without a significant increase in normal tissue com-

Stress Proteins in Biology and Medicine.

plications. Thus, it is felt that hyperthermia holds a great deal of promise for improving control rates for tumors with high local or regional failure probability. Despite the optimism, however, there are a number of critical questions that still need to be addressed regarding treatment optimization. For example: (1) How long does one need to heat in a given treatment session to achieve the maximum therapeutic benefit? (2) What temperatures are needed to obtain the maximum therapeutic gain? (3) How many heat treatments are needed? (4) How does one go about combining hyperthermia with radiation and/or chemotherapy to provide the maximum benefit? Answers to these questions will be complex, depending on tumor size, how well the tumor is heated, and the metabolic state of the tumor.

It is well known that temperature distributions in heated tissues are quite nonuniform when local heating techniques are used (Dewhirst and Sim 1984; Dewhirst et al. 1987; Sapozink et al. 1988). This is particularly true in tumors, since blood flow is so heterogeneous. The nonuniformity of the temperature distributions achieved significantly influences the therapeutic efficacy of the heat treatment. For example, Dewhirst and co-workers demonstrated in spontaneous tumors of pet animals that the probabilities for tumor response and duration of local control were dependent on minimum monitored temperature when between 3 and 10 intratumoral points were monitored (Fig. 1) (Dewhirst and Sim 1984; Dewhirst et al. 1984). Using multivariate analyses, they also showed that the effect was independent of tumor volume, although, in general, larger tumors were harder to heat. Subsequent analyses of several human series have shown similar results (Valdagni et al. 1988). The recent use of indwelling catheters for thermometry, wherein multijunction thermometers or mechanical mapping techniques have been used, has clearly shown, however, that simple descriptors of the temperature field, such as the minimum, will be inadequate to accurately predict response to therapy (Dewhirst et al. 1987). Clinical reports tend to support this idea, since the goodness of fit of multivariate models has often been poor despite strong correlations between temperature minima and response (Dewhirst and Sim 1984; Oleson et al. 1984). A number of the clinical reports have utilized thermal isoeffect dose calculations to convert complex time-temperature data into an equivalent number of minutes at a standard temperature (see below) (Sapareto 1987). Others have averaged temperatures at steady state (Valdagni et al. 1988). One problem with both approaches is that they could greatly over- or underestimate the amount of biologic effect if thermotolerance is not properly accounted for. Thus, one potential reason for the poor goodness of fit seen in some clinical data might have been lack of appropriate ac-

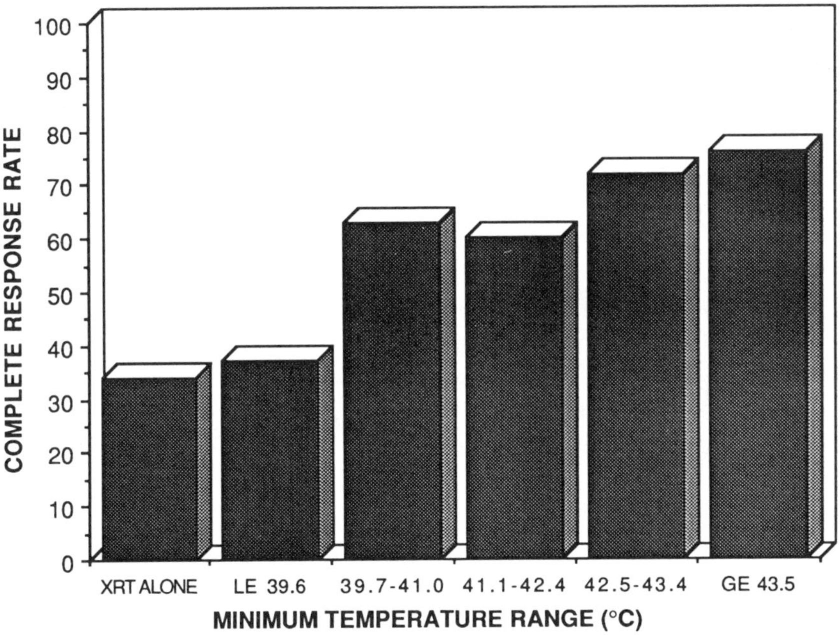

Figure 1 Correlation between complete response rate and minimum monitored temperature for pet animal tumors treated with heat and radiation. There was a significant correlation between the minimum monitored temperature and probability for complete response (p <0.001). The goodness of fit of the correlation was poor, however. This indicated that better descriptors of the temperature distribution might be needed. (Adapted from Dewhirst and Sim 1984.)

counting for thermotolerance, occurring either during treatment or between treatment sessions.

Since temperature distributions are typically nonuniform, the development and decay of thermotolerance may vary spatially throughout the tumor. Thus, it is critical to understand under what conditions thermotolerance develops in human tumors and the rate at which it decays, because such variability in biologic effect from heating could affect the answers to all of the previously mentioned questions regarding optimization of heat treatment schedules. Indices of thermotolerance that can be used clinically would help greatly in devising protocols to answer these and other important clinical questions. The purpose of this paper is to discuss those factors that influence thermotolerance, to review methods that are being developed for assessing thermotolerance, and, finally, to discuss how tumor morphologic and pathophysiologic heterogeneity influences the interpretation of such data.

II. FACTORS THAT INFLUENCE HYPERTHERMIC SENSITIVITY AND THERMOTOLERANCE

The biophysical relationship between hyperthermia and cell killing has been commonly depicted using an Arrhenius plot (Fig. 2) (Field and Morris 1984; Sapareto and Dewey 1984). This type of data presentation can be obtained by using the inverse of the slope of the in vitro cell survival curve versus temperature, or the treatment time to achieve an isoeffect can be used versus temperature. The latter sort of analysis is typically used for in vivo experiments. In either case, typically a biphasic plot is obtained that has a "breakpoint" around 43°C. Above this breakpoint, the amount of time needed to achieve an isoeffect can be halved for every 1°C rise. Below the breakpoint, the treatment time must be increased by a factor of 4–6 to achieve the same isoeffect. The break in the Arrhenius plot is thought to be due to thermotolerance development during heating when temperatures are less than 43°C. The slope of the Arrhenius plot above the breakpoint is quite consistent for a number of different cell lines studied in vitro (Sapareto 1987) and for in vivo systems (Field and Morris 1983). As a result, it is theoretically possible, at least, to convert

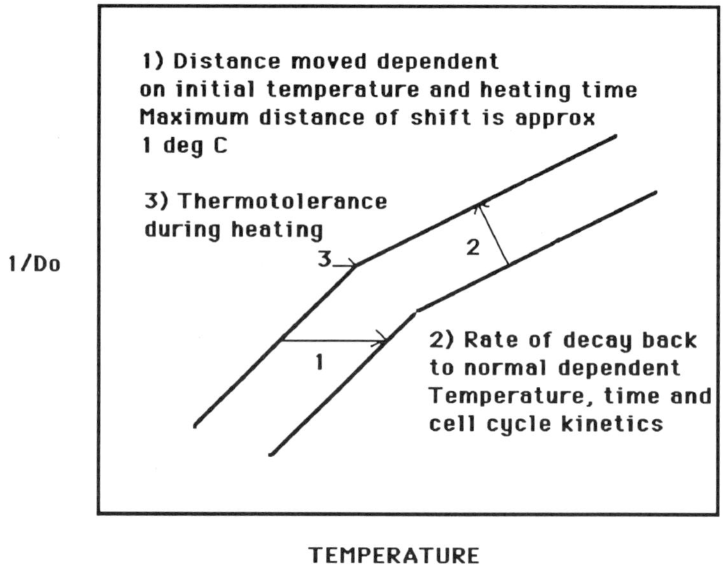

Figure 2 Typical Arrhenius plot, which relates the rate of cell killing ($1/D_o$) to temperature. The upper curve is typical of the shape of most Arrhenius plots from both in vitro and in vivo studies. The slope of the plot changes at about 43°C (*3*), because at temperatures below 43°C thermotolerance develops during the treatment. When thermotolerance develops between two heat treatments the curve shifts down to the right for the second heat (*1*, *2*).

any time-temperature combination into an equivalent number of minutes at 43°C. This type of conversion has been coined "thermal isoeffect dose" (Sapareto 1987) and has been advocated as a potential method to quantitate the effectiveness of hyperthermia therapy. The purpose of this paper is not to advocate the use of thermal isoeffect dose. However, it is a useful way to describe how factors that affect thermotolerance development and decay affect the overall thermal sensitivity of cells. The effectiveness of any treatment will depend on a number of factors, some of which are described below. Research is currently aimed at defining how these factors influence the Arrhenius plot, so that they can be accounted for when assessing the final outcome of therapy (Sapareto 1987).

When cells become thermally tolerant, the overall effect is to reduce the rate of cell killing at all temperatures. This shifts the Arrhenius plot (Fig. 2) down and to the right, in effect reducing the rate of cell kill as compared with nonthermally tolerant cells. The distance of the shift is positively related to the initial treatment temperature and time (Nielsen and Overgaard 1982). Thus, the degree of thermotolerance increases as the initial treatment time and temperature are increased. However, there is a maximum level of thermotolerance that can be achieved (Law et al. 1987). Typically, this translates into a maximum of 1°C temperature shift in the Arrhenius plot. The rate at which thermotolerance decays back to the pretreatment state is positively correlated with the initial treatment temperature and time (Nielsen and Overgaard 1982). In addition, the rate of decay may be dependent on cell cycle kinetics (Gerweck et al. 1983). The actual position of the breakpoint in the Arrhenius plot may be tissue dependent (Field and Morris 1983). Recent evidence suggests that human tumors have a breakpoint that is 0.5–1.0°C higher than that seen for murine cell lines (Roizin-Towle et al. 1986). A critical question that needs to be addressed in human tumors is whether sufficiently high temperatures are achieved clinically to induce thermotolerance. Second, if thermotolerance does develop, how long does it take to decay? Third, are there mechanisms that could be used to preferentially induce thermotolerance in normal tissues or inhibit its development in tumors? These considerations are important when trying to assess the overall efficacy of a given heat treatment, as well as when developing rationales for fractionating hyperthermia treatments.

A second factor that influences the shape of the Arrhenius plot is a phenomenon known as step-down heating (Henle 1980; Field 1987; Overgaard 1987). In this situation (which can be seen clinically; Fig. 3), temperatures initially rise above the breakpoint and subsequently drop to a temperature below the breakpoint. When this occurs, the breakpoint in the Arrhenius plot is lost. Thus, for every degree temperature drop below

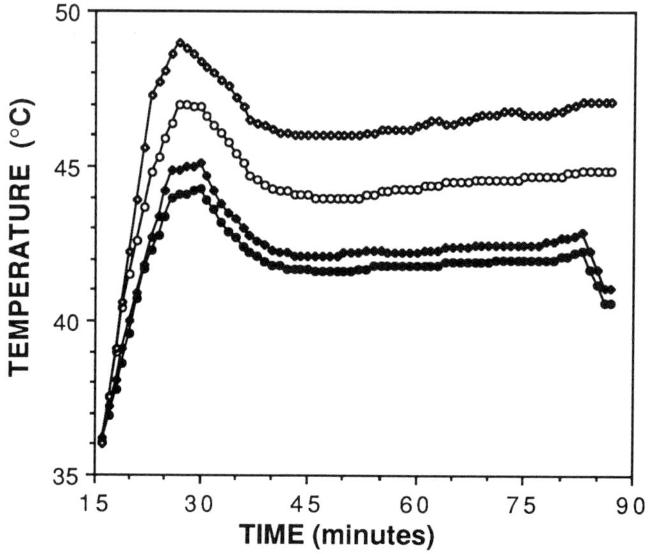

Figure 3 Clinical data demonstrating step-down heating. In this example, taken from a dog with a soft tissue sarcoma of the thigh, temperatures in all four monitored positions initially exceeded 43°C. After 30 min of heating using a 915-MHz microwave source at 80 W, power was reduced to 60 W. Temperatures in two of the sites dropped below 43°C after power reduction (step-down heating).

43°C, the time to achieve an isoeffect must be increased by a factor of 2, rather than by a factor of 4–6. The overall effect of step-down heating is to delay the onset of thermotolerance, thereby eliminating the break in the Arrhenius plot (Fig. 4).

A third factor that influences thermal sensitivity is pH. Initial studies on the effect of lowered pH on hyperthermic sensitivity of cells indicated that cells were much more sensitive at low pH and that low pH conditions inhibited the development of thermotolerance (Gerweck et al. 1982). The overall effect of an acute pH drop on the Arrhenius plot is to shift the curve up and to the left and to reduce the slope of the Arrhenius plot below the breakpoint (Fig. 5). Thus, the rate of cell killing is increased, relative to the cells at normal pH. Below the breakpoint, there is an inhibition or delay in the development of thermotolerance. Recent studies by Hahn and Shiu (1986) have shown that this effect is dependent on how long the cells have been conditioned at low pH. Chronic pH conditioning, that is, conditioning cells to low pH environmental conditions over 20–80 hours, causes the cells to become resistant to low pH heat

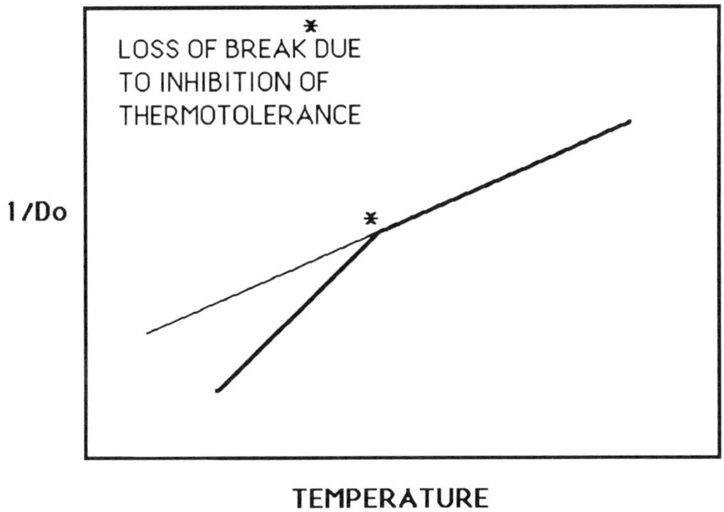

Figure 4 Arrhenius plot demonstrating the effect of step-down heating, as described in Fig. 3. Step-down and step-up heating delays the onset of thermotolerance, as shown by no change in slope near the breakpoint.

sensitization. In addition, they do not exhibit a delay or reduction in the amount of thermotolerance induction. This phenomenon may be important clinically. Acid pH conditions have been measured in human tumors in two independent laboratories using needle-type interstitial microelectrodes (Wike-Hooley et al. 1984; Thistlethwaite et al. 1987). Contradictory data have been obtained, however, when magnetic resonance spectroscopy (MRS) has been used to measure pH in tumors (Fig. 6). In those studies, tumor pH measurements have been largely in the normal-to-slightly-alkaline range (Oberhaensli et al. 1986; Sapareto et al. 1988; Semmler et al. 1988). These differences may result because microelectrode measurements sample primarily extracellular milieu, whereas MRS signals come from both intracellular and extracellular fluids (Hollis and Nunnally 1977). Furthermore, contamination of signals from adjacent normal tissues might make pH measurements fall more toward the normal range. Fox and co-workers have shown that cells can maintain a normal intracellular pH when placed in a chronically acidotic environment (Cook and Fox 1988). Thus, MRS signals could give normal pH values even when the tissues have an acidotic extracellular environment.

Numerous studies in murine systems have shown that it may be possible to acutely drop tumor pH by inducing transient hyperglycemia (Ward-Hartley and Jain 1987). Such results have been partly confirmed in human patients, although paradoxical results have been seen in pre-

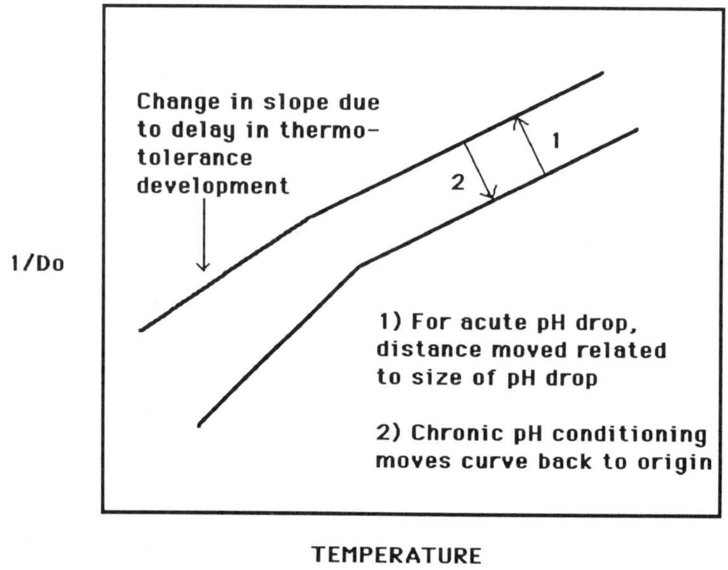

Figure 5 Effect of low pH on the Arrhenius plot. An acute pH drop increases the rate of cell kill at a given temperature and delays the onset of thermotolerance. Chronic pH conditioning, however, abrogates the sensitizing effect.

diabetics (Thistlethwaite et al. 1987). The magnitude of such pH reductions has ranged from 0.1 to 0.4 units. The results of Hahn and Shiu (1986) indicate that a drop of this magnitude would be sufficient to overcome the chronic pH conditioning effect. In summary, it is not known for certain whether chronically low pH conditions occur in human tumors. Second, aside from the data of Hahn and Shiu (1986) on one cell line, there are no data defining the time course of the low pH conditioning effect. There is a need for further work in this area to define more clearly in several model systems (in vivo and in vitro) the chronic pH conditioning effect as well as methods for overcoming it.

The development of thermotolerance and production of heat shock proteins (hsps) have also been delayed when cells are deprived of both oxygen and glucose (Gerweck et al. 1984; Lanks et al. 1988). Recent studies by Gerweck et al. (1989) have shown that such conditions may exist in vivo. Furthermore, tumors showing low PCr/P_i (phosphocreatine/inorganic phosphorus) ratios, which are a reflection of such conditions, are more sensitive to hyperthermic cell killing than tumors with higher PCr/P_i ratios. Some human and canine tumors have been shown to have lower PCr/P_i ratios than normal tissues, using ^{31}P MRS (Fig. 6) (Oberhaensly et al. 1986; Semmler et al. 1988). It is not known whe-

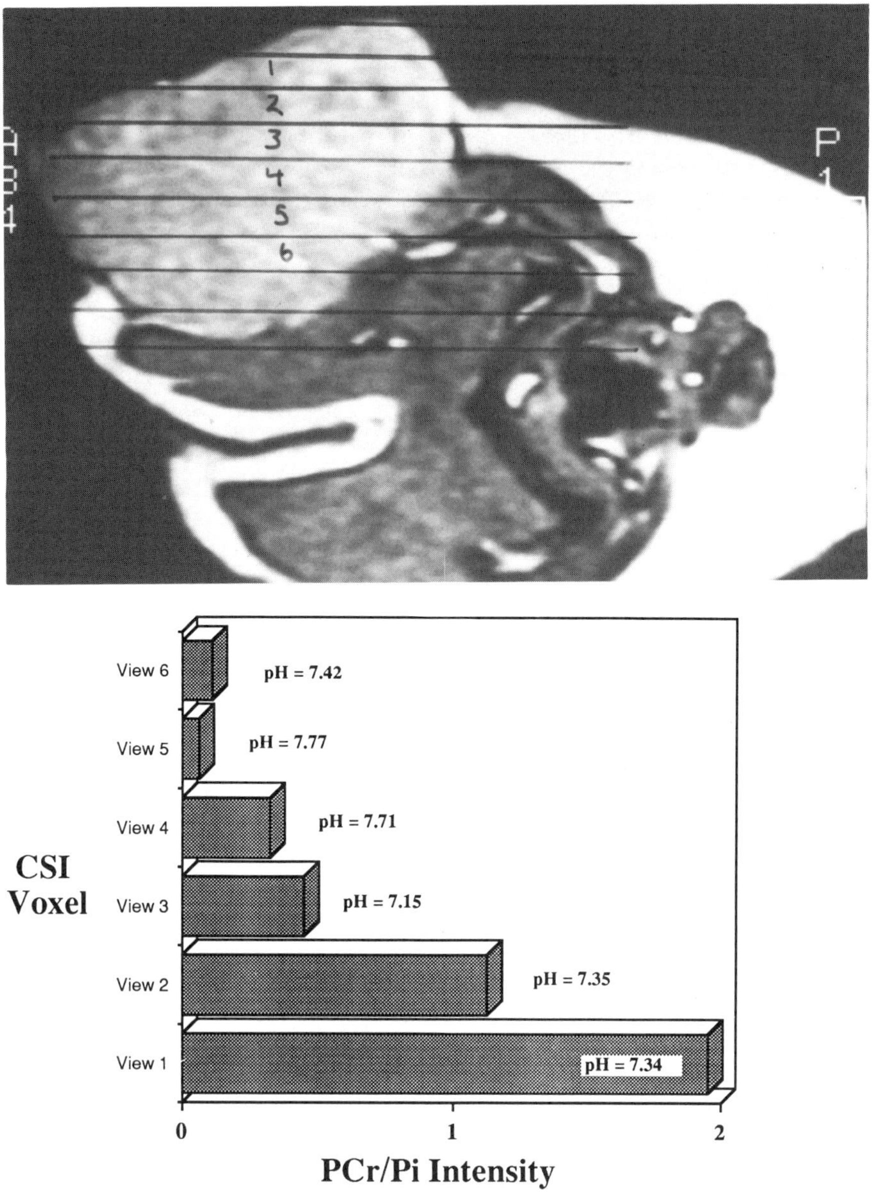

Figure 6 (*Top*) Magnetic resonance imaging scan of canine patient with a soft tissue sarcoma, depicting tissue slices in which high-energy phosphorus metabolites were monitored using dimensional chemical shift image MRS. (*Bottom*) The tumor was very heterogeneous in metabolite levels, with peripheral regions appearing more healthy (low PCr/P_i ratios) as compared with more centrally located regions, which had a lot of P_i and relatively little PCr. pH measurements ranged from acidotic to alkaline. (P_i) Inorganic phosphorus, (PCr) phosphocreatine. (View 1) Tumor surface, (view 6) 6–7 cm depth.

ther energy–O_2 depletion is sufficiently low to create enhanced thermocytoxicity and delay thermotolerance. Further work is needed to answer these questions.

Another factor that has been shown to influence the overall thermal sensitivity of cells is the absolute difference between the initial and final temperature. Culver and Gerner (1982) reported that cells conditioned to environmental temperatures below that at which cells are normally cultured are more sensitive to heat. When cells were grown at 35°C and heated for 2 hours at 43°C, survival was 50-fold lower than for cells grown at 39°C and heated at 43°C for the same period of time. An additional effect of this phenomenon is to inhibit the development of thermotolerance, similar to what is seen with step-down heating (D.B. Leeper, pers. comm.) (Fig. 4). There are a number of clinical situations in which step-up heating may be important. This may be particularly apparent in peripherally accessible tumor sites, such as the chest wall and head and neck regions, where the resting temperature of tumors is well below that of body core.

The overall shape of the Arrhenius plot may also be changed when hyperthermia is combined with other therapeutic modalities. Overgaard (1987) has shown that the Arrhenius plot for heat radiosensitization is essentially the same as that seen for heat alone when heat is given 3–4 hours after radiation. However, when heat is given simultaneously with radiation, the slope above the breakpoint changes. Since hyperthermia changes the reaction rate constants and pharmacokinetics of a number of chemotherapeutic drugs (Zakris et al. 1987; Riviere et al. 1986, 1989), it is probable that the Arrhenius relationship for heat-chemosensitization will be significantly different from that seen for heat alone. Probably it will be drug-dependent and possibly even sequence-dependent. Furthermore, thermotolerance has been shown to alter the cytotoxicity of some drugs, even when they are given at 37°C (Hahn 1987).

III. METHODS FOR ASSESSING TOLERANCE

It can be seen from the above discussion that a number of factors influence both the development and decay of thermotolerance. Thus, it becomes important to be able to assess whether thermotolerance is occurring in human tumors and normal tissues and, if so, to what degree it occurs. Li and associates have advocated the use of [^{35}S]methionine incorporation into heat-induced hsps as a way of monitoring the development and decay of thermotolerance (Li and Mak 1985). They have shown in in vitro and in vivo models that the kinetics of these two phenomena mirror each other quite closely. In a recent paper, Anderson et al. (1988) have looked at constitutive levels of hsps versus those that are heat inducible and have found that levels of both are important for interpreting the cytotoxic effects of heat, especially when comparing slow rates of heating with more rapid rates of heating. Furthermore, their data showed inconsistencies between the amount of hsps measured and the degree of thermotolerance expressed. Another technique that has been advocated as a method for measuring thermotolerance is dansyl lysine staining (Fisher et al. 1986). This technique involves challenging cells to a heat dose in vitro, following a priming dose (that could be given in vivo). The dansyl lysine stain is excluded by viable cells. Therefore, thermotolerant cells would show resistance to the stain, relative to nonpreheated controls. The disadvantage of all three of these techniques is that they require biopsies, both before and after heating. Thus, considerable clinical intuition as to where to biopsy both before and after heat would be needed to obtain the most prognostically important information (Woo et al. 1988). Possibly, strategies designed to maintain thermotolerance in normal tissues relative to tumor could be assessed in this way, although they would necessarily involve biopsies of adjacent normal tissues to assess whether thermotolerance persisted. In some normal tissues, it has been difficult to obtain single cell suspensions from such biopsies (Sekiguchi et al. 1988). Much more work needs to be done in this area to define protocols that will obtain information with the most prognostic significance. An alternative solution to the problem might be to use biologic probes for hsps that could be radiolabeled. Such probes would presumably have to enter cells in order to detect hsps. However, if this could be done, radiologic imaging techniques could be used to look for the presence of thermotolerance.

IV. EFFECT OF TUMOR MORPHOLOGIC AND PHYSIOLOGIC HETEROGENEITY ON ESTIMATION OF THERAPEUTIC EFFICACY FROM HEAT

The vascular morphology of tumors is very heterogeneous. For example, it is characterized by hypotension, regurgitant flow, vascular tortuosity, spontaneous hemorrhage, and low vascular density (Reinhold and Endrich 1986). This pathophysiologic state leads to a number of consequences that directly bear on the effectiveness of heat to kill tumor cells. First, circulatory heterogeneity effects the ability to deliver uniform temperatures within tumors. In general, the edge of tumors is more highly perfused than the center, thus, in a clinical situation, the center becomes much hotter than the periphery. As a result, in cases where more than one hyperthermia treatment is given, thermotolerance would persist longer in the center of the tumor than it would near the edge, if cells survive the higher temperatures in the center (Fig. 7). In the appended example, it is clear that certain regions of a tumor heated with 915-MHz microwaves were well within the therapeutic range, whereas other regions were well below it. Thus, optimization of the timing between hyperthermia fractions to avoid thermotolerance in the tumor becomes complicated, since the temperature distribution is nonuniform. As discussed above, other factors that may influence the development and decay of thermotolerance are pH and decreased nutrient availability. These factors are also likely to vary spatially (Fig. 6). Much more work needs to be done in spontaneous tumors to assess the impact of tumor physiologic environment on thermotolerance development and decay. Other factors that occur clinically include both step-up and step-down heating. These phenomena also vary throughout the treatment volume because of variations in blood flow and power deposition (Fig. 7). In conclusion, thermotolerance development and decay is likely to vary spatially

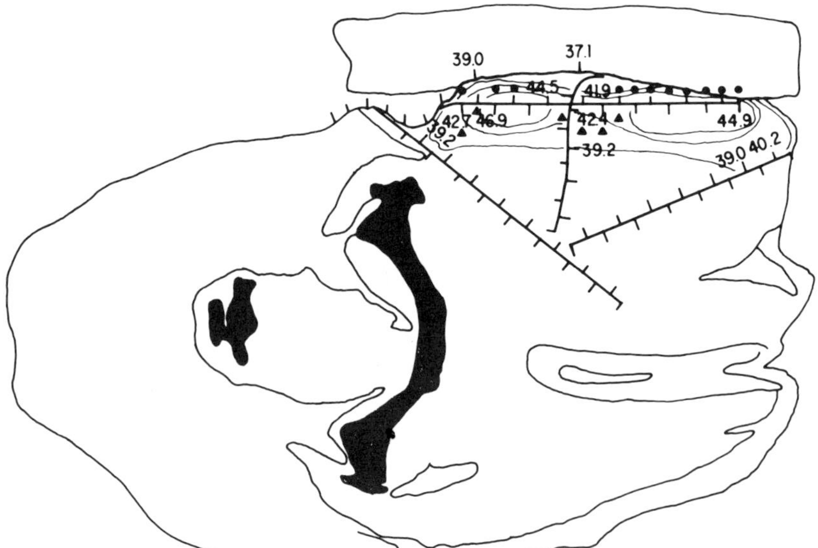

Figure 7 Estimate of two-dimensional temperature isotherms derived from multipoint thermometry in canine patient depicted in Figs. 3 and 6. Temperature distributions were very nonuniform. In addition, both step-up and step-down heating occurred in different parts of the tumor, as depicted by the symbols: (circles) step-down heating; (triangles) step-up heating. pH and phosphorus metabolite levels also varied significantly with depth. Because of the variable thermal history and energy status of different parts of the tumor, it is likely that thermotolerance induction and decay varied spatially within the tumor.

tially throughout the tumor because of multiple factors distributed heterogeneously throughout the tumor volume. As a result, estimation of therapeutic efficacy of heat treatment could be unfavorably biased unless the factors can be accounted for. One method to evaluate this problem would be to identify regions of tumor in which the potential for tolerance has the greatest prognostic significance and to biopsy those regions with thermotolerance assays. For example, it is possible that the peripheral regions have greater prognostic significance than the center, since peripheral regions contain the majority of the tumor mass (Oleson et al. 1989). A second potential solution would be to identify noninvasive techniques for evaluating thermotolerance, as suggested earlier with imaging techniques.

V. SUMMARY

In summary, considering the fact that the heat stress response with hsp production is conserved over virtually every living species, it seems inconceivable that human tumors would behave any differently in the face of heat shock. The questions that arise in the utilization of hyperthermia in human cancer therapy are more directed toward whether the heat treatments are sufficient to induce thermotolerance and whether the physiologic conditions in the tumor effect thermotolerance development and decay. These are critical questions that need to be addressed if hyperthermia is to achieve its greatest potential as an antitumor treatment.

ACKNOWLEDGMENTS

The author wishes to acknowledge Ms. Jeanne Forest for skilled typing assistance; Drs. Steve Sapareto, Dennis Leeper, and William Dewey for valuable scientific advice; and Drs. Dirk Sostman and Cecil Charles for providing MRS data shown in Figure 6. This work was supported in part by National Institutes of Health/National Cancer Institute grant 5 PO-1 CA-42745-02.

REFERENCES

Anderson, R.L., T.S. Herman, I. Kersen, and G.M. Hahn. 1988. Thermotolerance and heat shock protein induction by slow rates of heating. *Int. J. Rad. Oncol. Biol. Phys.* **15:** 717–725.

Cook, J.A. and M.H. Fox. 1988. Effects of chronic pH 6.6 on growth, intracellular pH and response to 42°C hyperthermia of Chinese hamster ovary cells. *Cancer Res.* **48:** 2147–2420.

Culver, P.S. and E.W. Gerner. 1982. Temperature acclimation and specific cellular components in the regulation of thermal sensitivity of mammalian cells. *Natl. Cancer Inst. Mongr.* **61:** 99–101.

Dewhirst, M.W. and D.A. Sim. 1984. The utility of thermal dose as a predictor of tumor and normal tissue responses to combined radiation and hyperthermia. *Cancer Res.* (suppl.) **44:** 4772s–4780s.

Dewhirst, M.W., D.A. Sim, S. Sapareto, and W.G. Connor. 1984. The importance of minimum tumor temperature in determining early and long term responses of pet animal tumors to heat and radiation. *Cancer Res.* **44:** 43–50.

Dewhirst, M.W., J.M. Winget, L. Edelstein-Keshet, I. Sylvester, M.J. Engler, D.E. Thrall, R.L. Page, and J.R. Oleson. 1987. Clinical application of thermal isoeffect dose. *Int. J. Hyperthermia* **3:** 307–318.

Field, S.B. 1987. Studies relevant to a means of quantifying the effects of hyperthermia. *Int. J. Hyperthermia* **3:** 291–296.

Field, S.B. and C.C. Morris. 1983. The relationship between heating time and temperature: Its relevance to clinical hyperthermia. *Radiother. Oncol.* **1:** 179–186.

———. 1984. Application of the relationship between heating time and temperature for use as a measure of thermal dose. In *Proceedings of the 4th International Symposium on Hyperthermic Oncology*, pp. 183–186. Aarhus, Denmark.

Fischer, G., G.C. Rice, and G.M. Hahn. 1986. Dansyl lysine, a new probe for assaying heat-induced cell killing and thermotolerance *in vitro* and *in vivo*. *Cancer Res.* **46:** 5064–5067.

Gerweck, L.E., B. Richards, and H.B. Michaels. 1982. Influence of low pH on the development and decay of 42°C thermotolerance in CHO cells. *Int. J. Rad. Oncol. Biol. Phys.* **8:** 1935–1941.

Gerweck, L.E., H. Majima, and T.F. Delany. 1983. Variability in kinetics of thermotolerance decay. In *Proceedings of the 7th International Congress of Radiation Research* (Abstract D6-13.) Amsterdam, The Netherlands.

Gerweck, L.E., W.K. Dahlberg, L.F. Epstein, and D.S. Shimm. 1984. Influence of nutrient and energy deprivation on cellular response to single and fractionated heat treatments. *Radiat. Res.* **99:** 573–581.

Gerweck, L.E., M. Urano, J. Kautcher, M.P. Fellenz, and J. Kahn. 1989. Relationship between energy status, hyponic cell fraction and hyperthermic sensitivity in a murine fibrosarcoma. *Radiat. Res.* **117:** 448–458.

Gillette, E.L., S.L. McChesney, M.W. Dewhirst, and R.J. Scott. 1987. Response of canine oral carcinomas to heat and radiation. *Int. J. Rad. Oncol. Biol. Phys.* **13:** 1861–1867.

Hahn, G.M. 1987. Thermotolerance and drugs. In *Thermotolerance* (ed. K.J. Henle), vol. I, pp. 9–112. CRC Press, Boca Raton, Florida.

Hahn, G.C. and E.C. Shiu. 1986. Adaptation to low pH modifies thermal and thermochemical responses of mammalian cells. *Int. J. Hyperthermia* **2:** 379–388.

Henle, K.J. 1980. Sensitization to hyperthermia below 43°C induced in Chinese hamster ovary cells by step down heating. *J. Natl. Cancer Inst.* **64:** 1479–1483.

Hollis, D.P. and R.L. Nunnally. 1977. Detection of regional ischemia in perfused beating hearts by phosphorus nuclear magnetic resonance. *Biochem. Biophys. Res. Commun.* **75:** 1086–1091.

Lanks, K.W., J. Gao, and F.J. Kasambalides. 1988. Nucleoside restoration of heat resistance and suppression of glucose-regulated synthesis by glucose-deprived L929 cells. *Cancer Res.* **48:** 1442–1445.

Law, M.P., R.G. Ahier, and S. Somaia. 1987. Thermotolerance induced by fractionated hyperthermia: Dependence on the interval between fractions. *Int. J. Hyperthermia* **3:** 433–440.

Li, G.C. and J.Y. Mak. 1985. Induction of heat shock protein synthesis in murine tumors during the development of thermotolerance. *Cancer Res.* **45:** 3816–3824.

Nielsen, O.S. and J. Overgaard. 1982. Importance of preheating temperature and time for the induction of thermotolerance in a solid tumour *in vivo*. *Br. J. Cancer* **46:** 894–903.

———. 1982. Influence of time and temperature on the kinetics of thermotolerance in L_1A_2 cells *in vitro*. *Cancer Res.* **42:** 4190–4196.

Oberhaensli, R.D., P.J. Bore, R.P. Rampling, D. Hilton-Jones, L.J. Hands, and G.K. Radda. 1986. Biochemical investigation of human tumours *in vivo* with phosphorous-31 magnetic resonance spectroscopy. *Lancet* **II:** 8–11.

Oleson, J.R., D.A. Sim, and M.R. Manning. 1984. Analysis of prognostic variables in hyperthermia treatment of 161 patients. *Int. J. Rad. Oncol. Biol. Phys.* **10:** 2231–2239.

Oleson, J.R., M.W. Dewhirst, J.M. Harrelson, K.A. Leopold, T.V. Samulski, and C.Y. Tso. 1989. Tumor temperature distributions predict hyperthermia effect. *Int. J. Rad. Oncol. Biol. Phys.* **16:** 559–570.

Overgaard, J. 1985. Rationale and problems in the design of clinical studies. In *Hyperthermia oncology* (ed. J. Overgaard), vol. 2. Taylor & Francis, London.

———. 1987. Some problems related to the clinical use of thermal isoeffect doses. *Int. J. Hyperthermia* **3:** 329–336.

Reinhold, H.S. and B. Endrich. 1986. Tumor microcirculation as a target for hyperthermia: A review. *Int. J. Hyperthermia* **2:** 111–137.

Riviere, J.E., R.L. Page, D.P. Aucoin, R.A. Rodgers, and P. Williams. 1989. Effect of heat on melphalan pharmacokinetics in normal dogs. *Int. J. Hyperthermia* (in press).

Riviere, J.E., R.L. Page, M.W. Dewhirst, K. Jyczkowska, and D.E. Thrall. 1986. Effect of hyperthermia, a cisplatin pharmacokinetics in normal dogs. *Int. J. Hypotherm.* **2:** 231–358.

Roizin-Towle, L., J.P. Pirro, and J. McDowell. 1986. A comparison of the heat and radiation sensitivity of rodent and human derived cells cultured *in vitro*. *Int. J. Rad. Oncol. Biol. Phys.* **12:** 647–653.

Sapareto, S.A. 1987. Thermal isoeffect dose: Addressing the problem of thermotolerance. *Int. J. Hyperthermia* **3:** 289–290.

Sapareto, S.A. and W.C. Dewey. 1984. Thermal dose determination in cancer therapy. *Int. J. Rad. Oncol. Biol. Phys.* **10:** 787–800.

Sapareto, S.A., J.L. Evelhock, W. Negendank, and J.J.H. Ackerman. 1988. Measuring physiological factors by NMR. In *Proceedings of the 5th International Congress of Hyperthermia*. Kyoto, Japan.

Sapozink, M.D., T.C. Cetas, P.M. Corry, M.J. Egger, P. Fessenden, and the NCI Equipment Evaluation Contractor's Group. 1988. Introduction to hyperthermia device evaluation. *Int. J. Hyperthermia* **4:** 1–16.

Sekiguchi, R., G.C. Rice, and G.M. Hahn. 1988. The use of dansyl lysine to assess heat damage and thermotolerance of normal tissues. *Int. J. Rad. Oncol. Biol. Phys.* **14:** 983–988.

Semmler, W., G. Gademann, P. Bachert-Baumann, H.J. Zabel, W.J. Lorenz, and G. van Kaick. 1988. Monitoring human tumor response to therapy by means of P-31 MR spectroscopy. *Radiology* **166:** 533–539.

Thistlethwaite, A.J., G.A. Alexander, D.J. Moylan, and D.B. Leeper. 1987. Modification of human tumor pH by elevation of blood glucose. *Int. J. Rad. Oncol. Biol. Phys.* **13:** 603–610.

Valdagni, R., M. Amichetti, and G. Paui. 1988. Radical radiation alone versus radical radiation plus microwave hyperthermia for N_3 (TNM-UICC) necknodes: A prospective randomized clinical trial. *Int. J. Rad. Oncol. Biol. Phys.* **15:** 13–24.

Valdagni, R., F.E. Liu, and D.S. Kapp. 1988. Important prognostic factors influencing outcome of combined radiation and hyperthermia. *Int. J. Rad. Oncol. Biol. Phys.* **15:** 959–972.

Ward-Hartley, K.A. and R.K. Jain. 1987. Effect of glucose and galactose on microcirculatory flow in normal and neoplastic tissues in rabbits. *Cancer Res.* **47:** 371–377.

Wike-Hooley, J.L., J. Haveman, and H.S. Reinhold. 1984. The relevance of tumour pH to the treatment of malignant disease. *Radiother. Oncol.* **2:** 343–366.

Woo, S., G.C. Rice, D. Kapp, and G.M. Hahn. 1988. A predictive assay for human tumor cellular response to hyperthermia using dansyl lysine and flow cytometry. *Int. J. Rad. Oncol. Biol. Phys.* **14:** 361–365.

Zakris, E.L., M.W. Dewhirst, J.E. Riviere, P.J. Hoopes, R.L. Page, and J.R. Oleson. 1987. Pharmacokinetics and toxicity of intraperitoneal cisplatin combined with regional hyperthermia. *J. Clin. Oncol.* **5:** 1613–1620.

6

Hyperthermia in Combination with Radiation in the Treatment of Cancers

Mitsuyuki Abe and Masahiro Hiraoka
Department of Radiology, Faculty of Medicine,
Kyoto University, Kyoto, 606, Japan

I. INTRODUCTION

Hyperthermia has been clinically investigated over the past two decades as a treatment modality for human malignant tumors. The biological rationale for the use of hyperthermia in the treatment of cancer is based on the following experimental findings:

1. Heat kills cells exponentially as a function of time at temperatures above 42°C.
2. Hypoxic cells are at least as sensitive to heat as oxic cells.
3. Tumors may contain cells with low pH and with nutrient deficiency, which make cells more sensitive to heat.
4. Tumors tend to become hotter than surrounding normal tissues during hyperthermia because of a more sluggish blood supply and a lesser capacity for increased blood flow upon heating.
5. The vasculature in tumors is damaged at lower temperatures than that in normal tissues, resulting preferentially in the secondary death of tumor cells and making the tumor temperature rise more easily upon the second heating.
6. Cells in the late S phase, known to be radioresistant, are sensitive to heat.
7. Under some conditions, heat interacts synergistically with radiation, partially as a result of inhibition of sublethal or potentially lethal damage repair.
8. Heat may potentiate the cytotoxic effects of some kinds of chemotherapeutic drugs either by increasing drug uptake or by enhancing sensitivity to the drug.

Stress Proteins in Biology and Medicine.

On the other hand, there are several problems in thermotherapy. The first is thermotolerance, which is defined as the development of resistance of cells to subsequent heat by prior heating. At lower temperatures of around 42°C, thermotolerance is induced during heating after an exposure of about 2 or 3 hours. This phenomenon should be taken into consideration in whole-body hyperthermia, in which target temperatures are kept at 41–42°C for 2–10 hours. At temperatures above 43°C, the cells appear not to develop thermotolerance during heating, but exhibit thermotolerance on subsequent heating if the time interval between two heat treatments is less than 72 hours or even longer. Loco-regional hyperthermia is clinically delivered in fractionated regimen. Thermotolerance is, therefore, important both in whole-body and loco-regional hyperthermia. This phenomenon has been well studied in both normal tissues and tumors in experimental animals. Few data, however, have been available for human tumors. Recently, two new methods, the dansyl lysine staining technique (Woo et al. 1988) and quantification of hsp70 (Birck et al. 1987), have been developed as a possible means to predict thermotolerance. With the introduction of these assays to clinics, the kinetics of thermotolerance of tumors and normal tissues in the human will be evaluated, leading to the establishment of an optimal fractionation regimen of hyperthermia.

The second problem is the lack of a unit of "thermal dose" to quantitatively assess the biological effect of heat. Just as the establishment of the unit of "radiation dose" has contributed greatly to advances in radiation therapy, it is essential to determine the thermal dose for advances in thermotherapy. Sapareto and Dewey (1984) introduced the equivalent-time concept, which is based on the assumption of a biological isoeffect relation of various combinations of time-temperature treatments. This concept is limited by the fact that other important biological parameters, such as thermotolerance, rate of heating, interaction between radiation and hyperthermia, effect of sequencing, and variations in fractionations, are not taken into consideration. Limitations of the existing thermometry system also hamper the use of the thermal dose concept in clinics.

In addition to above-mentioned biological problems, insufficient development of heating devices and thermometry systems still offers great limitations for clinical application of this new treatment modality. Nevertheless, clinical experience with combined loco-regional hyperthermia and radiation therapy has been rapidly accumulating over the past few decades. The purpose of this paper is to review briefly the current status of this combined modality in clinical cancer therapy.

II. HYPERTHERMIA FOR SUPERFICIAL TUMORS

A large number of clinical experiences with combined hyperthermia and radiation have been reported for superficial tumors. These tumors have several advantages over deep-seated tumors as a target of the combined treatment: relatively well-developed heating devices, accessibility of thermometry, and easy assessment of response to heat in both tumors and normal tissues. Therefore, they are considered to be a good model to investigate the combined effects of hyperthermia and radiation. Some superficial tumors, such as malignant melanoma, soft tissue tumors, locally advanced tumors, and recurrent tumors following radiation therapy, are still refractory to conventional treatments. The combined hyperthermia and radiation therapy, therefore, has potential clinical significance for the treatment of these tumors.

We treated 112 patients with a variety of radioresistant malignancies in superficial (less than 3 cm in depth) and subsurface (from 3 cm to 6 cm in depth) locations. Radiation with 4 Gy twice a week for a total of 40 Gy on the average, or with 2 Gy five times a week for a total of 50 Gy on the average, was delivered. After irradiation, hyperthermia at 42–43°C for 40–60 minutes was given twice a week for a total of 4.5 times on the average. Of 112 tumors treated, 55 (49.1%) showed complete regression. A greater response was observed in superficial tumors than in subsurface tumors (Table 1). Figure 1 shows a bulky neck tumor that responded well to combined radiation and hyperthermia.

Usefulness of local hyperthermia combined with radiation therapy was clearly demonstrated by several clinical trials using patients with two or more comparable tumors (matched tumors). We compared the

Table 1 Responses of superficial or subsurface tumors to radiation plus hyperthermia

Tumor site	No. of tumors	Tumor response			
		CR	PRa	PRb	NR
Superficial	39	26 (66.7%)	6 (15.4%)	4 (10.3%)	3 (7.7%)
Subsurface	73	29 (39.7%)	16 (21.9%)	21 (28.8%)	7 (9.6%)
Total	112	55 (49.1%)	22 (19.6%)	25 (22.3%)	10 (8.9%)

(CR) Complete regression; (PRa) partial regression with >80% decrease in tumor volume; (PRb) partial regression with 50–80% decrease in tumor volume; (NR) no regression.

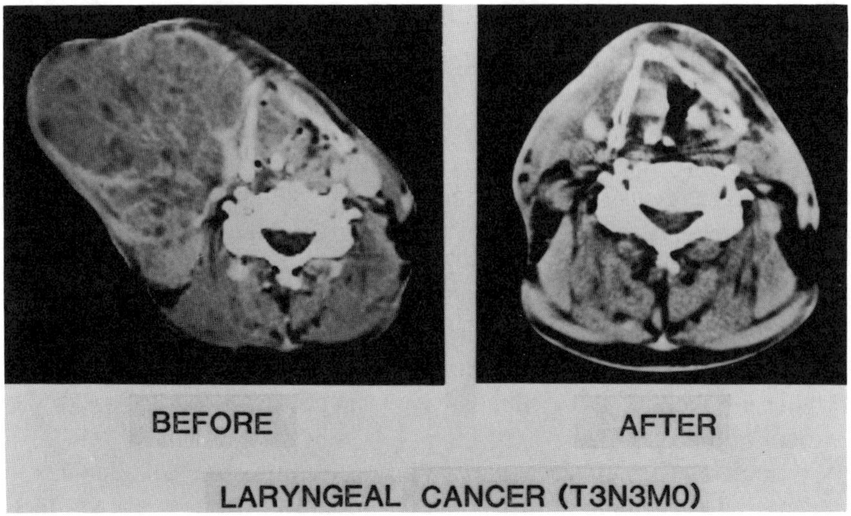

Figure 1 CT scans of a bulky neck tumor before and after thermoradiotherapy.

response of 11 matched tumors in the same patient treated by radiation alone or radiation combined with hyperthermia. All tumors were metastatic and superficial. In the thermoradiotherapy group, complete regression (CR) was obtained in 7 tumors, whereas in the radiation alone group, CR was seen in 2 tumors. These results indicate that the effects of radiation are enhanced by heat (Abe et al. 1986). A recent review of these trials demonstrates that the combined modality of hyperthermia and radiation showed a twofold or greater increase in local response rates than radiotherapy alone (Arcangeli et al. 1988).

In addition to these studies, recent prospective randomized trials indicated the superiority of combined thermoradiotherapy to radiotherapy alone in the response rate of superficial tumors. Valdagni et al. (1988a) reported a prospective randomized clinical trial for metastatic squamous cell cervical lymph nodes. Radiotherapy was given in conventional fractionation to a total dose of 64–70 Gy, and hyperthermia was delivered twice weekly. The complete response rate was 82.3% (14/17) for the combined treatment arm versus 36.8% (7/19) for the control irradiation arm, leading to an iso-dose thermal enhancement ratio value of 2.23. On the other hand, acute local toxicities were similar in irradiated and heated plus irradiated neck regions. A Japanese trial also demonstrated the usefulness of a combined treatment of hyperthermia and radiation for superficial tumors. The response rates (complete response plus partial response, more than 50% decrease in tumor volume) for thermoradiotherapy and radiotherapy alone were 81.8% and 62.5%, respectively, and this difference was statistically significant (S. Egawa et al., unpubl.). A trial carried out by the Radiation Therapy Oncology Group (RTOG) in the United States failed to show a difference in the response rate between radiation alone and radiation plus heat when the tumors treated were analyzed all together (Perez et al. 1987). When the tumor response was assessed according to the tumor size, a significantly higher response rate was achieved with the combined hyperthermia and radiotherapy for tumors with maximum diameter less than 3 cm, but not for tumors more than 3 cm in diameter. The combined treatment also showed a substantially higher response rate in breast tumors but not in head and neck tumors or other tumors in comparison with radiotherapy alone. Since smaller tumors and breast lesions are easier to heat, it is suggested that heating limitation may be the reason for the lack of enhanced effects of combined treatment for larger tumors and non-breast lesions.

Factors that would influence the response of tumors to combined radiation and hyperthermia have been analyzed (Valdagni et al. 1988b). Among several pretreatment factors, tumor volume seems to be most important. Most clinical reports show that tumor volume negatively correlates with tumor response, that is, as tumor volume increases, the tumor response decreases. It is reported that larger volumes are in turn correlated with lower minimum temperatures (Oleson et al. 1984). Therefore, the poor response of larger tumors to the combined treatment may result from greater intratumor thermal gradients, lower minimum temperatures, and the greater number of tumor cells. Among several thermal parameters, the minimum tumor temperatures seem to be mostly correlated to the response rate, which is consistent with experimental data using pet tumors (Dewhirst et al. 1984). Attempts to correlate the thermal dose to the tumor response have been successful in some clinical trials (Gonzalez et al. 1988).

Controversy exists concerning the optimal number and frequency of hyperthermia treatments. Most retrospective studies failed to show the superiority of a larger number of treatments to a smaller number in the response rate. Two prospective randomized trials comparing the two heat sessions versus six sessions both demonstrated a similar response rate (Kapp et al. 1987; Valdagni et al. 1988a). Alexander et al. (1987) reported a randomized trial in which hyperthermia treatments given once or twice a week were compared. Complete response rate was higher with treatments given once a week than twice a week. A possible explanation for this observation is the development of thermotolerance. As mentioned before, the kinetics of thermotolerance in human tumors has not been studied because of the lack of techniques for researching this phenomenon in the clinic. The dansyl lysine staining technique and quantification of hsp70 are expected to emerge as a method to explore the kinetics of thermotolerance in human tumors.

III. HYPERTHERMIA FOR DEEP-SEATED TUMORS

Major efforts have been devoted to the development of deep-heating equipment. Heating methods for deep-seated tumors are divided into external, intraluminal, and interstitial heating, all of which have inherent advantages and disadvantages.

External heating administers heat externally through the body surface. An ideal external heating device is capable of homogeneous heating of deep-seated tumors up to therapeutic temperatures without heating the adjacent normal tissues. None of the deep-heating devices available satisfy these requirements. One possible way for the selective heating of tumors is through the use of focused ultrasound. This method has been tested in the clinic with limited success (Fessenden et al. 1984). The main disadvantage of ultrasound is that it reflects at soft tissue/air and bone/soft tissue interfaces and attenuates rapidly in bone; both factors make ultrasound difficult to apply to tumors adjacent to or in the air or bones.

The most common method for external heating is regional heating using an annular phased array system (APAS) or radiofrequency (RF) capacitive heating. We have developed an RF capacitive heating machine (Thermotron RF-8) in cooperation with Yamamoto Vinyter Co. of Osaka. The machine is schematically illustrated in Figure 2. The RF generator has a self-excited oscillation circuit at 8 MHz and 1.5 kW maximum output power. The RF energy is transmitted from a generator via two coaxial cables to two disc electrodes. The RF is applied through a pair of electrodes placed on opposite sides of the body and power is distributed regionally through interaction of electric field produced between the parallel-opposed electrodes. The depth of heating can be controlled by changing the size of the paired electrodes. This technique deposits energy to deep-seated tumors in an unfocused manner. Therefore, energy is delivered to normal tissues as well as tumors. Under such conditions, selective heating of tumors over normal tissues is possible only when heat dissipation by blood flow is more predominant in the normal tissues than in tumors. Actually, a positive relationship between temperature elevation and the blood flow of tumors, as estimated by thermal washout curves, is demonstrated in deep heating by APAS (Samulski et al. 1987). Thermometry results of RF capacitive heating also show that small tumors and tumors considered to be hypervascular are difficult to heat (Hiraoka et al. 1987a).

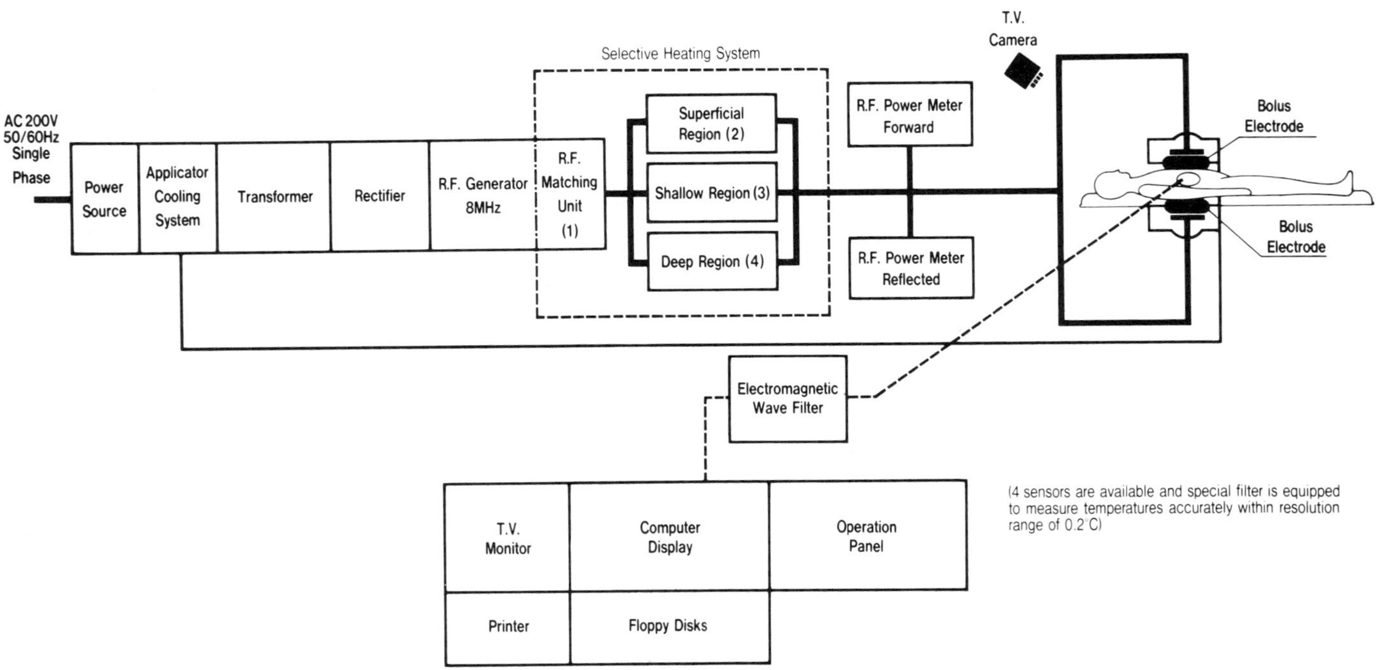

Figure 2 Schematic illustration of the Thermotron RF-8 hyperthermia system.

The second method for deep heating is an intraluminal or intracavitary heating using hot fluid, microwave, or RF. Intracavitary irrigation by hot fluid has been mostly applied to the carcinoma of the urinary bladder and, recently, to the peritoneal dissemination of various malignancies. Although the procedure is simple, it is only indicated for very superficial tumors because of its limited penetration of heat. Intraluminal heating is induced by inserting a microwave or an RF applicator into lumens in the human body, such as the esophagus, rectum, vagina, uterine canal, and biliary tract. Very localized heating is possible with this technique. It has a problem, however, in that heating distributions become inhomogeneous when the contact of the applicators with the tumor or the wall of the lumen is not sufficient.

The third method is interstitial heating, which is divided into RF current heating, microwave heating, and ferromagnetic heating. The advantages of interstitial heating are (1) selective heating of localized tumors and (2) feasibility of combined use of brachytherapy. On the other hand, the disadvantages are (1) invasiveness, (2) difficulty in repeated treatment, and (3) limitation of applicable sites.

In our hospital, 40 tumors in 40 patients were treated by RF capacitive hyperthermia using the Thermotron RF-8 in combination with radiation between 1981 and 1985. These tumors met all of the following criteria: (1) They were judged as refractory to radiotherapy from the tumor size or the histologic type; (2) they were located more than 7 cm in the maximum depth from the surface; and (3) they were tumors in which accurate thermometry was possible. Hyperthermia was given once or twice a week after irradiation, 4–13 sessions total. Radiotherapy was delivered in fractions of 170–250 cGy a day, 5 days a week, to 30–70 Gy to 33 patients, and the remaining 7 patients received a total dose of 28–60 Gy in fractions of 4 Gy, twice a week. Of the 40 tumors, 6 (15%) showed CR (Hiraoka et al. 1987b).

Since 1983, we have treated deep-seated abdominal or pelvic tumors that were considered to be refractory to conventional treatments by radiation in combination with RF capacitive heating. Of 33 tumors treated, 4 exhibited complete response, and 14 exhibited partial response (Nishimura et al. 1989). The tumor response is apparently dependent on the thermal parameters, such as average and minimum tumor temperatures, and the number of effective heat treatments.

Table 2 summarizes several clinical reports of thermoradiotherapy for deep-seated tumors. Most of them dealt with locally advanced or recurrent tumors refractory to conventional treatment modalities. The response rate obtained for deep-seated tumors is apparently lower than that in superficial tumors. There seem to be several reasons for this lower

Table 2 Clinical results of thermoradiotherapy for deep-seated tumors

Tumor	Heating device	Hyperthermia	Radiation therapy	Response rate of tumor	References
40 various tumors	capacitive RF	40–44°C 40–60 min 4–13 sessions	28–70 Gy	CR 15% PR 48%	Hiraoka et al. (1987b)
33 abdomino-pelvic tumors	capacitive RF	Tav >42°C in 17 tumors 40–60 min 3–14 sessions	20–70 Gy	CR 12% PR 42%	Nishimura et al. (1989)
44 various tumors	APAS	Tav 41°C	38 Gy (average)	CR 14% PR 11%	Shimm et al. (1988)
28 abdominal tumors	APAS	Tav >42°C in 13 tumors 4 sessions (av)	21 Gy (average)	CR 0 PR 18%	Sapozink et al. (1986a)
43 pelvic tumors	APAS	Tav >42°C in 31 tumors 4 sessions (av)	not specified	CR 12% PR 33%	Sapozink et al. (1986b)

(Tav) Average tumor temperature.

response rate in deep-seated tumors. First, thermal profiles shown in deep-seated tumors are obviously inferior to those in superficial tumors. The minimum tumor temperature mostly remains lower than 41°C. Second, the volume of deep-seated tumors treated is generally large, which will negatively affect the response rate. Third, local tumor response to therapeutic modalities is most commonly assessed by tumor regression. However, in hyperthermia, tumor response to heat is not fully assessed by tumor regression. Our current study has demonstrated that some tumors heated at high temperatures do not show regression but develop remarkable intratumor low-density areas that appear on posttreatment computed tomography (CT) scans (Hiraoka et al. 1987c). This finding is observed more frequently in deep-seated tumors with a large volume than in superficial tumors. These tumors do not regrow during long follow-up observations. Histopathologic examinations reveal that this low density corresponds to coagulation necrosis induced by heat (Jo et al. 1987). We classified the intratumor low-density area into three types according to its percent area in the biggest section of the tumor: type I, less than 50%; type II, 50-80%; and type III, more than 80%. The type of low-density area and tumor response were well correlated, as shown in Table 3. Furthermore, follow-up studies clearly demonstrated that type III tumors rarely regrow. Figure 3 shows a typical case of the type III change. The massive necrosis, which cannot be readily eliminated, explains the poor tumor regression. Destruction of microvasculature by hyperthermia may work to delay the tumor regression. The difficulty in eliminating necrosis via the overlying skin also explains the poor regression in deep-seated tumors. Therefore, new criteria based on the tumor size and the degree of intratumor low density must be established for the assessment of tumor response to heat.

Table 3 Results of the follow-up observation according to type of low density

Low density	No. of tumors	Tumor regrowth (+)	Tumor regrowth (−)	Removal after the treatment
Type I	14	10	1	3
Type II	7	6	1	0
Type III	5	1	4	0
Total	26	17	6	3

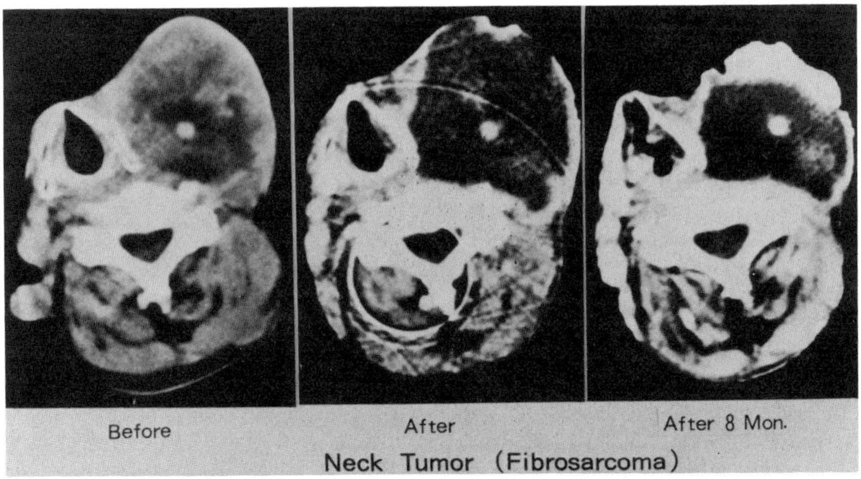

Figure 3 CT scans of a bulky neck tumor (fibrosarcoma) before and after thermoradiotherapy. After treatment, low density was found in the tumor. Although the tumor developed no regression, it did not regrow in the follow-up period of 8 months.

III. CONCLUSION

The use of heat in cancer therapy is the subject of an ever-broadening, multidisciplinary research effort involving the fields of biology, physics, engineering, and medicine. There are many questions still to be answered regarding heat shock protein, thermal dose, and thermometry. Continued development of heating devices that can heat tumors homogeneously at various sites is essential, and establishment of evaluation criteria for tumor response to heat is important in quantifying biological effects of hyperthermia. Although hyperthermia in cancer therapy is still at an early stage, it seems likely that this treatment modality will become a useful adjunct in the management of cancer.

ACKNOWLEDGMENTS

This work was supported in part by a grant-in-aid for scientific research (63010046) from the Ministry of Education, Science and Culture, Japan.

REFERENCES

Abe, M., M. Hiraoka, M. Takahashi, S. Egawa, T. Matsuda, Y. Onoyama, K. Morita, M. Kakehi, and T. Sugahara. 1986. Multi-institutional studies on hyperthermia using an 8

MHz radiofrequency capacitive heating device (Thermotron RF-8) in combination with radiation for cancer therapy. *Cancer* **58:** 1589–1595.

Alexander, G.A., D.J. Moylan, F.M. Waterman, R.E. Nerlinger, and D.B. Leeper. 1987. Randomized trial of 1 vs. 2 adjuvant hyperthermia treatments in patients with superficial metastases. In *Abstract Presented at the Thirty-fifth Annual Meeting of the Radiation Research Society*, February 21–26, p. 18. Atlanta.

Arcangeli, G., J. Overgaard, D.G. Gonzalez, and P.N. Shrivastava. 1988. Hyperthermia trials. *Int. J. Radiat. Oncol. Bios. Phys.* **14:** S93–S109.

Birck, E.M., J.Y. Mak, and G.C. Li. 1987. Possibility of using HSP 70 as a predictor of thermal response in tumors during fractionated hyperthermia. In *Abstract Presented at the Thirty-fifth Annual Meeting of the Radiation Research Society*, February 21–26, p. 47. Atlanta.

Dewhirst, M.W., D.A. Sim, S. Sapareto, and W.G. Connor. 1984. Importance of minimum tumor temperature in determining early and long-term responses of spontaneous carine and feline tumors to heat and radiation. *Cancer Res.* **44:** 43–50.

Fessenden, P., E.R. Lee, T.L. Anderson, J.W. Strohbehn, J. Meyer, T.V. Samulski, and J.B. Marmor. 1984. Experience with a multitransducer ultrasound system for localized hyperthermia of deep tissues. *IEEE Trans. Biomed. Eng.* **31:** 126–135.

Gonzalez, D., J.D.P. van Dijk, and L.E.C.M. Blank. 1988. Chest wall recurrences of breast cancer: Results of combined treatment with radiation and hyperthermia. *Radiother. Oncol.* **12:** 95–103.

Hiraoka, M., S. Jo, K. Akuta, Y. Nishimura, Y. Nagata, M. Takahashi, and M. Abe. 1987a. Radiofrequency capacitive hyperthermia for deep-seated tumors. I. Studies on thermometry. *Cancer* **60:** 121–127.

———. 1987b. Radiofrequency capacitive hyperthermia for deep-seated tumors. II. Effects of thermoradiotherapy. *Cancer* **60:** 128–135.

Hiraoka, M., K. Akuta, Y. Nishimura, Y. Nagata, S. Jo, M. Takahashi, and M. Abe. 1987c. Tumor response to thermoradiation therapy: Use of CT in evaluation. *Radiology* **164:** 259–262.

Jo, S., M. Hiraoka, K. Akuta, Y. Nishimura, H. Nishida, M. Furuta, M. Takahashi, and M. Abe. 1987. Histopathological studies on the effect of thermoradiotherapy for human malignant tumors. *Jpn. J. Hyperther. Oncol.* **3:** 49–61.

Kapp, D.S., T.V. Samulski, P. Fessenden, M.A. Bagshaw, E.R. Lee, A.W. Lohrbach, and R.S. Cox. 1987. Prognostic significance of tumor volume on response following local-regional hyperthermia (HT) and radiation therapy (XRT). In *Abstract Presented at the Thirty-fifth Annual Meeting of the Radiation Research Society*, February 21–26, p. 17. Atlanta.

Nishimura, Y., M. Hiraoka, S. Jo, K. Akuta, Y. Nagata, S. Masunaga, M. Takahashi, and M. Abe. 1989. Radiofrequency (RF) capacitive hyperthermia combined with radiotherapy in the treatment of abdominal and pelvic deep-seated tumors. *Radiother. Oncol.* (in press).

Oleson, J.R., D.A. Sim, and M.R. Manning. 1984. Analysis of prognostic variables in hyperthermia treatment of 161 patients. *Int. J. Radiat. Oncol. Biol. Phys.* **10:** 2231–2239.

Perez, C.A., B. Gillepie, N. Hornback, B. Emami, W. Moylan, W. Hans Baerwald, M.H. Seegenschmiedt, M. Bauer, and B. Pukris. 1987. Difficulties in assessing therapeutic results of clinical hyperthermia in an RTOG trial. In *The Twenty-ninth Annual Meeting of American Society for Therapeutic Radiology and Oncology*, Oct. 18–23, p. 164. Boston.

Samulski, T.V., P. Fessenden, R. Valdagni, and D.S. Kapp. 1987. Correlations of thermal washout rate, steady state temperatures and tissue type in deep seated recurrent or metastatic tumors. *Int. J. Radiat. Oncol. Biol. Phys.* **13:** 907–916.

Sapareto, S.A. and W.C. Dewey. 1984. Thermal dose determination in cancer therapy. *Int. J. Radiat. Oncol. Biol. Phys.* **10:** 787–800.

Sapozink, M.D., F.A. Gibbs, Jr., M.J. Egger, and J.R. Stewart. 1986a. Abdominal regional hyperthermia with an annular phased array. *J. Clin. Oncol.* **4:** 775–783.

———. 1986b. Regional hyperthermia for clinically advanced deep-seated pelvic malignancy. *Am. J. Clin. Oncol.* **9:** 162–169.

Shimm, D.S., T.C. Cetas, J.R. Oleson, E.R. Gross, D.N. Buechler, A.M. Fletcher, and S.E. Dean. 1988. Regional hyperthermia for deep-seated malignancies using the BSD annular array. *Int. J. Hyperthermia* **4:** 159–170.

Valdagni, R., M. Amichetti, and G. Pani. 1988a. Radical radiation alone versus radical radiation plus microwave hyperthermia for N_3 (TNM-UICC) neck nodes: A prospective randomized clinical trial. *Int. J. Radiat. Oncol. Biol. Phys.* **15:** 13–24.

Valdagni, R., F. Liu, and D.S. Kapp. 1988b. Important prognostic factors influencing outcome of combined radiation and hyperthermia. *Int. J. Radiat. Oncol. Biol. Phys.* **15:** 959–972.

Woo, S.Y., G.C. Rice, D.S. Kapp, and G.M. Hahn. 1988. Predictive assay for human tumor cellular response to hyperthermia using dansyl lysine staining and flow cytometry. *Int. J. Radiat. Oncol. Biol. Phys.* **14:** 361–365.

7

Stress Proteins and Infectious Diseases

Douglas B. Young and Angela Mehlert
MRC Tuberculosis and Related Infections Unit
Royal Postgraduate Medical School
Hammersmith Hospital, London W12 OHS
United Kingdom

Deborah F. Smith
Department of Biochemistry
Imperial College of Science, Technology and Medicine
London SW7 2AZ, United Kingdom

Stress Proteins in Biology and Medicine.

I. INTRODUCTION

Infectious diseases caused by bacterial and parasitic pathogens remain a major public health problem in most of the developing world. Improvement in the control programs for these diseases requires development of new diagnostic methods and effective prophylactic vaccines. Therefore, considerable effort has been invested in the analysis of pathogenic mechanisms and identification of the individual components of pathogens that are involved in interactions with the host immune system.

Analysis of stress proteins is an important part of two distinct aspects of such studies. First, major antigenic components recognized during a variety of bacterial and parasitic infections have been identified as members of stress protein families on the basis of sequence analysis. Stress proteins therefore have a role as antigens during the immune response to infection. Second, there is evidence that stress protein expression is upregulated in cells of the host and pathogen during the process of infection. Analysis of the differential expression of stress protein genes is therefore important in understanding factors involved in host/parasite interactions. Both of these aspects are addressed in this chapter.

II. STRESS PROTEINS AS ANTIGENS

A. The Immune Response to Infection

The immune system has a variety of mechanisms for killing and eliminating pathogens. The humoral immune response involves initial binding of antibody to the surface of a pathogen and subsequent killing by complement components in the serum. Cell-mediated immunity requires antigen recognition by regulatory T lymphocytes, which can then activate phagocytic cells capable of engulfing and destroying the pathogen. Regulatory T cells also activate a subset of cytotoxic T lymphocytes capable of lysing infected cells. Antibody recognition generally determines immunity to extracellular pathogens, whereas control of intracellular pathogens is more complex and often requires activation of phagocytic cells (for review, see Hahn and Kaufmann 1981). However, there is considerable overlap between the different arms of the immune system, and protective immunity usually involves the concerted action of different mechanisms of immune response. Activation of B lymphocytes for antibody production requires cooperation with activated T cells and, since antibody binding can enhance phagocytosis and B cells can present antigen to T cells, antibody recognition also plays a role in cell-mediated immunity. Thus, it is often important to analyze both antibody and T-cell recognition of antigens.

There is an important difference in the mechanisms involved in antigen recognition by antibodies and by T cells. Antibodies can bind to their target antigen when it is either free in solution or exposed on a surface, and they have no requirement for accessory cells or molecules. The antigen-specific receptor on T cells (which is a member of the immunoglobulin superfamily), however, will only bind antigen that is already attached to a molecule from the major histocompatibility complex (MHC) on the surface of an antigen-presenting cell. Attachment to the MHC molecule generally requires internalization of the antigen within the presenting cell and a "processing" event involving denaturation and partial degradation of the antigen. A number of cell types can function in antigen presentation, including the phagocytic cells responsible for degradation of pathogens, and B lymphocytes. Because of the different mechanisms of antigen recognition, antibodies are effective in protective immunity only if their antigen is exposed on the surface of the intact pathogen, but surface molecules and internal cell components can both be targets for T-cell recognition.

B. Identification of Antigens

Many studies have focused on identification of individual components involved in the immune response to pathogens. From the considerations discussed above, it is clear that identification of antigens recognized by antibodies is technically simpler than analysis of T-cell recognition. Sera from infected patients or from individuals who are immune to disease have been used in immunoprecipitation and immunoblotting experiments to identify important antigenic constituents of many pathogens. Further characterization of the antigens by biochemical purification from the organisms has been possible in some cases, but it has frequently proved technically difficult and has been limited by problems associated with culture of the organisms in vitro. Advances in molecular biology have allowed cloning of the relevant genes from the pathogens in recombinant DNA expression vectors suitable for screening with antibodies. The cloned gene can then be overexpressed for antigen purification and sequenced to derive the amino acid sequence of the protein. Immunoblotting techniques for the identification of antigens recognized by T cells have been developed (Lamb et al. 1988), and analysis of T-cell recognition has also been advanced by the use of recombinant antigens and synthesis of peptides predicted from derived amino acid sequences.

Following the approaches described above, prominent antigens from a wide variety of infectious agents have been identified as members of stress protein families. These are summarized in Table 1.

C. hsp70

1. Parasite Infections

The identification of parasite antigens related to the group of well-conserved eukaryotic proteins, the hsp70s, has come about from the

Table 1 Infectious agents that induce immune responses to members of stress protein families

Pathogen	Disease	Stress protein family	References
Protozoa			
P. falciparum	malaria	hsp70	1–5
		hsp90	6
T. cruzi	Chagas' disease	hsp70	7
		hsp90	8
L. donovani	visceral leishmaniasis	hsp70	9
Helminths			
S. mansoni	schistosomiasis	hsp70	10–12
		low m.w.	13
S. japonicum	schistosomiasis	hsp70	11, 14
B. malayi	lymphatic filariasis	hsp70	15, 16
O. volvulus	ocular filariasis	hsp70	17
Bacteria			
M. leprae	leprosy	hsp70	18, 19
		GroEL	20, 21
		low m.w.	22, 23
M. tuberculosis	tuberculosis	hsp70	18, 20, 24
		GroEL	20, 21
		GroES	25
C. burnetii	Q fever	GroEl	26
T. pallidum	syphilis	GroEL	27
L. pneumophila	Legionnaires' disease	GroEL	28
B. burgdorferi	Lyme disease	GroEL	29
C. trachomatis	blinding trachoma	GroEL	30

Key to references. [1]Bianco et al. (1986); [2]Ardeshir et al. (1987); [3]Yang et al. (1987); [4]Peterson et al. (1988); [5]Mattei et al. (1988); [6]Jendoubi and Bonnefoy (1988); [7]Engman et al. (1989); [8]Dragon et al. (1987); [9]Macfarlane et al. (1989); [10]Hedstrom et al. (1987); [11]Hedstrom et al. (1988); [12]Newport et al. (1988); [13]Nene et al. (1986); [14]Scallon et al. (1987); [15]Selkirk et al. (1987); [16]Selkirk et al. (1989); [17]Rothstein et al. (1989); [18]Britton et al. (1986); [19]Garsia et al. (1989); [20]Young et al. (1988a); [21]Shinnick et al. (1988); [22]Nerland et al. (1988); [23]Booth et al. (1988); [24]Mehlert and Young (1989); [25]Baird et al. (1988); [26]Vodkin and Williams (1988); [27]Hindersson et al. (1987); [28]Hoffman et al. (1989); [29]Hansen et al. (1988); [30]Morrison et al. (1989).

screening of expression libraries with sera derived from naturally infected humans or vaccinated laboratory animals. This approach, designed to identify dominant surface proteins of importance in immune recognition, has allowed the characterization of a number of hsp70-related molecules from a variety of organisms (Table 1). It is clear that the hsps rank among the most immunogenic of parasite antigens.

Several antigens with homology to eukaryotic hsp70 proteins have been identified in *Plasmodium falciparum* following the screening of cDNA expression libraries with sera from either adults living in areas endemic for malaria or immune owl monkeys, whose sera is known to be protective both in vitro and in vivo. One of these, Pfhsp 70-1, is an abundant soluble polypeptide of 75 kD, which shares 61% amino acid sequence identity with *Drosophila* hsp70 (Bianco et al. 1986; Ardeshir et al. 1987; Yang et al. 1987). An interesting feature of this antigen, which is expressed at all stages of the parasite's asexual cycle, is the presence of a tandemly repeated amino acid tetramer, Gly-Gly-Met-Pro, near the carboxyl terminus. This type of repeat is characteristic of immunogenic malarial antigens (Kemp et al. 1987) and may be of importance in other parasite hsp70 antigens (see below). A second antigen of 70 kD (Pfhsp 70-2) shows a similar pattern of expression and 64% amino acid identity with Pfhsp 70-1 (Peterson et al. 1988), but the two do not cross-react serologically, suggesting that the shared sequences are nonimmunogenic. Pfhsp70-2 does not share the amino acid tetramer repeat but does show a stretch of 20 amino acids at the carboxyl terminus with homology to the putative hepatocyte binding site of the circumsporozoite (CS) protein of *P. falciparum*. The significance of this homology is unclear, but it has been suggested that circulating antibodies to this sequence may block sporozoite invasion and thereby prevent hyperinfection. Alternatively, the presence of this sequence may favor sporozoite invasion by removing anti-hepatocyte binding site antibodies from the circulation (Peterson et al. 1988).

Another hsp-like antigen of *P. falciparum* shows only 55% amino acid identity with Pfhsp 70-1 but 72% identity with the rat 78-kD glucose-regulated protein, grp78 (Mattei et al. 1988). Therefore, it is clear that this parasite possesses a diverse family of hsp70-like proteins, the functions of which are at present unknown.

The identification of similar hsp70-like peptides in other protozoan parasites has progressed more slowly, but it is evident that such proteins are just as immunogenic in these organisms. In the African trypanosomes *(Trypanosoma brucei)*, which evade the host immune response by the sophisticated genetic mechanism of antigenic variation (for review, see Boothroyd 1985; Borst 1986), hsp70 proteins have not yet been implicated as antigens, although their genes have been fully characterized (section C.1). Conversely, the etiological agent of Chagas' disease, *Trypanosoma cruzi,* expresses antigens related to both hsp70 and hsp90. The screening of λgt11 expression libraries with rabbit serum against *T. cruzi* yielded a cognate *hsp70* sequence, TC1, whose deduced amino acid sequence shows 46% identity with the *Escherichia coli dnaK* gene product but only 38% identity with the *T. cruzi* heat-inducible *hsp70* (Engman et al. 1989). The gene encoding the latter protein was isolated from a genomic library (using a heterologous probe), has 65% identity with *Drosophila hsp70* and, like the *Plasmodium* antigens, has the same tetrameric amino acid repeat (Gly-Gly-Met-Pro) at the carboxyl terminus of its deduced open reading frame (Requena et al. 1988).

Kinetoplastid protozoa of the genus *Leishmania* also express immunogenic hsp70-like antigens. In *L. donovani*, the causative agent of visceral leishmaniasis, a cDNA expression clone isolated from a screen with infection serum shows 75% amino acid identity with the *T. brucei* hsp70 sequence (Macfarlane et al. 1989). The protein encoded by this gene shows stage-specific expression and is recognized by sera from approximately 50% of patients tested. *L. major* also differentially expresses several hsp70-related proteins during its life cycle, but sera from Tunisian patients with cutaneous leishmaniasis do not reproducibly recognize these molecules (Searle et al. 1989; Smith et al. 1989; S. Searle and D.F. Smith, unpubl.).

hsp70-related antigens are also produced by the parasitic helminths, which include trematode worms of the genus *Schistosoma* and filarial species of *Brugia* and *Onchocerca.* Nearly all individuals infected with *Schistosoma mansoni* react to two schistosome antigens of 70 and 38 kD (Newport et al. 1988). The 70-kD protein is an hsp70 homolog (sharing 80% amino acid identity with the human protein and 71% with *Drosophila)* that has its immunogenic domain at the carboxyl terminus, a region of low conservation in hsp70 proteins generally. A similar protein is found in *S. japonicum* (Scallon et al. 1987), which does not, however, cross-react with the *S. mansoni* antigen (Hedstrom et al. 1988). This recalls the lack of cross-reactivity seen between the hsp70 antigens of *P. falciparum* and is further discussed in sections II.F and II.G below. Antibodies to the schistosome hsp70s are not detectable in subjects infected with other parasitic helminths or in uninfected individuals, suggesting a potential diagnostic role for these molecules (Newport et al. 1988).

In the parasitic helminths of *Brugia* species, an hsp70-related protein has been defined as a prominent immunogen (Selkirk et al. 1987, 1989). This antigen, present in all life cycle stages examined, shows strongest amino acid homology (85%) to the constitutively expressed rat hsc protein. Western blotting analysis has suggested that antibody generated by infection with *B. malayi* is directed at filarial-specific determinants of the hsc molecule. Some infected individuals do produce antibodies with cross-reactivity to the hscs of *P. falciparum* and *S. mansoni*, however, thereby reducing the suitability of this protein for the diagnosis of Brugian filariasis. The causative agent of ocular filariasis, *Onchocerca volvulus*, also produces an hsp70 that is the major immunogen recognized by amicrofilaremic individuals in an endemic area (Rothstein et al. 1989). Again, this antigen shows strongest amino acid homology to the rat hsc protein.

2. *Bacterial Infections*

To identify the proteins involved in immune recognition of mycobacteria, the causative agents of leprosy and tuberculosis, panels of monoclonal antibodies were generated following immunization of mice with *Mycobacterium leprae* and *Mycobacterium tuberculosis* (Engers et al. 1985, 1986). A striking finding of this approach was that, although the crude mycobacterial extracts used to immunize the mice contained a large number of proteins, many antibodies generated independently in separate laboratories were directed to a limited set of apparently dominant antigens. A 70-kD antigen was identified by a monoclonal antibody raised to an *M. leprae* extract (Britton et al. 1985), and analysis of recombinant DNA clones subsequently showed that this was closely related to a 71-kD antigen independently identified by a monoclonal antibody to *M. tuberculosis* (Engers et al. 1986). Britton and his colleagues (1986) showed that this protein was involved in both antibody and T-cell recognition of mycobacteria during leprosy infection. The amino acid sequence derived by analysis of the cloned gene for the *M. tuberculosis* protein indicated that it belonged to the *hsp70* gene family (Young et al. 1988a), and it has subsequently been shown that this protein has a high affinity for ATP and that its synthesis is induced during the hs response of mycobacteria (Mehlert and Young 1989). Similarly, partial sequence analysis of the gene coding for the *M. leprae* 70-kD antigen has shown that it belongs to the *hsp70* family (Garsia et al. 1989).

Other bacterial pathogens clearly express hsp70 proteins that are cross-reactive with the mycobacterial antigen, but the role of these proteins in antibody and T-cell responses during infection has not been analyzed.

D. GroEL

1. *Mycobacteria*

A high proportion of monoclonal antibodies raised to mycobacterial extracts recognized a protein referred to as the 65-kD antigen (Buchanan et al. 1987; Young et al. 1987b). This antigen was first identified by antibodies that were specific for *M. leprae* (Gillis and Buchanan 1982), but it was subsequently shown that other epitopes on the protein were cross-reactive with a wide range of bacteria (Young et al. 1987b; Thole et al. 1988b). In addition to the monoclonal antibody approach, the same protein (referred to as the 64-kD antigen) was independently identified in studies using polyclonal antisera (Thole et al. 1985). Using purified recombinant antigen, it has been shown that the 65-kD protein has a prominent role in T-cell recognition of mycobacteria, both in human and mouse systems (Emmrich et al. 1985; Boom et al. 1987; Lamb et al. 1987). When mice were immunized with *M. tuberculosis*, 20% of the resulting mycobacteria-reactive T cells were directed toward this one protein (Kaufmann et al. 1987).

The genes for 65-kD antigens of *M. tuberculosis* and *M. leprae* were cloned by R. Young and his colleagues (1985a,b) using the λgt11 vector, and the gene from the vaccine strain *M. bovis* BCG was cloned by Thole and van Embden using an EMBL expression library (Thole et al. 1987). The genes from *M. tuberculosis* and *M. bovis* BCG were found to be identical, and between these and the *M. leprae* gene, there was 95% amino acid identity (Shinnick et al. 1987). Screening of sequence data banks initially showed no homology with other proteins, but a subsequent entry revealed a striking relationship with an open reading frame described as the *ams* gene of *E. coli* (Young et al. 1987b). The *ams* sequence had been derived from a DNA fragment capable of suppressing a temperature-sensitive mutation in mRNA half-life (Chanda et al. 1985). Renewed analysis of this fragment showed that it coded for two proteins, with molecular weights of 15 kD and 60 kD, and that the larger protein was antigenically cross-reactive with the mycobacterial 65-kD antigen. The suppression of the *ams* mutation was analogous to reports of the extragenic suppression of temperature-sensitive mutations in the *dnaA* gene by overexpression of the GroEL and GroES proteins of *E. coli* (Fayet et al. 1986; Jenkins et al. 1986). The plasmids capable of suppressing the two mutations were then compared and found to be identical, thus identifying the *E. coli* equivalent of the mycobacterial 65-kD antigen as the GroEL heat shock protein (Young et al. 1988a). The homology between the two proteins was later confirmed by comparison of the derived amino acid sequences, and many of the monoclonal antibodies originally described as being directed to the mycobacterial 65-kD protein have been shown to bind to the purified *E. coli* GroEL (Mehlert et al. 1988). The relationship between the mycobacterial 65-kD antigen and *E. coli* GroEL was confirmed by Shinnick and his colleagues (1988), who went on to show that the mycobacterial homolog behaves as an hs protein in a rapidly growing mycobacterium *(M. smegmatis)*.

2. *Other Bacteria*

A library of genomic DNA from *Coxiella burnetii*, the causative agent of Q fever, was constructed in a cosmid vector (pHC79) and screened with sera from a patient with Q fever and from a vaccinated individual. Both sera reacted strongly with a clone that expressed a 62-kD immunoreactive protein and a 14-kD nonimmunoreactive protein. Sequence analysis showed that these corresponded to the GroEL and GroES homologs of *Coxiella* (Vodkin and Williams 1988).

A "common antigen," originally identified in *E. coli* (Kaijser 1975) and *Pseudomonas aeruginosa* (Hoiby 1975), has been documented in extracts from more than 50 different bacterial species. Hindersson and his colleagues (1984) at the Statens Seruminstitut in Copenhagen showed that a major antigen (Tp-4) from *Treponema pallidum*, the causative agent of syphilis, is a member of the common antigen family. A combined interest in both mycobacteria and *T. pallidum* in van Embden's laboratory resulted in the demonstration that the mycobacterial 65-kD antigen also belongs to the common antigen family (Thole et al. 1988b), and it is now clear that the common antigen is in fact the GroEL hsp. The gene coding for the *T. pallidum* GroEL has been cloned and sequenced, and corresponding genes from *Borrelia burgdorferi* and *Legionella micdadei* have also been cloned in Hindersson's laboratory (Hindersson et al. 1987; P. Hindersson, pers. comm.).

A dominant 60-kD antigen has been identified by immunoblot analysis using sera from patients in the convalescent phase of legionellosis. The gene coding for this antigen has recently been cloned from *Legionella pneumophila* and has been identified as a member of the *groEL* family on the basis of sequence analysis and response to hs (Hoffman et al. 1989). Western blot analysis of sera from mice infected with *Salmonella typhimurium* again indicates the presence of a prominent GroEL-like antigen (C. Hormaeche, pers. comm.), and a major antigen from *Chlamydia trachomatis* also appears to belong to the same protein family (Morrison et al. 1989).

These studies show that GroEL is a major antigenic component in an extraordinarily wide variety of bacteria and that elevated antibody levels to species-specific and cross-reactive epitopes on GroEL are a common feature of bacterial infections.

3. Eukaryotic GroEL

It has recently been established that proteins related to GroEL are present in a wide diversity of organisms. A major plant chloroplast protein involved in assembly of the RuBisCO enzyme shows extensive sequence homology with bacterial GroEL proteins (Hemmingsen et al. 1988). Studies on hsps of *Tetrahymena thermophila* demonstrated the presence of a GroEL-related protein in the mitochondria of eukaryotic cells (McMullin and Hallberg 1987, 1988), and the corresponding yeast protein has been analyzed with regard to structure and function (Cheng et al. 1989; Reading et al. 1989). Sequence analysis of a gene coding for a 63-kD human mitochondrial protein that is altered in cell lines resistant to antimitotic drugs has revealed 50% amino acid identity with the mycobacterial 65-kD antigen, indicating that it is a human member of the *groEL* gene family (Gupta and Austin 1987; Jindal et al. 1989). When analyzing the immune response to GroEL antigens from pathogens, it is important to bear in mind that homologous proteins are present in the diet and also in the host cells themselves.

There are as yet no published reports of the expression of a GroEL protein in parasitic pathogens, but a polyclonal antiserum raised against the 65-kD antigen of *M. leprae* recognizes a protein of a similar size in promastigotes of *Leishmania major* (S. Searle and D.F. Smith, unpubl.). An oligonucleotide complementary to a region conserved between the human, plant, and *M. leprae* GroEL is currently being used to isolate the *Leishmania* gene(s).

E. Other Stress Proteins

In *Trypanosoma cruzi,* an 85-kD antigen with homology to hsp83 of *Drosophila* (64% amino acid identity) and to hsp90 of *Saccharomyces cerevisiae* (62% identity) has been characterized. The protein is constitutively expressed and was identified by immunoprecipitation, with infection serum, of translation products of mRNA that had been selected using clones from a cDNA library (Dragon et al. 1987). It has been suggested that recognition of this host-related protein may be of importance in the autoimmunity associated with Chagas' disease. A heat-shock-like antigen related to hsp90 has also been identified in *P. falciparum* (Jendoubi and Bonnefoy 1988). The *E. coli* protein C62.5 (HtpG) also belongs to the hsp90 gene family (Bardwell and Craig 1987), and there is a corresponding cross-reactive protein present in mycobacterial extracts (D.B. Young, unpubl.). It is possible, therefore, that further examples of hsp90 proteins acting as antigens may be found in other parasitic and bacterial infections.

In *Schistosoma mansoni*, an immunodominant antigen of 40 kD shows homology to the α-crystallins and the *Drosophila* small hsps (Nene et al. 1986). This takes the form of a duplicated region of amino acid homology to the α-crystallin/small hsp domain, which probably reflects an intragenic duplication (de Jong et al. 1988). The antigen is differentially regulated during the parasite life cycle, and it elicited a strong immune response in over 90% of infected patients. An 18-kD protein of *M. leprae* that is recognized by monoclonal antibodies (Britton et al. 1985) and by human T-cell clones (Mustafa et al. 1986) also shares some sequence elements with the low-molecular-weight hsps (Nerland et al. 1988). The sequence identities in these cases are very much less than those observed for the hsp70, hsp90, and GroEL families and, although they may represent conservation of certain structural features of these proteins, it is unlikely that they will result in antigenic cross-reactivity.

Immunoscreening of an expression library from *T. cruzi* with serum from a chronically infected mouse has yielded a ubiquitin cDNA that encodes a precursor protein of this ubiquitous protein (Kirchhoff et al. 1988). No stress-induced changes in the expression of this antigen have yet been demonstrated.

A 10-kD antigen identified in *M. bovis* BCG and *M. tuberculosis* by monoclonal antibody analysis (Minden et al. 1984) was recognized by antibodies and by T cells of immunized animals and showed a high degree of species specificity. Sequence analysis reveals 45% identity between this protein and the product of the *groES* gene of *E. coli* (Baird et al. 1988). The GroES homolog of *Coxiella burnetii* was found to be nonimmunogenic (Vodkin and Williams 1988) and, in contrast to GroEL, immunization with purified *E. coli* GroES does not induce a strong antibody response (A. Mehlert and D.B. Young, unpubl.). Further analysis will show whether or not the GroES family represents a frequent target of the immune response.

F. Mapping of Antigenic Determinants

A variety of studies have been undertaken to investigate the antigenic determinants involved in antibody and T-cell recognition of stress proteins. This approach has been stimulated by several considerations: (1) The strong immune response to stress protein antigens has led to speculation about their potential as immunodiagnostic probes; development of this approach would require identification of species-specific determinants. (2) The presence of homologous proteins in host and pathogen cells suggests that stress proteins could have a role in the induction of autoimmune responses; analysis of this possibility requires identification of cross-reactive determinants. (3) Human T cells that proliferate in response to stress proteins can readily be isolated from many individuals; these antigens therefore represent a convenient model for detailed analysis of the effects of genetic restriction on T-cell recognition in human populations.

1. GroEL: Antibody Epitopes

Antibody responses to GroEL proteins characteristically include recognition of both species-specific and cross-reactive determinants. A set of monoclonal antibodies recognizing different epitopes on GroEL was generated as a result of the studies on mycobacterial antigens, and systematic efforts to identify the epitopes recognized by such antibodies have been undertaken. Recombinant DNA sublibraries based on expression of fragments of the mycobacterial 65-kD antigen fused to β-galactosidase were prepared using λgt11 and pEX2 vectors. Mehra et al. (1986) defined the epitopes recognized by 6 of the monoclonal antibodies to sequences of the protein ranging in size from 13 to 35 amino acids and a further 20 monoclonal antibody binding sites were mapped to 8 distinct regions of the protein by Thole et al. (1988a).

A surprising finding from these studies was that all of the antibodies mapped to linear determinants, which were found throughout the whole sequence of the protein. This was unexpected, since antibodies usually recognize epitopes on the surface of soluble protein antigens (Benjamin et al. 1984). The degraded nature of the antigen used for immunization in these experiments could have contributed to the high proportion of antibodies generated to linear epitopes, but these same antibodies also bind to intact protein preparations under nondenaturing conditions.

Having identified the region of the protein containing an antibody epitope, it is possible to further refine the analysis using synthetic peptide chemistry. The *M. leprae* specific monoclonal antibody (IIIE9) binds to a 15-amino-acid peptide, and detailed analysis showed that the amino acids in positions 4 and 8 (lysine and threonine) were key residues involved in antibody recognition (Brown et al. 1986). The corresponding protein from *M. tuberculosis* has glutamate residues in these two positions and is not recognized. Monoclonal antibody IIC8 binds to the 65-kD antigens of *M. leprae* and *M. tuberculosis* but not to the corresponding *E. coli* GroEL. The epitope recognized by IIC8 has been mapped in detail using synthetic peptides (Anderson et al. 1986), and sequence comparison shows that only 2 of the 9 residues required for antibody recognition are conserved in the *E. coli* protein (Table 2). Antibody IIIC8, on the other hand, does recognize *E. coli* GroEL, suggesting that the conserved Gly-Gly-Met motif at the carboxyl terminus of GroEL proteins may have a role in antibody recognition. Oxidation of methionine residues of the IIIC8 epitope led to a loss of antigenicity (Anderson et al. 1986). A comparison of amino acid sequences in regions of GroEL proteins corresponding to four antibody epitopes is shown in Table 2. Mapping of additional monoclonal antibody epitopes using synthetic peptides is described by Anderson et al. (1988a).

Table 2 Sequence analysis of epitopes from the mycobacterial GroEL protein involved in recognition by monoclonal antibodies

Antibody	Amino acid sequence	Recognition
IIH9 (104-123)		
M. leprae	RNVAAGANPLGLKRGIEKAV	yes
M. tuberculosis	********************	yes
E. coli	KA****M**MD*****D***	yes
IIIE9 (423-437)		
M. leprae	ALDKLKLTGDEATGA	yes
M. tuberculosis	T**E***E*******	no
E. coli	K*AD*RGQNEDQNVV	no
IIC8 (482-490)		
M. leprae	YEDLLKAGV	yes
M. tuberculosis	*****A***	yes
E. coli	*GNMIDM*I	no
IIIC8 (531-541)		
M. leprae	DPTGGMGGMDF	yes
M. tuberculosis	PGGGD******	yes
E. coli	GAA******GG	yes

Sequence analysis of epitopes from the mycobacterial GroEL protein involved in recognition by monoclonal antibodies. The epitopes recognized by four monoclonal antibodies directed to *M. leprae* GroEL (Gillis and Buchanan 1982) were mapped using a recombinant DNA sublibrary and synthetic peptides (Anderson et al. 1986; Brown et al. 1986; Mehra et al. 1986). The species specificity of each antibody was tested by Western blot experiments using GroEL proteins purified from *M. tuberculosis* and *E. coli* (Mehlert et al. 1988). Amino acid sequences are shown using the single letter code with an asterisk to denote residues identical to those in the *M. leprae* protein; residue numbers refer to their position in the *M. leprae* sequence (data from Mehra et al. 1986; Shinnick 1987; Hemmingsen et al. 1988). Alteration of one or two key residues resulted in loss of recognition by some antibodies, but in other cases, several residues could be changed without affecting antigenicity.

Despite the success achieved in mapping the epitopes recognized by monoclonal antibodies, there are no reports describing the mapping of epitopes recognized by sera from patients or infected animals. It is possible that the natural antibody response during infection involves recognition of conformational determinants on GroEL proteins rather than the linear determinants recognized by monoclonal antibodies. This point is important if serological tests based on responses to GroEL-related antigens are to be developed.

2. *GroEL: T-cell Epitopes*

Analysis of antigen recognition by T cells is complicated by the need for antigen processing and presentation with MHC molecules. However, polymorphism of MHC molecules underlies the genetic restriction of immune responses, and investigation of the pathways by which different MHC molecules "select" different regions of antigen for presentation to T cells is an important area of current research on immune mechanisms. Thus, in addition to the identification of species-specific and cross-reactive determinants that may be of diagnostic value, mapping of stress protein epitopes seen by human T cells represents an important study in basic immunology.

As in the case of antibody epitopes, a number of T-cell epitopes on the mycobacterial 65-kD antigen have been mapped using recombinant DNA sublibraries (Lamb et al. 1987; de Vries et al. 1988; Thole et al. 1988a). A complementary approach has been the empirical prediction of potential T-cell epitopes by analyzing the primary sequence for the presence of structural elements, or motifs, previously identified in other known T-cell epitopes (DeLisi and Berzofsky 1985; Rothbard and Taylor 1988). Preliminary results do indeed indicate that different epitopes are seen by individuals with different MHC haplotypes, possibly because of the specificity associated with the antigen-binding site of MHC molecules (Sette et al. 1987). This type of "determinant selection" by MHC molecules may have a role in the observed differences in genetic susceptibility to infection.

Generally, T-cell epitopes have been localized on peptides other than those carrying antibody epitopes. However, in experiments that used defined peptides from the mycobacterial 65-kD protein to provide T-cell help for responses to adjacent antibody epitopes, it was shown that antibodies to these T-cell-epitope-bearing peptides can be generated (Cox et al. 1988). Therefore, there appears to be no strict discrimination between those areas of the protein involved in antibody and T-cell recognition.

At present, all of the T-cell epitopes mapped on the 65-kD antigen have been those recognized by helper T cells bearing the CD4 phenotype and interacting with class II MHC molecules. A role for cytotoxic T cells—CD8 cells restricted by class I MHC molecules—in the immune response to intracellular infections has been proposed (Kaufmann 1988), and it would be of interest to determine whether such cells also recognize GroEL proteins. The mycobacterial 65-kD antigen has been expressed in recombinant vaccinia virus vectors (J.D.A. van Embden, pers. comm.), and such constructs, in combination with synthetic peptide antigens, may allow detailed mapping of class-I-restricted T-cell recognition (Townsend et al. 1986). Class-II-restricted CD4 T cells that are functionally cytotoxic have recently been described and found to recognize a determinant in the amino-terminal portion of the 65-kD antigen (Ottenhoff et al. 1988).

It has recently been reported that the mycobacterial 65-kD antigen is recognized by a class of T cells that are negative for both the CD4 and CD8 molecules and that express an antigen receptor made up of γ and δ subunits, in contrast to the conventional α/β receptor (O'Brien et al. 1989; Holoshitz et al. 1989). The antigen specificity of γ/δ T cells appears to be much more restricted than that allowed by the α/β receptor, and the ability of the 65-kD antigen to trigger proliferation of these cells may indicate some fundamental role of hsps in development of the immune system (Raulet 1989). Further analysis and detailed mapping of the specificity of γ/δ T cells will be of considerable interest.

The results of mapping of antibody and T-cell determinants on the mycobacterial GroEL protein are summarized in Figure 1.

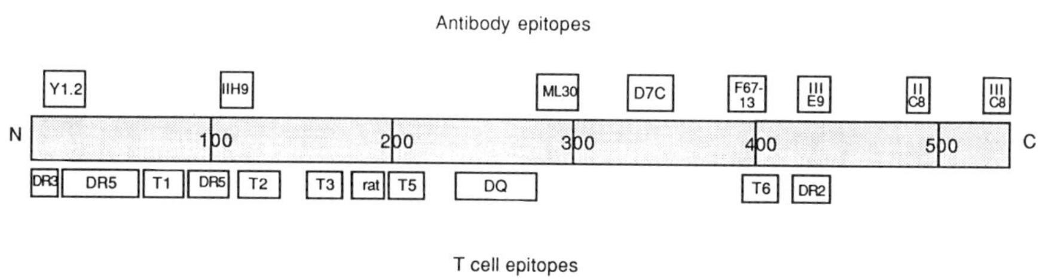

Figure 1 Mapping of antigenic determinants on mycobacterial GroEL. Epitopes involved in recognition of mycobacterial GroEL by monoclonal antibodies and by human T lymphocytes have been mapped using recombinant DNA sublibraries and synthetic peptides. Antigenic determinants have been found throughout the molecule. (Data from Mehra et al. 1986; Lamb et al. 1987, 1989; Anderson et al. 1988a,b; deVries et al. 1988; Thole et al. 1988a; van Eden et al. 1988.)

3. *hsp70*

Epitope mapping of parasite stress-related antigens has been relatively limited so far and is confined to a few examples. However, it is already apparent that the most immunogenic regions of these hsp70-like peptides are near the carboxyl termini, which represent the least well conserved domains of hsp70s in general.

In *Plasmodium falciparum*, the amino acid tetramer, Gly-Gly-Met-Pro, found at the carboxyl terminus of the 75-kD antigen, Pfhsp 70-1, may be the dominant epitope, given that much of the antibody response induced by malarial infections is directed against repeat epitopes (Kemp et al. 1987). Using carrier-free synthetic peptides against this region, antiserum with good reactivity against the native protein has been raised in rabbits (Richman and Reese 1988).

The *Schistosoma mansoni* 70-kD antigen also has its immunodominant epitope located near the carboxyl terminus, as judged by the ability of a cDNA expression clone (encoding 516 nucleotides from the 3′ end of the gene) to remove essentially all of the relevant antibodies from infection sera (Hedstrom et al. 1987). The predicted amino acid sequence of this clone also shows a single Gly-Gly-Met-Pro element as described above. Comparisons of the nucleotide sequences of the 70-kD schistosome antigens and the human hsp70 show very few predicted amino acid differences between the parasite proteins and the human host protein (Hedstrom et al. 1988). It is striking that this limited diversity is sufficient to elicit a discriminatory antibody response to these parasite host-like antigens.

A similar location for the immunodominant epitope has been demonstrated in the *Brugia pahangi* hsc protein. Comparison of the immunological reactivity of the products of two cDNA expression clones showed that the absence of the 23 most carboxy-terminal amino acids drastically reduced the antigenicity of the fusion protein (Selkirk et al. 1989). These missing residues include a repeat of the Gly-Gly-Met-Pro element (found in the hsp70-related antigens of both *P. falciparum* and *S. mansoni*), which presumably contributes to the antigenicity of this region. The presence of this tetramer, or elements of it, near the carboxyl termini of both the rat hsc70 protein (O'Malley et al. 1985) and the *E. coli* GroEL protein (Hemmingsen et al. 1988) shows that it is not confined to infectious organisms. Given the lack of cross-reactive antibodies that have been generated naturally against these proteins from different species, it seems likely that it is the tertiary structure of their carboxy-terminal domains, rather than particular amino acid motifs, that gives rise to distinctive, non-cross-reactive epitopes.

Under appropriate immunization conditions, antibody responses to conserved determinants on hsp70 proteins can be induced (Kelly and Schlesinger 1982; Chappell et al. 1986), but it seems that parasites present their hsp70 proteins to the immune system in such a way as to avoid an autoimmune response. The manner of presentation of this antigenic region is evidently tightly defined and may be an important feature in pathogenesis, both for survival of the parasite and the course of the disease.

Preliminary analysis of the specificity of antibody responses to the mycobacterial hsp70 protein indicates a situation analogous to that observed with the parasite antigens. Mice infected with *M. tuberculosis* have a strong antibody response to the mycobacterial hsp70 but, although this response cross-reacts with the *E. coli* homolog, there is little or no recognition of determinants conserved with the self (murine) hsp70 protein (S.J. Brett, pers. comm.). Immunization of mice with the mycobacterial protein in adjuvant does, however, induce an antibody response that is cross-reactive with the self-protein. CD8 T-cell clones isolated from tuberculosis patients have been shown to proliferate in response to hsp70 proteins from mycobacteria, *E. coli*, and human cells (Rees et al. 1988).

G. Stress Proteins and Autoimmunity

When highly conserved stress proteins are the target for immune recognition, the immune system must guard against mounting a response that could be directed to the homologous self-proteins. The mechanisms by which the immune system distinguishes between "self" and "nonself" have not been fully elucidated but are generally thought to involve control at the level of T-cell recognition. The frequent occurrence of immune responses to stress proteins provides an opportunity to use these antigens as a model for detailed investigation of these mechanisms. Induction of responses to self stress proteins could play a role in the immunopathology associated with chronic infections and could also provide a focus for autoimmune disease.

As discussed above, there is evidence to suggest that the antibody response to stress proteins during infection is predominantly directed toward determinants that are not conserved with the homologous self-protein (Hedstrom et al. 1988). However, antibody responses to self stress proteins have been reported in patients suffering from several autoimmune diseases. Patients with systemic lupus erythematosus develop autoantibodies to a wide variety of self-proteins, and this has been shown to include hsp90 and hsp70 (Minota et al. 1988a,b). Autoantibodies to self-hsp70 have also been found in sera from rheumatoid arthritis patients (G. Tsoulfa et al., in prep.). It is not known whether these antibodies were originally induced as part of the immune response to a previous infection or as a secondary consequence of the tissue breakdown and failure in immune regulation occurring during autoimmune disease.

T-cell recognition of stress proteins, including human hsp70, has recently been shown using lymphocytes obtained from the synovial effusion of patients with rheumatoid arthritis and reactive arthritis (Res et al. 1988; Gaston et al. 1989; Lamb et al. 1989). The ability to carry out mapping of antigenic determinants allows detailed analysis of the specificity of the human T-cell response to the mycobacterial GroEL stress protein and T-cell lines, and clones directed to five separate epitopes were tested for species specificity (Young et al. 1988b; Lamb et al. 1989). It was shown that, whereas the response to three of the epitopes was specific for bacterial GroEL, two of the determinants were conserved in the homologous human protein (Fig. 2). The mechanisms of self/nonself discrimination are, therefore, not sufficiently stringent to completely exclude the generation of autoreactive responses to stress proteins.

Although these experiments demonstrate the occurrence of autoimmune recognition of stress proteins, there is currently no evidence that this plays an active role in the pathology of human autoimmune or infectious disease. In a rat model of adjuvant-induced arthritis, however, it has been demonstrated that transfer of an autoreactive T-cell clone directed to a determinant on the mycobacterial 65-kD antigen can cause autoimmune pathology and that the disease can be prevented by previous immunization with the purified 65-kD protein (van Eden et al. 1988). If the findings from the rat model are considered applicable to human autoimmune disease, then analysis of T-cell responses to self stress proteins can be seen as an important area of investigation.

Models for "molecular mimicry," in which a foreign antigen induces an immune response that cross-reacts with a host component, have been proposed to explain relationships between infection and autoimmunity (Oldstone 1987). Stress proteins have several features that make them particularly attractive in such a model. Their ubiquitous occurrence and immunogenicity in a broad spectrum of infections would provide a satisfactory explanation for the wide diversity of organisms that have been associated with the etiology of autoimmune disease (Saag and Bennett 1987). In addition, the fact that synthesis of stress proteins is subject to differential regulation suggests a potential mechanism to account for the localization of autoimmune responses to particular lesions—possibly corresponding to sites of particularly high levels of stress protein accumula-

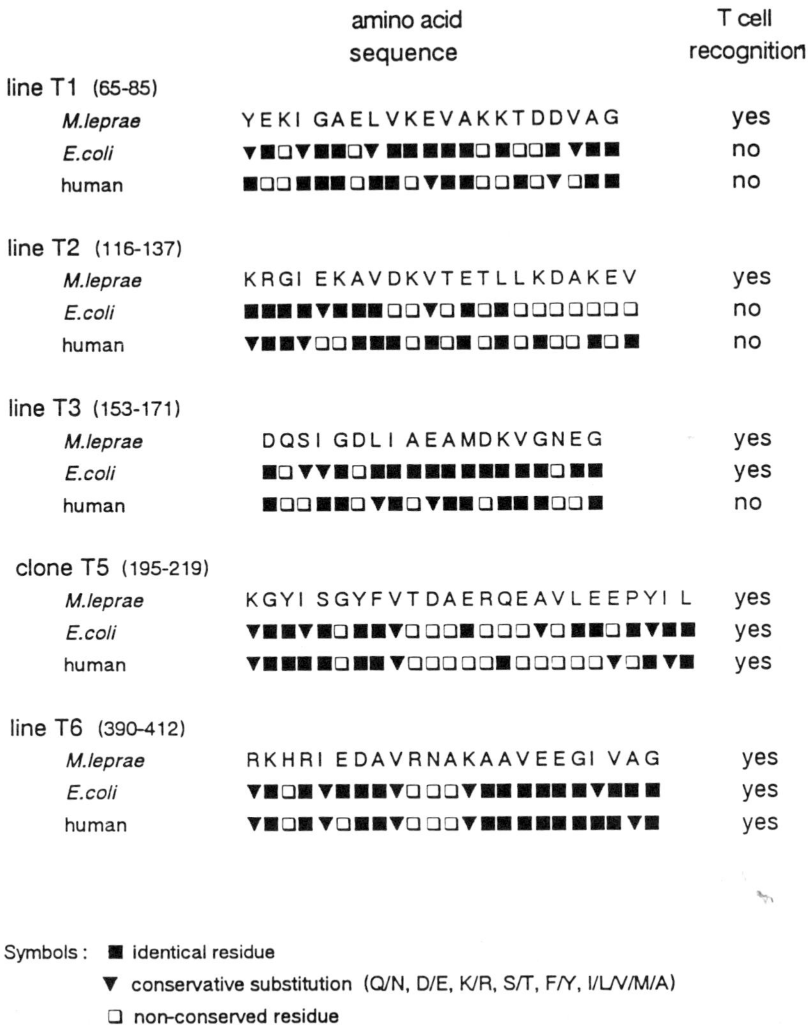

Figure 2 Defined T-cell epitopes of proteins belonging to the GroEL family. Peptides corresponding to T-cell epitopes on *M. leprae* GroEL were predicted and synthesized as described by Lamb et al. (1987). Human T-cell lines and clones recognizing these peptides were tested for recognition of corresponding GroEL proteins from *E. coli* and from human lymphocytes (Lamb et al. 1989). As in the case of the antibody epitopes (Table 2), conservation of particular key residues is important in determining immune recognition. A T-cell clone (T5) and a line (T6) are capable of autoimmune recognition of the "self" GroEL protein. Sequence data for the bacterial proteins is from Mehra et al. (1986) and Hemmingsen et al. (1988). Sequence of the human protein was derived by Jindal et al. (1989). Amino acid residue numbers refer to the *M. leprae* sequence.

tion. Bacterial or parasitic infection could induce a T-cell population reactive with a self stress protein, and a subsequent stress stimulus, such as viral infection, could set up a focus for generation of autoimmune pathology. Testing of such a model will involve a combination of detailed mapping of immune responses to stress proteins and analysis of the regulation of stress protein genes at sites of autoimmune lesions.

III. EXPRESSION OF STRESS PROTEINS DURING INFECTION

A. Parasites

The often complex life cycles of many parasites of medical and veterinary importance frequently include a major upshift in ambient temperature for the organism. Examples include ingestion by a warm-blooded host of a cyst or free-living larval form and the transfer of a parasite from a poikilothermic insect vector to its homeothermic mammalian host. This sudden stress to the parasite might be expected to give rise to the expression of hsps, as is found in other biological systems. This does indeed occur but, increasingly, it seems likely that these changes are not a true hs response but may be part of a normal developmental program, evolutionarily related to the stress response (Newport et al. 1988). The study of parasite hsps and the regulation and expression of their genes has intensified recently, particularly in view of the role of these proteins as immunodominant antigens.

1. hsp70 Expression

Many parasites constitutively express hsps in one or more stages of their life cycle, some at a level comparable to the most abundant of their cellular products (Newport et al. 1988). Constitutive expression is a particular feature of parasites in poikilothermic vectors, with the exception of malarial sporozoites (Bianco et al. 1986). Additionally, an increase in ambient temperature induces the synthesis of hsps in insect forms of *Leishmania* (Hunter et al. 1984; Lawrence and Robert-Gero 1985; van der Ploeg et al. 1985; Shapira et al. 1988; Smith et al. 1989) and of *Trypanosoma* (Glass et al. 1986), trophozoites of *Giardia* (Lindley et al. 1988), L_3 larvae of *Brugia* (Selkirk et al. 1987), and schistosomulae of *Schistosoma* (Blanton et al. 1987; Yuckenberg et al. 1987; Hedstrom et al. 1988). This increase in parasite hsp synthesis, which is generally regulated at the level of transcription, is usually accompanied by the inhibition of synthesis of other proteins, typical of a classic heat shock response. An increase in hsp production has been correlated with an increase in infectivity in species of *Leishmania* and is associated with morphological and biochemical changes that are characteristic of the transformation from the insect to the mammalian form of the parasite (Smejkal et al. 1988; Smith et al. 1989). Indeed, a hypothesis has been proposed that hsps could be involved in triggering this cell transformation (van der Ploeg et al. 1985), but recent evidence strongly suggests that differential expression of hsp genes between the two forms is merely concomitant with this event (Lee et al. 1988). The increase in hsp production is independent of a temperature change, however, suggesting that the stimulus for hsp production may be a nutritional or other type of cue. This is supported by the observation that trypomastigotes of *Trypanosoma brucei* show a "heat shock" response when maintained in an environment similar to conditions within the mammalian host, even at an ambient temperature of less than 30°C (van der Ploeg et al. 1985).

The occurrence of host fever during parasitic infection might be expected to increase hsp synthesis and, although this topic has been neglected so far, it has been recorded that stumpy form trypanosomes (which dominate during fever) synthesize more hsp70 than other mammalian forms (Davis et al. 1987).

2. *Parasite hsp Genes*

Parasite hsps are typically encoded by multigene families that tend to be clustered in head-to-tail tandem arrays (Glass et al. 1986; Dragon et al. 1987; Hedstrom et al. 1987; Lee et al. 1988; Requena et al. 1988; Swindle et al. 1988). Exceptions to this arrangement are seen in *Plasmodium falciparum*, which possesses at least five different hsp70 genes at four different chromosomal locations (Peterson et al. 1988), and *Leishmania major*, in which there are at least four separate genes of different copy number dispersed within the genome (Lee et al. 1988; Searle et al. 1989). One of the malarial hsps shows greater homology to the mammalian glucose-regulated proteins and immunoglobulin heavy-chain binding proteins than to the mammalian heat-inducible hsp70 (Mattei et al. 1988), whereas one of the hsps from *Leishmania* is the homolog of the mitochondrial hsp70-like antigen found in *T. cruzi* (Engman et al. 1989; Searle et al. 1989). It is likely that most parasites express more than one hsp70-like protein, of as yet unknown function.

Transcription of the hsp70 genes in trypanosomes shows a number of features characteristic of the Kinetoplastida. A single polycistronic mRNA molecule is the primary transcription product, which is processed before the addition of a 39-bp 5′ leader RNA by *trans*-splicing (for review, see Boothroyd 1985; Borst 1986). Processing of gene transcripts is disrupted during hs, with the resulting accumulation of high-molecular-weight precursors (Muhich and Boothroyd 1988). Steady-state levels of hsp70 mRNA actually increase after hs, however, and it has been shown that *trans*-splicing of this message is unaffected by stress, presumably due to the evolution of some compensatory mechanism (Muhich and Boothroyd 1989).

The sequence elements required for temperature-dependent regulation of the transcription of eukaryotic hs genes have also been demonstrated in the tandem arrays of parasite genes: Consensus hs elements (HSEs) and putative TATA boxes have been identified within the highly conserved intergenic regions of the *Trypanosoma brucei* tandem repeat (Glass et al. 1986) and 5′ of the polyubiquitin coding regions (Swindle et al. 1988). The functional significance of these elements remains to be determined.

B. Bacteria

1. hsp Expression

In *E. coli,* each stress protein family is represented by a single gene that regulates constitutive and stress-induced expression. If this pattern is repeated in other bacteria, demonstration of sequence homology with stress proteins will be an indication that expression of a particular antigen is likely to be regulated by stress stimuli. This has been verified for the 65-kD and 70-kD antigens from mycobacteria, for the 62-kD antigen from *Coxiella,* and for the *Legionella* 60-kD antigen (Shinnick et al. 1988; Vodkin and Williams 1988; Hoffman et al. 1989; Mehlert and Young 1989). In each case, expression of the antigen was induced by hs.

No direct analysis has been carried out to determine whether the environmental stimuli encountered by bacterial pathogens during infection will cause induction of stress responses. Consideration of the factors involved in the process of infection, particularly by intracellular pathogens, does suggest that this is likely to be the case. In addition to a possible hs associated with the body temperature of the host, a variety of potential stress stimuli can be envisaged for bacteria that adopt an intracellular lifestyle (for review, see Hahn and Kaufmann 1981; Moulder 1985). An important feature of the host defense involves generation of reactive oxygen intermediates that can kill bacteria, and survival under such conditions will involve induction of the oxidative stress response. The response to hydrogen peroxide has been well characterized in *Salmonella typhimurium* and was found to include induction of several heat shock proteins (Christman et al. 1985; Morgan et al. 1986). Consistent with the proposed role of the stress response during infection, a mutant strain of *S. typhimurium* that was unable to mount an oxidative stress response was found to be impaired in its ability to resist macrophage killing (Fields et al. 1986; Morgan et al. 1986). The hsp70 proteins of *S. typhimurium* and *E. coli* are induced by hydrogen peroxide, but although induction of GroEL in *E. coli* has been shown (Reuter and Shapiro 1987), other reports have indicated that this is not a component of the oxidative stress response (Morgan et al. 1986; Van Bogelen et al. 1987). Nutrient deprivation may be another common stress stimulus capable of inducing hsps in bacteria during adaptation to growth within the infected host (Jenkins et al. 1988).

2. *Bacterial hsp Genes*

The *groES-groEL* operon structure of *E. coli* is conserved in *C. burnetii* (Vodkin and Williams 1988) and in *Legionella* (Hoffman et al. 1989; P. Hindersson, pers. comm.), but comparison of restriction maps of clones expressing the 65-kD and 10-kD antigens of mycobacteria indicates that there is a significant gap between the *groEL* and *groES* genes in these organisms. The high degree of conservation of the *groE* genes is indicative of a fundamental role for these proteins in bacterial physiology, and it will be of interest to determine whether the dissociation of the two genes in mycobacteria represents a basic functional difference between these organisms and *E. coli*. An open reading frame corresponding to a DnaJ homolog has been identified in *M. tuberculosis* (Lathigra et al. 1988). The mycobacterial *dnaK* and *dnaJ* genes are separated by approximately 500 bp, and analysis of the regulation of these two genes provides another opportunity for detailed comparison of the control of mycobacterial and *E. coli* stress responses.

During sequence analysis of the genes for GroEL proteins of mycobacteria and *C. burnetii*, it has been pointed out that potential σ^{32} recognition sites are found upstream from the structural genes (Vodkin and Williams 1988; Young 1988). In contrast to findings with other mycobacterial proteins (Clark-Curtiss et al. 1985), the stress protein antigens can be expressed from their own promoters in *E. coli* (Thole et al. 1985; Young et al. 1987a; Shinnick et al. 1988). It is attractive to speculate that expression of these genes in *E. coli* is mediated by σ^{32} and that perhaps, like the stress proteins themselves, the mechanisms for induction of the stress response may be highly conserved among bacteria. It would be predicted then that promoter regions from stress proteins will be suitable for expression of genes in a wide variety of different prokaryotes. Further study of stress protein genes from different bacteria and detailed analysis of mRNA transcripts will allow testing of this hypothesis.

C. Host Cells

The expression of stress proteins in the host cells infected by intracellular parasites has been largely neglected, despite the likelihood that the invasion of a mammalian cell by a foreign organism could evoke such a stress response. It has been proposed that hsps may have a general role in the inflammatory response, and evidence suggesting a link between immune mediators (IL-1, IL-2, etc.) and stress responses has been reviewed (Polla 1988). Phagocytosis of a pathogen induces an oxidative respiratory burst in macrophages that results in superoxide generation, with the spontaneous production of hydrogen peroxide and the formation of free radicals (Fridovich 1978). It can be anticipated that induction of host cell oxidative stress proteins may be necessary to protect the phagocyte against the effects of its own oxygen radicals. This effect might be particularly pronounced during infection with parasites, such as *Leishmania* and mycobacteria, which persist and multiply within the phagocytic cells. In addition, accumulation of abnormal proteins resulting from degradation of a phagocytosed organism could provide a signal for induction of stress proteins.

More information is available concerning hsp expression in virally infected mammalian cells. Host-encoded proteins have a critical role in determining the effect that a virus exerts on its host cell, and this has been well established for both RNA and DNA tumor viruses (Lane and Crawford 1979; Bishop 1981; Hayward et al. 1981; Sarnow et al. 1982). The synthesis of cellular proteins related to hsp has been demonstrated following infection by a number of lytic viruses, including paramyxoviruses (Collins and Hightower 1982), adenoviruses (Nevins 1982), herpesviruses (Notarianni and Preston 1982; LaThangue et al. 1984), papovaviruses (Khandjian and Turler 1983), α viruses and rhabdoviruses (Garry et al. 1983). Different viruses may preferentially induce different stress proteins: Paramyxoviral infection induces the synthesis of glucose-regulated proteins of 99 and 78 kD, as well as hsps of 88, 72, and 71 kD (Collins and Hightower 1982), whereas host proteins of 90, 70, and 25 kD are synthesized following infection with herpes simplex virus-1 (HSV-1) (Notarianni and Preston 1982). The mechanisms involved in regulation of hsp expression during viral infection are discussed in detail by Morimoto and Milarski (this volume).

Bacterial hsps are induced by phage infection and perform essential functions in promoting phage replication. It is possible that viruses utilize eukaryotic heat shock proteins in an analogous fashion, but no specific examples of such behavior have been reported. The fact that hs can induce reactivation of some latent viral infections also suggests that viral growth cycles may be linked in some way to the host stress response (Zerbini et al. 1985).

D. Cellular Location of Stress Proteins

Given the constitutive, often high-level, expression of the hsps in many parasites and their ability to act as antigens, it must be assumed that their cellular location allows access to the host immune system during infection. This could occur by leakage into domains surveyed by the immune system or following parasite death. Current evidence suggests that most hsp70 proteins are not surface molecules (Peterson et al. 1988), with the exception of the malarial merozoite 75-kD surface antigen (Ardeshir et al. 1987). A unique mitochondrial location has been demonstrated for the hsp70-related antigen of *T. cruzi* (Engman et al. 1989).

The GroEL and DnaK heat shock proteins of *E. coli* are found predominantly in soluble cytoplasmic extracts. Immunoelectron microscopy has indicated a cytoplasmic location for the GroEL homolog in *M. bovis* BCG (Thole et al. 1988b), but it has been reported that the protein sediments with the insoluble pellet following cell lysis of *M. leprae* and *C. burnetii* (Gillis et al. 1985; Vodkin and Williams 1988). Growth of mycobacteria in zinc-deficient media results in accumulation of GroEL as the major culture filtrate protein (deBruyn et al. 1987), and in *Neisseria gonorrhoeae*, GroEL can be detected in lithium acetate outer membrane preparations following exposure to environmental stress (J.P.M. van Putten, pers. comm.). The mycobacterial hsp70 homolog has also been found in culture supernatants (Abou-Zeid et al. 1988). The absence of a signal peptide in the derived sequences for the bacterial stress proteins suggests that their normal location will be cytoplasmic. It remains possible, however, that these proteins may become membrane-associated at particular stages of growth, perhaps reflecting a functional involvement in intracellular cytoskeletal structures or in protein secretion. Subsequent membrane damage could then account for the preferential accumulation of stress proteins in the extracellular medium.

In host cells, the 90-kD hsp90-related protein that accumulates as a result of HSV-2 infection is also atypical. This protein is found on the cell surface following viral infection, but it is located in the cytoplasm of uninfected cells (LaThangue and Latchman 1988).

IV. CONCLUDING REMARKS

A substantial body of evidence indicates that members of the highly conserved stress protein families are frequently recognized as major antigens during infection. These observations indicate a number of interesting areas for further research.

What makes stress proteins attractive to the immune system? Stress proteins are a major cellular component of every pathogenic bacterium or parasite, and it can be proposed that repeated challenge to the immune system during a diversity of infections results in an enhanced frequency of lymphocytes able to recognize a familiar target in a new pathogen. Against this argument, however, are the data demonstrating a remarkable species specificity in the antibody recognition of hsp70 homologs. In addition, the ubiquitous occurrence of these proteins—in host cells and as constituents of the diet and normal gut flora—would suggest that the immune system might become tolerized by repeated exposure to such antigens. An attractive hypothesis to account for the ability of stress proteins to induce T-cell recognition would be that they are able to gain preferential access to processing mechanisms within the antigen-presenting cell. This could result from the ability of stress proteins to bind to abnormal or denatured polypeptides, which are themselves likely targets for antigen processing. Alternatively, by analogy with the role of hsp70-related proteins in assembly of antibody molecules, perhaps assembly of MHC molecules requires interaction with an hsp70 protein, and homologous "foreign" proteins could inadvertently be incorporated into such a pathway. A recent report describing localization of two *hsp70* genes within the region of human chromosome 6 encoding the MHC lends support to this hypothesis (Sargent et al. 1989). Recognition of the mycobacterial 65-kD antigen by γ/δ T cells may also be an indication of some intimate relationship between hsps and components of the immune system. It has been suggested that γ/δ T cells may originally have evolved in order to recognize and destroy self cells that are defective or stressed and that the ability to recognize foreign stress proteins may have developed during subsequent evolution (Raulet 1989). Further investigation of the immunogenicity of stress proteins may provide new insights into mechanisms of antigen processing and may help in understanding the steps involved in the selection and development of T-cell repertoires.

Are stress proteins potential vaccine candidates? Antibody-based immunity is dependent on recognition of antigens exposed on the surface of a pathogen. Although the possibility of surface expression has not been ruled out, it is generally agreed that stress proteins are intracellular components, and it is unlikely, therefore, that they will be suitable targets for protective antibodies. On the other hand, the ability of stress proteins to stimulate helper T cells (allowing subsequent macrophage or B-cell activation) could play a role in protective immunity. The possible induction of autoimmune responses would, however, be a factor in any consideration of the use of these proteins as vaccines, and it is interesting to note that the only examples of vaccination experiments with stress proteins are in animal models of arthritis (van Eden et al. 1988; van der Broek et al. 1989).

Are stress proteins involved in autoimmunity? Stress proteins are attractive candidates for models of autoimmunity based on molecular mimicry. Current evidence indicates that the normal immune response to these antigens is dominated by recognition of variable determinants, but experiments involving selection of particular T-cell populations, or immunization in the presence of adjuvant, show that immune recognition of conserved determinants can occur. Analysis of the factors that determine the region of the antigen selected for immune recognition may reveal mechanisms that are important in the pathogenesis of infectious and autoimmune diseases.

Is virulence a stress regulon? For several bacterial pathogens it has been shown that factors involved in interactions with host cells (toxins, adhesins, etc.) are coregulated at the transcriptional level in the form of "virulence regulons," which are presumed to be induced in response to environmental signals during infection (Finlay and Falkow 1989). It is possible that the hs protein antigens are part of a virulence regulon, and this concept may provide a link between the immunogenicity and virulence of bacterial pathogens. Further overlap between hs and virulence regulons is suggested by the recent finding that important virulence factors (including the hemolysin) are induced by hs in pathogenic isolates of *Listeria monocytogenes* (Sokolovic and Goebel 1989).

ACKNOWLEDGMENTS

We are grateful to the following individuals for helpful discussion and communication of unpublished results: Ted Bianco, Sara Brett, Jo Cox, Jan Bradley, Peter Hindersson, Jon Lamb, Peter Lydyard, Graham Rook, Murray Selkirk, Jan van Embden, and Rick Young. The support of the Wellcome Trust to D.F.S. is acknowledged.

REFERENCES

Abou-Zeid, C., I. Smith, J.M. Grange, T.L. Ratliff, J. Steele, and G.A.W. Rook. 1988. The secreted antigens of *Mycobacterium tuberculosis* and their relationship to those recognised by the available antibodies. *J. Gen. Microbiol.* **134:** 531–538.

Anderson, D.C., M.E. Barry, and T.M. Buchanan. 1988a. Exact definition of species-specific and cross-reactive epitopes of the 65-kilodalton protein of *Mycobacterium leprae* using synthetic peptides. *J. Immunol.* **141:** 607–613.

Anderson, D.C., R.A. Young, and T.M. Buchanan. 1986. Solid phase peptide synthesis of epitopes that react with monoclonal antibodies to the 65,000 protein of *Mycobacterium leprae. Leprosy Rev.* (suppl.) **57:** 169–175.

Anderson, D.C., W.C.A. van Schooten, M.E. Barry, A.A.M. Janson, T.M. Buchanan, and

R.R.P. de Vries. 1988b. A *Mycobacterium leprae* specific human T cell epitope cross-reactive with an HLA-DR2 peptide. *Science* **242:** 258–260.

Ardeshir, F., J.E. Flint, S.J. Richman, and R.T. Reese. 1987. A 75 kd merozoite surface protein of *Plasmodium falciparum* which is related to the 70 kd heat-shock proteins. *EMBO J.* **6:** 493–499.

Baird, P.N., L.M.C. Hall, and A.R.M. Coates. 1988. A major antigen from *Mycobacterium tuberculosis* which is homologous to the heat shock proteins groES from *Escherichia coli* and htpA gene product of *Coxiella burnetii. Nucleic Acid Res.* **16:** 9047.

Bardwell, J.C.A. and E.A. Craig. 1987. Eukaryotic M_r 83,000 heat shock protein has a homologue in *Escherichia coli. Proc. Natl. Acad. Sci.* **84:** 5177–5181.

Benjamin, D.C., J.A. Berzofsky, and I.J. East. 1984. The antigenic structure of proteins: A reappraisal. *Annu. Rev. Immunol.* **2:** 67–101.

Bianco, A.E., J.M. Favaloro, T.R. Burkot, J.G. Culvenor, P.E. Crewther, G.V. Brown, R.F. Anders, R.L. Coppel, and D.J. Kemp. 1986. A repetitive antigen of *Plasmodium falciparum* that is homologous to heat shock protein 70 of *Drosophila melanogaster. Proc. Natl. Acad. Sci.* **83:** 8713–8717.

Bishop, J.M. 1981. Enemies within: The genesis of retrovirus oncogenes. *Cell* **23:** 5–6.

Blanton, R., E.C. Loula, and J. Parker. 1987. Two heat-induced proteins are associated with transformation of *Schistosoma mansoni* cercariae to schistosomula. *Proc. Natl. Acad. Sci.* **84:** 9011–9014.

Boom, W.F., R.N. Husson, R.A. Young, J.R. David, and W.F. Piessens. 1987. In *vivo* and *in vitro* characterisation of murine T-cell clones reactive to *Mycobacterium tuberculosis. Infect. Immun.* **55:** 2223–2229.

Booth, R.J., D.P. Harris, J.M. Love, and J.D. Watson. 1988. Antigenic proteins of *Mycobacterium leprae:* Complete sequence of the gene for the 18kDa protein. *J. Immunol.* **140:** 597–601.

Boothroyd, J.C. 1985. Antigenic variation in African trypanosomes. *Annu. Rev. Microbiol.* **39:** 475–502.

Borst, P. 1986. Discontinuous transcription and antigenic variation in trypanosomes. *Annu. Rev. Biochem.* **55:** 701–732.

Britton, W.J., L. Hellqvist, A. Basten, and A.S. Inglis. 1986. Immunoreactivity of a 70kD protein purified from *Mycobacterium bovis* BCG by monoclonal antibody affinity chromatography. *J. Exp. Med.* **164:** 695–708.

Britton, W.J., L. Hellqvist, A. Basten, and R.L. Raison. 1985. *Mycobacterium leprae* antigens involved in human immune responses. Identification of four antigens by monoclonal antibodies. *J. Immunol.* **135:** 4171–4177.

Brown, W.B., W.B. Larrabee, and P.S. Kim. 1986. Analysis of a leprosy-specific antibody epitope. *Leprosy Rev.* (suppl.) **57:** 157–162.

Buchanan, T.M., H. Nomaguchi, D.C. Anderson, R.A. Young, T.P. Gillis, W.J. Britton, J. Ivanyi, A.H.J. Kolk, O. Closs, B.R. Bloom, and V. Mehra. 1987. Characterization of antibody-reactive epitopes on the 65-kilodalton protein of *Mycobacterium leprae. Infect. Immun.* **55:** 1000–1004.

Chanda, P.K., M. Ono, K. Michihiko, and H.F. Kung. 1985. Cloning, sequence analysis and expression of alteration of the mRNA stability gene (ams+) of *Escherichia coli. J. Bacteriol.* **161:** 446–449.

Chappell, T.G., W.J. Welch, D.M. Schlossman, K.B. Palter, M.J. Schlesinger, and J.E. Rothman. 1986. Uncoating ATPase is a member of the 70kD family of stress proteins. *Cell* **45:** 3–13.

Cheng, M.Y., F.-U. Hartl, J. Martin, R.A. Pollock, F. Kalousek, W. Neupert, E.M. Halberg, R.L. Halberg, and A.L. Horwich. 1989. Mitochondrial heat-shock protein hsp60 is essential for assembly of proteins imported into yeast mitochondria. *Nature* **337:** 620–625.

Christman, M.F., R.W. Morgan, F.S. Jacobson, and B.N. Ames. 1985. Positive control of a regulon for defences against oxidative stress and some heat shock proteins in *Salmonella typhimurium*. *Cell* **41:** 753–762.

Clark-Curtiss, J.E., W.R. Jacobs, M.A. Docherty, L.R. Ritchie, and R. Curtiss III. 1985. Molecular analysis of DNA and construction of genomic libraries of *M. leprae*. *J. Bacteriol.* **161:** 1093–1102.

Collins, P.L. and L.E. Hightower. 1982. Newcastle disease virus stimulates the cellular accumulation of stress (heat shock) mRNAs and proteins. *J. Virol.* **44:** 703–707.

Cox, J.H., J. Ivanyi, D.B. Young, J.R. Lamb, A.D. Syred, and M.J. Francis. 1988. Orientation of epitopes influences the immunogenicity of synthetic peptide dimers. *Eur. J. Immunol.* **18:** 2015–2019.

Davis, C.E., D.G. Guiney, and M.E.M. Colmerauer. 1987. Molecular strategies of parasite invasion. *UCLA Symp. Mol. Cell. Biol.* **42:** 169–179.

deBruyn, J., R. Bosmans, M. Turneer, M. Weckx, J. Nyabenda, J.-P. van Voorten, P. Falmagne, H.G. Wiker, and M. Harboe. 1987. Purification, partial characterisation, and identification of a skin-reactive protein antigen from *Mycobacterium bovis* BCG. *Infect. Immun.* **55:** 245–252.

de Jong, W.W., J.A.M. Leunissen, P.J.M. Leenan, A. Zweers, and M. Versteeg. 1988. Dogfish alpha-crystallin sequences: Comparison with small heat shock proteins and *Schistosoma* egg antigen. *J. Biol. Chem.* **263:** 5141–5149.

DeLisi, C. and J.A. Berzofsky. 1985. T-cell antigenic sites tend to be amphipathic structures. *Proc. Natl. Acad. Sci.* **82:** 7048–7052.

de Vries, R.R.P., T.H.M. Ottenhoff, and W.C.A. van Schooten. 1988. Human leukocyte antigens (HLA) and mycobacterial disease. *Springer Semin. Immunopathol.* **10:** 305–318.

Dragon, E.A., S.R. Sias, E.A. Kato, and J.D. Gabe. 1987. The genome of *Trypanosoma cruzi* contains a constitutively expressed tandemly arranged multicopy gene homologous to a major heat shock protein. *Mol. Cell. Biol.* **7:** 1271–1275.

Emmrich, F., J. Thole, J. van Embden, and S.H.E. Kaufmann. 1985. A recombinant 64 kDa protein of *Mycobacterium bovis* BCG specifically stimulates human T 4 clones reactive to mycobacterial antigens. *J. Exp. Med.* **163:** 1024–1029.

Engers, H.D., M. Abe, B.R. Bloom, V. Mehra, W. Britton, T.M. Buchanan, S.R. Khanolkar, D.B. Young, O. Closs, T. Gillis, M. Harboe, J. Ivanyi, A.H.J. Kolk, and C.C. Shepard. 1985. Results of a World Health Organisation sponsored workshop on monoclonal antibodies to *Mycobacterium leprae*. *Infect. Immun.* **48:** 603–605.

Engers, H.D., V. Houba, J. Bennedsen, T.M. Buchanan, S.D. Chaparas, G. Kadival, O. Closs, J.R. David, J.D.A. van Embden, T. Godal, A.S. Mustafa, J. Ivanyi, D.B. Young, S.H.E. Kaufmann, A.G. Khomenko, M. Kubin, J.A. Louis, P. Minden, T.M. Shinnick, L. Trnka, and R.A. Young. 1986. Results of a WHO sponsored workshop to characterise antigens recognised by mycobacteria-specific monoclonal antibodies. *Infect. Immun.* **51:** 718–720.

Engman, D.M., L.V. Kirchoff, and J.E. Donelson. 1989. Molecular cloning of mpt70, a mitochondrial member of the hsp70 family. *Mol. Cell. Biol.* **9:** 5163–5167.

Fayet, O., J.-M. Louarn, and C. Georgopoulos. 1986. Suppression of the *Escherichia coli dnaA46* mutation by amplification of the *groES* and *groEL* genes. *Mol. Gen. Genet.* **202:** 435–445.

Fields, P.I., R.V. Swanson, C.G. Haidaris, and F. Heffron. 1986. Mutants of *Salmonella typhimurium* that cannot survive within the macrophage are avirulent. *Proc. Natl. Acad. Sci.* **83:** 5189–5193.

Finlay, B.B. and S. Falkow. 1989. Common themes in microbial pathogenicity. *Microbiol. Rev.* **53:** 210–230.

Fridovich, I. 1978. The biology of oxygen radicals. *Science* **201:** 875–880.
Garry, R.F., E.T. Ulug, and H.R. Bose. 1983. Induction of stress proteins in Sindbis virus- and vesicular stomatitis virus-infected cells. *Virology* **129:** 319–332.
Garsia, R.J., L. Hellqvist, R.J. Booth, A.J. Radford, W.J. Britton, L. Astbury, R.J. Trent, and A. Basten. 1989. Homology of the 70-kilodalton antigens from *Mycobacterium leprae* and *Mycobacterium bovis* with the *Mycobacterium tuberculosis* 71-kilodalton antigen and with conserved heat shock protein 70 of eucaryotes. *Infect. Immun.* **57:** 204–212.
Gaston, J.S.H., P.F. Life, L.C. Bailey, and P.A. Bacon. 1989. *In vitro* responses to a 65-kilodalton mycobacterial protein by synovial T cells from inflammatory arthritis patients. *J. Immunol.* **143:** (in press).
Gillis, T.P. and T.M. Buchanan. 1982. Production and partial characterisation of monoclonal antibodies to *Mycobacterium leprae. Infect. Immun.* **37:** 172–178.
Gillis, T.P., R.A. Miller, D.B. Young, S.R. Khanolkar, and T.M. Buchanan. 1985. Immunochemical characterisation of a protein associated with *Mycobacterium leprae* cell wall. *Infect. Immun.* **49:** 371–377.
Glass, D.J., R.I. Polvere, and L.H.T. van der Ploeg. 1986. Conserved sequences and transcription of the hsp70 gene family in *Trypanosoma brucei. Mol. Cell. Biol.* **6:** 4657–4666.
Gupta, R.S. and R.C. Austin. 1987. Mitochondrial matrix localisation of a protein altered in mutants resistant to the microtubule inhibitor podophyllotoxin. *Eur. J. Cell Biol.* **45:** 170–176.
Hahn, H. and S.H.E. Kaufmann. 1981. The role of cell mediated immunity in bacterial infections. *Rev. Infect. Dis.* **2:** 1221–1250.
Hansen, K., J.M. Bangsborg, H. Fjordvang, N.S. Pedersen, and P. Hindersson. 1988. Immunochemical characterisation of, and isolation of the gene for a *Borrelia burgdorferi* immunodominant 60kDa antigen common to a wide range of bacteria. *Infect. Immun.* **56:** 2047–2053.
Hayward, W.S., B.C. Neel, and S.M. Astrin. 1981. Activation of a cellular *onc* gene by promoter insertion in ALV-induced lymphoid leukosis. *Nature* **290:** 475–480.
Hedstrom, R., J. Culpepper, R.A. Harrison, N. Agabian, and G. Newport. 1987. A major immunogen in *Schistosoma mansoni* infections is homologous to the 70kD heat-shock protein. *J. Exp. Med.* **165:** 1430–1435.
Hedstrom, R., J. Culpepper, V. Schinski, N. Agabian, and G. Newport. 1988. Schistosome heat-shock proteins are immunologically distinct host-like antigens. *Mol. Biochem. Parasitol.* **29:** 275–282.
Hemmingsen, S.M., C. Woolford, S.M. van der Vies, K. Tilly, D.T. Dennis, C.P. Georgopoulos, R.W. Hendrix, and R.J. Ellis. 1988. Homologous plant and bacterial proteins chaperone oligomeric protein assembly. *Nature* **333:** 330–334.
Hindersson, P., J.D. Knudsen, and N.H. Axelsen. 1987. Cloning and expression of *Treponema pallidum* common antigen (Tp-4) in *Escherichia coli* K-12. *J. Gen. Microbiol.* **133:** 587–596.
Hindersson, P., C.S. Petersen, N.S. Pedersen, N. Hoiby, and N.H. Axelsen. 1984. Immunological cross reaction between antigen Tp-4 of *Treponema pallidum* and an antigen common to a wide range of bacteria. *Acta Pathol. Microbiol. Immunol. Scand.* **92:** 183–188.
Hoffman, P.S., C.A. Butler, and F.D. Quinn. 1989. Cloning and temperature-dependent expression in *Escherichia coli* of a *Legionella pneumophila* gene coding for a genus-common 60-kilodalton antigen. *Infect. Immun.* **57:** 1731–1739.
Hoiby, N. 1975. Cross reactions between *Pseudomonas aeruginosa* and 36 other bacterial species. *Scand. J. Immunol.* (suppl.) **4:** 187–196.
Holoshitz, J., F. Koning, J.E. Coligan, J. de Bruyn, and S. Strober. 1989. Isolation of $CD4^-CD8^-$ mycobacteria-reactive T lymphocyte clones from rheumatoid arthritis synovial fluid. *Nature* **339:** 226–229.

Hunter, K.W., C.L. Cook, and E.G. Hayunga. 1984. Leishmanial differentiation *in vitro*: Induction of heat shock proteins. *Biochem. Biophys. Res. Commun.* **125:** 755–760.

Jendoubi, M. and S. Bonnefoy. 1988. Identification of a heat shock like antigen in *P. falciparum,* related to the heat shock protein 90 family. *Nucleic Acids Res.* **16:** 10928.

Jenkins, A.J., J.B. March, I.R. Oliver, and M. Masters. 1986. A DNA fragment containing the *groE* genes can suppress mutations in the *Escherichia coli dnaA* gene. *Mol. Gen. Genet.* **202:** 446–454.

Jenkins, D.E., J.E. Schultz, and A. Matin. 1988. Starvation-induced cross protection against heat or hydrogen peroxide challenge in *Escherichia coli. J. Bacteriol.* **170:** 3910–3914.

Jindal, S., A.K. Dudani, B. Singh, C.B. Harley, and R.S. Gupta. 1989. Primary structure of a human mitochondrial protein homologous to the bacterial and plant chaperonins and to the 65-kilodalton mycobacterial antigen. *Mol. Cell. Biol.* **9:** 2279–2283.

Kaijser, B. 1975. Immunological studies of an antigen common to many gram negative bacteria with special reference to *Escherichia coli. Int. Arch. Allergy Appl. Immunol.* **48:** 72–81.

Kaufmann, S.H.E. 1988. CD8+ T lymphocytes in intracellular microbial infections. *Immunol. Today* **9:** 168–174.

Kaufmann, S.H.E., U. Vath, J.E.R. Thole, J.D.A. Van Embden, and F. Emmrich. 1987. Enumeration of T cells reactive with *Mycobacterium tuberculosis* organisms and specific for the recombinant 64kDa protein. *Eur. J. Immunol.* **17:** 351–357.

Kelly, P.M. and M.J. Schlesinger. 1982. Antibodies to two major chicken heat shock proteins cross-react with similar proteins in widely divergent species. *Mol. Cell. Biol.* **2:** 267–274.

Kemp, D.J., R.L. Coppel, and R.F. Anders. 1987. Repetitive proteins and genes of malaria. *Annu. Rev. Microbiol.* **41:** 181–208.

Khandjian, E.W. and H. Turler. 1983. Simian virus 40 and polyoma virus induce synthesis of heat shock proteins in permissive cells. *Mol. Cell. Biol.* **3:** 1–8.

Kirchhoff, L.V., K.S. Kim, D.M. Engman, and J.E. Donelson. 1988. Ubiquitin genes in Trypanosomatidae. *J. Biol. Chem.* **263:** 12698–12704.

Lamb, J.R., R.E. O'Hehir, and D.B. Young. 1988. The use of nitrocellulose immunoblots for the analysis of antigen recognition by T lymphocytes. *J. Immunol. Methods* **110:** 1–10.

Lamb, J.R., J. Ivanyi, A.D.M. Rees, J.B. Rothbard, K. Holland, R.A. Young, and D.B. Young. 1987. Mapping of T cell epitopes using recombinant antigens and synthetic peptides. *EMBO J.* **6:** 1245–1249.

Lamb, J.R., V. Bal, P. Mendez-Samperio, A. Mehlert, A. So, J.B. Rothbard, S. Jindal, R.A. Young, and D.B. Young. 1989. Stress proteins may provide a link between the immune response to infection and autoimmunity. *Int. Immunol.* **1:** 191–196.

Lane, D.P. and L.V. Crawford. 1979. T antigen is bound to a host protein in SV40-transformed cells. *Nature* **278:** 261–263.

LaThangue, N.B. and D.S. Latchman. 1988. A cellular protein related to heat-shock protein 90 accumulates during herpes simplex virus infection and is overexpressed in transformed cells. *Exp. Cell. Res.* **178:** 169–179.

LaThangue, N.B., K. Shriver, C. Dawson, and W.L. Chan. 1984. Herpes simplex virus infection causes the accumulation of a heat-shock protein. *EMBO J.* **3:** 267–277.

Lathigra, R., D.B. Young, D. Sweetser, and R.A. Young. 1988. A gene from *M. tuberculosis* which is homologous to the DnaJ heat shock protein of *Escherichia coli. Nucleic Acids Res.* **16:** 1636.

Lawrence, F. and M. Robert-Gero. 1985. Induction of heat shock and stress proteins in promastigotes of three *Leishmania* species. *Proc. Natl. Acad. Sci.* **82:** 4414–4417.

Lee, M.G.S., B.L. Atkinson, S.H. Giannini, and L.H.T. Van der Ploeg. 1988. Structure and expression of the hsp 70 gene family of *Leishmania major*. *Nucleic Acids Res.* **40:** 9569–9585.

Lindley, T.A., P.R. Chakraborty, and T.D. Edlind. 1988. Heat shock and stress response in *Giardia lamblia*. *Mol. Biochem. Parasitol.* **28:** 135–144.

Macfarlane, J., M.L. Blaxter, R.P. Bishop, M.A. Miles, and J.M. Kelly. 1989. Characterisation of a *Leishmania donovani* antigen homologous to heat shock protein 70. *Biochem. Soc. Trans.* **17:** 168–169.

Mattei, D., L.S. Ozaki, and L.P. da Silva. 1988. A *Plasmodium falciparum* gene encoding a heat shock-like antigen related to the rat 78 Kd glucose-regulated protein. *Nucleic Acids Res.* **16:** 5204.

McMullin, T.W. and R.L. Hallberg. 1987. A normal mitochondrial protein is selectively synthesised and accumulated during heat shock in *Tetrahymena thermophila*. *Mol. Cell. Biol.* **7:** 4414–4423.

———. 1988. A highly evolutionarily conserved mitochondrial protein is structurally related to the protein encoded by the *Escherichia coli groEL* gene. *Mol. Cell. Biol.* **8:** 371–380.

Mehlert, A. and D.B. Young. 1989. Biochemical and antigenic characterization of the *Mycobacterium tuberculosis* 71kD antigen, a member of the 70kD heat-shock protein family. *Mol. Microbiol.* **3:** 125–130.

Mehlert, A., J.R. Lamb, and D.B. Young. 1988. Analysis of stress-related proteins involved in the immune response to mycobacterial infection. *Biochem. Soc. Trans.* **16:** 721–722.

Mehra, V., D. Sweetser, and R.A. Young. 1986. Efficient mapping of protein antigenic determinants. *Proc. Natl. Acad. Sci.* **83:** 7013–7017.

Minden, P., P.J. Kelleher, and J.H. Freed. 1984. Immunological evaluation of a component isolated from *Mycobacterium bovis* BCG with a monoclonal antibody to *Mycobacterium bovis*. *Infect. Immun.* **46:** 519–525.

Minota, S., B. Cameron, W.J. Welch, and J.B. Winfield. 1988a. Autoantibodies to the constitutive 73-kD member of the hsp70 family of heat shock proteins in systemic lupus erythematosus. *J. Exp. Med.* **168:** 1475–1480.

Minota, S., S. Koyasu, I. Yahara, and J. Winfield. 1988b. Autoantibodies to the heat shock protein hsp 90 in systemic lupus erythematosus. *J. Clin. Invest.* **81:** 106–109.

Morgan, R.W., M.F. Christman, F.S. Jacobson, G. Storz, and B.N. Ames. 1986. Hydrogen peroxide-inducible proteins in *Salmonella typhimurium* overlap with heat shock and other stress proteins. *Proc. Natl. Acad. Sci.* **83:** 8059–8063.

Morrison, R.P., K. Lyng, and H.D. Caldwell. 1989. Chlamydial disease pathogenesis: Ocular hypersensitivity elicited by a genus-specific 57-kD protein. *J. Exp. Med.* **169:** 663–675.

Moulder, J.W. 1985. Comparative biology of intracellular parasitism. *Microbiol. Rev.* **49:** 298–337.

Muhich, M.L. and J.C. Boothroyd. 1988. Polycistronic transcripts in trypanosomes and their accumulation during heat shock: Evidence for a precursor role in mRNA synthesis. *Mol. Cell. Biol.* **8:** 3837–3846.

———. 1989. Synthesis of trypanosome hsp70 mRNA is resistant to disruption of trans-splicing by heat shock. *J. Cell. Biochem.* **13E:** 85.

Mustafa, A.S., H.K. Gill, A. Nerland, W.J. Britton, V. Mehra, B.R. Bloom, R.A. Young, and T. Godal. 1986. Human T cell clones recognise a major *M. leprae* protein antigen expressed in *Escherichia coli*. *Nature* **319:** 63–66.

Nene, V., D.W. Dunne, K.S. Johnson, D.W. Taylor, and J.S. Cordingley. 1986. Sequence and expression of a major egg antigen from *S. mansoni*. Homologies to heat shock proteins and alpha crystallins. *Mol. Biochem. Parasitol.* **21:** 179–188.

Nerland, A.H., A.S. Mustafa, D. Sweetser, T. Godal, and R.A. Young. 1988. A protein antigen of *Mycobacterium leprae* is related to a family of small heat shock proteins. *J. Bacteriol.* **170:** 5919–5921.

Nevins, J.R. 1982. Induction of the synthesis of a 70,000 dalton mammalian heat shock protein by the adenovirus E1A gene product. *Cell* **29:** 913–919.

Newport, G.R., J. Culpepper, and N. Agabian. 1988. Parasite heat-shock proteins. *Parasitol. Today* **4:** 306–312.

Newport, G.R., R.C. Hedstrom, J. Kallestad, P. Tarr, S. Klebanoff, and N. Agabian. 1988. Identification, molecular cloning, and expression of a schistosome antigen displaying diagnostic potential. *Am. J. Trop. Med. Hyg.* **38:** 540–546.

Notarianni, E.L. and C.M. Preston. 1982. Activation of cellular stress protein genes by Herpes Simplex virus temperature-sensitive mutants which overproduce immediate early polypeptides. *Virology* **123:** 113–122.

Oldstone, M.B. 1987. Molecular mimicry and auto-immune disease. *Cell* **50:** 819–820.

O'Brien, R.L., M.P. Happ, A. Dallas, E. Palmer, R. Kubo, and W.K. Born. 1989. Stimulation of a major subset of lymphocytes expressing T cell receptor *gamma/delta* by an antigen derived from *Mycobacterium tuberculosis*. *Cell* **57:** 667–674.

O'Malley, K., A. Mauron, J.D. Barchas, and L. Kedes. 1985. Constitutively expressed rat mRNA encoding a 70-kilodalton heat shock-like protein. *Mol. Cell. Biol.* **5:** 3476–3483.

Ottenhoff, T.H.M., B.K. Ab, J.D.A. van Embden, J.E.R. Thole, and R. Kiessling. 1988. The recombinant 65kD heat shock protein of *Mycobacterium bovis* BCG/ *Mycobacterium tuberculosis* is a target molecule for CD4+ cytotoxic T lymphocytes that lyse human monocytes. *J. Exp. Med.* **168:** 1947–1952.

Peterson, M.G., P.E. Crewther, J.K. Thompson, L.M. Corcoran, R.L. Coppel, G.V. Brown, R.F. Anders, and D.J. Kemp. 1988. A second antigenic heat shock protein of *Plasmodium falciparum*. *DNA* **7:** 71–78.

Polla, B.S. 1988. A role for heat-shock proteins in inflammation? *Immunol. Today* **9:** 134–137.

Raulet, D.H. 1989. Antigens for *gamma/delta T* cells. *Nature* **339:** 342–343.

Reading, D.S., R.L. Halberg, and A.M. Myers. 1989. Characterization of the yeast hsp60 gene coding for a mitochondrial assembly factor. *Nature* **337:** 655–659.

Rees, A., A. Scoging, A. Mehlert, D.B. Young, and J. Ivanyi. 1988. Specificity of proliferative response of human CD8 clones to mycobacterial antigens. *Eur. J. Immunol.* **18:** 1881–1887.

Requena, J.M., M.C. Lopez, A. Jimenez-Ruiz, J.C. de la Torre, and C. Alonso. 1988. A head-to-tail tandem organization of hsp70 genes in *Trypanosoma cruzi*. *Nucleic Acids Res.* **16:** 1393–1406.

Res, P.C.M., C.G. Schaar, F.C. Breedveld, W. van Eden, J.D.A. van Embden, I.R. Cohen, and R.R.P. de Vries. 1988. Synovial fluid T cell reactivity against 65 kDa heat shock protein of mycobacteria in early chronic arthritis. *Lancet* **II:** 478–480.

Reuter, S.H. and L. Shapiro. 1987. Asymmetric segregation of heat-shock proteins upon cell division in *Caulobacter crescentus*. *J. Mol. Biol.* **194:** 653–662.

Richman, S.J. and R.T. Reese. 1988. Immunologic modeling of a 75-kDa malarial protein with carrier-free synthetic peptides. *Proc. Natl. Acad. Sci.* **85:** 1662–1666.

Rothbard, J.B. and W.R. Taylor. 1988. A sequence pattern common to T cell epitopes. *EMBO J.* **7:** 93–100.

Rothstein, N.M., G. Higashi, J. Yates, and T.V. Rajan. 1989. *Onchocerca volvulus* heat shock protein 70 is a major immunogen in amicrofilaremic individuals from a filariasis-endemic area. *Mol. Biochem. Parasitol.* **33:** 229–236.

Saag, M.S. and J.C. Bennett. 1987. The infectious etiology of chronic rheumatic diseases. *Semin. Arthritis Rheum.* **17:** 1–23.

Sargent, C.A., I. Dunham, J. Trowsdale, and R.D. Campbell. 1989. Human major histocompatibility complex contains genes for the major heat shock protein hsp70. *Proc. Natl. Acad. Sci.* **86:** 1968–1972.

Sarnow, P., Y.S. Ho, J. Williams, and A.J. Levine. 1982. Adenovirus Elb-58kd tumor antigen and SV40 large tumor antigen are physically associated with the same 54kd cellular protein in transformed cells. *Cell* **28:** 387–394.

Scallon, B.J., B.J. Bogitsh, and C.E. Carter. 1987. Cloning of a *Schistosoma japonicum* gene encoding a major immunogen recognized by hyperinfected rabbits. *Mol. Biochem. Parasitol.* **24:** 237–245.

Searle, S., A.J.R. Campos, R.M.R. Coulson, T.W. Spithill, and D.F. Smith. 1989. A family of heat shock protein 70-related genes are expressed in the promastigotes of *Leishmania major*. *Nucleic Acids Res.* **17:** 5081–5095.

Selkirk, M.E., D.A. Denham, F. Partono, and R.M. Maizels. 1989. Heat shock cognate 70 is a prominent immunogen in Brugian filariasis. *J. Immunol.* **143:** 299–308.

Selkirk, M.E., P.J. Rutherford, D.A. Denham, F. Partano, and R.M. Maizels. 1987. Cloned antigen genes of *Brugia* filarial parasites. *Biochem. Soc. Symp.* **53:** 91–102.

Sette, A., S. Buus, S. Colon, J.A. Smith, C. Miles, and H.M. Grey. 1987. Structural character of an antigen required for its interaction with Ia and recognition by T cells. *Nature* **328:** 395–399.

Shapira, M., J.C. McEwan, and C.L. Jaffe. 1988. Temperature effects on molecular processes which lead to stage differentiation in *Leishmania*. *EMBO J.* **7:** 2895–2901.

Shinnick, T.M. 1987. The 65kDa antigen of *M. tuberculosis*. *J. Bacteriol.* **169:** 1080–1088.

Shinnick, T.M., M.H. Vodkin, and J.L. Williams. 1988. The *Mycobacterium tuberculosis* 65kDa antigen is a heat shock protein which corresponds to common antigen and to the *E.coli* GroEL protein. *Infect. Immun.* **56:** 446–451.

Shinnick, T.M., D. Sweetser, J.E.R. Thole, J.D.A. van Embden, and R.A. Young. 1987. The etiological agents of leprosy and tuberculosis share an immunoreactive protein antigen with the vaccine strain *Mycobacterium bovis* BCG. *Infect. Immun.* **55:** 1932–1935.

Smejkal, R.M., R. Wilff, and J.G. Olenick. 1988. *Leishmaniasis braziliensis panamensis*: Increased infectivity resulting from heat shock. *Exp. Parasitol.* **65:** 1–9.

Smith, D.F., P.D. Ready, R.M.R. Coulson, S. Searle, and A.J.R. Campos. 1989. Gene expression in the infective promastigotes of *Leishmania major*. *NATO-ASI Ser. Ser A* **163:** 569–580.

Sokolovic, Z. and W. Goebel. 1989. Synthesis of listeriolysin in *Listeria monocytogenes* under heat shock conditions. *Infect. Immun.* **57:** 295–298.

Swindle, J., J. Ajioka, H. Eisen, B. Sanwal, C. Jacquemot, Z. Browder, and G. Buck. 1988. The genomic organisation and transcription of the ubiquitin genes of *Trypanosoma cruzi*. *EMBO J.* **7:** 1121–1127.

Thole, J.E.R., H.G. Dauwerse, P.K. Das, D.G. Groothius, L.M. Schouls, and J.D.A. van Embden. 1985. Cloning of *Mycobacterium bovis* BCG DNA and expression of antigens in *Escherichia coli*. *Infect. Immun.* **50:** 800–806.

Thole, J.E.R., W.J. Keulen, A.H.J. Kolk, D.G. Groothius, L.G. Berwald, R.H. Tiesjema, and J.D.A. van Embden. 1987. Characterisation, sequence determination and immunogenicity of a 64 kD protein of *Mycobacterium bovis* BCG, expressed in *Escherichia coli* K-12. *Infect. Immun.* **55:** 1466–1475.

Thole, J.E.R., W.C.A. van Schooten, W.J. Keulen, P.W.M. Hermans, A.A.M. Janson, R.R.P. de Vries, A.H.J. Kolk, and J.D.A. van Embden. 1988a. Use of recombinant antigens expressed in *Escherichia coli* K-12 to map B cell and T cell epitopes on the immunodominant 65 kDa protein of *Mycobacterium bovis* BCG. *Infect. Immun.* **56:** 1633–1640.

Thole, J.E.R., P. Hindersson, J. de Bruyn, F. Cremers, J. van der Zee, H. de Cock, J. Tommassen, W. van Eden, and J.D.A. van Embden. 1988b. Antigenic relatedness of a strongly immunogenic 65 kDa mycobacterial protein antigen with a similarly sized ubiquitous bacterial common antigen. *Microb. Pathog.* **4:** 71–83.

Townsend, A.R.M., J. Rothbard, F.M. Gotch, G. Bahadur, D. Wraith, and A.J.

McMichael. 1986. The epitopes of influenza nucleoprotein recognized by cytotoxic T lymphocytes can be defined with short synthetic peptides. *Cell* **44:** 959–968.

van Bogelen, R.A., P.M. Kelly, and F.C. Neidhardt. 1987. Differential induction of heat shock, SOS and oxidation stress regulons and accumulation of nucleotides in *Escherichia coli*. *J. Bacteriol.* **169:** 26–32.

van der Broek, M.F., M.C.J. van Bruggen, E.J.M. Hogervorst, W. van Eden, R. van der Zee, and W.B. van der Berg. 1989. Protection against streptococcal cell wall induced arthritis by pretreatment with the mycobacterial 65kD heat shock protein. *J. Exp. Med.* **170:** 449–466.

van der Ploeg, L.H.T., S.H. Giannini, and C.R. Cantor. 1985. Heat shock genes: Regulatory role for differentiation in parasitic protozoa. *Science* **228:** 1443–1446.

van Eden, W., J.E.R. Thole, R. van der Zee, A. Noordzij, J.D.A. van Embden, E.J. Hensen, and I.R. Cohen. 1988. Cloning of the mycobacterial epitope recognised by T lymphocytes in adjuvant arthritis. *Nature* **331:** 171–173.

Vodkin, M.H. and J.C. Williams. 1988. A heat shock operon in *Coxiella burnetii* produces a major antigen homologous to a protein in both mycobacteria and *Escherichia coli*. *J. Bacteriol.* **170:** 1227–1234.

Yang, Y.F., P. Tan-ariya, Y.D. Sharma, and A. Kilejian. 1987. The primary structure of a *Plasmodium falciparum* polypeptide related to heat shock proteins. *Mol. Biochem. Parasitol.* **26:** 61–67.

Young, D.B. 1988. Stress-induced proteins and the immune response to leprosy. *Microbiol. Sci.* **5:** 143–146.

Young, D.B., L. Kent, and R.A. Young. 1987a. Screening of a recombinant mycobacterial DNA library with polyclonal antiserum and molecular weight analysis of expressed antigens. *Infect. Immun.* **55:** 1421–1425.

Young, D.B., J. Ivanyi, J.H. Cox, and J.R. Lamb. 1987b. The 65 kDa antigen of mycobacteria—a common bacterial protein? *Immunol. Today* **8:** 215–219.

Young, D.B., R. Lathigra, R. Hendrix, D. Sweetser, and R.A. Young. 1988a. Stress proteins are immune targets in leprosy and tuberculosis. *Proc. Natl. Acad. Sci.* **85:** 4267–4270.

Young, D.B., A. Mehlert, V. Bal, P. Mendez-Semperio, J. Ivanyi, and J.R. Lamb. 1988b. Stress proteins and the immune response to mycobacteria—antigens as virulence factors? *Antonie van Leeuwenhoek J. Microbiol.* **54:** 431–439.

Young, R.A., B.R. Bloom, C.M. Grosskinsky, J. Ivanyi, and R.W. Davis. 1985a. Dissecting *Mycobacterium tuberculosis* antigens using recombinant DNA. *Proc. Natl. Acad. Sci.* **82:** 2583–2587.

Young, R.A., V. Mehra, D. Sweetser, T.M. Buchanan, J.E. Clark-Curtiss, R.W. Davis, and B.R. Bloom. 1985b. Genes for the major protein antigens of the leprosy parasite. *Nature* **316:** 450–452.

Yuckenberg, P.D., F. Poupin, and T.E. Mansour. 1987. *Schistosoma mansoni:* Protein composition and synthesis during early development; evidence for early synthesis of heat shock proteins. *Exp. Parasitol.* **63:** 301–311.

Zerbini, M., M. Musiani, and M. La Placa. 1985. Effect of heat-shock on Epstein-Barr virus and cytomegalovirus expression. *J. Gen. Virol.* **66:** 633–636.

8

The Function and Regulation of Heat Shock Proteins in *Escherichia coli*

Carol A. Gross, David B. Straus, and James W. Erickson
Department of Bacteriology
University of Wisconsin
Madison, Wisconsin, 53706

Takashi Yura
Institute for Virus Research
Kyoto University
Kyoto, Japan 606

I. INTRODUCTION

The heat shock response in *E. coli* was first identified by monitoring the response of individual proteins to temperature upshift (Lemaux et al. 1978; Yamamori et al. 1978; for review, see Neidhardt and VanBogelen 1987). Most *E. coli* proteins do not change their rate of synthesis significantly when the temperature is suddenly increased. However, a small group of proteins show a large but transient increase in their rate of synthesis. These proteins have been called the heat shock proteins (hsps). hsps are induced by several stimuli other than heat, including DNA-damaging agents, coumermycin, amino acid analogs, starvation for

Stress Proteins in Biology and Medicine.

amino acids, bacteriophage λ infection, and ethanol. hsps may serve to protect the cell against a variety of adverse conditions.

The heat shock regulon includes about 20 of the approximately 2000 proteins that *E. coli* synthesizes (Neidhardt and VanBogelen 1987). Two of the major *E. coli* hsps are homologous to eukaryotic hsps. DnaK is homologous to the eukaryotic hsp70 family (Bardwell and Craig 1984), and GroEL is related to eukaryotic mitochondrial hsps (McMullin and Hallberg 1988), as well as the plant RuBisCO binding protein (Hemmingsen et al. 1988). HptG, another *E. coli* hsp, is homologous to hsp83 (Bardwell and Craig 1987). This conservation of hsps suggests that the functions required to protect the cell from thermal stress are evolutionarily conserved among nearly all organisms. Thus, a study of the function of the *E. coli* hsps will not only contribute to an understanding of the physiology of eubacteria, but will also add to our understanding of how eukaryotic organisms respond to temperature.

Genetic evidence indicates that the hsps are positively regulated by the *rpoH* gene (also called *htpR* or *hin*) (Neidhardt and VanBogelen 1981; Yamamori and Yura 1982). Mutations in *rpoH* prevent the transient increase in synthesis of hsps following temperature upshift, but have little effect on the expression of other proteins. The heat shock regulon is defined by the *rpoH* gene, whose product, a 32-kD sigma factor (σ^{32}), directs core RNA polymerase to recognize the promoters for heat shock genes (Grossman et al. 1984; Landick et al. 1984; Cowing et al. 1985). The studies of how the cell uses σ^{32} to regulate the heat shock response have provided a great deal of information about the molecular mechanisms involved in mounting a transient, global response to an environmental stimulus.

In this chapter, we describe work on the function of hsps in *E. coli* and investigations into the mechanism and regulation of heat shock gene expression. First we discuss the evidence that RNA polymerase containing σ^{32} ($E\sigma^{32}$) is responsible for nearly all the transcription of the heat shock genes and summarize our current understanding of the details of the $E\sigma^{32}$ heat shock promoter interaction. In the next section, we address the function of hsps in *E. coli*. We present evidence indicating that hsps are required for growth at most temperatures, discuss the role of the hsps in thermotolerance, and integrate the results of the many investigations into the function of the hsps. We then discuss the regulation of the heat shock response and how the transcription and translation of *rpoH* mRNA and the stability of σ^{32} are regulated to provide the amount of σ^{32} required for heat shock gene expression. In the final section, we discuss the emerging evidence for the existence of a second set of heat shock genes that may be under the control of a recently discovered minor σ factor.

II. THE HEAT SHOCK GENES ARE TRANSCRIBED BY RNA POLYMERASE HOLOENZYME CONTAINING A 32-KD SUBUNIT (Eσ^{32})

The first indication that the *rpoH* locus encoded a positive regulator of the heat shock response came from an analysis of an *rpoH* mutant (Neidhardt and VanBogelen 1981; Yamamori and Yura 1982). Strains carrying *rpoH165*, a non-sense mutation in *rpoH*, and a Ts suppressor tRNA expressed hsps at low temperature but did not show heat-inducible transcription of heat shock genes. When the factor permitting transcription of the heat shock genes was purified, it turned out to be a 32-kD σ factor encoded by *rpoH* (Grossman et al. 1984). σ^{32} directs core RNA polymerase to recognize the heat-inducible promoters preceding heat shock genes (Cowing et al. 1985; Bloom et al. 1986; Fujita et al. 1987). The fact that the promoters for heat shock genes are recognized by Eσ^{32} rather than by the predominant form of RNA polymerase holoenzyme, Eσ^{70}, allows their regulation to be distinct from the vast majority of *E. coli* genes.

A combination of promoter cloning, S1 mapping of in vivo RNAs, and in vitro transcription with Eσ^{32} was used to identify the heat-inducible promoters preceding heat shock genes (Cowing et al. 1985). To date, there is information concerning seven heat shock promoters. The map locations and sequences of these seven promoters are summarized in Table 1 (Cowing et al. 1985; Cowing 1988). The sequence of the promoters has been used to derive a consensus Eσ^{32} promoter: TCTC - CCCTTGAA - 15 bp - CCCCAT - TA. Like Eσ^{70} promoters, the two conserved regions are centered at −35 and −10 bp from the start point of transcription, but the sequence of both regions differs substantially from the consensus Eσ^{70} promoter.

There are strong indications that heat shock promoters are recognized only by Eσ^{32} in vivo. The most compelling evidence comes from an

Table 1 Heat shock promoters

Map position	Promoter	−35	Spacing	−10
94′	*groE*	TTTCCCCCTTGAA	15	CCCCATTTC
.5′	*dnaK P1*	TCTCCCCCTTGAT	14	CCCCATTTA
.5′	*dnaK P2*	TTGGGCAGTTGAA	13	CCCCTATTA
11′	*htpG P1*	GCTCTCGCTTGAA	15	CCCCATCTC
67′	*rpoD Phs*	TGCCACCCTTGAA	15	GACGATATA
10′	*lon*	TCTCGGCGTTGAA	14	CCCCATATA
57′	*grpE*	TGCTTCCCTTGAA	14	CCCCATAAT
	consensus	TCTC-CCCTTGAA	13–15	CCCCAT-TA

analysis of the RNAs present in strains lacking σ^{32} either because of an insertion at the 5′ end of *rpoH* or deletion of *rpoH* (Zhou et al. 1988). Strains without σ^{32} have no detectable transcription initiating at the *dnaKP1*, *dnaKP2*, *groE*, or *htpG* heat shock promoters (<1% of wild-type levels), although initiation at $E\sigma^{70}$ promoters occurs at normal rates. This experiment establishes that $E\sigma^{70}$ does not recognize heat shock promoters in vivo. It also indicates that strains without σ^{32} have no additional σ factors with a promoter specificity overlapping that of σ^{32}. If other σ factors recognize heat shock promoters, their activity either requires a functional σ^{32} or is repressed under the growth conditions of the experiments.

In vitro transcription studies are consistent with the conclusion that $E\sigma^{70}$ does not recognize heat shock promoters. Two different groups have found that $E\sigma^{70}$ is unable to initiate transcription at the *groE* promoter, the strongest heat shock promoter in vitro (Cowing et al. 1985; Fujita et al. 1987). The *dnaKP1*, *dnaKP2*, *rpoDpHS*, and *htpG* heat shock promoters were also not recognized by $E\sigma^{70}$. A report that the *groE* and *dnaK* promoters were transcribed by $E\sigma^{70}$ (Bloom et al. 1986), albeit poorly, most likely reflects both the existence of a minor $E\sigma^{70}$-controlled promoter for *groE* (Zhou et al. 1988) and low-level contamination of that enzyme preparation with $E\sigma^{32}$ (Skelly et al. 1988). Conversely, $E\sigma^{32}$ does not recognize any of the 15 $E\sigma^{70}$ promoters tested to date (Cowing et al. 1985; Bloom et al. 1986; Fujita et al. 1987). Thus, $E\sigma^{32}$ and $E\sigma^{70}$ do not appear to have overlapping promoter specificities.

The interaction of $E\sigma^{32}$ with several promoters has recently been examined in vitro by footprinting (Cowing and Gross 1989). These studies indicate that the contacts made by $E\sigma^{32}$ at heat shock promoters are similar to those made by $E\sigma^{70}$ at its cognate promoters. The DNase I footprint of $E\sigma^{32}$ on the *groE* promoter is almost indistinguishable from that of $E\sigma^{70}$ on the *lacUV5* promoter. In both cases, sequences from about −60 to +20 are protected, and almost identical positions between the −10 and −35 region, and upstream of the −35 region, show enhanced cleavage by DNase I. More precise structural information was obtained by footprinting using hydroxyl radical cleavage of the promoter DNA (J. Mecsas and C. Gross, in prep.). These studies indicate that $E\sigma^{32}$ protects both faces of the DNA helix from cleavage in the −10 region but only one face of the DNA helix in the −35 region. Ethylation interference experiments have indicated that the same geometry exists for $E\sigma^{70}$ binding to *lacUV5* (Siebenlist et al. 1980). Therefore, despite differences in the size of the σ factors and the conserved promoter sequences, both $E\sigma^{70}$ and $E\sigma^{32}$ appear to form homologous open complexes with their promoters.

Although similar contacts are made by Eσ^{70} and Eσ^{32} with their cognate promoters, the complexes formed by the two RNA polymerases can be distinguished from each other. Eσ^{32} open complexes at the *groE* promoter are similar in stability to Eσ^{70} promoter complexes ($T_{1/2}$ about 50 min) but are very labile to competitors that do not destabilize Eσ^{70} open complexes (Cowing 1988). Addition of heparin, poly[d(AT)], or poly[d(IC)] as competitor destabilized these Eσ^{32} open complexes in a concentration-dependent fashion. Although open complexes at some Eσ^{70} promoters are destabilized by heparin, sensitivity to poly[d(AT)] or poly[d(IC)] has not been reported. This suggests that the Eσ^{70} and Eσ^{32} open complexes are structurally distinct. The RNA polymerase site(s) required for destabilization by competitors are apparently accessible in Eσ^{32} open complexes, but not in Eσ^{70} open complexes.

III. FUNCTION OF THE HEAT SHOCK PROTEINS

A. hsps Are Required for *E. coli* Growth

It is a mistake to believe that the functions provided by the hsps are required only for cell survival at high temperature. In fact, most evidence suggests that hsps provide function(s) that are needed for cell growth at all temperatures. The increased intracellular concentration of hsps found at higher temperatures probably reflects an increased requirement for hsp function as the growth temperature rises.

The first indication that more hsps are required at higher temperature came from a systematic study of the expression of hsps as a fraction of total protein synthesis (Herendeen et al. 1979). The rate of synthesis of hsps is relatively invariant at temperatures up to 30°C. However, at temperatures greater than 30°C, hsps gradually become a greater fraction of total cell protein. At 46°C, which is near the maximum growth temperature of *E. coli*, the steady-state rate of synthesis of the GroEL hsp is tenfold greater than that at 30°C. hsps represent a major fraction (about 20%) of total cell protein at this high temperature.

The correlation between the intracellular concentration of hsps and growth temperature is not just circumstantial. The first experimental evidence that cells growing at higher temperatures required more hsps came from a study of the *rpoH165* strain described earlier. This *rpoH* nonsense mutation was suppressed with a series of suppressor tRNAs of increasing efficiency, each of which inserted the same amino acid into the mutant σ^{32} protein. Strains with better suppressors made more hsps and thus were able to grow at higher temperatures than those with poorer suppressors (Yamamori and Yura 1982).

More recently, the role of hsps in normal cell growth has been pursued by characterizing strains in which *rpoH* is inactivated, either by an insertion into the 5′ end of the gene or by deletion of the entire coding sequence (Zhou et al. 1988). Such strains lack σ^{32} and are viable only at temperatures below 20°C, suggesting that hsps are required for growth over most of the normal temperature range. It is also likely that hsps are required for growth even below 20°C. In fact, genetic analysis has shown that the GroE hsps (GroES and GroEL) and the GrpE hsp are required for growth at temperatures below 20°C, since neither the *groE* genes nor *grpE* may be inactivated without the presence of an additional copy of the appropriate gene (Fayet et al. 1989; Ang and Georgopoulos 1989). Strains lacking σ^{32} survive at temperatures below 20°C, despite the absence of detectable transcription from heat shock promoters, probably as a result of heat shock gene transcription initiating at $E\sigma^{70}$ promoters. For example, GroEL is present at 30% of its normal level in these cells because of the presence of a minor $E\sigma^{70}$ promoter that is equally active in both mutant and wild-type cells at 16°C.

The *rpoH* deletion strain is an ideal starting point for a genetic analysis designed to determine the functions required for growth at high temperature. Revertants were selected at a variety of higher temperatures and obtained at temperatures up to 40°C (Kusukawa and Yura 1988). Each of these revertants proved to have an insertion upstream of the *groE* operon, which led to overexpression of the GroEL and GroES proteins. The greater the expression of *groE*, the higher the permissible growth temperature, suggesting that among the hsps, GroEL and GroES are uniquely required in large amounts for cell growth up to 40°C. To confirm that the GroE proteins were sufficient for restoration of growth, it was shown that the *rpoH* deletion strain carrying a multicopy plasmid with the *groE* operon could grow at temperatures up to 40°C. To achieve growth above 40°C, an additional mutation was required. Analysis of revertants growing at 42°C indicated the presence of a mutation allowing overexpression of *dnaK*. The requirement for DnaK was confirmed by showing that the *rpoH* deletion strain carrying plasmids with the *groE* operon and the *dnaK* gene could grow at temperatures up to 42°C. These results suggest that among the hsps, abundant amounts of only the GroES and GroEL proteins are required for growth up to 40°C. For growth above 40°C, abundant amounts of DnaK are also required.

These findings do not exclude a role for small amounts of hsps other than GroEL and GroES in cell growth at temperatures below 40°C. Such a role is in fact likely given that an intact *grpE* heat shock gene is essential for growth at all temperatures (Ang and Georgopoulos 1989), and a *dnaK* null mutant quickly accumulates extragenic suppressor mutations during growth at 30°C (Bukau and Walker 1989). Small amounts of

DnaK and GrpE are probably supplied from weak Eσ^{70} promoters. This expression would allow the growth of the *rpoH* deletion strain at temperatures up to 40°C in the presence of abundant amounts of GroEL and GroES proteins.

B. GroE and DnaK Function

The studies described above identify GroEL, GroES, and DnaK as the key hsps required in abundant amounts for cell growth in the normal temperature range. In this section, we analyze the available data on the functions of these proteins to try to arrive at an understanding of why they are required for cell growth.

GroEL, GroES, and DnaK are among the most abundant proteins in the cell. Together GroES and GroEL constitute about 1% of total cell protein at 30°C and 12% at 46°C (Herendeen et al. 1979). DnaK constitutes about 4% of total cell protein at 46°C. The abundance of these proteins suggests that they play a structural role in the cell. Two of these proteins, GroEL and DnaK, exhibit a weak ATPase activity (Hendrix 1979; Zylicz et al. 1983). In addition, DnaK can autophosphorylate (Zylicz et al. 1983) and has been reported to be involved in the phosophorylation of the threonyl- and glutaminyl-tRNA synthetases (Wada et al. 1986). Both GroEL and DnaK are homologous to eukaryotic proteins. DnaK is homologous to the hsp70 class of proteins (Bardwell and Craig 1984), whereas GroEL is homologous to the mitochondrial hsp60 proteins (McMullin and Hallberg 1988) as well as the chloroplast RuBisCO binding protein (Hemmingsen et al. 1988).

GroEL and GroES purify as homopolymeric particles of 14 and 7 subunits, respectively (Hendrix 1979; R. Hendrix and T. Zeigelhoffer, pers. comm.). GroES and GroEL particles associate in an Mg-ATP-dependent manner in vitro (Chandrasekhar et al. 1986). Furthermore, genetic analysis suggests that they interact in vivo; mutations in *groEL* can partially suppress the phenotype of mutations in *groES* (Tilly and Georgopoulos 1982). DnaK is purified as a monomer, but there is evidence that it interacts with two other hsps. There is strong genetic and biochemical evidence that GrpE and DnaK interact. A mutation in *grpE* is suppressed by mutations in *dnaK* (Johnson et al. 1989). Moreover, GrpE binds to a DnaK affinity column in the presence of 2 M KCl and can be eluted by ATP (Zylicz et al. 1987). The evidence for direct interaction of DnaJ with the other two proteins is weaker. However, DnaJ works together with DnaK and GrpE to promote the same step in λ replication, suggesting at least a functional, if not structural, interaction between these three proteins (Dodson et al. 1989; Mensa-Wilmot et al. 1989; Zylicz et al. 1989).

The *groES, groEL*, and *dnaK* loci were originally identified because mutations in these genes affected bacteriophage growth. In all cases, the defects result from alterations in protein-protein interactions in the mutant cells. Mutations in *groES* and *groEL* lead to defects in protein assembly and proteolytic processing. These mutations affect a variety of morphogenetic processes in bacteriophage growth, including tail assembly in bacteriophage T5 (Zweig and Cummings 1973) and head assembly in bacteriophages T4 and λ (Georgopoulos et al. 1972, 1973; Coppo et al. 1973; Sternberg 1973). Mutations in *dnaK* impair the ability of DnaK (in concert with GrpE and DnaJ) to disrupt the interaction between the P protein of λ and the DnaB protein. Consequently, *dnaK* mutant cells are unable to activate the DnaB helicase required for DNA synthesis and are defective in λ replication (Zylicz et al. 1983).

More recently, the cellular phenotypes of *groES, groEL*, and *dnaK* mutations have been investigated. Strains containing mutations in these genes exhibit a wide variety of alterations in cellular metabolism, including temperature-sensitive growth, impaired DNA and RNA synthesis, altered cell division, and possibly altered permeability (Sternberg 1973; Itikawa and Ryu 1979; Wada and Itikawa 1984). Mutations in *dnaK* and *groEL* lead to defects in ATP-dependent proteolysis of puromycyl fragments (Keller and Simon 1988; Straus et al. 1988), and *dnaK* mutant cells also show impaired degradation of the X90 nonsense fragment of β-galactosidase (Straus et al. 1988). A genetic analysis has also suggested that GroEL is involved in the monomer-tetramer equilibrium of Ssb (single-strand DNA binding protein). A *groEL* mutation was obtained as a suppressor of the temperature-sensitive defect of *ssb-1* (Ruben et al. 1988). This strain is probably temperature-sensitive because the mutant Ssb protein is defective in tetramer formation at high temperature (Williams et al. 1984). The *groEL411* suppressor mutation may restore the appropriate tetramer concentration either by preventing their dissociation or promoting association of mutant Ssb1 monomers. Recent biochemical work indicates that GroEL plays an important role in the transport of pre-β-lactamase into membrane vesicles, probably by maintaining the protein in an unfolded state, which is thought to be essential for passage through the membrane (Bochkareva et al. 1988).

Both genetic and biochemical investigations of the roles of GroEL, GroES, and DnaK proteins are consistent with the proposal that the general function of hsps is to control the higher-order structure of proteins present either as individual polypeptides or in multicomponent assemblies by controlling aggregation and refolding (Pelham 1986). Certainly this general description correctly describes the function of each of these proteins in bacteriophage growth: GroEL and GroES facilitate protein-protein interactions involved in morphogenesis, whereas DnaK facilitates protein-protein interactions involved in DNA replication (Dodson et al. 1986). The involvement of these proteins in proteolysis can also be understood in this context. As abundant cellular proteins, it is highly unlikely that these proteins are proteinases. Instead, they may facilitate proteolysis by binding either to the substrate or to the proteinase. Consequent alterations in conformation could facilitate the substrate-enzyme interaction and enhance the rate of proteolysis. The other diverse cellular phenotypes reported for *dnaK* and *groE* mutants could result from other instances in which protein-protein interactions mediated by GroE and DnaK are disrupted. The proposed function of GroE and DnaK in controlling higher-order protein structure and other protein-protein interactions may explain their essential role in cell growth.

What is needed now is a good biochemical system in which to explore such protein-protein interactions. The proposed interaction of Ssb1 and GroEL may provide an excellent model system in which to study these novel protein-protein interactions.

C. Role of the hsps in Thermotolerance

When *E. coli* cells are shifted directly from a low growth temperature to a lethal temperature, e.g., 50°C, the cells are rapidly killed. However, if the cells are first "preadapted" by growth at 42°C for 30 minutes, the rate of killing upon shift to lethal temperature is dramatically decreased (Yamamori and Yura 1982). Conditions that increase survival when cells are shifted to lethal temperatures are said to induce thermotolerance. Pretreatment with ethanol, $CdCl_2$, and H_2O_2 also induce thermotolerance.

Several observations suggest that hsps are involved in thermotolerance. There is a good correlation between conditions that induce thermotolerance and the induction of hsps. *rpoH* mutant strains that are defective in the heat shock response do not acquire thermotolerance when shifted to 42°C prior to shift to lethal temperature, and in wild-type strains, protein synthesis during the 42°C adaptation phase is required for the acquisition of thermotolerance. These observations imply that the transient overproduction of hsps after shift to 42°C is involved in protecting cells from killing at 55°C (Yamamori and Yura 1982). Additionally, by 15–20 minutes after cells are shifted to a lethal temperature, hsps are the predominant proteins synthesized, suggesting that they may have a major role in protecting the cell from thermal killing.

Increased amounts of hsps alone, however, are not a sufficient condition to create thermotolerance. The new steady-state concentration of hsps is rapidly established following shift to 42°C, but the acquisition of thermotolerance is transient, peaking 30 minutes after temperature upshift (Yamamori and Yura 1982). The availability of a strain in which hsps could be induced simply by increasing the amount of σ^{32} in the cell permitted a direct determination of whether increased amounts of hsps were sufficient to create thermotolerance (VanBogelen et al. 1987). hsp induction was accomplished using a strain carrying a plasmid with *rpoH* under control of the isopropylthiogalactoside (IPTG)-inducible P*tac* promoter. Addition of IPTG results in overexpression of σ^{32} and the consequent overexpression of hsps to an extent comparable to that observed after temperature upshift. Nonetheless, overexpression of hsps without a temperature upshift was not sufficient to induce thermotolerance. Although the hsps are likely to be involved in thermotolerance, either additional proteins, not under $E\sigma^{32}$ control, or other cellular changes must also be required to induce thermotolerance. Proteins encoded by transcripts initiated by the newly identified form of RNA polymerase, $E\sigma^{E}$ (see next section), may be involved in thermotolerance.

D. Role of hsps in F-plasmid Maintenance

It was recently shown that expression of an F-plasmid function essential for replication is controlled by σ^{32}. It was initially found that the *rpoH* mutants are defective in F-plasmid maintenance and segregate F^- cells at high frequency (Wada et al. 1986). Mini-F plasmids that essentially consist of the origin of replication and the gene encoding replication initiator protein (*repE*) exhibited similar instability. Promoter cloning and analysis of in vivo transcripts indicated that *repE* transcription depends on *rpoH* function (Wada et al. 1987). The amount of *repE* mRNA increased transiently upon shift from 30°C to 42°C in the *rpoH*$^+$ (mini-F) strain, but decreased drastically at high temperature in an *rpoH* nonsense mutant carrying a Ts suppressor tRNA. In fact, the *repE* promoter sequence is similar to one of the heat shock promoters *rpoD* Phs (Table 1). Studies of in vitro transcription using purified RNA polymerase showed that *repE* is transcribed primarily by $E\sigma^{32}$, although minor RNAs may be transcribed by $E\sigma^{70}$ (Wada et al. 1987). Furthermore, mini-F cannot transform *rpoH* deletion strains unless the host bacteria contain a multicopy plasmid that can provide RepE function in *trans*. These experiments establish that the supply of RepE protein is necessary and sufficient for the replication of mini-F in the absence of σ^{32}. The synthesis of Rep protein of another plasmid, mini-Rts1, was also shown to be under σ^{32} control (C. Wada and T. Yura, unpubl.). It thus appears that the synthesis of replicator initiator proteins of F and related plasmids are at least partially dependent on the function of σ^{32}.

In additon to the role of σ^{32} in *repE* transcription, several heat shock proteins appear to be required for RepE protein to function normally. Strains of *E. coli* carrying a missense (or deletion) mutation in *dnaK*, *dnaJ*, or *grpE* are virtually unable to be transformed by mini-F DNA (Ezaki et al. 1989; Kawasaki et al. 1989). Interestingly, the requirement for DnaK, DnaJ, and GrpE disappears when RepE protein is provided in *trans* from a multicopy plasmid, explaining why excess RepE suffices for mini-F replication in the absence of σ^{32}. Thus, σ^{32} appears to control replication of F at two different levels: first, σ^{32} controls the synthesis of RepE and second, three heat shock proteins transcribed by σ^{32} are required for the function of RepE.

A requirement for DnaK, DnaJ, and GrpE was also found for the replication of plasmid mini-P1 (Tilly and Yarmolinsky 1989; Bukau and Walker 1989) and mini-Rts1 (Y. Kawasaki and T. Yura, unpubl.). Although plasmids like F are not required for vegetative growth of the cell, they play a vital role in evolution/differentiation of bacterial species by providing a means for genetic recombination. Further studies may contribute to an understanding of the relationship between plasmid replication and cell physiology.

IV. REGULATION OF HEAT SHOCK GENE EXPRESSION

A. Transcription of the *rpoH* Gene

The *rpoH* gene has one of the most complex promoter structures known in *E. coli*. There are at least four promoters (P1, P3, P4, and P5), the activity of which varies with the temperature and growth conditions. The promoters P1 and P4 are recognized by $E\sigma^{70}$ and account for the majority of *rpoH* transcription under most growth conditions. Neither promoter appears to be highly regulated, although transcription from P4 increases two- to threefold after shift to 42°C (Erickson et al. 1987; Fujita and Ishihama 1987). The minor promoter P5 is recognized by $E\sigma^{70}$ and is active only when the growth medium lacks glucose or when ethanol is present in the medium. In vitro CAP-cAMP stimulates transcription from P5, but the mechanism of induction by ethanol is unknown (H. Nagai and T. Yura, unpubl.). The other minor promoter P3 is strongly induced by shift to high temperature. This induction is particularly strong when the cells are shifted to the lethal temperature of 50°C—a temperature at which σ^{70} appears to be inactive (Erickson et al. 1987). The factor that recognizes P3 has been purified and shown to be a form of RNA polymerase containing a 24-kD σ factor called σ^E (Erickson and Gross 1989). It is not known how many genes are under the control of σ^E, but the existence of a second alternate σ factor that is especially active at high temperature suggests that there may be a second set of heat shock genes controlled by σ^E (see below).

Despite the many levels of regulation, the transcription of *rpoH* is not the primary factor that controls the synthesis of σ^{32} immediately after shift to high temperature (see the next section). Instead, transcription of *rpoH* is probably controlled so as to provide the proper amount of σ^{32} expression under a variety of steady-state conditions.

B. Transient Accumulation of σ^{32} Mediates the Heat Shock Response

When cells are shifted from 30°C to 42°C, the rate of synthesis of the hsps increases about 10- to 20-fold by 5 minutes after temperature upshift and then declines to a new steady-state rate of synthesis. The actual amount of increase is dependent on the particular protein and the final temperature. The higher the temperature, the greater the increase in synthesis of hsps, and the higher the final steady-state rate of synthesis.

The first indication that the increased synthesis of hsps following temperature upshift resulted from increased synthesis of heat shock gene mRNAs came from a study examining the effect of rifampicin, an inhibitor of transcription, on the heat shock response (Yamamori and Yura 1980). The addition of rifampicin prior to temperature upshift prevents increased synthesis of hsps, suggesting that increased synthesis of heat shock mRNAs is a requisite for the increased synthesis of hsps. Subsequent studies examining either the rate of synthesis (Yamamori and Yura 1982) or the accumulation (Taylor et al. 1984; Cowing et al. 1985) of discrete heat shock mRNAs substantiated the idea that the heat shock response was regulated at the transcriptional level. Cloning of several heat shock promoters upstream of reporter genes established that increased synthesis of heat shock mRNA resulted from increased transcription initiation at promoters recognized by $E\sigma^{32}$ (Taylor et al. 1984; Cowing et al. 1985; Yano et al. 1987).

The mechanism underlying the rapid, but transient, increase in the rate of transcription initiation at heat shock promoters has been intensively investigated. It was initially speculated that the activity of σ^{32} increased after temperature upshift as a result of an alteration in the intracellular environment, because the magnitude and rapidity of the response seemed to prohibit the simple model that changes in the concentration of σ^{32} regulated heat shock gene expression (e.g., see Grossman et al.

1984). However, using immunological techniques capable of detecting the low level of σ^{32} present in the cell, it was discovered that changes in the level of σ^{32} are directly responsible for increases in transcription initiation from heat shock promoters (Lesley et al. 1987; Skelly et al. 1987; Straus et al. 1987). The intracellular concentration of σ^{32} increases 15- to 20-fold by 5 minutes after temperature upshift and then declines to a new level, severalfold higher than at 30°C. The kinetics and magnitude of this increase are sufficient to account for the increased rate of transcription initiation from heat shock promoters after temperature upshift.

A transient change in the stability of σ^{32} contributes to the accumulation observed after temperature upshift (Straus et al. 1987). During steady-state growth at either 30°C or 42°C, σ^{32} is rapidly degraded ($T_{1/2}$ = 1 min). However, for the first 4–5 minutes after temperature upshift, σ^{32} is stabilized, permitting its accumulation. After this initial period, rapid degradation of σ^{32} resumes. The extreme instability of σ^{32} allows the rapid adjustments in its concentration that regulate the heat shock response.

Changes in the rate of σ^{32} synthesis also contribute to the accumulation of σ^{32} after temperature upshift (Straus et al. 1987). The rate of σ^{32} synthesis increases about tenfold within 3–4 minutes after shift to high temperature and then declines to a new rate of synthesis by about 6 minutes after temperature upshift. Experiments with protein and operon fusions of *rpoH* to a *lacZ* reporter gene suggest that the transient increase in synthesis of σ^{32} results from altered translation of *rpoH* mRNA. When both the transcriptional and translational signals of *rpoH* were fused to *lacZ*, the rate of synthesis of β-galactosidase increased after temperature upshift in a manner similar to the increase in synthesis exhibited by authentic σ^{32}. When only the *rpoH* transcriptional signals are fused to *lacZ*, little increase in synthesis of β-galactosidase was observed following temperature upshift. Additional evidence that transcription of *rpoH* does not increase following temperature upshift was obtained by measuring the rate of synthesis of *rpoH* mRNA before and after temperature upshift using filter hybridization (Erickson et al. 1987). This study indicated that the rate of synthesis of *rpoH* mRNA increased less than twofold immediately after temperature upshift. The fivefold accumulation of *rpoH* mRNA following temperature upshift reported by several investigators (Tilly et al. 1986; Erickson et al. 1987) occurs after the increase in σ^{32} synthesis and is probably due to the effect of increased translation on the stability of *rpoH* mRNA.

We believe that σ^{32} synthesis is negatively regulated. This conclusion is based on the observation that *rpoH* mRNA is relatively abundant, but σ^{32} is normally synthesized at a very low rate (Straus et al. 1987). Furthermore, there is no obvious secondary structure in the message that would mask the ribosome binding site, and preferred codons are used to encode σ^{32} (Landick et al. 1984; Yura et al. 1984). Apparently the synthesis of σ^{32} is normally repressed, but immediately following temperature upshift, repression is lost and σ^{32} synthesis is induced.

A relatively simple picture thus emerges to explain the transient induction of hsps after shift to high temperature. σ^{32}, the direct regulator of heat shock gene expression, is present in limiting amounts in the cell. Following temperature upshift, both the stability and synthesis of σ^{32} transiently increase, leading to a transient 20-fold increase in the intracellular concentration of σ^{32}. The consequent changes in $E\sigma^{32}$ concentration lead to a transient increase in transcription initiation at heat shock promoters and increased levels of hsps.

At present, it is not known how a change in temperature is translated into alterations in the synthesis and stability of σ^{32}. Under steady-state growth conditions, the concentration of σ^{32} is regulated through rapid degradation and translational repression. The signal produced upon temperature upshift must interfere with this negative regulation. σ^{32} could be transiently stabilized because of altered activity of the regulatory proteinase or a modification of σ^{32} that prevents its degradation. In addition, the rate of translation of *rpoH* mRNA could transiently increase due to inactivation of a repressor. In the simplest case, shutoff of the response would occur by a removal of the signal and a reversal of the alterations in σ^{32} degradation and *rpoH* mRNA translation. In this model, the signal for the heat shock response transiently interrupts the normal negative regulation that governs σ^{32} synthesis and stability, thereby transiently increasing the intracellular concentration of σ^{32}.

C. A Regulatory Loop Governs Expression of Heat Shock Genes

At least three hsps, DnaK, DnaJ, and GrpE, are involved in the negative regulation of σ^{32} synthesis and stability, and thereby regulate their own synthesis and the synthesis of the other hsps. This regulatory loop allows the expression of heat shock genes to be coupled to the activity of hsps. The involvement of the DnaK, DnaJ, and GrpE proteins was determined by analyzing the effect of mutations in these genes on the heat shock response. *dnaK* mutant strains overexpress hsps at 30°C, and fail to shut off the heat shock response after shift to 42°C (Tilly et al. 1983). Mutations in *dnaJ* and *grpE* have similar effects on heat shock gene expression (D. Straus and C. Gross, unpubl.). A *dnaK* deletion strain has a phenotype similar to that of strains carrying a point mutation in either gene (Paek and Walker 1987), suggesting that the phenotypes observed in these mutants are due to a loss of gene product function.

The molecular basis for the alteration in the heat shock response in these mutants has recently been established (D. Straus and C. Gross, unpubl.). Strains carrying mutations in *dnaK*, *dnaJ*, or *grpE* are defective in degrading σ^{32} at 30°C, and the *dnaK* and *grpE* strains are also defective in degrading σ^{32} after shift to 42°C. Tilly et al. (1989) have also shown that degradation of σ^{32} is defective in a *dnaK* mutant at 30°C. All three mutants fail to shut off σ^{32} synthesis following its induction after temperature upshift (Table 2). The products of these three genes, then, are part of a negative regulatory loop controlling the concentration of σ^{32}. Their normal function is to increase the rate of σ^{32} degradation and to permit the reestablishment of translational repression following temperature upshift.

The phenotypes of the mutants may provide us with some information about the mechanisms by which the hsps affect σ^{32} synthesis and stability. The *dnaK*, *dnaJ*, and *grpE* mutants fail to reduce σ^{32} synthesis during the shutoff of the heat shock response, but they do not show altered rates of σ^{32} synthesis at 30°C. Since deletions of *dnaK* also have no effect on σ^{32} synthesis at low temperature, this phenotype is not the result of the mutant proteins retaining function at low temperature (D. Straus et al., unpubl.). If a single mechanism governs σ^{32} synthesis during steady-state growth at 30°C and during shutoff of the heat shock response, the heat shock gene mutants should affect either both processes or neither one. One explanation is that hsps are required for repression of σ^{32} synthesis only at high temperature. The repression mechanism could be temperature sensitive, requiring hsp function for activity after temperature upshift. The involvement of hsps in proteolysis or the control of protein structure could couple hsp function to the repression of σ^{32} synthesis during the shutoff of the heat shock response.

Table 2 Heat shock gene mutants are defective in σ^{32} synthesis and degradation

	σ^{32} synthesis			σ^{32} degradation		
	30°C	42°C (4 min)	42°C (8 min)	30°C	42°C (4 min)	42°C (8 min)
wild type	+	++++	++	+	–	+
dnaK	+	++++	++++	–	–	–
grpE	+	++++	++++	–	–	–
grpE	+	++++	++++	–	–	–
dnaJ	+	++++	+++	–	–	+

The mechanism by which the DnaK, GrpE, and DnaJ hsps influence σ^{32} stability is not known, but the altered degradation of σ^{32} in the mutant strains may be the result of defects in hsp function discussed earlier. hsps are required for wild-type rates of protein degradation, probably because they influence the structure of the substrate and/or proteinase. Similarly, the DnaK, GrpE, and DnaJ proteins may enhance the rate of σ^{32} degradation by interacting with the proteinase degrading σ^{32}, the proteinase/σ^{32} complex, or σ^{32} itself. Mutations in any one of these hsps would disrupt the proposed interactions, thereby decreasing the rate of proteolysis. The observed rate of σ^{32} degradation in *dnaK*, *dnaJ*, or *grpE* mutant strains reflects either degradation of σ^{32} by its normal proteinase, without the benefit of interactions provided by hsps, or degradation of σ^{32} by an entirely different proteolytic pathway. Georgopoulos et al. (this volume) have specifically proposed that hsps control the stability of σ^{32} by regulating its association with core RNA polymerase. The DnaK, DnaJ, and GrpE hsps could act to dissociate σ^{32} from core RNA polymerase, which would make the σ factor more susceptible to proteolysis.

It is fair to say that some, but not all, of the players in the negative regulatory loop governing the concentration of σ^{32} are known. Three hsps, DnaK, DnaJ, and GrpE, are involved in negatively regulating the stability and synthesis of σ^{32}. However, the proteinase degrading σ^{32} and the putative translational repressor of σ^{32} synthesis have not been identified. An understanding of how hsps are involved in governing the concentration of σ^{32} will require the identification of these direct regulators.

D. What Are the Signals That Induce the Heat Shock Response?

What are the signal(s) that induce the synthesis and stabilization of σ^{32} following temperature upshift, and how might they act? In light of the involvement of hsps in the control of σ^{32} synthesis and stability, it is possible that hsps themselves are the sensors of signal(s) generated by a change in the cell environment. Alterations in hsp function resulting from changes in the cell environment would transmit the signal(s) to the direct regulators of σ^{32} synthesis and stability. Goff and Goldberg (1985) proposed a specific model linking hsp function to the regulation of heat shock gene expression. They suggested that after heat shock, increased numbers of abnormal proteins would titrate the function of the La proteinase, an hsp required for the degradation of abnormal proteins, resulting in the transient stabilization of σ^{32}. Our results indicate that σ^{32} is not degraded by La (D. Straus, unpubl.), but they do show that other hsps, which are involved in proteolysis, are required for normal degradation of σ^{32}. The DnaK, DnaJ, and GrpE hsps could play the role that was originally assigned to La.

Excess proteolytic substrates may not be the only means of titrating of hsp function. Substrates other than, or in addition to, unstable proteins could be responsible for titration of hsp function, since hsps almost certainly carry out several functions in addition to proteolysis. For example, the role of DnaK, DnaJ, and GrpE in dissociation of protein complexes during the initiation of λ replication suggests that protein assemblies or aggregates may also be substrates of these hsps. If titration of hsps does provide the signal for the heat shock response, it indicates that hsp function is extremely sensitive to changes in substrate concentration. Presumably, the increase in substrate concentration following a shift from 30°C to 42°C is small, if hsps function near their capacity, since the concentration of major hsps increases only twofold. The proposed sensitivity of hsp function to substrate concentration may not be surprising if hsps are involved primarily in polymeric interactions.

The regulation of σ^{32} stability after a temperature upshift could be explained by a transient titration of DnaK and GrpE function, since analysis of *dnaK* and *grpE* mutants indicates that the hsps encoded by these genes are required for the normal degradation of σ^{32} both at low and high temperature. In contrast, the phenotype of a *dnaJ* mutant indicated that DnaJ is only required for the normal degradation of σ^{32} at low temperature. This conditional phenotype is not the result of residual DnaJ function at high temperature, since a *dnaJ* deletion mutant has the same conditional phenotype (D. Straus et al., unpubl.). Since DnaJ is not required for σ^{32} degradation at high temperature, saturation of DnaJ function cannot provide an explanation for increased σ^{32} stability during heat shock. However, DnaJ function could be a target of stimuli that induce hsps at low temperature by stabilizing σ^{32}.

Alternatively, induction of hsps might result from the direct modification of DnaK, DnaJ, or GrpE activity instead of the titration of their function. For example, it is known that DnaK can autophosphorylate (Zylicz et al. 1983) and that both DnaK and DnaJ affect the phosphorylation of at least one cellular protein (Wada et al. 1986). The phosphorylation state of DnaK might be altered upon temperature upshift, modifying the spectrum of proteins with which it interacts. In this scenario, the form of DnaK appearing immediately following temperature upshift would be unable to interact with the σ^{32}/proteinase complex. During this time, degradation of σ^{32} could not be facilitated by DnaK. A return to the normal state of DnaK phosphorylation would restore the rapid degradation of σ^{32} and shutoff of the heat shock response.

Since hsps are apparently not required for the regulation of σ^{32} synthesis at low temperature, it is unlikely that they are involved in the induction of σ^{32} synthesis after temperature upshift. In this case, it is more likely that the signal for heat shock acts by directly altering the regulator of *rpoH* mRNA translation rather than by titrating or modifying hsp function. If the mechanism for inducing σ^{32} synthesis is distinct from the mechanism for stabilizing σ^{32}, then some treatments that induce hsps could work by primarily altering σ^{32} synthesis, whereas others could primarily alter σ^{32} stability. At present, only the induction of hsps by heat and ethanol has been studied in detail, and in these two cases, σ^{32} synthesis and stability are co-regulated. Unlike induction, shutoff of σ^{32} synthesis does require hsp function which, as discussed for regulation of σ^{32} stability, may reflect changes in substrate concentration or direct modification of hsp function.

E. A Second System of Heat Shock Regulation?

The extensive studies of the σ^{32}-controlled hsps have been extremely productive but have tended to obscure the fact that there are a number of *E. coli* proteins induced after shift to high temperature that are not part of the σ^{32}-controlled regulon (Herendeen et al. 1979; Yamamori and Yura 1982; Neidhardt and VanBogelen 1987). The newly discovered minor σ, σ^{E}, which is active at high temperature, may control the expression of a number of these σ^{32}-independent heat shock genes. If this is the case, there may be a second heat-inducible regulon in *E. coli*.

The role of σ^{E} is not known, nor can we guess at the number of genes that it regulates. So far, two *E. coli* genes are known to be transcribed by $E\sigma^{E}$: *rpoH* (Erickson and Gross 1989) and the *htrA* gene (Lipinska et al. 1988a,b). *htrA* has recently been shown to be the same as *degP* (Lipinska et al. 1988b; Strauch et al. 1989), which encodes a periplasmic or membrane proteinase (Strauch et al. 1989). *htrA*/*degP* is required for viability only at high temperature since *htrA*::Tn*10* insertion mutants grow well at or below 37°C but do not survive at 42°C (Lipinska et al. 1988b; Strauch et al. 1989).

We are currently attempting to identify other genes under σ^{E} control and to obtain mutations affecting σ^{E} in order to identify the functions of this new set of heat-inducible genes. One possibility is that these proteins are the missing component of thermotolerance, but there are other possibilities, and the study of these genes should bring new insight into the molecules necessary for growth and survival at elevated temperatures.

V. PROSPECTS

A study of the heat shock response is poised to provide new insights concerning both the function and regulation of the heat shock response. An intensive study of the spectrum of protein-protein interactions mediated by GroEL, GroES, DnaK, DnaJ, and GrpE is now possible. At the present time, investigation of the role of DnaK and cohorts in λ replication provides the best mechanistic description of the role of these proteins. However, this system contains so many interacting proteins that the development of additional model systems is desirable. Most of these will come from a biochemical analysis of interactions inferred from in vivo studies. For example, a study of the interaction of GroEL with wild-type and mutant Ssb1 proteins, and the interaction of DnaK, DnaJ, and GrpE with the proteinase that degrades σ^{32} would be very informative. It is certainly likely that such studies will be carried out by a number of groups.

There is also a sufficient base of information concerning the regulation of the heat shock response to anticipate rapid progress in understanding how signal transduction occurs. Genetic approaches are being used to find additional genes involved in signal transduction. In particular, several groups are pursuing the identification of the translational repressor of σ^{32} synthesis and the identification of the proteinase that degrades σ^{32}, using both genetic and biochemical means. In addition, the tools exist to determine whether the phosphorylation state of DnaK influences its function in the regulation of the heat shock response. Taken together, these investigations should go a long way in refining our knowledge of how σ^{32} is regulated.

Finally, studies are just beginning on defining the role and regulation of Eσ^E. Future investigations will answer the question of whether Eσ^E-mediated gene expression is the missing link in thermotolerance. In addition, future experiments should help to identify other functions that are required for growth and survival at high temperature.

REFERENCES

Ang, D. and C. Georgopoulos. 1989. The heat shock-regulated *grpE* gene of *Escherichia coli* is required for bacterial growth at all temperatures but is dispensable in certain mutant backgrounds. *J. Bacteriol.* **171:** 2748–2755.

Bardwell, J.C.A. and E.A. Craig. 1984. Major heat shock gene of *Drosophila* and the *Escherichia coli* heat-inducible *dnaK* gene are homologous. *Proc. Natl. Acad. Sci.* **81:** 848–852.

———. 1987. Eukaryotic M_r-83,000 heat shock protein has a homologue in *Escherichia coli*. *Proc. Natl. Acad. Sci.* **84:** 5177–5181.

Bloom, M., S. Skelly, R. VanBogelen, F. Neidhardt, N. Brot, and H. Weissbach. 1986. *In vitro* effect of the *Escherichia coli* heat shock regulatory protein on expression of heat shock genes. *J. Bacteriol.* **166:** 380–384.

Bochkareva, E., N. Lissin, and A. Girshovich. 1988. Transient association of newly synthesized unfolded proteins with the heat-shock GroEL protein. *Nature* **336:** 254–257.

Bukau, B. and G. Walker. 1989. *ΔdnaK52* mutants of *Escherichia coli* have defects in chromosome segregation and plasmid maintenance at normal growth termperatures. *J. Bacteriol.* **171:** 6030–6038.

Chandrasekhar, G., K. Tilly, C. Woolford, R. Hendrix, and C. Georgopoulos. 1986. Purification and probes of the groES morphogenetic protein of *Escherichia coli. J. Biol. Chem.* **261:** 12414–12419.

Coppo, A., A. Manzi, J. Pulitzer, and H. Takahashi. 1973. Abortive bacteriophage T4 head assembly in mutants of *Escherichia coli. J. Mol. Biol.* **76:** 61–87.

Cowing, D.C. 1988. "Interaction of *Escherichia coli* RNA polymerase holoenzyme containing sigma 32 with promoters for heat shock genes." Ph.D. thesis, University of Wisconsin, Madison.

Cowing, D.W. and C.A. Gross. 1989. The interaction of *E. coli* RNA polymerase holoenzyme containing σ^{32} with heat shock promoters: DNase I footprinting and methylation protection. *J. Mol. Biol.* (in press).

Cowing, D.W., J.C.A. Bardwell, E.A. Craig, C. Woolford, R.W. Hendrix, and C. Gross. 1985. Consensus sequence for *Escherichia coli* heat-shock gene promoters. *Proc. Natl. Acad. Sci.* **80:** 2679–2683.

Dodson, M., R. McMacken, and H. Echols. 1989. Specialized nucleoprotein structures at the origin of replication of bacteriophage λ. *J. Biol. Chem.* **264:** 10719–10725.

Dodson, M., H. Echols, S. Wickner, C. Alfano, K. Mensa-Wilmont, B. Gomes, I. LeBowitz, J. Roberts, and R. McMaken. 1986. Specialized nucleoprotein structures at the origin of bacteriophage lambda: Localized unwinding of DNA by a six-protein reaction. *Proc. Natl. Acad. Sci.* **83:** 7638–7642.

Erickson, J. and C. Gross. 1989. Identification of the σ^E subunit of *Escherichia coli* RNA polymerase: A second alternate σ factor involved in high temperature gene expression. *Genes Dev.* **3:** 1462–1471.

Erickson, J.W., V. Vaughn, W. Walter, F.C. Neidhardt, and C.A. Gross. 1987. Transcription of the *rpoH* (*htpR*) gene in *Escherichia coli.* Identification of the promoters and evidence that the amount of *rpoH* mRNA increases after heat shock by a post-transcriptional mechanism. *Genes Dev.* **1:** 419–432.

Ezaki, B., T. Ogura, H. Mori, H. Niki, and S. Hiraga. 1989. Involvement of DnaK protein in mini-F plasmid replication: Temperature-sensitive *seg* mutations are located in the *dnaK* gene. *Mol. Gen. Genet.* **218:** 183–189.

Fayet, O., T. Ziegelhoffer, and C. Georgopoulos. 1989. The *groES* and *groEL* heat shock genes of *Escherichia coli* are essential for bacterial growth at all temperatures. *J. Bacteriol.* **171:** 1379–1385.

Fujita, N. and A. Ishihama. 1987. Heat shock induction of RNA polymerase sigma-32 synthesis in *Escherichia coli:* Transcriptional control and a multiple promoter system. *Mol. Gen. Genet.* **210:** 10–15.

Fujita, N., T. Nomura, and A. Ishihama. 1987. Promoter selectivity of *Escherichia coli* RNA polymerase. *J. Biol. Chem.* **262:** 1855–1859.

Georgopoulos, C.G., R.W. Hendrix, S.R. Casjens, and A.D. Kaiser. 1973. Host participation in bacteriophage lambda head assembly. *J. Mol. Biol.* **76:** 45–60.

Georgopoulos, C., R. Hendrix, A. Kaiser, and W. Wood. 1972. Role of the host cell in bacteriophage morphogenesis: Effects of a bacterial mutation on T4 head assembly. *Nature New Biol.* **239:** 38–41.

Goff, S.A. and A.L. Goldberg. 1985. Production of abnormal proteins in *E. coli* stimulates transcription of *lon* and other heat shock genes. *Cell* **41:** 587–595.

Grossman, A.D., J.W. Erickson, and C.A. Gross. 1984. The *htpR* gene product of *E. coli* is a sigma factor for heat-shock promoters. *Cell* **38:** 383–390.

Hemmingsen, S., C. Woolford, S. van der Vies, K. Tilly, D. Dennis, C. Georgopoulos, R. Hendrix, and R.J. Ellis. 1988. Homologous plant and bacterial proteins chaperone oligomeric protein assembly. *Nature* **333:** 330–334.

Hendrix, R.W. 1979. Purification and properties of GroE, a host protein involved in bacteriophage assembly. *J. Mol. Biol.* **129:** 375–392.

Herendeen, S.L., R.A. VanBogelen, and F.C. Neidhardt. 1979. Levels of major proteins of *Escherichia coli* during growth at different temperatures. *J. Bacteriol.* **139:** 185–194.

Itikawa, H. and I. Ryu. 1979. Isolation and characterization of a temperature-sensitive *dnaK* mutant of *Escherichia coli B. J. Bacteriol.* **138:** 339–344.

Johnson, C., G.N. Chandrasekhar, and C. Georgopoulos. 1989. The DnaK and GrpE heat shock proteins in *Escherichia coli* interact both *in vivo* and *in vitro*. *J. Bacteriol.* **171:** 1590–1596.

Kawasaki, Y., C. Wada, and T. Yura. 1989. Roles of *Escherichia coli* heat shock proteins DnaK, DnaJ and GrpE in mini-F plasmid replication. *Mol. Gen. Genet.* (in press).

Kelly, J. and L. Simon. 1988. Divergent effects of a *dnaK* mutation on abnormal protein degradation in *Escherichia coli*. *Mol. Microbiol.* **2:** 31–41.

Kusukawa, N. and T. Yura. 1988. Heat shock protein GroE of *Escherichia coli*: Key protective roles against thermal stress. *Genes Dev.* **2:** 874–882.

Landick, R., V. Vaughn, E.T. Lau, R.A. VanBogelen, J.W. Erickson, and F.C. Neidhardt. 1984. Nucleotide sequence of the heat shock regulatory gene of *E. coli* suggests its protein product may be a transcription factor. *Cell* **38:** 175–182.

Lemaux, P.G., S.L. Herendeen, P.L. Boch, and F.C. Neidhardt. 1978. Transient rates of synthesis of individual polypeptides in *E. coli* following temperature shifts. *Cell* **13:** 427–434.

Lesley, S., N. Thompson, and R. Burgess. 1987. Studies of the role of the *Escherichia coli* heat shock regulatory protein σ32 by the use of monoclonal antibodies. *J. Biol. Chem.* **262:** 5404–5407.

Liberek, K., C. Georgopoulos, and M. Zylicz. 1988. Role of the *Escherichia coli* DnaK and DnaJ heat shock proteins in the initiation of bacteriophage λ DNA replication. *Proc. Natl. Acad. Sci.* **85:** 6632–6636.

Lipinska, B., S. Sharma, and C. Georgopoulos. 1988a. Sequence analysis and regulation of the *htrA* gene of *Escherichia coli*: A σ^{32}-independent mechanism of heat-inducible transcription. *Nucleic Acids Res.* **16:** 10053–10066.

Lipinska, B., O. Fayet, L. Baird, and C. Georgopoulos. 1988b. Identification, characterization and mapping of the *Escherichia coli htrA* gene, whose product is essential for bacterial viability only at elevated temperatures. *J. Bacteriol.* **171:** 1547–1584.

McMullin, T. and R. Hallberg. 1988. A highly evolutionarily conserved mitochondrial protein is structurally related to the protein encoded by the *Escherichia coli groEL* gene. *Mol. Cell. Biol.* **8:** 371–380.

Neidhardt, F.C. and R.A. VanBogelen. 1981. Positive regulatory gene for temperature-controlled proteins in *Escherichia coli*. *Biochem. Biophys. Res. Commun.* **100:** 894–900.

———. 1987. The heat shock response. In Escherichia coli *and* Salmonella typhimurium (ed. F.C. Neidhardt), pp. 1334–1345. American Society for Microbiology, Washington, D.C.

Paek, K.H. and G.C. Walker. 1987. *Escherichia coli dnaK* null mutants are inviable at high temperature. *J. Bacteriol.* **169:** 283–290.

Pelham, H.R.B. 1986. Speculations on the functions of the major heat shock and glucose-regulated proteins. *Cell* **46:** 959–961.

Ruben, S., S. VanDenBrink-Webb, D. Rein, and R. Meyer. 1988. Suppression of the *Escherichia coli ssb-1* mutation by an allele of *groEL*. *Proc. Natl. Acad. Sci.* **85:** 3767–3771.

Siebenlist, U., R. Simpson, and W. Gilbert. 1980. *E. coli* RNA polymerase interacts homologously with two different promoters. *Cell* **20:** 269–281.

Skelly, S., T. Coleman, C.F. Fu, N. Brot, and H. Weissbach. 1987. Correlation between the 32-kDa σ factor levels and *in vitro* expression of *Escherichia coli* heat shock genes. *Proc. Natl. Acad. Sci.* **84:** 8365–8369.

Skelly, S., C.F. Fu, B. Dalie, B. Redfield, T. Coleman, N. Brot, and H. Weissbach. 1988. Antibody to σ32 cross-reacts with DnaK: Association of DnaK protein with *Escherichia coli* RNA polymerase. *Proc. Natl. Acad. Sci.* **85:** 5497–5501.

Sternberg, N. 1973. Properties of a mutant of *Escherichia coli* defective in bacteriophage λ head formation (*groE*). I. Initial characterization. *J. Mol. Biol.* **67:** 1–24.

Strauch, K., K. Johnson, and J. Beckwith. 1989. Characterizations of *degP*, a gene required for proteolysis in the cell envelope and essential for growth of *Escherichia coli* at high temperature. *J. Bacteriol.* **171:** 2689–2696.

Straus, D.B., W.A. Walter, and C.A. Gross. 1987. The heat shock response of *E. coli* is regulated by changes in the concentration of σ^{32}. *Nature* **329:** 348–351.

———. 1988. *Escherichia coli* heat shock gene mutants are defective in proteolysis. *Genes Dev.* **2:** 1851–1858.

Taylor, W.E., D.B. Straus, A.D. Grossman, Z.F. Burton, C.A. Gross, and R. Burgess. 1984. Transcription from a heat-inducible promoter causes heat shock regulation of the sigma subunit of *E. coli* RNA polymerase. *Cell* **38:** 371–381.

Tilly, K. and C. Georgopoulos. 1982. Evidence that the two *Escherichia coli groE* morphogenetic gene products interact *in vivo*. *J. Bacteriol.* **149:** 1082–1088.

Tilly, K. and M. Yarmolinsky. 1989. Participation of *Escherichia coli* heat shock proteins DnaJ, DnaK, and GrpE in P1 plasmid replication *J. Bacteriol.* **171:** 6025–6029.

Tilly, K., J. Spence, and C. Georgopoulos. 1989. Modulation of the stability of *Escherichia coli* heat shock regulatory factor σ^{32}. *J. Bacteriol.* **171:** 1585–1589.

Tilly, K., J. Erickson, S. Sharma, and C. Georgopoulos. 1986. Heat shock regulatory gene *rpoH* mRNA level increases after heat shock in *Escherichia coli*. *J. Bacteriol.* **168:** 1155–1158.

Tilly, K., N. McKittrick, M. Zylicz, and C. Georgopoulos. 1983. The dnaK protein modulates the heat shock response of *Escherichia coli*. *Cell* **34:** 641–646.

VanBogelen, R.A., M.A. Acton, and F.C. Neidhardt. 1987. Induction of the heat shock regulon does not produce thermotolerance in *Escherichia coli*. *Genes Dev.* **1:** 525–531.

Wada, M. and H. Itikawa. 1984. Participation of *Escherichia coli* K-12 *groE* gene products in the synthesis of cellular DNA and RNA. *J. Bacteriol.* **157:** 694–696.

Wada, C., M. Imai, and T. Yura. 1987. Host control of plasmid replication: Requirement for the σ factor σ^{32} in transcription of mini-F replication initiator gene. *Proc. Natl. Acad. Sci.* **84:** 8849–8853.

Wada, M., K. Sekine, and H. Itikawa. 1986. Participation of the *dnaK* and *dnaJ* gene products in phosphorylation of glutaminyl-tRNA synthetase and threonyl-tRNA synthetase in *Escherichia coli K12*. *J. Bacteriol.* **168:** 213–220.

Wada, C., Y. Akiyama, K. Ito, and T. Yura. 1986. Inhibition of F plasmid replication in *htpR* mutants of *Escherichia coli* deficient in sigma 32 protein. *Mol. Gen. Genet.* **203:** 208–213.

Williams, K., J. Murphy, and J. Chase. 1984. Characterization of the structural and functional defect in the *Escherichia coli* single-stranded DNA binding protein encoded by the *ssb-1* mutant gene. *J. Biol. Chem.* **259:** 11804–11811.

Yamamori, T. and T. Yura. 1980. Temperature-induced synthesis of specific proteins in *Escherichia coli*: Evidence for transcriptional control. *J. Bacteriol.* **142:** 843–851.

———. 1982. Genetic control of heat-shock protein synthesis and its bearing on growth and thermal resistance in *Escherichia coli K12*. *Proc. Natl. Acad. Sci.* **79:** 860–864.

Yamamori, T., K. Ito, Y. Nakamura, and T. Yura. 1978. Transient regulation of protein synthesis in *Escherichia coli* upon shift-up of growth temperature. *J. Bacteriol.* **134:** 1133–1140.

Yano, R., M. Imai, and T. Yura. 1987. The use of operon fusions in studies of the heat shock response: Effects of altered sigma 32 on heat shock promoter function in *Escherichia coli*. *Mol. Gen. Genet.* **207:** 24–28.

Yura, T., T. Tobe, K. Ito, and T. Osawa. 1984. Heat shock regulatory gene (*htpR*) of *Escherichia coli* is required for growth at high temperature but is dispensable at low temperature. *Proc. Natl. Acad. Sci.* **81:** 6803–6807.

Zhou, Y.N., N. Kusukawa, J. Erickson, C. Gross, and T. Yura. 1988. Isolation and characterization of *Escherichia coli* mutants that lack the heat shock sigma factor σ32. *J. Bacteriol.* **170:** 3640–3649.

Zweig, M. and D. Cummings. 1973. Cleavage of head and tail proteins during bacteriophage T5 assembly: Selective host involvement in the cleavage of a tail protein. *J. Mol. Biol.* **80:** 505–518.

Zylicz, M., D. Ang, and C. Georgopoulos. 1987. The grpE protein of *Escherichia coli*. *J. Biol. Chem.* **262:** 17437–17442.

Zylicz, M., D. Ang, K. Liberek, and C. Georgopoulos. 1989. Initiation of λ DNA replication with purified host- and bacteriophage-encoded proteins: The role of the DnaK, DnaJ., and GrpE heat shock proteins. *EMBO J.* **8:** 1601–1608.

Zylicz, M., J. LeBowitz, R. McMaken, and C. Georgopoulos. 1983. The dnaK protein of *Escherichia coli* possesses an ATPase and autophosphorylating activity and is essential in an *in vitro* DNA replication system. *Proc. Natl. Acad. Sci.* **80:** 6431–6435.

Zylicz, M., T. Yamamoto, N. McKittrick, S. Sell, and C. Georgopoulos. 1985. Purification and properties of the DnaJ replication protein of *Escherichia coli*. *J. Biol. Chem.* **260:** 7591–7598.

9

Properties of the *Escherichia coli* Heat Shock Proteins and Their Role in Bacteriophage λ Growth

Costa Georgopoulos and Debbie Ang
Department of Cellular, Viral and Molecular Biology
University of Utah School of Medicine
Salt Lake City, Utah 84132

Krzysztof Liberek and Maciej Zylicz
Division of Biophysics
Department of Molecular Biology
University of Gdansk
80-822 Kladki
Gdansk, Poland

I. INTRODUCTION

The history of the discovery of the heat shock response in *E. coli* in the laboratories of F. Neidhardt and T. Yura, as well as its regulation, has been ably outlined in the reviews by Neidhardt et al. (1985), Neidhardt and VanBogelen (1987), and Gross et al. (this volume). In this chapter, we concentrate on (1) a description of the biochemical properties of the known heat shock proteins, (2) an analysis of the roles that the heat shock proteins play in bacteriophage λ growth, (3) a summary of the known protein-protein interactions among the heat shock proteins, and (4) a model of heat shock gene regulation, based on the known biochemical properties of the heat shock proteins.

Stress Proteins in Biology and Medicine.

II. KNOWN HEAT SHOCK GENES AND THEIR PRODUCTS

Of the approximately 20 *E. coli* heat shock proteins under Eσ^{32} RNA polymerase transcriptional control, the identity of ten of the corresponding genes has been established during the past few years. Surprisingly, 5 of them were previously known in a different context, because certain mutations in their corresponding genes interfered with bacteriophage λ growth (for review, see Friedman et al. 1984; Neidhardt et al. 1985; Neidhardt and VanBogelen 1987; Lindquist and Craig 1988; Gross et al., this volume). Figure 1 shows the locations on the *E. coli* genetic map of the ten known Eσ^{32}-regulated heat shock genes. In addition, the *rpoH* (coding for σ^{32}) and the *htrA* genes are indicated because they have been shown to belong to a new heat shock regulon, expressed best at 50°C, and whose expression is under the control of the Eσ^{E} (σ^{24}) holoenzyme (Erickson et al. 1987; Lipinska et al. 1988a, 1989; Erickson and Gross 1989).

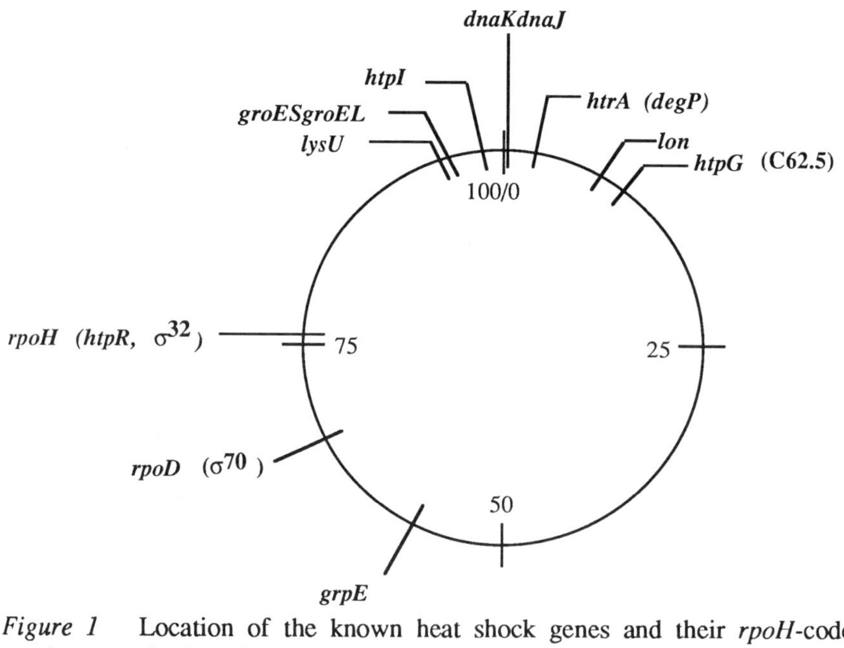

Figure 1 Location of the known heat shock genes and their *rpoH*-coded regulator on the *E. coli* genetic map. The *htrA (degP)* gene is included because its transcription recently has been shown to be also temperature-inducible, but under the regulation of the σ^{E} polypeptide (Lipinska et al. 1988a; 1989; C. Georgopoulos et al., with J. Erickson and C. Gross, unpubl.; for further details, see Gross et al., this volume).

Table 1 summarizes the current data on the biochemical properties of the products of the known Eσ^{32}-regulated heat shock genes. Although on the surface there seems to be no obvious unity of purpose for these gene products, there are some proteins that contribute synergistically to common functions, such as the autoregulation of the heat shock response, or to the control of similar intracellular processes, such as intracellular proteinase activity, protein folding, and protein assembly and disassembly (Gross et al., this volume; see below). Below, we summarize those properties of the heat shock proteins that we consider pertinent to this section.

A. The *dnaK*, *dnaJ*, and *grpE* Gene Products

The *dnaK, dnaJ*, and *grpE* genes were originally discovered because mutations in them were shown to block bacteriophage λ DNA replication (for review, see Friedman et al. 1984). Mutations in the λ*P* gene enable bacteriophage λ to replicate on hosts carrying these mutations, suggesting that the λP protein interacts with each of these three *E. coli* heat shock proteins (Friedman et al. 1984). This conclusion has been substantiated in vitro by demonstrating that the pairs of DnaK-λP and GrpE-λP purified proteins do indeed interact (Zylicz et al. 1983, 1987, 1989; D. Ang et al., unpubl.; see below).

Mutations in the *dnaK, dnaJ*, and *grpE* genes have also been shown to lead to similar phenotypes in bacterial physiology and regulation. These include (1) overproduction of all of the heat shock proteins, even at permissive temperatures (Tilly et al. 1983, 1989; Sell 1987; Straus et al. 1987; D. Straus and C. Gross, pers. comm.; C. Georgopoulos et al., unpubl.). This heat shock protein overproduction correlates with the stabilization of the otherwise unstable σ^{32} polypeptide (Tilly et al. 1989; Gross et al., this volume; D. Straus and C. Gross, pers. comm.). They also include (2) a block on host DNA and RNA syntheses at nonpermissive temperatures (Saito and Uchida 1977, 1978; Itikawa and Ryu 1979; Wada et al. 1982; Ang et al. 1986) and (3) a generalized defect in proteolysis (Keller and Simon 1988; Straus et al. 1988). For further discussion of the properties of this subset of the heat shock proteins, the reader is referred to the article by Gross et al. (this volume).

B. The DnaK Protein

The *dnaK* gene forms an operon with *dnaJ*, located at 0.3 minutes on the *E. coli* map (Fig. 1), with the order being promoter-*dnaK-dnaJ*. The

Table 1 Summary of the properties of the known $E\sigma^{32}$-transcribed heat shock gene products

Gene	Product	Relevant properties
dnaJ	dimer[a] of 40,973-M_r subunits (Bardwell et al. 1986; Ohki et al. 1986)	binds to DNA nonspecifically (Zylicz et al. 1985); interacts with DnaB[a] and DnaK[a]; found mostly in membrane fraction (Zylicz et al. 1985); necessary for cell division (Ohki et al. 1987); essential for growth but extragenic suppressors accumulate rapidly[a]
dnaK	mostly monomer[a] of 69,121-M_r (Bardwell and Craig 1984); some higher forms exist[a]	5′-nucleotidase activity (Bochner et al. 1986); autophosphorylates (Zylicz et al. 1983); interacts with λO, λP, GrpE, and DnaJ[a]; 48% identical to hsp70 of eukaryotes (Bardwell and Craig 1984); essential for growth; necessary for cell division (Bukau et al. 1989)
grpE	dimer[a] of 21,668-M_r subunits (Lipinska et al. 1988b)	interacts with DnaK (Zylicz et al. 1987; Johnson et al. 1989) and λP[a]; inhibits autophosphorylation of DnaK[a]; essential for *E. coli* growth at all temperatures (Ang and Georgopoulos 1989)
groES (*mopB*)	heptamer (Chandrasekhar et al. 1986; R. Hendrix and T. Ziegelhoffer, pers. comm.) of 10,368-M_r subunits	interacts with GroEL (Tilly and Georgopoulos 1982; Chandrasekhar et al. 1986); necessary for λ head-tail connector assembly[a] and T5 tail assembly (Zweig and Cummings 1973); assembly and disassembly of oligomeric proteins; essential for growth (Fayet et al. 1989)
groEL (*mopA*)	decatetramer (Hendrix 1978; Hohn et al. 1978) of 57,259 M_r subunits (Hemmingsen et al. 1988); sevenfold axis of symmetry (Hendrix 1979; Hohn et al. 1979)	46% identical to RuBisCO binding protein (Hemmingsen et al. 1988) and 54% identical to hsp60 of eukaryotes (Reading et al. 1989); identical to common antigen in bacterial infections (Young et al. 1988); necessary for assembly of λ, T4, and T5 (see above); interacts with GroES; binds to nascent polypeptide chains (Bochkareva et al. 1988) and unfolded proteins (Lecker et al. 1989); essential for growth (Fayet et al. 1989)
htpG	C62.5; dimer (Spence and Georgopoulos 1989) of 71,429-M_r (Bardwell and Craig 1987)	42% identical to eukaryotic hsp90 (Bardwell and Craig 1987); facilitates growth at high temperatures (Bardwell and Craig 1988)
htpI	D48.5 gene partly sequenced[b]	not essential for *E. coli* growth, except slightly at 46°C[b]
lon	tetramer of 87,000-M_r subunits (Chin et al. 1988)	ATP-dependent proteinase that degrades selected proteins such as SulA (Mizusawa and Gottesman 1983); not absolutely essential for bacterial growth (Maurizi et al. 1985)
lysU	lysyl-tRNA-synthetase (VanBogelen et al. 1983); gene partly sequenced[b]	cryptic form, not expressed at low temperatures (VanBogelen et al. 1983); not essential for growth[b]
rpoD	σ^{70}; monomer of 70,263-M_r (Burton et al. 1981)	binds to RNA polymerase core and helps transcribe the majority of *E. coli* genes; essential for growth

[a]For details, see Table 2 and text.
[b]F.C. Neidhardt (pers. comm.)

dnaK gene product has been shown to be approximately 50% identical, at the amino acid sequence level, to the eukaryotic hsp70 family of proteins (Bardwell and Craig 1984; Lindquist and Craig 1988). Mutations in the *dnaK* gene have been shown to result in a variety of interesting phenotypes. For example, *dnaK*⁻ mutants overexpress the heat shock proteins, suggesting that the DnaK protein is a negative modulator of the heat shock response (Tilly et al. 1983, 1989; Paek and Walker 1987; Johnson et al. 1989; C. Gross, pers. comm.). In agreement with this, it has been shown that overproduction of DnaK protein leads to a dampening of the heat shock response (Tilly et al. 1983). It has been shown that in the *dnaK*756 mutant background, the half-life of σ^{32} is increased approximately fivefold (Tilly et al. 1989; D. Straus and C. Gross, pers. comm.). This stabilization of σ^{32} (whose half-life fluctuates from approximately 45 sec at 42°C to 15 min at 22°C [Straus et al. 1987; Tilly et al. 1989]) results in higher intracellular levels of σ^{32}, leading to higher levels of heat shock gene expression (Grossman et al. 1987; Tilly et al. 1989). *dnaK*⁻ mutants also express a temperature-sensitive phenotype for both RNA and DNA syntheses (Saito and Uchida 1977; Itikawa and Ryu 1979; Sakakibara 1988). These phenotypes can be occasionally separated; e.g., Sakakibara (1988) has shown that the DNA replication defect of bacteria carrying the *dnaK*111 mutation is alleviated at the nonpermissive temperature by inactivating the *rnh* gene, coding for the RNase H enzyme. This result provides genetic evidence that intact DnaK protein is required at 42°C for the *initiation* of DNA replication, but not for *elongation* (Sakakibara 1988). Possibly in vivo the DnaK protein is required for one or more of the steps leading to initiation at the *E. coli* DNA replication origin, *oriC*. Such an involvement of DnaK could be at the level of assembly of the correct DNA/protein complex at *oriC*, stabilization of an unstable DNA replication protein at 42°C, or even assisting the conversion of the DnaA-ADP form to the active DnaA-ATP form (Sekimizu et al. 1987). Last, excessive filamentation at 30°C has been observed in bacteria carrying the *dnaK*52 deletion mutation (Paek and Walker 1987; Bukau and Walker 1989; Bukau et al. 1989). These filaments do not possess septa, suggesting that cell division is blocked at an early step (Bukau and Walker 1989; Bukau et al. 1989). Interestingly, overproduction of the FtsZ cell division protein leads to decreased filamentation (Bukau and Walker 1989; Bukau et al. 1989). It is not known whether the DnaK protein affects the activity or intracellular levels of the FtsZ protein.

Insight about DnaK protein function has come from studies with the purified protein, since a number of its properties have been identified. It has a nonspecific 5′-nucleotidase activity, with a preference for tri- and monophosphates and for purines over pyrimidines (Bochner et al. 1986). This 5′-nucleotidase activity is inhibited by the alarmone ApppppA, which accumulates following a temperature shift (Lee et al. 1983; Bochner et al. 1986). The mutant DnaK756 protein has been shown to be (1) less active against triphosphates, (2) more active against di- and monophosphates, and (3) not inhibited by ApppppA (Bochner et al. 1986). These results may explain why *dnaK*756 bacteria overproduce the heat shock proteins at all temperatures tested (Tilly et al. 1983). The ability of the $DnaK^+$ and DnaK756 proteins to bind to ATP-agarose has considerably aided their purification (Zylicz et al. 1987). The ATPase activity of the DnaK protein is (1) extremely stable to heating (Zylicz and Georgopoulos 1983), (2) inhibited by the λP protein (Zylicz et al. 1983), and (3) stimulated by the λO protein (Zylicz et al. 1983).

The DnaK protein also possesses the ability to autophosphorylate. It has been shown that a small fraction (1–5%) of the purified DnaK protein autophosphorylates at a threonine residue in vitro following incubation with ATP (Zylicz et al. 1983). In addition, the autophosphorylation activity is stimulated at least tenfold by Ca^{++} ions (A. Cegielska and C. Georgopoulos, in prep.), and the mutant DnaK756 protein exhibits substantially reduced levels of autophosphorylating activity (A. Cegielska and C. Georgopoulos, in prep.). The in vitro autophosphorylation activity of DnaK protein is inhibited by purified GrpE protein, whereas autophosphorylation of the DnaK756 protein appears to be refractory to the inhibition of GrpE (unpublished data of C. Georgopoulos with A. Cegielska and C. Johnson). Recently, it has been shown that the DnaK protein is capable of in vitro autophosphorylation using GTP, this occurring also at a threonine residue (H. Weissbach, pers. comm.; C. Georgopoulos with A. Cegielska, unpubl.). The mutant DnaK756 protein, which is also stimulated by Ca^{++} (unpublished data of C. Georgopoulos with A. Cegielska), can also be autophosphorylated by GTP, albeit at reduced levels. Previous work has shown that a small percentage of the total DnaK protein in vivo is also phosphorylated (Zylicz et al. 1983; Zylicz and Georgopoulos 1984). Surprisingly, a more recent report has shown that the in vivo phosphorylation of the DnaK protein occurs primarily at a serine residue, but, following M13 phage infection, the extent of phosphorylation is not only increased but is primarily detected at a threonine residue (Rieul et al. 1987). The relationship of the low level of DnaK phosphorylation in vivo to any of its biological properties remains elusive. In this regard, it is interesting to note that Wada et al. (1986) have shown that in *dnaK*⁻ or *dnaJ*⁻ bacteria, as opposed to wild type, the threonyl- and glutaminyl-tRNA synthetases are not phosphorylated. The direct or indirect role that the DnaK and DnaJ proteins play in this phosphorylation is not clear. However, purified DnaK protein has not been shown to phosphorylate the purified threonyl- and glutaminyl-tRNA synthetases in vitro (C. Georgopoulos with C. Johnson, unpubl.).

A potential role for DnaK is the protection or renaturation of other proteins during heat shock (Pelham 1986). This is supported by the preferential binding of DnaK to the unfolded form of bovine pancreatic trypsin inhibitor (BPTI) as opposed to its native form (Johnson 1988; C. Georgopoulos with C. Johnson and A. Cegielska, unpubl.). In this regard, it is interesting that eukaryotic homologs of DnaK can also bind to aberrant polypeptides (Kassenbrock et al. 1988). In addition, they help proteins translocate across membranes, presumably by maintaining them in an unfolded state (Chirico et al. 1988; Deshaies et al. 1988a,b). This activity of the DnaK protein is elegantly manifested in its role in λ DNA replication (see below), and the DnaK protein binds specifically to the same eukaryotic protein, p53, as its eukaryotic hsc70 homolog (Clarke et al. 1988).

The DnaK protein appears to be essential for *E. coli* growth, at least at high temperature (Georgopoulos 1977). The *dnaK* gene can be deleted in *E. coli* (Paek and Walker 1987), but the resulting strains grow very poorly and only within a narrow temperature range centered around 30–37°C. These strains also often accumulate extragenic suppressors (Bukau and Walker 1989; Bukau et al. 1989). We have also shown that many missense mutations in the *dnaK* gene, isolated as unable to propagate phage λ (G. Gaitanaris and M. Gottesman, pers. comm.), rapidly accumulate extragenic suppressors that allow them to grow better in the 30–37°C temperature range (our unpublished data). Surprisingly, we have recently found that many of these *dnaK*⁻ missense mutations, as well as the *dnaK*52 deletion (Paek and Walker 1987), can be readily transferred into an otherwise wild-type background, provided that the transductions are performed in M9 minimal media (C. Georgopoulos and D. Ang, unpubl.). The *dnaK*⁻ bacterial strains constructed in this way readily accumulate extragenic suppressor mutations that allow them to grow on rich media (C. Georgopoulos and D. Ang, unpubl.).

The DnaK protein is also essential for cell viability during starvation for carbon. Analysis of proteins induced under starvation conditions, such as during stationary phase, shows that the *dnaK*103 mutant is defective for the synthesis of a subset of the proteins induced by the wild-type parent strain under these conditions, with the concomitant induction of other proteins not seen with the parent strain (Spence 1988). These results suggest a more fundamental role for the DnaK protein in cellular physiology than previously suspected.

C. The DnaJ Protein

Mutations in the *dnaJ* gene result in pleiotropic effects on cell growth and metabolism, in many ways similar to those exhibited by mutations in the *dnaK* gene. It is perhaps not surprising, then, that these two genes constitute an operon (Yochem et al. 1978). The *dnaJ* gene can be insertionally inactivated or deleted in *E. coli*, resulting in growth only at low temperatures. *E. coli* with an inactivated *dnaJ* gene can grow at high temperatures (above 30°C) only in the presence of rapidly accumulating extragenic suppressors (Sell 1987; C. Georgopoulos et al., unpubl.). P. Kang and E. Craig (pers. comm.) have shown that the temperature-sensitive phenotype of a *dnaKdnaJ*-deleted *E. coli* strain can be suppressed by the acquisition of multiple copies of a gene(s) located at 3.7 minutes on the *E. coli* genetic map. The mechanism by which this overproduction compensates for bacterial viability is not known. In principle, this gene could code for a product that performs a function analogous to those of the DnaK and DnaJ proteins.

The *dnaJ* gene has been sequenced (Bardwell et al. 1986; Ohki et al. 1986). It has been shown that similar proteins are made in both gram-positive and other gram-negative bacteria (Lathigra et al. 1988). The predicted amino acid sequence of the protein includes a 16-amino-acid segment duplicated tandemly four times. The duplicated sequence contains the segment Cys-X-X-Cys-X-Gly-X-Gly (Bardwell et al. 1986; Ohki et al. 1986). The biological significance, if any, of this conserved sequence is not known at present. The *dnaJ* mutants exhibit a plethora of phenotypes including (1) overproduction of the heat shock proteins (Sell 1987), presumably as a consequence of stabilization of the unstable σ^{32} polypeptide (D. Straus and C. Gross, pers. comm.), (2) extensive filamentation, even at permissive temperatures (Sell 1987; P. Kang and E. Craig, pers. comm.) (it is not known whether this filamentation can be suppressed by high levels of FtsZ protein, as in the case of the *dnaK*-deleted strain [Bukau and Walker 1989; Bukau et al. 1989]), (3) a block to both host DNA and RNA syntheses at the nonpermissive temperature (Saito and Uchida 1978; Wada et al. 1982), and (4) an arrest of the cell cycle at a defined stage, prior to the synthesis of a specific subset of proteins, similar to the effect of a *divE* mutation (Ohki et al. 1987).

The DnaJ protein has been purified to homogeneity and shown to possess the following properties: (1) It is a basic, dimeric protein (Zylicz et al. 1985); (2) it is localized mostly in the bacterial membrane (Zylicz et al. 1985); (3) it binds to both single- and double-stranded DNA with no apparent sequence specificity; (4) it interacts with the DnaK protein, since it binds to a DnaK-affinity column, and mutations at or near the *dnaK* gene compensate for the temperature-sensitive phenotype of the *dnaJ*259 mutation (Sell 1987); (5) it interacts with the *E. coli* DnaB replication protein, as judged by affinity chromatography (K. Liberek et al., unpubl.); (6) it interacts weakly with the λP protein (Table 2), as judged by both affinity chromatography (M. Zylicz et al., unpubl.) and genetic studies (Sunshine et al. 1977; Saito and Uchida 1978); and (7) it associates specifically with the prophage P1 replication protein, RepA (S. Wickner, pers. comm.).

D. The GrpE Protein

The *grpE* gene was originally defined because the only known missense mutation in it, *grpE*280, blocks λ DNA replication (Saito and Uchida 1977). Subsequently, it was shown that the GrpE protein is not only a heat shock protein, but also essential for bacterial growth at all temperatures (Ang et al. 1986; Ang and Georgopoulos 1989). The phenotype at the nonpermissive temperature includes a block to both DNA and RNA syntheses and overproduction of heat shock proteins (Ang et al. 1986; Sell 1987). These phenotypes are reminiscent of those found in *dnaK* and *dnaJ* bacterial mutants. D. Straus and C. Gross (pers. comm.) have shown that the *grpE*280 mutation results in an increase in the half-life of the σ^{32} polypeptide, as well as other aberrant polypeptides (see Gross et al., this volume).

The essentiality of the *grpE* gene product for *E. coli* viability has recently been addressed (Ang and Georgopoulos 1989). It has been found that a deleted version of the *grpE* gene can be tolerated in *E. coli* only in the presence of another copy of the *grpE* gene located elsewhere in the cell. This result has been obtained at temperatures ranging from 18° to 43°C, as well as on minimal M9 medium (Ang and Georgopoulos 1989), demonstrating that the *grpE* gene product is essential for wild-type *E. coli* viability. Surprisingly, we have found that the *grpE* gene can be deleted in certain mutant *E. coli* strains. These *E. coli* strains possess extragenic suppressors, enabling them to tolerate the presence of certain *dnaK*$^-$ mutations (Ang and Georgopoulos 1989). The loss of *grpE* in these strains was demonstrated by showing the complete absence of GrpE antigen in these cells. It appears, then, that the same mutations that enable *E. coli* to tolerate certain defects in the *dnaK* gene also suppress the need for GrpE protein. Since the DnaK and GrpE proteins functionally interact both in vitro (Zylicz et al. 1987; Johnson et al. 1989) and in vivo (Johnson et al. 1989), this result is perhaps not surprising. The identity or nature of these extragenic suppressors is not known.

The in vivo evidence supporting a GrpE/DnaK interaction consists of

Table 2 Interactions among λ- and *E. coli*-encoded proteins required for initiation of λ DNA replication in vitro

	*ori*λ	λO	λP	DnaB	DnaK	DnaJ	GrpE
*ori*λ		1					
λO		2	3		4		5
λP			6	7	8	9	10
DnaB				11		12	
DnaK					13	14	15
DnaJ						16	
GrpE							17

1. The amino terminus of λO binds preferentially to DNA fragments containing the 19-bp repeats found in *ori*λ (Tsurimoto and Matsubara 1981).

2. λO sediments as a dimer under native conditions (Tsurimoto and Matsubara 1981; Zylicz et al. 1984).

3. Mutations in λP can suppress a λ*O*ts mutation (Tomizawa 1971). In hybrid λO proteins, the carboxy-terminal region determines which lambdoid P protein can be used for replication (Furth and Yates 1978). A λO-λP complex can be detected in metrizamide gradients (Zylicz et al. 1984) and can be isolated by gel filtration (Wickner and Zahn 1986).

4. λO protein stimulates DnaK 5′-nucleotidase activity in vitro (Zylicz et al. 1983; unpubl.). λO binds to a DnaK-affinity column and is eluted at 0.6 M KC1 (Liberek et al. 1988 and unpubl.). A λO/DnaK complex can be isolated by glutaraldehyde cross-linking and by gel filtration (Liberek et al. 1988; M. Zylicz et al., unpubl.).

5. Preliminary experiments suggest that a λO/GrpE complex can be isolated by glutaraldehyde cross-linking (M. Zylicz et al., unpubl.)

6. λP sediments as a dimer in metrizamide gradients under native conditions, but sediments as a monomer in glycerol gradients (Zylicz et al. 1984).

7. Active λP protein can be copurified with DnaB as a complex (Wickner 1978; Klein et al. 1980). λP inhibits the DNA-dependent ATPase activity of DnaB in vitro (Wickner 1978). λπ mutants that suppress the GroP⁻ phenotype in *E. coli dnaB*⁻ mutants can be isolated (Georgopoulos and Herskowitz 1971; Saito and Uchida 1977).

8. λP inhibits DnaK 5′-nucleotidase activity in vitro (Zylicz et al. 1983). λπ mutants can be isolated that propagate on the *E. coli dnaK*756 mutant (Georgopoulos 1977). λP binds to a DnaK-affinity column and is partially eluted in 10 mM ATP (Liberek et al. 1988; M. Zylicz et al., unpubl.). λP/DnaK complexes can be isolated by glutaraldehyde cross-linking and by gel filtration (M. Zylicz et al., unpubl.).

9. λπ mutants able to propagate on *E. coli dnaJ*259 can be isolated (Sunshine et al. 1977). In preliminary experiments, λP binds to a DnaJ-affinity column, and is eluted in 50 mM KC1, suggesting a very weak interaction (M. Zylicz et al., unpubl.).

10. λπ mutants able to propagate on *E. coli grpE*280 can be isolated (Saito and Uchida 1977). A λP/GrpE complex can be isolated by glutaraldehyde cross-linking (M. Zylicz et al., unpubl.).

11. DnaB forms a hexamer under native conditions (Reha-Krantz and Hurwitz 1978; Arai et al. 1981).

12. Preliminary experiments suggest that DnaB binds to a DnaJ-affinity column and is eluted by 0.2 M KCl (M. Zylicz et al., unpubl.).

13. DnaK behaves mostly like a monomer in solution. However, larger forms can be detected in size columns and following cross-linking with glutaraldehyde when the concentration of DnaK protein is increased (M. Zylicz et al., unpubl.).

14. DnaJ binds to a DnaK-affinity column and is eluted by 0.5 M KCl (Sell 1987). Some extragenic suppressors of *dnaJ*259 map in or near the *dnaK* gene (Sell 1987).

15. DnaK and GrpE are co-immunoprecipitated by either anti-DnaK or anti-GrpE antibodies (Ang et al. 1987; Johnson et al. 1989). GrpE binds to a DnaK-affinity column, is not eluted in 2 M KC1, but is eluted in 10 mM ATP. GrpE cosediments with DnaK in glycerol gradients except in the presence of ATP (Zylicz et al. 1987). A DnaK/GrpE complex can be isolated by glutaraldehyde cross-linking (Ang 1988; M. Zylicz et al., unpubl.). A DnaK/GrpE280 complex can also be isolated, but this complex is less stable (Ang 1988). Some Tr⁺ revertants of *E. coli grpE*280 have mutations that map in the *dnaK* gene (Johnson et al. 1989). Purified GrpE inhibits autophosphorylation of purified DnaK (C. Georgopoulos et al., with A. Cegielska and C. Johnson, unpubl.).

16. Under native conditions, DnaJ acts as a dimer (Zylicz et al. 1985).

17. In glycerol gradients, GrpE behaves as a monomer (Zylicz et al. 1987). However, dimers can be isolated by glutaraldehyde cross-linking in the absence of glycerol (Ang 1988; M. Zylicz et al., unpubl.). Under native conditions, ^{35}S-labeled grpE behaves as an aggregate of undetermined size by gel filtration (M. Zylicz et al., unpubl.).

the isolation of mutations in the *dnaK* gene that suppress the temperature-sensitive phenotype of the *grpE*280 mutation (Johnson et al. 1989). The in vitro evidence supporting this interaction consists of (1) the specific binding of the GrpE protein to a DnaK-affinity column (this interaction is resistant to 2 M KCl, but can be disrupted by ATP [Zylicz et al. 1987]); (2) the coprecipitation of wild-type GrpE and DnaK proteins in the presence of either anti-GrpE or anti-DnaK antibodies. Interestingly, this coprecipitation does not occur between the $DnaK^+$/GrpE280 and DnaK756/$GrpE^+$ pairs of proteins but is reinstated in the DnaK9/GrpE280 pair of proteins (the *dnaK*9 mutation was isolated as suppressing the *grpE*280 temperature-sensitive phenotype [Johnson et al. 1989]); (3) isolation of a $GrpE^+$/$DnaK^+$ complex following glutaraldehyde cross-linking or in glycerol gradients (Zylicz et al. 1987; D. Ang et al., unpubl.); and (4) inhibition of the $DnaK^+$, but not DnaK756, autophosphorylation activity by purified GrpE protein (C. Georgopoulos with A. Cegielska and C. Johnson, unpubl.).

The GrpE and λP proteins have been shown to interact both in vivo and in vitro. The in vivo evidence consists of the isolation of mutations in the λ*P* gene that compensate for the λ DNA replication block exerted by the *grpE*280 mutation (Saito and Uchida 1977; Ang et al. 1986). The in vitro evidence consists of glutaraldehyde cross-linking experiments between the $GrpE^+$/λP pair of proteins (D. Ang et al., unpubl.). The formation of the $GrpE^+$/λP complex is resistant to the presence of ATP. This property is important in understanding the role that GrpE plays in λ DNA replication (see below, section on λ DNA replication).

The purified GrpE protein behaves like a monomer in glycerol gradient centrifugation experiments (Zylicz et al. 1987), but both $GrpE^+$ and GrpE280 proteins behave like dimers in glutaraldehyde cross-linking experiments in the absence of glycerol (Ang 1988). The GrpE280 protein also forms a complex with the DnaK protein, but, unlike the $GrpE^+$/DnaK complex, this GrpE280/DnaK complex appears to be extremely unstable (Ang 1988; D. Ang et al., unpubl.).

Under normal growth conditions, the bulk of the *grpE* gene transcription is carried out by the $E\sigma^{32}$ holoenzyme in vivo (Lipinska et al. 1988b). However, a substantial amount of *grpE* transcription must be carried out by other forms of the RNA polymerase holoenzyme, since substantial amounts of GrpE protein antigen (up to 50%) can be detected in an *E. coli* strain carrying a deletion in the *rpoH* gene (D. Ang and C. Georgopoulos, unpubl.). Transcription may start at an $E\sigma^{70}$ holoenzyme consensus promoter that is found within the $E\sigma^{32}$-recognized *grpE* promoter (Lipinska et al. 1988b). For the *groE* operon, there is a similar arrangement of $E\sigma^{32}$- and $E\sigma^{70}$-recognized promoters, and the $E\sigma^{32}$-

recognized promoter is again almost exclusively used under normal growth conditions (Zhou et al. 1988).

E. The GroE Proteins

The *groE* locus of *E. coli* was originally identified because mutations in it block phage λ and T4 head morphogenesis (Georgopoulos et al. 1972, 1973; Takano and Kakefuda 1972; Coppo et al. 1973; Sternberg 1973b). It was subsequently shown that mutations in the *groE* locus block phage T5 morphogenesis (Georgopoulos et al. 1973), but at the level of tail assembly (Zweig and Cummings 1973). Following this early work, a careful genetic analysis showed that the *groE* locus is made up of two genes, the order being promoter-*groES*-*groEL* (Tilly et al. 1981; Tilly 1982). The *groE* genes form an operon under $E\sigma^{32}$ regulation (Cowing et al. 1985), although a second promoter, located immediately downstream, can be efficiently utilized by the $E\sigma^{70}$ holoenzyme in the absence of the σ^{32} polypeptide (Zhou et al. 1988). A recent genetic study has demonstrated that both the *groES* and *groEL* gene products are essential for *E. coli* viability at all temperatures (Fayet et al. 1989).

Mutations in either the *groES* or *groEL* genes exert similar phenotypes with respect to *E. coli* physiology and phage infection. For example, mutations in either gene can block λ head assembly and T5 tail assembly. However, only mutations in the *groEL* gene appear to block T4 head assembly, implying that the *groES* gene product may not be essential for T4 head morphogenesis (Tilly and Georgopoulos 1982; C. Georgopoulos, unpubl.). A block to λ head assembly by mutations in either the *groES* or *groEL* gene can be bypassed by mutations in the λ*B* gene (Georgopoulos et al. 1973; Sternberg 1973b), suggesting that the λB protein and the GroE proteins functionally interact. In the case of T4, mutations in gene *31* can bypass the block to phage head assembly imposed by the *groEL*⁻ mutations (Georgopoulos et al. 1972; Takano and Kakefuda 1972; Coppo et al. 1973; Revel et al. 1980), implying that the 31 and GroEL proteins functionally interact. However, no physical evidence for such interaction exists yet. Mutations in either the *groEL* or *groES* gene also result in (1) a block to both DNA and RNA syntheses at the nonpermissive temperature (Sternberg 1973a; Wada and Itikawa 1984); (2) a block to cell division, resulting in the formation of long filaments without septa (Georgopoulos and Eisen 1974; Miki et al. 1988); and (3) an overall reduction in generalized proteinase activity in *E. coli* (Straus et al. 1988). This result suggests that the GroE proteins are indirectly involved in proteinase activity, either by properly "presenting" the polypeptides to proteinases, by interacting with the proteinases directly, or both. Such a regulation of protease activity is also exhibited by the DnaK, DnaJ, and GrpE proteins (Straus et al. 1988; see above and Gross et al., this volume).

The ample genetic and biochemical evidence that the GroES and GroEL proteins functionally interact helps explain why mutations in either the *groES* or *groEL* gene often result in similar phenotypes. The genetic evidence consists of the isolation of missense mutations in the *groEL* gene that compensate for the temperature-sensitive phenotype of certain *groES*$^{-}$ mutations in an allele-specific manner (Tilly and Georgopoulos 1982). The biochemical evidence consists of (1) the demonstration that the two purified proteins cosediment in the presence of ATP (Chandrasekhar et al. 1986), (2) the inhibition of the weak ATPase activity of the GroEL protein by the GroES protein (Chandrasekhar et al. 1986), and (3) the specific retention of GroES protein on a GroEL affinity column (Chandrasekhar et al. 1986).

F. The GroEL Protein

The GroEL protein and its homologs appear not only to be conserved throughout evolution, but also to perform analogous functions in all organisms tested so far. The GroEL protein is 54% identical at the amino acid sequence level to yeast hsp60, a heat shock protein coded by the nucleus but found associated with mitochondria (McMullin and Hallberg 1987, 1988; Reading et al. 1989). It is also 48% identical to the RuBisCO (ribulose-1,5-bisphosphate carboxylase-oxygenase) binding protein, coded also by the nucleus but present in plant chloroplasts (Hemmingsen et al. 1988). Because of the demonstrated ability of the GroEL-like proteins to promote oligomerization and assembly of other proteins, the term "chaperonin" has been proposed to describe their function (Hemmingsen et al. 1988). The RuBisCO binding protein has been shown to be transiently required for the assembly of the RuBisCO enzyme in chloroplasts (for review, see Hemmingsen et al. 1988). The GroEL protein of *E. coli* has also been shown to promote the assembly of a prokaryotic RuBisCO enzyme (composed of 16 subunits) in *E. coli* (Goloubinoff et al. 1989). For this RuBisCO assembly to be effective, the GroES protein of *E. coli* is required (Goloubinoff et al. 1989), again demonstrating that the two GroE proteins function synergistically in *E. coli*. It has been demonstrated that the hsp60 protein, which is imported into mitochondria, is the product of an essential gene in *Saccharomyces* required for the correct assembly of other proteins that are transported into the mitochondria. For example, the β subunit of the F_1-ATPase enzyme is properly translocated into mitochondria and processed to its mature form, but fails to assemble into the proper complex in the absence of functional hsp60 protein (Cheng et al. 1989). Thus, the GroEL-like proteins of prokaryotes and eukaryotes carry out similar functions.

The GroEL protein is made up of 14 identical subunits arranged in a sevenfold axis of symmetry (Hendrix 1979; Hohn et al. 1979). The dimensions of the ring-shaped particle are 125–130 Å in diameter and 100–115 Å in height (Hendrix 1979; Hohn et al. 1979). The dissociation of the GroEL protein into dimers in the presence of metrizamide and ATP suggests that the first step in the assembly of the mature enzyme may be the formation of a GroEL dimer whose subsequent oligomerization gives rise to the final decatetrameric structure. It may be relevant that the building block of the octameric large subunit of RuBisCO is also a dimer (Goloubinoff et al. 1989).

A second function for the GroEL protein appears to be the binding of other unfolded polypeptide chains. Perhaps such binding is an integral first step in the chaperonin function discussed above. For example, Bochkareva et al. (1988) have shown, through cross-linking experiments, that the *E. coli* GroEL protein associates with the unfolded forms (as the nascent chains emerge from the ribosome) of both the chloramphenicol acetyltransferase and β-lactamase enzymes. This association is disrupted in the presence of ATP, but not in the presence of its nonhydrolyzable derivatives (Bochkareva et al. 1988). The GroEL protein does not appear to associate with the folded forms of these polypeptides (Bochkareva et al. 1988). Recently, it has been demonstrated that the GroEL protein is a "trigger" factor, binding to the pro-ompA and pre-phoE polypeptides. Such binding presumably maintains the polypeptides in an unfolded state, thus allowing them to pass through the *E. coli* cytoplasmic membrane (Lecker et al. 1989). Assuming that this role of the GroEL protein has been conserved during evolution, it may explain why the eukaryotic GroEL analog, hsp60, is imported into mitochondria, e.g., it could be that the hsp60 protein not only helps to refold and oligomerize proteins imported into mitochondria, but may also be involved in the subsequent transport of proteins from the mitochondrial lumen into the intermembrane space.

Clues about the role of GroEL in assembly of other *E. coli* proteins are also provided by genetic studies. (1) A *groEL*411 mutation has been shown to suppress the temperature-sensitive defect of the *ssb*1 (*ssb* codes for the single-stranded DNA binding protein essential for *E. coli* DNA replication), but not that of the *ssb*113, mutation (Ruben et al. 1988). A likely explanation is that the GroEL411 protein helps the mutant Ssb1 polypeptides to tetramerize properly at the nonpermissive temperature. This property is not found for the wild-type GroEL protein, even at overproducing levels (Ruben et al. 1988). Determining the mechanism by which the *groELA*11 mutation causes the protein to acquire a new function will contribute toward our understanding of GroEL's mechanism of action. (2) The finding that mutations in the *rpoA* gene, coding for the α subunit of RNA polymerase, can suppress the temperature-sensitive phenotype of certain *groES*⁻ mutations suggests a potential role for the GroE proteins in the assembly of the RNA polymerase core enzyme (Wada et al. 1987). (3) Overproduction of both GroE proteins in *E. coli* has been shown to suppress the temperature-sensitive phenotype of certain *dnaA*⁻ mutations of *E. coli* (Fayet et al. 1986; Jenkins et al. 1986). The temperature-sensitive phenotype of an amber *dnaA* mutation is not suppressed, so the GroE suppression is not simply a bypass of DnaA protein's function in the initiation of DNA replication. The intracellular concentration of the mutant DnaA polypeptides does not change under the GroE-suppressing conditions (C. Georgopoulos et al., with O. Fayet, unpubl.), so it probably simply allows the correct assembly of the mutant DnaA polypeptides at *oriC*. In support of this interpretation, it has been recently shown that overproduction of both GroES and GroEL results in the suppression of the Tr⁻ phenotype of 20–30% of random, temperature-sensitive alleles in various genes (R. LaRossa, pers. comm.). (4) Mutations in either of the *groE* genes lead to a nonmutable phenotype under SOS-inducible conditions (Bukau et al. 1989), suggesting that the GroE proteins may be important in assembling the DNA replication machinery operating under SOS-inducible conditions.

G. The GroES Protein

The GroES protein in *E. coli* consists of a heptamer of 97-amino-acid polypeptides (Chandrasekhar et al. 1986; C. Georgopoulos et al., with R. Hendrix and T. Ziegelhoffer, unpubl.). This conclusion is based on the chromatographic and sedimentation behavior of GroES, as well as cross-linking experiments with the purified protein. Mutations in the *groES* and *groEL* genes cause many of the same phenotypes, so their corresponding gene products may act together to carry out many of their intracellular functions. Since the GroES and GroEL proteins interact only in the presence of ATP (Chandrasekhar et al. 1986), it could be that GroES is important in removing from GroEL the attached, denatured polypeptide chains observed by Bockhareva et al. (1988). Recent experiments by Miki et al. (1988) suggest that the GroES protein of *E. coli* may be one of the targets of the lethal action of the F-factor-coded CcdB protein. The evidence consists of the isolation of mutations in the *groES* gene that dampen the lethal effect of the CcdB protein (Miki et al. 1988).

The GroES protein appears to be essential for *E. coli* growth (Fayet et al. 1989), for the proper assembly of the λ head (see below), and for the assembly of a functional T5 tail structure (Zweig and Cummings 1973). It does not appear to be essential for the correct assembly of T5 proheads (Zweig and Cummings 1973) or T4 proheads (C. Georgopoulos, unpubl.).

Although the GroES protein appears to have been conserved among prokaryotic organisms (Baird et al. 1988; Vodkin and Williams 1988), there is no evidence yet that it has also been conserved in eukaryotes.

III. ROLE OF THE HEAT SHOCK PROTEINS IN PHAGE λ GROWTH

A. Role in Morphogenesis

A key role of the host in phage λ morphogenesis was discovered through the isolation of mutants, called *groE*$^-$, that allow λ DNA injection, DNA replication, late gene expression, and cell lysis, but block phage assembly (Georgopoulos et al. 1973; Sternberg 1973b). The block was shown to be specifically at the level of proper prohead assembly since (1) no functional proheads accumulated, (2) abnormal head structures were clearly visible under the electron microscope, and (3) phage mutations able to bypass the block map either in the *E* gene (which codes for the major capsid protein) or *B* gene (Georgopoulos et al. 1973; Sternberg 1973b, 1976; Murialdo 1979). Infection of *groE*$^-$ host bacteria by wild-type phage λ was shown, in many respects, to be equivalent to infection of wild-type bacteria by phage λ*B*$^-$ mutants (for review, see Georgopoulos et al. 1983; Friedman et al. 1984).

The specific step at which the GroEL protein is required for λ prohead assembly was shown by Kochan and Murialdo (1983) to be the correct assembly of the λ head-tail connector structure, a dodecamer composed entirely of λB polypeptides. This structure is essential for the correct assembly of the λ prohead, but remains attached to the tail following mild dissociation of the λ virion (Tsui and Hendrix 1980). The head-tail connector has a donut-like structure when viewed from the top, with an overall diameter of 15 nm and a 2-nm central hole (Kochan et al. 1984). The hole is large enough to allow passage of the λ DNA chromosome.

Figure 2 shows an outline of the pathway used to generate the λB head-tail connector structure. The participation of the host-coded GroEL protein and the phage-coded λNu3 protein has been proven by Kochan and Murialdo (1983). The exact role that λNu3 plays in the correct assembly of the head-tail structure is not known. The λNu3 protein is also needed at a subsequent morphogenetic step, in which it forms the

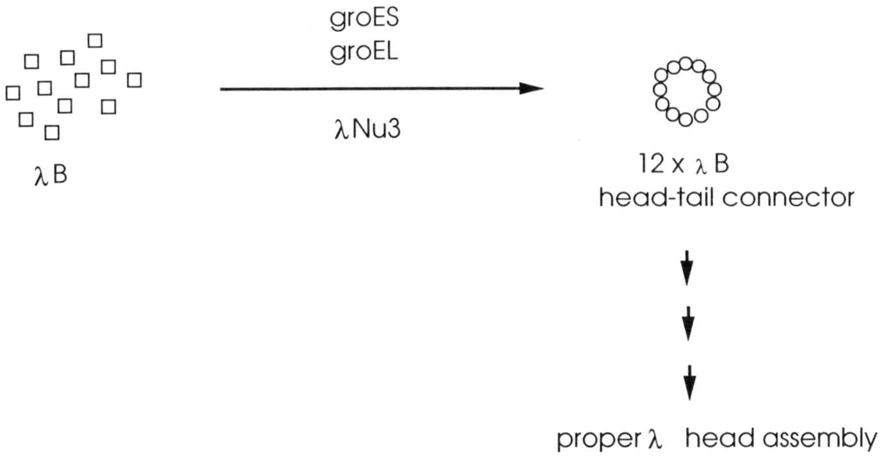

Figure 2 A pathway for the assembly of the λ head-tail connector. For details, see the text and the article by Georgopoulos et al. (1983).

prohead "core" around which the λE capsid protein polymerizes (for review, see Georgopoulos et al. 1983). Upon formation of the mature prohead, the λNu3 protein exits from the prohead and at that time is degraded (Ray and Murialdo 1975). For more details of the events leading to the formation of a mature prohead, the reader is referred to the review by Georgopoulos et al. (1983).

Because Kochan and Murialdo (1983) used only *groEL*⁻ mutant bacteria in their reported studies, it has not yet been determined whether or not the GroES protein is also essential for the formation of the λ head-tail connector (Fig. 2). It is likely that GroES is involved in the assembly of the connector because mutations in either the *groES* or *groEL* genes exhibit identical phenotypes for λ infection (Georgopoulos et al. 1973) and because the GroES and GroEL proteins intimately interact in vitro (see above).

The studies of Kochan and Murialdo (1983) strongly suggest that the involvement of the GroES and GroEL proteins in λ prohead assembly is *solely* at the level of the assembly of the head-tail connector. No evidence to suggest involvement of the GroE proteins in the subsequent prohead assembly steps has been observed. However, Hohn et al. (1979) reported the association of six copies of GroEL with immature proheads assembled in a specific *groEL*⁻ mutant. One possible way to explain this association is that in this particular *groE*⁻ mutant, the mutated form of GroEL binds more tightly to the λB protein and so is trapped within or attached to the immature proheads.

A specific protein-protein interaction between GroEL and λB was observed during glycerol gradient centrifugation (Murialdo 1979). However, later studies from the same laboratory concluded that a GroEL-λB interaction is transient at best (Kochan and Murialdo 1983). One possibility, not shown in Figure 2, is that the GroE proteins help λB protein to dimerize, and that the λB dimers spontaneously associate to form the final dodecameric head-tail connector structure. If this is true, the decatetrameric GroEL and dodecameric λB structures need not interact directly, so no explanation of how a structure with a sevenfold axis of symmetry helps assemble one with a sixfold axis of symmetry would be needed. Studies of the *trans*-dominant λ*B*dc10 mutation which leads to a block in λ head assembly (Georgopoulos et al. 1979) may contribute to determining the assembly pathway. One possibility is that the λBdc10 mutant protein unproductively binds to and sequesters the GroE proteins. Alternatively, the λBdc10 mutant protein could oligomerize properly but fail to carry out its functions.

Since none of the morphogenetic protein cleavages associated with proper head or tail assemblies were seen during infections of *groE*$^-$ hosts, it was hypothesized that the GroE proteins could play a proteolytic role in assembly (Georgopoulos and Eisen 1974). However, no evidence exists for a direct proteolytic role of the GroE proteins in phage morphogenesis. Rather, the failure to build the *proper* morphogenetic structure is more likely to be the cause for the lack of proper protein cleavages.

B. Role in λ DNA Replication

The role of the *E. coli* DnaK, DnaJ, and GrpE proteins in λ DNA replication has been reviewed (Friedman et al. 1984; McMacken et al. 1987, 1988; Keppel et al. 1988; Zylicz et al. 1988; 1989). Here, we briefly outline the steps leading to λ DNA replication and highlight some of the conclusions based on recent work.

The key to both *E. coli* and λ DNA replication appears to be the positioning of the *E. coli* DnaB protein on single-stranded DNA near the appropriate origin of replication. This is important, both because DnaB is a helicase (LeBowitz and McMacken 1986), and because the *E. coli* DnaG primase needs to recognize a DnaB single-stranded DNA complex in order to synthesize the primer RNA molecules that are elongated into DNA by DNA polymerase III holoenzyme (for review, see McMacken et al. 1987). Phage λ must solve several problems before it can replicate. First, λ must displace the host DnaC initiation protein from the few molecules of DnaB (10–20 per cell). Second, λ must direct the DnaB helicase to *ori*λ, its origin of replication. Third, λ must "position" the DnaB protein onto single-stranded DNA, thus enabling the protein to utilize its helicase activity. Following the positioning of the DnaB helicase, all subsequent steps could be carried out by the host DNA replication proteins. In essence, a λ DNA replication fork on λ DNA could be exactly analogous to an *E. coli* DNA replication fork.

Figure 3 shows the steps that lead to λ DNA replication in the "GrpE-dependent" and the "GrpE-independent" purified protein systems (Zylicz et al. 1989). An analogous GrpE-independent system has been established independently in the laboratory of R. McMacken (McMacken et al. 1988; Mensa-Wilmot et al. 1989). As shown, the steps are complex and involve the interaction of many different components. Table 2 summarizes our current knowledge of the many identifiable specific protein-protein and DNA-protein interactions that lead to the assembly and disassembly of this complex structure.

Step 1. The λO protein is analogous to the *E. coli* DnaA protein, since it specifically recognizes the *ori*λ DNA segment located inside the *O* gene. The λO dimers not only bind to and occupy all four binding sites but also form a larger, "O-some" structure (Dodson et al. 1985) through specific λO-λO dimer interactions.

It has been shown that binding of λO to *ori*λ, in an ATP-independent, superhelicity-dependent fashion, causes destabilization of the duplex within the AT-rich region adjacent to *ori*λ (Schnos et al. 1988), resulting in the formation of an "open complex." The *ori*λ-λO complex is stable, since it can be purified by gel filtration and shown to be active in λ DNA replication in the absence of further λO protein addition (Dodson et al. 1985; Yamamoto et al. 1987; McMacken et al. 1988).

Step 2. The λP protein is a dimer, capable of binding effectively to DnaB (Table 2). Since it binds better than DnaC to DnaB, λP effectively "sequesters" the DnaB helicase for specific use by the λ DNA replication machinery (McMacken et al. 1987, 1988).

The λP-DnaB complex is attracted to the *ori*λ site through a specific protein-protein interaction between λP and λO (Table 2). It is possible that some of the λP protein first binds to the *ori*λ-λO complex and then attracts the DnaB helicase. However, since the *ori*λ-λO-λP complex is relatively unstable compared to that of *ori*λ-λO-λP-DnaB (Yamamoto et al. 1987; Liberek et al. 1988; McMacken et al. 1988), it is more likely that the λP-DnaB complex forms first and then finds its way to *ori*λ. The *ori*λ-λO-λP-DnaB complex can assemble efficiently on relaxed circular and linear templates (Alfano and McMacken 1988). The resulting *ori*λ-λP-λO-DnaB structure appears to be larger than the *ori*λ-λO structure (Dodson et al. 1985) and is very stable since it can be purified by gel fil-

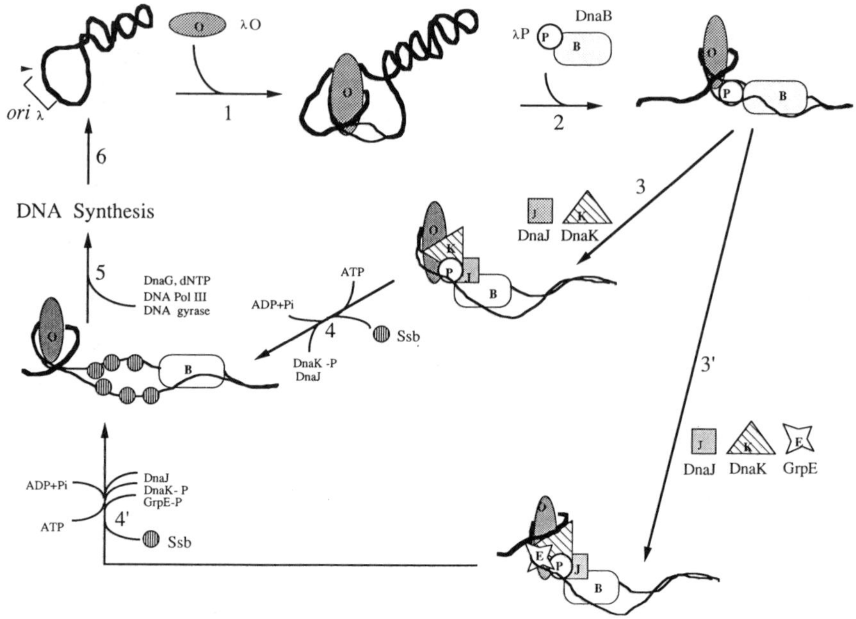

Figure 3 Steps in the circle-to-circle mode of replication of phage λ. The identity of each step has been established by in vitro purified DNA replication systems. For details, see the text.

tration and shown to be active in λ DNA replication in the absence of further λO, λP, and DnaB addition (Dodson et al. 1985; Yamamoto et al. 1987; McMacken et al. 1988; Zylicz et al. 1988; 1989). At this point, phage λ has managed to attract the key replication protein, DnaB, to the proper place, *ori*λ. However, the strong λP-DnaB interaction, which assures the transport of DnaB to the *ori*λ site in the first place, now maintains the DnaB helicase in an inactive state. The next problem for λ, then, is to dissociate DnaB from the λP protein, thus activating its helicase activity.

Step 3. The in vivo genetic data has shown that the DnaK, DnaJ, and GrpE heat shock proteins are essential for the initiation of λ DNA replication (Furth and Wickner 1983; Friedman et al. 1984; McMacken et al. 1987, 1988; Keppel et al. 1988). Nevertheless, the λ DNA replication process can proceed in vitro up to this stage without the active involvement of the DnaK, DnaJ, and GrpE proteins. Both our groups and that of R. McMacken have used an in vitro system, made up of purified components, to show that the DnaB helicase can be liberated from its inactive state by the DnaK and DnaJ heat shock proteins and ATP hydrolysis (McMacken et al. 1988; Zylicz et al. 1988, 1989). In addition, we have purified ^{14}C- or ^{35}S-labeled proteins or have used antibodies raised against these proteins to monitor their presence or absence from *ori*λ DNA (Liberek et al. 1988; Ang 1988). The antibody procedure has also been used by R. McMacken's laboratory (Alfano and McMacken 1988; McMacken et al. 1988).

In both step 3 and step 3′ depicted in Figure 4, the DnaJ protein can be added alone to the *ori*λ-λO-λP-DnaB complex, and a stable DNA replication intermediate can be isolated (McMacken et al. 1988; Zylicz et al. 1988, 1989). The DnaJ protein may be attracted to the complex through its interaction with λP and DnaB (Table 2) and its ability to bind DNA nonspecifically (Table 2). This complex can be purified by gel filtration and shown to be active for replication in the absence of additional λO, λP, DnaB, and DnaJ proteins (McMacken et al. 1988; Zylicz et al. 1988; 1989). We have hypothesized that the DnaJ protein assists in positioning the DnaK protein near λP protein by binding to the λP protein. The subsequent addition of DnaK protein results in the formation of an even larger stable complex, which is also capable of DNA replication (Dodson et al. 1986; Liberek et al. 1988; McMacken et al. 1988; Mensa-Wilmot et al. 1989; Zylicz et al. 1989). The addition of ATP or dGTP results in the dissociation of the λP protein from the *ori*λ complex (Liberek et al. 1988; McMacken et al. 1988; Mensa-Wilmot et al. 1989). However, the released λP protein, when not properly sequestered, can reassociate with DnaB present at *ori*λ or the replication fork, thus again inhibiting the helicase activity of DnaB and creating an unfavorable situation for the initiation of λ DNA replication. The DnaK protein is probably responsible for the sequestration of λP protein in this system. However, in the presence of ATP, the DnaK-λP interaction is extremely weak (Liberek et al. 1988; Zylicz et al. 1989; Table 2). We postulate that in this GrpE-independent system, a high concentration of DnaK protein is needed to shift the equilibrium of the λP-DnaK $\xrightarrow{\text{ATP}}$ λP + DnaK reaction in favor of formation of the λP-DnaK complex. This reaction is important at two points of the prepriming reaction: (1) recognition of λP that is bound to DnaB protein at *ori*λ and (2) efficient sequestration of λP protein when released from the *ori*λ-λO-λP-DnaB complex. However, sequestration of λP protein by DnaK engages the DnaK molecules, effectively preventing them from releasing additional λP protein complexed with DnaB at *ori*λ. This could justify the need for the high concentration of DnaK (600 monomers per *ori*λ; tenfold higher than the number of λO, λP, DnaJ, and DnaB monomers required [Liberek et al. 1988; McMacken et al. 1988; Zylicz et al. 1988, 1989]). As expected from previous in vitro work (Zylicz and Georgopoulos

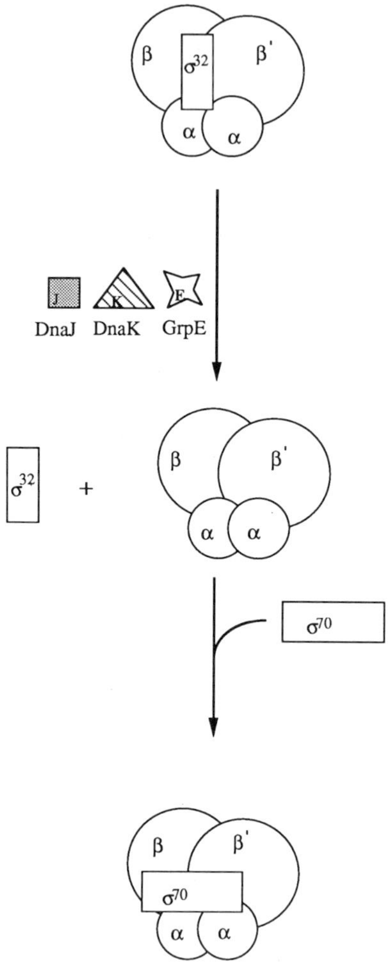

Figure 4 Hypothetical model of heat shock gene regulation. The proposed model is analogous to that functioning in λ DNA replication. See text for details.

1984), purified DnaK756 mutant protein is not able to substitute for the wild-type DnaK protein (Zylicz et al. 1989).

Step 3′. The GrpE protein, purified using a DnaK-affinity column (Zylicz et al. 1987), efficiently stimulates the purified in vitro system for λ DNA replication. The presence of GrpE protein in the reconstitution assay allows an approximately tenfold reduction in the molar concentration of DnaK protein required for maximal DNA synthesis (Zylicz et al. 1988, 1989). The optimal concentration of GrpE used in this GrpE-dependent assay has a rather narrow range, since the DNA replication reaction is inhibited when the level of GrpE protein exceeds that of DnaK (Zylicz et al. 1989; M. Zylicz et al., unpubl.). It has been established that the GrpE protein does not *substitute* for any of the other proteins in the purified in vitro λ DNA replication system (Zylicz et al. 1989).

The mutant GrpE280 protein, purified from cells transformed with a plasmid that overproduces GrpE280 protein (Ang 1988), does not substitute for the wild-type GrpE protein in the purified in vitro system when added in comparable concentrations. This is consistent with the observation that the *E. coli grpE*280 mutant does not support λ DNA replication in vivo at any temperature (Saito and Uchida 1977; D. Ang and C. Georgopoulos, unpubl.).

A hypothesis to explain the ability of GrpE to appreciably lower the concentration of DnaK protein required for efficient *ori*λ-dependent replication is to assume that GrpE protein itself sequesters the λP protein following DnaK action, thus alleviating the requirement for high concentrations of DnaK protein. In support of this, a λP-GrpE complex can be seen following glutaraldehyde cross-linking (Table 2). The DnaK protein, free from its complex with soluble λP protein, could then recycle and help release additional λP protein, which could be complexed to DnaB at *ori*λ or which could have "recaptured" DnaB at the replication fork(s). The inhibition of both DNA replication and DNA unwinding by excess GrpE (Zylicz et al. 1989) could be due to the demonstrable GrpE-DnaK interaction (Table 2). The high levels of GrpE could shift the equilibrium toward the formation of a GrpE-DnaK complex, thereby interfering with the ability of DnaK to release λP.

It could be argued that the GrpE-independent system is due to an artifact of GrpE protein contaminating our DnaK preparations. After all, DnaK protein is known to form a stable, hydrophobic complex with GrpE (Zylicz et al. 1987). Three control experiments eliminate this possibility: (1) Immunoblot assays of our DnaK protein preparation using anti-GrpE antibodies do not detect any GrpE contamination (D. Ang et al., unpubl.), (2) our DnaK protein preparations fail to stimulate a crude in vitro λ DNA replication system derived from *grpE*280 mutant bacteria (Zylicz et al. 1987), and (3) wild-type DnaK protein purified from *E. coli* strain DA259, which carries a deletion of the *grpE* gene (Ang and Georgopoulos 1989), has the same specific activity in the GrpE-independent purified system as DnaK protein purified from a wild-type DnaK-overproducing strain. From these control experiments, we conclude that, in the absence of GrpE protein, DnaK protein is needed in large quantity in the purified system to stimulate *ori*λ-dependent DNA synthesis.

Step 4 and Step 4′. The DnaB helicase, liberated from λP, begins to unwind the DNA, starting at the partially denatured AT-rich DNA region adjacent to the origin (Schnos et al. 1988). Replication is unidirectional, toward gene *Q* (Dodson et al. 1986), probably because the λO protein still bound at *ori*λ inhibits DnaB migration toward the *c*I gene. The addition of Ssb protein stabilizes the single-stranded DNA regions resulting from the action of the DnaB helicase (LeBowitz and McMacken 1986; Alfano and McMacken 1988). Following its liberation from λP, the DnaB helicase can completely unwind the *ori*λ-containing DNA duplex in the presence of Ssb, gyrase, and ATP (Zylicz et al. 1989).

Step 5. The DnaG primase enzyme recognizes the DnaB-single-stranded DNA complex and synthesizes RNA primers at apparently random locations (McMacken et al. 1987). The DNA polymerase III holoenzyme extends these short RNA primers into DNA.

Step 6. Following completion of the replication cycle, ligation of the nicks by DNA ligase, and separation of the catenated circles by gyrase (McMacken et al. 1987), the DNA-protein structure at *ori*λ reassembles, allowing the beginning of a new replication cycle.

IV. A SPECIFIC MODEL FOR HEAT SHOCK REGULATION

The overproduction of the heat shock proteins resulting from mutations in either the *dnaK, dnaJ*, or *grpE* genes can be explained by a model analogous to that proposed for the action of these three heat shock proteins in disassembling the λP protein from a protein-DNA complex assembled at *ori*λ. According to this model, the σ^{32} polypeptide is more stable when complexed with the *E. coli* RNA polymerase core than when "free." Free σ^{32} polypeptide has an extremely short half-life because it is subject to proteolysis by one or more proteinase systems. It could be that its relatively long half-life at 22°C (~15 min) versus 42°C (~45 sec) (Straus et al. 1987; Tilly et al. 1989) is due to (1) an altered σ^{32} conformation, leading to resistance to proteinase action, (2) lower activity of the proteinase(s) degrading σ^{32} at low temperature, or (3) longer association with the polymerase core. In this model, we explain the *dnaK*⁻, *grpE*⁻, and *dnaJ*⁻ effects on σ^{32} by assuming that these three heat shock gene products (and potentially the GroE proteins as well) help to disassemble $E\sigma^{32}$ (Fig. 4). As in the case with λP disassembly, a mutation in any of these three genes can lead to inability to remove σ^{32} from the core, thus resulting in higher intracellular levels of $E\sigma^{32}$ holoenzyme. Alternatively, DnaK, DnaJ, and GrpE could help to assemble the $E\sigma^{70}$ holoenzyme. The efficient assembly of $E\sigma^{70}$ could limit the formation of $E\sigma^{32}$, perhaps through direct competition for the core. Mutations in *dnaK, dnaJ*, or *grpE* would slow down $E\sigma^{70}$ formation, thus indirectly favoring $E\sigma^{32}$ assembly and, hence, stabilization of the $E\sigma^{32}$ polypeptide.

ACKNOWLEDGMENTS

We thank Jeni Urry for cheerful and excellent editorial assistance. This work was supported by National Institutes of Health grants GM-23917 and AI-21029 and Polish Academy of Sciences grant CPBR 3.13.

REFERENCES

Alfano, C. and R. McMacken. 1988. The role of template superhelicity in the initiation of bacteriophage λ DNA replication. *Nucleic Acids Res.* **16:** 9611–9630.

Ang, D. 1988. "The role of the *Escherichia coli* heat shock protein, grpE, in *Escherichia coli* growth and λ DNA replication." Ph.D. thesis, University of Utah, Salt Lake City.

Ang, D. and C. Georgopoulos. 1989. The heat shock-regulated *grpE* gene of *Escherichia coli* is required for bacterial growth at all temperatures, but is dispensable in certain mutant backgrounds. *J. Bacteriol.* **171:** 2748–2755.

Ang, D., G.N. Chandrasekhar, M. Zylicz, and C. Georgopoulos. 1986. *Escherichia coli grpE* gene codes for heat shock protein B25.3, essential for both λ DNA replication at all temperatures and host growth at high temperature. *J. Bacteriol.* **167:** 25–29.

Ang, D., G.N. Chandrasekhar, C. Johnson, M. Zylicz, and C. Georgopoulos. 1987. A genetic and biochemical analysis to the *Escherichia coli grpE* gene and its product. *UCLA Symp. Mol. Cell. Biol. New Ser.* **47:** 521–532.

Arai, K., S. Yasuda, and A. Kornberg. 1981. Mechanism of DnaB protein action. I. Crystalization and properties of DnaB protein, an essential replication protein in *Eschierchia coli*. *J. Biol. Chem.* **256:** 5247–5252.

Baird, P.N., L.M.C. Hall, and A.R.M. Coates. 1988. A major antigen from *Mycobacterium tuberculosis* which is homologous to the heat shock proteins groES from *E. coli* and the *htpA* gene product of *Coxiella burnetii*. *Nucleic Acids Res.* **16:** 9047–9051.

Bardwell, J.C.A. and E.A. Craig. 1984. Major heat shock gene of *Drosophila* and the *Escherichia coli* heat-inducible *dnaK* gene are homologous. *Proc. Natl. Acad. Sci.* **81:** 848–852.

———. 1987. Eukaryotic M_r 83,000 heat shock protein has a homologue in *Escherichia coli*. *Proc. Natl. Acad. Sci.* **84:** 5177–5181.

———. 1988. Ancient heat shock gene is dispensable. *J. Bacteriol.* **170:** 2977–2983.

Bardwell, J.C.A., K. Tilly, E. Craig, J. King, M. Zylicz, and C. Georgopoulos. 1986. The nucleotide sequence of the *Escherichia coli* K12 *dnaJ*+ gene: A gene that encodes a heat shock protein. *J. Biol. Chem.* **261:** 1782–1785.

Bochkareva, E.S., N.M. Lisson, and A.S. Girshovich. 1988. Transient association of newly synthesized unfolded proteins with the heat-shock GroEL protein. *Nature* **336:** 254–257.

Bochner, B.R., M. Zylicz, and C. Georgopoulos. 1986. *Escherichia coli* dnaK protein possesses a 5′-nucleotidase activity that is inhibited by AppppA. *J. Bacteriol.* **168:** 931–935.

Bukau, B. and G.C. Walker. 1989. Cellular defects caused by deletion of the *Escherichia coli dnaK* gene indicate roles for heat shock protein in normal metabolism. *J. Bacteriol.* **171:** 2337–2346.

Bukau, B., C. E. Donnelly, and G. C. Walker. 1989. The role of the *Escherichia coli* heat shock proteins. *UCLA Symp. Mol. Cell. Biol.* **96:** 27–36.

Burton, Z., R.R. Burgess, J. Lin, D. Moore, S. Holder, and C.A. Gross. 1981. The nucleotide sequence of the cloned *rpoD* gene for the RNA polymerase sigma subunit from *E. coli* K-12. *Nucleic Acids Res.* **9:** 2889–2903.

Chandrasekhar, G.N., K. Tilly, C. Woolford, R. Hendrix, and C. Georgopoulos. 1986. Purification and properties of the groES morphogenetic protein of *Escherichia coli. J. Biol. Chem.* **261:** 12414–12419.

Cheng, MY., F.-U. Hartl, J. Martin, R.A. Pollock, F. Kalousek, W. Neupert, E.M. Hallberg, R.L. Hallberg, and A.L. Horwich. 1989. Mitochondrial heat shock protein hsp60 is essential for assembly of proteins imported into yeast mitochondria. *Nature* **337:** 620–625.

Chin, D.T., S.A. Goff, T. Webster, T. Smith, and A.L. Goldberg. 1988. Sequence of the *lon* gene in *Escherichia coli. J. Biol. Chem.* **263:** 11718–11728.

Chirico, W.J., M.G. Waters, and G. Blobel. 1988. 70 K Heat shock related proteins stimulate protein translocation into microsomes. *Nature* **332:** 805–810.

Clarke, C.F., K. Cheng, A.B. Frey, R. Stein, P.W. Hinds, and A.J. Levine. 1988. Purification of complexes of nuclear oncogene p53 with rat and *Escherichia coli* heat shock protein: *In vitro* dissociation of hsc70 and dnaK from murine p53 by ATP. *Mol. Cell. Biol.* **8:** 1206–1215.

Coppo, A., A. Manzi, J.F. Pulitzer, and H. Takahashi. 1973. Abortive bacteriophage T4 head assembly in mutants of *Escherichia coli. J. Mol. Biol.* **76:** 61–87.

Cowing, D.W., J.C.A. Bardwell, E.A. Craig, C. Woolford, R.W. Hendrix, and C.A. Gross. 1985. Consensus sequence for *Escherichia coli* heat shock gene promoters. *Proc. Natl. Acad. Sci.* **82:** 2679–2683.

Deshaies, R.J., B.D. Koch, and R. Schekman. 1988a. The role of stress proteins in membrane biogenesis. *Trends Biochem. Sci.* **13:** 384–388.

Deshaies, R.J., B.D. Koch, M. Werner-Washburne, E.A. Craig, and R. Schekman. 1988b. A subfamily of stress proteins facilitates translocation of secretory and mitochondrial precursor polypeptides. *Nature* **332:** 800–805.

Dodson, M., J. Roberts, R. McMacken, and H. Echols. 1985. Specialized nucleoprotein structures at the origin of replication of bacteriophage λ: Complexes with λO, λP and *Escherichia coli* DnaB proteins. *Proc. Natl. Acad. Sci.* **82:** 4678–4682.

Dodson, M., H. Echols, S. Wickner, C. Alfano, K. Mensa-Wilmot, B. Gomes, J.H. LeBowitz, J.D. Roberts, and R. McMacken. 1986. Specialized nucleoprotein structures at the origin of replication of bacteriophage λ: Localized unwinding of duplex DNA by six-protein reaction. *Proc. Natl. Acad. Sci.* **83:** 7638–7642.

Erickson, J.W. and C.A. Gross. 1989. Identification of the σE subunit of *Escherichia coli* RNA polymerase; a second alternative sigma involved in high temperature gene expression. *Genes Dev.* **3:** 1462–1471.

Erickson, J.W., V. Vaughn, W.A. Walter, F.C. Neidhardt, and C.A. Gross. 1987. Regulation of the promoters and transcripts of *rpoH,* the *Escherichia coli* heat shock regulatory gene. *Genes Dev.* **1:** 419–432.

Fayet, O., J.-M. Louarn, and C. Georgopoulos. 1986. Suppression of the *Escherichia coli dnaA46* mutation by amplification of the *groES* and *groEL* genes. *Mol. Gen. Genet.* **202:** 435–445.

Fayet, O., T. Ziegelhoffer, and C. Georgopoulos. 1989. The *groES* and *groEL* heat shock genes of *Escherichia coli* are essential for bacterial growth at all temperatures. *J. Bacteriol.* **171:** 1379–1385.

Friedman, D.E., E.R. Olson, C. Georgopoulos, K. Tilly, I. Herskowitz, and F. Banuett. 1984. Interactions of bacteriophage and host macromolecules in the growth of bacteriophage λ. *Microbiol. Rev.* **48:** 299–325.

Furth, M.E. and S.H. Wickner. 1983. Lambda DNA replication. In *Lambda II* (ed. R.W. Hendrix et al.) pp. 145–173. Cold Spring Harbor Laboratory, Cold Spring Harbor, New York.

Furth, M. and J. Yates. 1978. Specificity determinants for bacteriophage lambda DNA replication. II. Structure of O protein of λ-ϕ80 and λ-82 hybrid phages and of a λ mutant defective in the origin of replication. *J. Mol. Biol.* **126:** 227–240.

Georgopoulos, C.P. 1977. A new bacterial gene *(groPC)* which affects λ DNA replication. *Mol. Gen. Genet.* **151:** 35–39.

Georgopoulos, C.P. and H. Eisen. 1974. Bacterial mutants which block phage assembly. *J. Supramol. Struct.* **2:** 349–359.

Georgopoulos, C.P. and I. Herskowitz. 1971. *Escherichia coli* mutants blocked in lambda DNA synthesis. In *The bacteriophage lambda* (ed. A.D. Hershey), pp. 553–564. Cold Spring Harbor Laboratory, Cold Spring Harbor, New York.

Georgopoulos, C., K. Tilly, and S. Casjens. 1983. Lambdoid phage head assembly. In *Lambda II* (ed. R.W. Hendrix et al.) pp. 279–304, Cold Spring Harbor Laboratory, Cold Spring Harbor, New York.

Georgopoulos, C. P., R.W. Hendrix, S.R. Casjens, and A.D. Kaiser. 1973. Host participation in bacteriophage lambda head assembly. *J. Mol. Biol.* **76:** 45–60.

Georgopoulos, C P., R.W. Hendrlx, A.D. Kaiser, and W.B. Wood. 1972. Role of the host cell in bacteriophage morphogenesis: Effects of a bacterial mutation on T4 head assembly. *Nature New Biol.* **239:** 38–42.

Georgopoulos, C., R. Bisig, M. Magasin, H. Eisen, and D. Court. 1979. A novel mutation affecting λ head assembly. *J. Virol.* **29:** 782–788.

Goloubinoff, P., A.A. Gatenby, and G.H. Lorimer. 1989. GroE heat shock proteins promote assembly of foreign prokaryotic ribulose bisphosphate carboxylase oligomers in *Escherichia coli*. *Nature* **337:** 44–47.

Grossman, A.D., D.B. Straus, W.A. Walter, and C.A. Gross. 1987. σ^{32} synthesis can regulate the synthesis of heat shock proteins in *Escherichia coli*. *Genes Dev.* **1:** 179–184.

Hemmingsen, S.M., C. Woolford, S.M. van der Vies, K. Tilly, D.T. Dennis, C.P. Georgopoulos, R.W. Hendrix, and R.J. Ellis. 1988. Homologous plant and bacterial proteins chaperone oligomeric protein assembly. *Nature* **333:** 330–334.

Hendrix, R.W. 1979. Purification and properties of GroE, a host protein involved in bacteriophage assembly. *J. Mol. Biol.* **129:** 375–392.

Hohn, T., B. Hohn, A. Engel, M. Wurtz, and P.R. Smith. 1979. Isolation and characterization of the host protein GroE involved in bacteriophage lambda assembly. *J. Mol. Biol.* **129:** 359–373.

Itikawa, H. and J.-I. Ryu. 1979. Isolation and characterization of a temperature-sensitive *dnaK* mutant of *Escherichia coli* B. *J. Bacteriol.* **138:** 339–344.

Jenkins, A.J., J.B. Marsh, I.R. Oliver, and M. Master. 1986. A DNA fragment containing the *groE* genes can suppress mutations in the *Escherichia coli dnaA* gene. *Mol. Gen. Genet.* **202:** 446–454.

Johnson, C. 1988. " The *Escherichia coli* dnaK protein: Aspects of its function during heat stress." Ph.D. thesis, University of Utah, Salt Lake City.

Johnson, C., G.N. Chandrasekhar, and C. Georgopoulos. 1989. The dnaK and grpE heat shock proteins in *Escherichia coli* interact both *in vivo* and *in vitro*. *J. Bacteriol.* **171:** 1590–1596.

Kassenbrock, C.K., P.D. Garcia, P. Walter, and R.B. Kelly. 1988. Heavy-chain binding protein recognized aberrant polypeptides translocated *in vitro*. *Nature* **333:** 90–93.

Keller, J.A. and L.D. Simon. 1988. Divergent effects of a *dnaK* mutation on abnormal protein degradation in *Escherichia coli*. *Mol. Microbiol.* **2:** 31–41.

Keppel, F., O. Fayet, and C. Georgopoulos. 1988. Strategies of bacteriophage DNA replication. In *The bacteriophages* (ed. R. Calendar), vol. 2, pp. 145–262, Plenum Press, New York.

Klein, A., E. Lanka, and E. Schuster. 1980. Isolation of a complex between the *P* protein of phage λ and the *dnaB* protein of *Escherichia coli*. *Eur. J. Biochem.* **105:** 1–6.

Kochan, J. and H. Murialdo. 1983. Early intermediates in bacteriophage lambda prohead assembly. II. Identification of biologically active intermediates. *Virology* **131:** 100–115.

Kochan, J., J.L. Carrascosa, and H. Murialdo. 1984. Bacteriophage lambda preconnectors: Purification and structure. *J. Mol. Biol.* **174:** 433–447.

Lathigra, R.B., D.B. Young, D. Sweetser, and R.A. Young. 1988. A gene from *Mycobacterium tuberculosis* which is homologous to the DnaJ heat shock protein of *E. coli*. *Nucleic Acids Res.* **16:** 1636.

LeBowitz, J.H. and R. McMacken. 1986. The *Escherichia coli* DnaB replication protein is a DNA helicase. *J. Biol. Chem.* **261:** 4738–4748.

Lecker, S., R. Lill, T. Ziegelhoffer, C. Georgopoulos, P.J. Bassford, Jr., C.A. Kumamoto, and W. Wickner. 1989. Three pure chaperone proteins of *Escherichia coli*, secB, trigger factor and groEL, form soluble complexes with precursor proteins *in vitro*. *EMBO J.* **8:** 2703–2709.

Lee, P.C., B.R. Bochner, and B.N. Ames. 1983. AppppA, heat-shock stress, and cell oxidation. *Proc. Natl. Acad. Sci.* **80:** 7496–7500.

Liberek, K., C. Georgopoulos, and M. Zylicz. 1988. The role of the *Escherichia coli* DnaK and DnaJ heat shock proteins in the initiation of bacteriophage λ DNA replication. *Proc. Natl. Acad. Sci.* **85:** 6632–6636.

Lindquist, S. and E.A. Craig. 1988. The heat-shock proteins. *Annu. Rev. Genet.* **22:** 631–677.

Lipinska, B., S. Sharma, and C. Georgopoulos. 1988a. Sequence analysis and transcriptional regulation of the *htrA* gene of *Escherichia coli:* A σ^{32}-independent mechanism of heat-inducible transcription. *Nucleic Acids Res.* **16:** 10053–10067.

Lipinska, B., O. Fayet, L. Baird, and C. Georgopoulos. 1988b. Identification, characterization and mapping of the *Escherichia coli htrA* gene, whose product is essential for bacterial viability only at elevated temperatures. *J. Bacteriol.* **171:** 1574–1584.

Lipinska, B., J. King, D. Ang. and C. Georgopoulos. 1989. Sequence analysis and transcriptional regulation of the *Escherichia coli grpE* gene, encoding a heat shock protein. *Nucleic Acids Res.* **16:** 7545–7562.

Maurizi, M.R., P. Trisler, and S. Gottesman. 1985. Insertional mutagenesis of the *lon* gene in *Escherichia coli*: *lon* is dispensable. *J. Bacteriol.* **164:** 1124–1135.

McMacken, R., L. Silver, and C. Georgopoulos. 1987. DNA replication. In Escherichia coli *and* Salmonella typhimurium: *Cellular and molecular biology* (ed. F.C. Neidhardt et al.), pp. 564–612, American Society for Microbiology, Washington, D.C.

McMacken, R., K. Mensa-Wilmot, C. Alfano, R. Seaby, K. Carroll, B. Gomes, and K. Stephens. 1988. Reconstitution of purified protein systems for the initiation and regulation of bacteriophage λ DNA replication. *Cancer Cells* **6:** 25–34.

McMullin, T.W. and R.L. Hallberg. 1987. A normal mitochondrial protein is selectively synthesized and accumulated during heat shock in *Tetrahymena thermophila*. *Mol. Cell. Biol.* **7:** 4414–4423.

———. 1988. A highly evolutionary conserved mitochondrial protein is structurally related to the protein encoded by the *Escherichia coli groEL* gene. *Mol. Cell. Biol.* **8:** 371–180.

Mensa-Wilmot, K., R. Seaby, C. Alfano, M.S. Wold, B. Gomes, and R. McMacken. 1989. Reconstitution of a nine-protein system that initiates bacteriophage λ DNA replication. *J. Biol. Chem.* **264:** 2853–2861.

Miki, T., T. Orita, M. Funruno, and T. Horiuchi. 1988. Control of cell division by sex factor F in *Escherichia coli*. *J. Mol. Biol.* **201:** 327–338.

Mizusawa, S. and S. Gottesman. 1983. Protein degradation in *Escherichia coli*: The *lon* gene controls the stability of *sulA* protein. *Proc. Natl. Acad. Sci.* **80:** 358–362.

Murialdo, H. 1979. Early intermediates in bacteriophage lambda prohead assembly. *Virology* **96:** 341–367.

Neidhardt, F.C. and R.A. VanBogelen. 1987. Heat shock response. In Escherichia coli *and* Salmonella typhimurium: *Cellular and molecular biology* (ed. F.C. Neidhardt et al.), pp. 1334–1345, American Society for Microbiology, Washington, D.C.

Neidhardt, F.C., R.A. VanBogelen, and V. Vaughn. 1985. The genetics and regulation of heat shock proteins. *Annu. Rev. Genet.* **18:** 295–329.

Ohki, M., F. Tamura, S. Nishimura, and H. Uchida. 1986. Nucleotide sequence of the *Escherichia coli dnaJ* gene and purification of the gene product. *J. Biol. Chem.* **261:** 1778–1781.

Ohki, M., H. Uchida, F. Tamura, R. Ohki, and S. Nishimura. 1987. The *Escherichia coli dnaJ* mutation affects biosynthesis of specific proteins, including those of the *lac* operon. *J. Bacteriol.* **169:** 1917–1922.

Paek, K.-H. and G.C. Walker. 1987. *Escherichia coli dnaK* null mutants are inviable at high temperature. *J. Bacteriol.* **169:** 283–290.

Pelham, H.R.B. 1986. Speculations on the functions of the major heat shock and glucose-regulated proteins. *Cell* **46:** 959–961.

Ray, P.N. and H. Murialdo. 1975. The role of gene *Nu3* in bacteriophage lambda morphogenesis. *Virology* **64:** 247–263.

Reading, D.S., R.L. Hallberg, and A.M. Myers. 1989. Characterization of the yeast hsp60 gene coding for a mitochondrial assembly factor. *Nature* **337:** 655–659.

Reha-Krantz, L.J. and J. Hurwitz. 1978. The *dnaB* gene product of *Escherichia coli*. II. Single-stranded DNA-dependent ribonucleoside triphosphatase activity. *J. Biol. Chem.* **253:** 4051–4057.

Revel, H.R., B.L. Stitt, I. Lielausis, and W.B. Wood. 1980. Role of the host cell in bacteriophage T4 development. I. Characterization of host mutants that block T4 head assembly. *J. Virol.* **33:** 366–376.

Rieul, C., J.-C. Cortay, F. Bleicher, and A.J. Cozzone. 1987. Effect of bacteriophage M13 infection on phosphorylation of dnaK protein and other *Escherichia coli* proteins. *Eur. J. Biochem.* **168:** 621–627.

Ruben, S.M., S.E. VanDenBrink-Webb, D.C. Rein, and R.R. Meyer. 1988. Suppression of the *ssb-1* mutation by an allele of *groEL*. *Proc. Natl. Acad. Sci.* **85:** 3767–3771.

Saito, H. and H. Uchida. 1977. Initiation of the DNA replication of bacteriophage lambda in *Escherichia coli* K12. *J. Mol. Biol.* **113:** 1–25.

———. 1978. Organization and expression of the *dnaJ* and *dnaK* genes of *Escherichia coli* K12. *Mol. Gen. Genet.* **164:** 1–8.

Sakakibara, Y. 1988. The *dnaK* gene of *Escherichia coli* functions in initiation of chromosome replication. *J. Bacteriol.* **170:** 972–979.

Schnos, M., K. Zahn, R.B. Inman, and F.R. Blattner. 1988. Initiation protein induced helix destabilization at the λ origin: A prepriming step in DNA replication. *Cell* **52:** 385–395.

Sekimizu, K., D. Bramhill, and A. Kornberg. 1987. ATP activates dnaA protein in initiating replication of plasmids bearing the origin of the *E. coli* chromosome. *Cell* **50:** 259–265.

Sell, S. 1987. "Studies of bacterial genes whose products are involved in heat shock regulation and bacteriophage lambda DNA replication." Ph.D. thesis, University of Utah, Salt Lake City.

Spence, J. 1988. " The role of the C62.5 and DnaK heat shock proteins in the response to nutritional deprivation and characterization of the C62.5 protein." Ph.D. thesis, University of Utah, Salt Lake City.

Spence, J. and C. Georgopoulos. 1989. Purification and properties of the *Escherichia coli* heat shock protein HtpG. *J. Biol. Chem.* **264:** 4398–4403.

Sternberg, N. 1973a. Properties of a mutant of *Escherichia coli* defective in bacteriophage λ head formation *(groE)*. I. Initial characterization. *J. Mol. Biol.* **67:** 1–24.

———. 1973b. Properties of a mutant of *Escherichia coli* defective in bacteriophage λ head formation (*groE*). II. The propagation of phage λ. *J. Mol. Biol.* **76:** 25–44.

———. 1976. A genetic analysis of bacteriophage λ head assembly. *Virology* **71:** 568–582.

Straus, D.B., W.A. Walter, and C.A. Gross. 1987. The heat shock response of *E. coli* is regulated by changes in the concentration of σ^{32}. *Nature* **329:** 348–351.

———. 1988. *Escherichia coli* heat shock gene mutants are defective in proteolysis. *Genes Dev.* **2:** 1851–1858.

Sunshine, M., M. Feiss, J. Stuart, and J. Yochem. 1977. A new host gene *(groPC)* necessary for lambda DNA replication. *Mol. Gen. Genet.* **151:** 27–34.

Takano, T. and T. Kakefuda. 1972. Involvement of a bacterial factor in morphogenesis of bacteriophage capsid. *Nat. New Biol.* **239:** 34–37.

Tilly, K. 1982. "Studies on bacteriophage and host genes involved in morphogenesis." Ph.D. thesis, University of Utah, Salt Lake City.

Tilly, K. and C. Georgopoulos. 1982. Evidence that the two *Escherichia coli groE* morphogenetic gene products interact *in vivo*. *J. Bacteriol.* **149:** 1082–1088.

Tilly, K., H. Murialdo, and C. Georgopoulos. 1981. Identification of a second *Escherichia coli groE* gene whose product is necessary for bacteriophage morphogenesis. *Proc. Natl. Acad. Sci.* **78:** 1629–1633.

Tilly, K., J. Spence, and C. Georgopoulos. 1989. Modulation of the stability of *Escherichia coli* heat shock regulatory factor σ^{32}. *J. Bacteriol.* **171:** 1585–1589.

Tilly, K., N. McKittrick, M. Zylicz, and C. Georgopoulos. 1983. The *dnaK* protein modulates the heat-shock response of *Escherichia coli*. *Cell* **34:** 641–646.

Tomizawa, J. 1971. Functional cooperation of genes O and P. In *The bacteriophage lambda* (ed. A.D. Hershey), pp. 549–552. Cold Spring Harbor Laboratory, Cold Spring Harbor, New York.

Tsui, L. and R. Hendrix. 1980. Head-tail connector of bacteriophage lambda. *J. Mol. Biol.* **142:** 419–438.

Tsurimoto, T. and K. Matsubara. 1981. Purified bacteriophage λ*O* protein binds to four repeating sequences at the λ replication origin. *Nucleic Acids Res.* **9:** 1789–1799.

VanBogelen, R.A., V. Vaughn, and F.C. Neidhardt. 1983. Gene for heat-inducible lysyl-tRNA synthetase *(lysU)* maps near *cadA* in *Escherichia coli*. *J. Bacteriol.* **153:** 1066–1068.

Vodkin, M.H. and J.C. Williams. 1988. A heat shock operon in *Coxiella burnetii* produces a major antigen homologous to a protein in both mycobacteria and *Escherichia coli*. *J. Bacteriol.* **170:** 1227–1234.

Wada, M. and H. Itikawa. 1984. Participation of *Escherichia coli* K-12 *groE* gene products in the synthesis of cellular DNA and RNA. *J. Bacteriol.* **157:** 694–696.

Wada, M., H. Fujita, and H. Itikawa. 1987. Genetic suppression of a temperature-sensitive *groES* mutation by an altered subunit of RNA polymerase of *Escherichia coli* K-12. *J. Bacteriol.* **169:** 1102–1106.

Wada, M., Y. Kadokami, and H. Itikawa. 1982. Thermosensitive synthesis of DNA and RNA in *dnaJ* mutants of *Escherichia coli* K-12. *Jpn. J. Genet.* **57:** 407–413.

Wada, M., K. Sekine, and H. Itikawa. 1986. Participation of the *dnaK* and *dnaJ* gene products in phosphorylation of glutaminyl-tRNA synthetase and threonyl-tRNA synthetase of *Escherichia coli* K-12. *J. Bacteriol.* **168:** 213–220.

Wickner, S. 1978. DNA replication proteins of *Escherichia coli* and phage λ. *Cold Spring Harbor Symp. Quant. Biol.* **43:** 303–310.

Wickner, S.H. and K. Zahn. 1986. Characterization of the DNA binding domain of bacteriophage λ O protein. *J. Biol. Chem.* **261:** 8537–8543.

Yamamoto, T., J. McIntyre, S.M. Sell, C. Georgopoulos, D. Skowyra, and M. Zylicz. 1987. Enzymology of the pre-priming steps in lambda DNA replication *in vitro*. *J. Biol. Chem.* **262:** 7996–7999.

Yochem, J., H. Uchida, M. Sunshine, H. Saito, C. Georgopoulos, and M. Feiss. 1978. Genetic analysis of two genes, *dnaJ* and *dnaK,* necessary for *Escherichia coli* and bacteriophage lambda DNA replication. *Mol. Gen. Genet.* **164:** 9–14.

Young, D., R. Lathigra, R. Hendrix, D. Sweetser, and R.A. Young. 1988. Stress proteins are immune targets in leprosy and tuberculosis. *Proc. Natl. Acad. Sci.* **85:** 4267–4270.

Zhou, Y., N. Kusukawa, J.W. Erickson, C.A. Gross, and T. Yura. 1988. Isolation and characterization of *Escherichia coli* mutants that lack the heat shock sigma factor (σ^{32}). *J. Bacteriol.* **170:** 3640–3649.

Zweig, M. and D.J. Cummings. 1973. Cleavage of head and tail proteins during bacteriophage T5 assembly: Selective host involvement in the cleavage of a tail protein. *J. Mol. Biol.* **80:** 505–518.

Zylicz, M. and C. Georgopoulos. 1983. Properties of the purified *dnaK* gene product of *Escherichia coli*. *UCLA Symp. Mol. Cell. Biol.* **10:** 317–326.

———. 1984. Purification and properties of the *Escherichia coli* dnaK replication protein. *J. Biol. Chem.* **259:** 8820–8825.

Zylicz, M., D. Ang, and C. Georgopoulos. 1987. The grpE protein of *Escherichia coli:* Purification and properties. *J. Biol. Chem.* **262:** 17437–17442.

Zylicz, M., D. Ang, K. Liberek, and C. Georgopoulos. 1989. Initiation of λ DNA replication with purified host- and bacteriophage-encoded proteins: The role of the dnaK, dnaJ and grpE heat shock proteins. *EMBO J.* **8:** 1601–1608.

Zylicz, M., I. Gorska, K. Taylor, and C. Georgopoulos. 1984. Bacteriophage λ replication proteins: formation of a mixed oligomer and binding to the origin of λ DNA. *Mol. Gen. Genet.* **196:** 401–406.

Zylicz, M., J.H. LeBowitz, R. McMacken, and C. Georgopoulos. 1983. The dnaK protein of *Escherichia coli* possesses an ATPase and autophosphorylating activity and is essential in an *in vitro* DNA replication system. *Proc. Natl. Acad. Sci.* **80:** 6431–6435.

Zylicz, M., D. Ang, K. Liberek, T. Yamamoto, and C. Georgopoulos. 1988. Initiation of lambda DNA replication reconstituted with purified lambda and *Escherichia coli* replication proteins. *Biochem. Biophys. Acta* **951:** 344–350.

Zylicz, M., T. Yamamoto, N. McKittrick, S. Sell, and C. Georgopoulos. 1985. Purification and properties of the DnaJ replication protein of *Escherichia coli*. *J. Biol. Chem.* **260:** 7591–7598.

10

The Mammalian Stress Response: Cell Physiology and Biochemistry of Stress Proteins

William J. Welch
Departments of Medicine and Physiology
Lung Biology Center, University of California
San Francisco, California 94143

I. INTRODUCTION

Organisms exhibit specific homeostatic-like responses when confronted with sudden changes in their environmental circumstance. As was dis-

Stress Proteins in Biology and Medicine.

cussed in some of the earlier chapters in this volume, the ability of an organism to adapt or acclimate to its new environment may be an integral driving force for evolution. At the cellular level, there are also a number of mechanisms by which the cell can respond and/or acclimate to changes in its normal growth environment. Examples of this include temporary or reversible changes in enzyme structure or function by which certain metabolic pathways are modified. Alternatively, there may be more long-term changes that involve permanent alterations in the pattern of gene expression, thereby facilitating the acclimation of the cell to its new environmental circumstance. Finally, all organisms appear to have developed a rather specific and remarkably similar strategy by which to confront abrupt changes in their surroundings, such as that which occurs following increases in their normal growth temperature. This latter response, historically referred to as the heat shock response, involves the coordinated increased expression of a group of proteins, the heat shock proteins (hsps), whose collective function appears to provide the cell protection during the insult and/or to facilitate the restoration of basic metabolic pathways that were perturbed as a consequence of the environmental insult (see also reviews by Ashburner and Bonner 1979; Nover 1984; Craig 1985; Lindquist 1986; Subjeck and Shyy 1986).

That the heat shock response is involved in the protection and/or enhanced survival of the cell has been amply documented. For example, cells given a mild, sublethal heat shock treatment (sufficient to increase the levels of the hsps or to perhaps "activate" the pre-existing hsps) and then returned to their normal growth temperature exhibit significantly higher survival rates to a second and what would otherwise be a lethal heat shock challenge (Gerner and Scheider 1975; Gerner et al. 1976; Henle and Leeper 1976). This phenomenon, referred to as acquired thermotolerance, appears to be partly dependent on the levels and/or activities of the hsps (Henle and Leeper 1982; Landry et al. 1982; Li and Werb 1982; Li et al. 1982; Subjeck et al. 1982; Li and Laszlo 1985). Moreover, cellular thermotolerance appears transient, with the decay of tolerance correlated with the half-life of one or more of the hsps (Landry et al. 1982; Li and Werb 1982; Subjeck et al. 1982). As discussed below, I suspect that the key to the thermotolerance is simply the ability of the insulted cell to continue functioning normally; i.e., to maintain essential metabolic pathways under conditions (e.g., high temperature) that are typically toxic or inhibitory to such pathways. Therefore, one would surmise that the general function of many of the hsps is the overall protection, maintenance, and/or restoration of various cellular machineries that are inactivated or compromised by the temperature insult. However, it is probably naive to suggest that the hsps are solely responsible for the maintenance or protection of all essential metabolic functions perturbed in the environmentally stressed cell.

When considering the response of different organisms to increases in their local environmental temperature, it is important to point out that a major difference exists between many unicellular versus multicellular organisms. For example, whereas most prokaryotes are capable of acclimating to rather large increases in their external temperature, cells from homeothermic animals appear to tolerate only relatively slight increases in their local environmental temperature and, for the most part, rarely exhibit complete acclimation to a higher nonphysiological temperature (i.e., temperatures much greater than 41°C). Instead, with prolonged exposure to the higher temperature, most homeothermic cells eventually expire. Nevertheless, cells from both poikilothermic and homeothermic organisms, following the temperature upshift, increase the expression of the hsps. In the case of the former, however, the cells, while still at the higher temperature, eventually shift back to their normal patterns of gene expression and resume typical growth and proliferation. In contrast, most homeothermic cells, following a temperature upshift, remain in a heat shock response, continue to express high levels of the hsps, but eventually die if left at the higher temperature. Thus, although the heat shock response may play an integral role in the ability of some organisms to confront and eventually adapt to higher temperatures, it is apparently not sufficient to allow for similar temperature acclimation in homeothermic cells.

In this paper, I discuss a number of aspects concerning the heat shock or stress response of mammalian cells. Particular emphasis is given to describing pertinent changes in the physiology of the stressed cell, most notably those alterations that occur in cell morphology, the integrity of intracellular organelles, and alterations occurring in different metabolic pathways. By understanding the specific lesions faced by the cell under stress, one can better approach the possible function of the hsps, especially as it pertains to the possible repair of such lesions and the mechanisms by which the cell eventually returns to its normal lifestyle. Finally, considerable discussion is given to our current understanding of the properties of the individual stress proteins, including their identity, posttranslational modifications, intracellular locale, and their possible biochemical function.

II. INDUCTION OF THE RESPONSE

Numerous studies over the past 15 years have demonstrated that all organisms, be they prokaryotes, plants, yeast, or higher eukaryotes, respond in a similar fashion to abrupt increases in their local environmental temperature by increasing the synthesis of a group of proteins, the hsps. Moreover, a heat-shock-like response can also be elicited upon exposure of the cell to a wide variety of other agents. Although too extensive to list here in entirety, such agents include amino acid analogs, various heavy metals, certain ionophores, ethanol, inhibitors of mitochondrial function, and, in some cases, viral infections (Ashburner and Bonner 1979; Thomas et al. 1982). Consequently, because of the variety of different agents that elicit the increased expression of the same group of proteins, the heat shock response is often referred to more generally as the stress response, with the proteins that are expressed at increased levels referred to as the stress proteins. As discussed below, there are some minor differences with respect to the changes in gene expression elicited by these different agents. For example, whereas most of these agents elicit the increased expression of the classically defined stress proteins, there also occurs an induction of a few other proteins unique to the particular stress agent employed. Consequently, by examining the proteins whose synthesis increases in the cell under stress, it is often possible to ascertain the nature of agent responsible for the activation of the response (e.g., heat vs. metals).

The actual mechanism by which the cell recognizes and responds to a particular stress agent is still unclear. One obvious common denominator shared by the many different agents that induce the response is their ability to promote, at least in vitro, perturbations in protein conformation. However, not all of the stress agents fall into this category. Nevertheless, it has been shown that the introduction into oocytes of denatured or cross-linked proteins will result in the activation of a stress response (Ananthan et al. 1986). Consequently, the accumulation in the cell of denatured proteins (either in general or perhaps specific proteins) might be the triggering mechanism by which the response is activated. Considering the extremely rapid kinetics by which the genes encoding the stress proteins are activated following temperature upshift (i.e., within minutes), this general model for the induction of the response is probably not the whole story. Moreover, whether all of the various agents that induce the response act via independent pathways or converge within a single point along one pathway is entirely unclear and, at present, difficult to approach experimentally. The isolation and characterization of the heat shock transcription factor that regulates the expression of the genes encoding the stress proteins should greatly facilitate further studies regarding the mechanism by which the response is initiated. Until then, it will be difficult to delineate the so-called "thermometer" of the cell involved in the recognition and activation of the stress response.

III. PHYSIOLOGY OF THE STRESSED CELL

When discussing the complex biochemistry of the stress proteins and their role in the cell, it is important to keep in mind that most of these proteins are rather abundant components of the normal, unstressed cell. Consequently, it seems reasonable to assume that the stress proteins participate in processes essential to the lifestyle of the normal, unstressed cell. As discussed in detail below, recent data have indeed implicated a role for many of the stress proteins in a number of basic and essential biochemical pathways. One suspects that after exposure to physiologic stress, the increased production (or activation) of the stress proteins occurs as a consequence of perturbations in these biochemical pathways in which the stress proteins normally participate. However, it is possible that some of the stress proteins also are recruited to perform new tasks that arise because of unique changes that occur in the cell as a result of the stress event. Consequently, to understand the biochemistry of the stress proteins and their putative role in protecting and/or facilitating recovery of normal cellular processes, it is imperative to also understand the physiological and biochemical changes that occur in the cell experiencing stress. Once defined, the question then becomes whether one or more of the stress proteins are involved in the repair and/or restoration of metabolic pathways that have been perturbed as a consequence of the stress event. Although this seems to be a reasonable approach to the problem, we know relatively little concerning which metabolic pathways are compromised in the cell experiencing stress. Moreover, whether the observed alteration in a metabolic pathway is a primary event or a secondary event arising from the change in cellular outlook from one of maintenance and/or growth to one of cellular defense is difficult to resolve. For example, if a particular biochemical pathway is inhibited in the stressed cell, is it due to the inadvertent inactivation of a key enzyme or to an intentional change made by the cell to provide for its protection?

An obvious consequence of physiologic stress in mammalian cells is the abrupt arrest of cell growth and proliferation. In cells from homeothermic organisms, induction of the stress response is accompanied by an arrest of the cell cycle and the temporary cessation of proliferation (Rao and Engelberg 1965; Westra and Dewey 1971; Zeuthen 1971). However, as was discussed earlier, in the case of cells from most poikilotherms, such growth arrest is temporary. For example, after a period of time at the elevated temperature, usually sufficient to increase the expression of the stress proteins, both bacteria and yeast apparently acclimate to their new temperature (or environment) and slowly recover their normal proliferative activities (see Lindquist 1986). This is evidenced by their return to normal gene expression (i.e., expression similar to that prior to the insult) and re-entry into and progression through the cell cycle. In contrast, in cells from higher organisms, where body temperature is tightly regulated, acclimation to higher temperatures (i.e., >41°–42°C) is not typically observed. Instead, with prolonged exposure at the higher temperature, most mammalian cells will continue to express the stress proteins up to the point of cell death.

A. Changes Occurring in the Nucleus

Using the electron microscope, a number of morphological changes occurring in the nucleus of mammalian cells placed under stress have been described. In general, many of these changes may be a consequence of the increased formation of insoluble protein (and/or nucleic acid) complexes resulting from the exposure of the cell to the various protein denaturants that activate the stress response (Littlewood et al. 1987). One obvious thermal effect is the accumulation and/or aggregation of so-called perichromatin granules dispersed throughout the nucleus (Fig. 1A, B) (Heine et al. 1971; Welch and Suhan 1985). Because hnRNA processing (i.e., "splicing"), but not transcription per se, appears compromised after heat shock treatment, it is likely that such granules represent accumulated aggregates of unprocessed mRNA (Mayrand and Pederson 1983; Yost and Lindquist 1986). Indirect immunofluorescence techniques have demonstrated that some of the components involved in RNA processing are perturbed after heat shock. For example, using antibodies specific to some of the components involved in hnRNA processing, the so-called small nuclear ribonucleoprotein complexes (snRNPs), one observes a change in the normal intranuclear distribution of the snRNPs following heat shock treatment. In contrast, in the cell first made

Figure 1 Heat shock treatment results in a number of morphological changes within the nucleus and cytoplasm of rat fibroblasts. Rat fibroblasts growing in 35-mm plastic dishes were incubated at either 37°C or 42°C for 3 hr. Immediately after the heat shock treatment, both the normal and stressed cells were fixed and stained with aqueous uranyl acetate/lead citrate. After embedding in EPON, thin sections through the nucleus or the cytoplasm were taken and analyzed by electron microscopy. (*A*, *C*, *E*) Sections from the 37°C cells; (*B*, *D*, *F*) sections taken from the heat-shock-treated cells. (*A*, *B*) Thin sections of the nucleus showing the perturbations in nucleolar (Nu) integrity and the appearance of intranuclear actin filaments (AF). (*C*, *D*) Section near the perinuclear region of the cells illustrating the fragmentation of the Golgi (G) complex. (*E*, *F*) Sections taken further out in the cytoplasm showing the heat-induced changes in mitochondrial morphology.

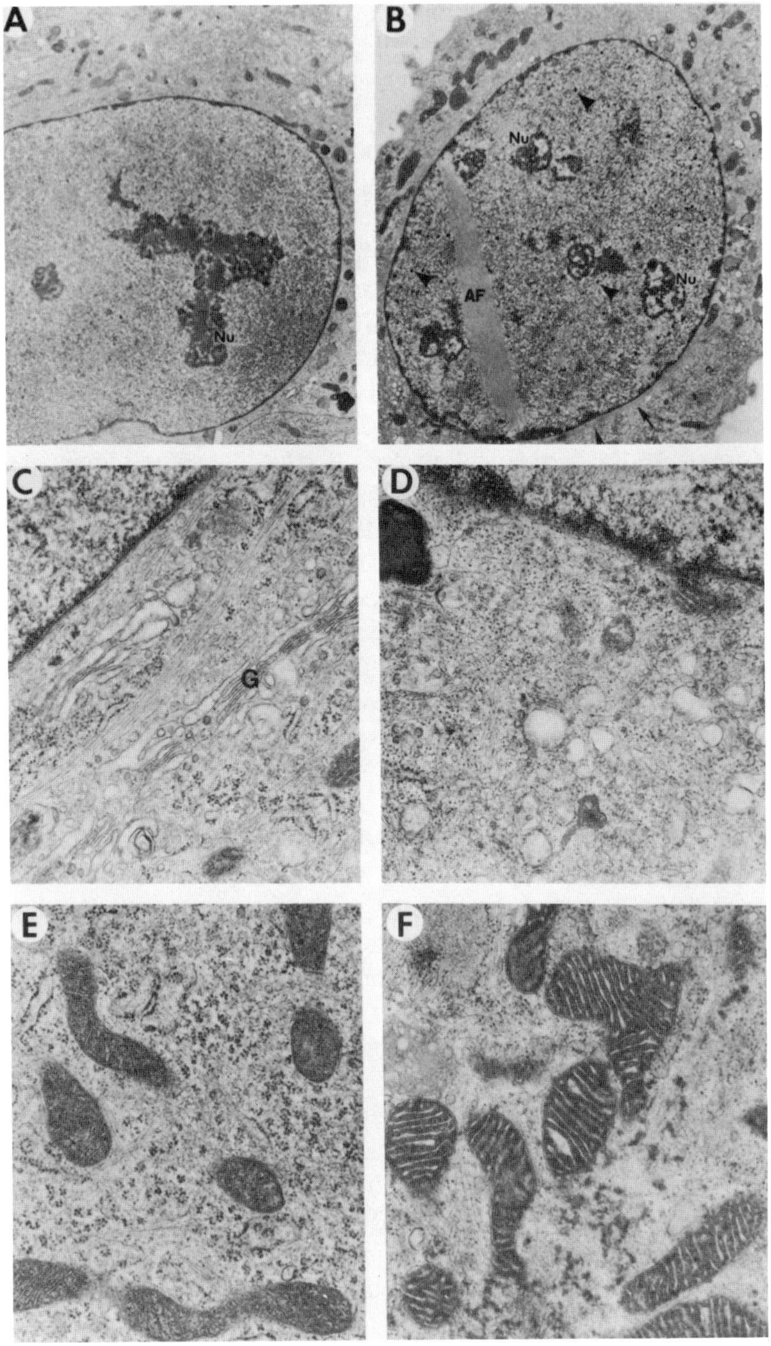

Figure 1 (See facing page for legend.)

thermotolerant, significantly less disruption in normal snRNPs staining pattern is observed after the heat challenge (Welch and Mizzen 1988). Consistent with this latter observation are studies demonstrating less inhibition of normal RNA processing in cells first made thermotolerant and then exposed to a subsequent heat shock challenge (Yost and Lindquist 1986).

Hyperthermia also results in rather significant changes in the integrity and morphology of the nucleolus, that site within the cell involved in ribosome biogenesis. In the electron microscope, one observes a loss and/or aggregation of the so-called granular components of the nucleolus, these representing preribosomes and/or other maturing ribonucleoprotein complexes (Fig. 1A, B) (Simard and Bernhard 1967; Amalric et al. 1969; Cervera 1978; Neumann et al. 1984; Welch and Suhan 1985). Consistent with these observed morphological alterations are biochemical studies demonstrating an inhibition of proper ribosomal RNA processing and a corresponding cessation of ribosome biogenesis following heat shock (Amalric et al. 1969; Ellgaard and Clever 1971; Rubin and Hogness 1975; Bouche et al. 1979). Thus, the nucleolus appears to be an extremely sensitive organelle and, for the most part, is rendered inactive in most cells after heat shock treatment. Again, nucleolar morphology and function appear less compromised in cells first made thermotolerant and then challenged with a second hyperthermic event (Pelham 1984; Welch and Mizzen 1988). As discussed below, a major target for some of the hsps is the damaged nucleolus and in particular the aggregated and/or denatured preribosomal particles.

Finally, a third and rather curious abnormality occurring in the nucleus of the heat-treated cell concerns the appearance of rod-like filaments comprised, at least in part, of the cytoskeletal component, actin (Pekkala et al. 1984; Welch and Suhan 1985; Iida et al. 1986). Following a 43°C/90-minute heat shock treatment of rat fibroblasts, approximately 50% of the cells exhibit densely packed and parallel-running fibers of actin filaments that traverse the nucleus, but which never appear to extend through the nuclear envelope into the cytoplasm (Fig. 1A, B). To date, similar intranuclear actin rods have not been observed in cells placed under stress by agents other than heat. Other investigators, however, have reported the appearance of similar intranuclear actin inclusion bodies in certain cells exposed to high levels of dimethylsulfoxide or to calcium ionophores (Fukui and Katsumura 1979; Osborn and Weber 1980; Sanger et al. 1980). Both the source of the intranuclear actin and the physiological relevance of the actin rods remain unclear. These intranuclear actin bundles could arise from preexisting nonfilamentous nuclear actin or via recruitment of actin from the cytoplasm. Regarding their possible function, it has been suggested that nuclear actin may serve some role in transcription and/or hnRNA processing activities (Scheer et al. 1984). Further studies are necessary to delineate the significance of these very unusual intranuclear actin structures and whether they are involved in the changing patterns of gene expression that occur in the cell experiencing stress.

B. Changes Occurring in the Cytoplasm

As mentioned earlier, one consequence of stress in mammalian cells is an abrupt arrest in cell growth. Providing that the heat shock treatment is not too severe, proliferating fibroblasts will exhibit increased growth arrest and a corresponding increased spreading on their substratum. Accompanying their increased spreading is an apparent increase in the number of actin microfilaments and stress fibers spanning the cytoplasm of the stressed cell (Thomas et al. 1982). Similar increases in the number of actin filaments are often observed in cells rendered growth-arrested by other means (e.g., upon density arrest, removal of growth factors). With time of recovery from the stress event and as the cells regain their normal pattern of gene expression and reenter the cell cycle, there is a corresponding decreased number of actin stress fibers. Some investigators have reported a diminishment in the number of actin filaments following severe heat shock treatment (Glass et al. 1985). Thus, depending on the cell type, hyperthermia may result in completely different effects on the actin cytoskeleton. However, heat shock treatment must not be so severe as to result in the cell's detaching from its substratum, thereby leading to a reduction in the number of actin microfilaments.

Alteration of a second cytoskeletal network, the intermediate filaments, has been observed in a number of different mammalian cells placed under physiologic stress. The intermediate or so-called 10-nm filaments comprise primarily a subunit protein of 50–60,000 daltons whose exact identity differs slightly depending on cell type. For example, there are five major classes of intermediate filaments: (1) vimentin filaments present in mesenchymal cells; (2) desmin filaments of smooth, skeletal, and cardiac muscle cells; (3) keratin filaments of epithelial cells; (4) neurofilaments found in neuronal cells; and (5) glial filaments present in all types of glial cells. It is not uncommon that a particular cell type will contain two classes of intermediate filaments. Within the normal, unstressed cell, the intermediate filaments exhibit a well-spread distribution throughout the cytoplasm, extending from the nuclear envelope out to the plasma membrane. For the most part, the filaments appear closely aligned with another cytoskeletal network, the tubulin-containing microtubules. Following exposure of cells to a variety of agents that induce the stress response, the intermediate filaments rapidly redistribute (or "collapse") from their cytoplasmic locale into a meshwork of filaments wrapped around the nucleus (Fig. 2) (Falkner et al. 1981; Thomas et al. 1982; Welch et al. 1985; Collier and Schlesinger 1986). Collapse of the filaments after stress still occurs in cells treated with inhibitors of transcription and translation and, therefore, their redistribution appears to be independent of the changes in gene expression associated with the activation of the stress response (Welch and Feramisco 1985a; Collier and Schlesinger 1986). Unlike the intermediate filaments, little or no obvious change occurs with respect to the organization of the microtubules after stress. As the cell recovers from the stress event, the intermediate fila-

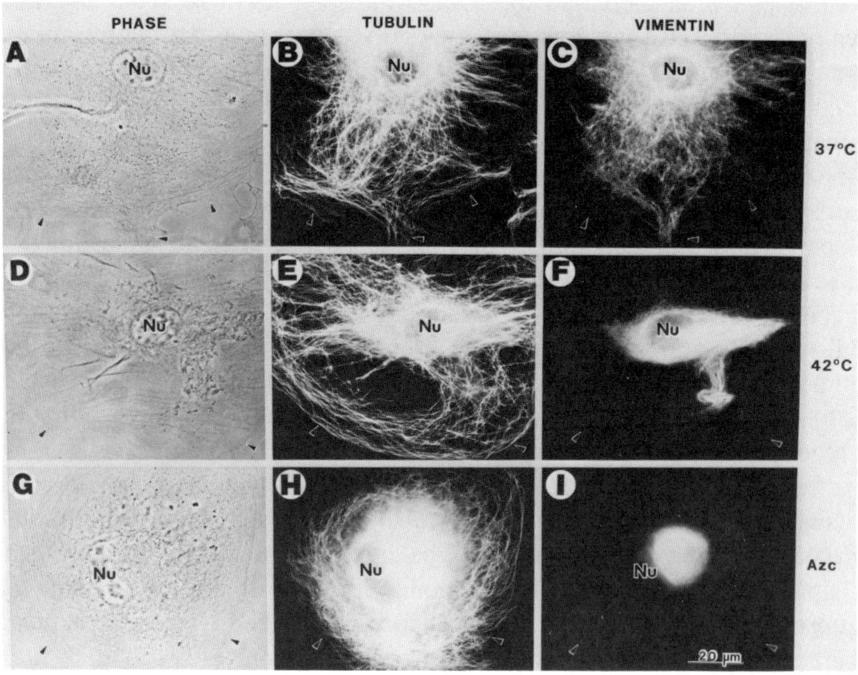

Figure 2 Distribution of microtubles and intermediate filaments in normal and stressed rat embryo fibroblasts. Rat embryo fibroblasts were grown at 37°C, 42°C, or 37°C in the presence of 5 mM L-azetidine 2-carboxylic acid (AzC), an amino acid analog of proline. The cells were processed and analyzed by double-label indirect immunofluorescence using a mouse monoclonal antibody specific for α-tubulin and a rabbit polyclonal antibody specific for vimentin. For reference, the cell borders are indicated by arrowheads in the phase and fluorescence micrographs. (Nu) Nucleus. (*A–C*) 37°C; (*D–F*) heat-shock-treated cells; (*G–I*) AzC-treated cells.

ments slowly regain their normal well-spread distribution throughout the cytoplasm. Such recovery, in contrast to their collapse, does appear to require new transcription/translation events (Collier and Schlesinger 1986). Moreover, the kinetics of recovery of normal intermediate filament morphology appears to correlate with the cessation of stress protein synthesis and the return of normal patterns of protein synthesis (W.J. Welch, unpubl.).

Although most agents that induce the stress response similarly elicit the collapse of the intermediate filaments, substantial data indicate that collapse can occur in cells as a consequence of other treatments that do not result in an activation of the stress response. For example, exposure of cells to agents that lead to the depolymerization of the microtubles (e.g., vinblastine, colcemide, colchicine) also initiates a collapse of the intermediate filaments but no discernible increase in stress protein synthesis (Welch and Feramisco 1985a). Microinjection studies have demonstrated that subtle perturbations in the organization and/or the structure of the microtubules can initiate the collapse of the neighboring intermediate filaments. Specifically, introduction into the cell of antibodies against tubulin, the microtubule subunit component, results in a collapse of the intermediate filaments but, again, little or no change in microtubule organization nor any increased expression of the stress proteins (Blose et al. 1984). Finally, a prior collapse of the intermediate filaments (e.g., by microtubule poisons) does not compromise the ability of the cell to activate the stress response and increase the expression of the stress proteins (Welch and Feramisco 1985a).

Although it is clear that the actin-containing microfilaments and the tubulin-containing microtubules are involved in basic cellular events such as cell locomotion, adhesion, and mitosis, the exact biological role of the intermediate filaments is not known. Consequently, the relevance of intermediate filament collapse in the cell experiencing stress is unclear. The redistribution of the filaments into a tight cage enveloping the nucleus of the stressed cell may provide some type of protection for the nuclear contents. However, despite this envelopment of the nucleus with the collapsed intermediate filaments, new mRNAs apparently can pass from the nucleus into the cytoplasm, and proteins made in the cytoplasm appear capable of migrating back into the nucleus (Welch and Feramisco 1985a). Recent studies have implicated a role for the cAMP-dependent protein kinase (A-kinase) in the regulation of intermediate filament collapse (Escribano and Rozengurt 1988; Lamb et al. 1989). For example, treatment of mammalian cells with agents that elevate A-kinase activities result in a rapid collapse of the 10-nm filaments. Accompanying such collapse by these means is the increased phosphorylation of two sites along the vimentin molecule. In the case of heat shock or other forms of physiological stress, increased phosphorylation of these same two sites within vimentin is observed. Moreover, a temporal correlation exists between the dephosphorylation of these sites on vimentin and the recovery of normal intermediate filament morphology (Lamb et al. 1989). These results raise the possibility that in the stressed cell there is an increased activity (or "redirection" of activity) of the cAMP-dependent protein kinase resulting, in part, in the increased phosphorylation of vimentin and the concomitant collapse of the filaments. A number of other studies have also implicated a role for A-kinase in the stress response. First, a number of mutations in yeast that result in abnormally high levels of cAMP (and presumably a corresponding elevated activity of A-kinase) render the cells thermosensitive (Iida and Yahara 1984; Shin et al. 1987). Conversely, cells that display low levels of cAMP appear more thermoresistant. Second, in yeast, the heat shock transcription factor, a pre-existing cellular protein, appears to be activated via a phosphorylation event, perhaps mediated either directly or indirectly by A-kinase (Sorger et al. 1987). Hence, further studies analyzing the activity of A-kinase and its possible role in mediating one or more of the many events occurring in the stressed cell appear warranted.

Concomitant with the collapse of the intermediate filaments, a relocalization of both the mitochondria and ribosomes/polysomes occurs in the stressed cell. For example, after heat shock, both the mitochondria and much of the translation machinery are observed within the perinuclear region of the cell in close association with the collapsed intermediate filaments (Welch and Suhan 1986). As the cells recover from the stress event and regain their normal distribution of intermediate filaments, the mitochondria and ribosomes slowly regain their normal distribution throughout the cytoplasm. On the basis of these and other observations, it has been suggested that the mitochondria and translation machinery are in fact "anchored" to the intermediate filaments, and as the filaments collapse, there is a corresponding redistribution of these two organelles (Fulton et al. 1980; Cerveza et al. 1981). Possibly, however, due to their large size, the mitochondria and polysomes are simply entrapped in a nonspecific fashion within the collapsing intermediate filaments.

Analysis of mammalian cells has revealed a number of perturbations occurring in certain cytoplasmic organelles following heat shock treatment. For example, high-temperature treatment of rat fibroblasts results in subtle alterations in the structure of the endoplasmic reticulum (ER) and more obvious and severe changes in the organization of the Golgi complex. In contrast to its well-defined and highly organized collection of membrane stacks in the normal cell, the Golgi complex appears fragmented and dispersed as a collection of membrane vesicles throughout the cytoplasm of the cell following hyperthermic shock (Fig. 1C, D) (Welch and Suhan 1985). Surprisingly, little is known concerning whether these alterations in the integrity of the Golgi complex have any direct consequence on the ability of the stressed cell to continue proper protein glycosylation and secretory activities. It is of interest that two of the major stress proteins, the glucose-regulated 74–80-kD and 94–100-kD proteins, are components of the ER and ER/Golgi complex, respectively. Moreover, and as discussed below, the smaller glucose-regulated protein has been shown to participate in protein trafficking events in the ER. With respect to the functionality of these two organelles after stress, recent studies have described the increased levels of a heavily glycosylated 14-kD protein in human melanoma cells exposed to various heavy metals, thiol reactive agents, or hyperthermia (Niewiarowski et al. 1987). Increases in this 14-kD glycosylated molecule were observed even when the cells were placed under stress in the presence of transcription and translation inhibitors, indicating it to be a preexisting protein. In contrast, tunicamycin treatment of the cells before the heat stress prevented the accumulation of this small glycoprotein, a result that led the investigators to suggest that the protein is a lipid-linked intermediate involved in N-linked protein glycosylation. These studies may be indicative of an inhibition in normal protein glycosylation events following stress and a corresponding increase in the levels of this 14-kD glycoprotein possibly involved in N-linked protein glycosylation. With the recent excitement concerning a role for some of the stress proteins in protein trafficking events, we can expect more studies examining the ability of the stressed cell to properly glycosylate and/or transport proteins through both the ER and Golgi complex.

Striking alterations occur with respect to the structure and function of the mitochondria following induction of the stress response. Whereas the mitochondria of 37°C fibroblasts display a typical cylindrical shape with well-packed and organized cristae, after heat shock the mitochondria appear enlarged or "swollen" with the cristae now very prominent and the intracristal spaces enlarged (Fig. 1E, F) (Welch and Suhan 1985). In addition, as mentioned above, after stress the mitochondria also show a redistribution into the perinuclear region of the cells, presumably due to the collapsed intermediate filaments. These changes observed in the integrity of the mitochondria following heat shock are similar to those observed when cells are exposed to agents that inhibit electron transport or uncouple oxidative phosphorylation (Weinbach and Garbus 1968; Buffa et al. 1970). Moreover, similar to the case of cells treated with mitochondrial poisons, heat shock treatment results in a reduction of cellular ATP levels (Leenders et al. 1974a; Ellgaard and Maxwell 1975; Findly et al. 1983). Because so many mitochondrial poisons elicit the stress response in isolated *Drosophila* salivary glands, Leenders et al. (1974b) investigated whether a compromise in overall mitochondrial function and a reduction in cellular ATP levels might be the common trigger by which the heat shock response is activated. No definitive data were obtained to support this notion. Moreover, some agents, such as sodium arsenite, will result in an induction of the stress response but render no obvious changes in the levels of endogenous ATP. Nevertheless, considering the fact that the mitochondria (as well as the ER) are the major sites within the cell where calcium is sequestered and the fact that intracellular calcium levels transiently increase in cells following hyperthermic treatment (discussed below), further studies are needed to evaluate whether changes in mitochondrial integrity/function might be one mechanism by which the stress response is initiated.

IV. EARLY BIOCHEMICAL CHANGES IN THE STRESSED CELL

Although the events that lead to the transcription and translation of the stress proteins have been actively investigated, we still know relatively little concerning the early biochemical events that accompany the induction of the stress response. Specifically, only a few studies have focused on the possible changes in various metabolic pathways that may accompany the early part of the stress response. Although there are some definitive data in this regard, we still do not have a complete picture of what occurs, biochemically, during the early stages of the response. Moreover, even when changes in certain metabolic pathways are observed, it is difficult to ascertain whether such changes are intimate with respect to the initiation of the stress response or represent more secondary effects that arise as a consequence of the stress event and/or the redirection of cellular outlook from that of cell growth and maintenance to one of cell defense.

Because of the known effects of heat on phospholipids and fatty acids, a number of studies examined possible changes in membrane integrity and/or function following hyperthermic treatment of mammalian cells. Some investigators have reported little or only subtle change in membrane fluidity following heat shock treatment (Anderson et al. 1981; Lepock 1982; Lepock et al. 1983). However, there are conflicting reports concerning membrane lipid composition and the sensitivity to hyperthermic killing. For example, Cress and Gerner (1980) have reported that cells more resistant to hyperthermia (as assayed by cell survival) exhibit higher cholesterol content than those cells more sensitive to the same hyperthermic treatment. In addition, these investigators observed an overall increase in cholesterol content in cells made thermotolerant (Cress et al. 1982). Such an increase in membrane cholesterol would likely result in a decreased fluidity of cellular membranes. In contrast, Anderson and Parker (1982) concluded that thermotolerance was not accompanied by any obvious changes in membrane lipid composition and in particular, cholesterol levels. Similarly, Lepock and colleagues (1981) could find no evidence that membrane fluidity played an important role in hyperthermic killing nor that decreased membrane fluidity was correlated with increased thermoresistance. Although most workers in the field would agree that subtle changes do occur in the organization of the plasma membrane in cells after heat shock, the relative contributions of lipid versus protein thermal transitions and their effect on cell survival are still difficult to resolve.

Although potential changes in plasma membrane fluidity and the role of lipid composition in the sensitivity of cells to hyperthermia remain a controversial issue, there seems little doubt that heat shock or other forms of stress do effect marked changes in the activity and/or regulation of a number of plasma membrane components. Examples of changes occurring at the plasma membrane in the stressed cell include (1) a compromise in Na^+/K^+ ATPase activities with the net result being an increased level of K^+ and a corresponding decreased level of Na^+ inside the cell (Yi 1979; Burdon and Cutmore 1982; Anderson and Hahn 1985); (2) an overall decrease of epidermal growth factor binding activities, apparently due to a reduced affinity of the receptor for its ligand (Magun and Fennie 1981); (3) a decrease in insulin binding activities, apparently as a consequence of reduced numbers of receptor (rather than a change in affinity) (Calderwood and Hahn 1983); (4) a reduced number of cell-surface histocompatability antigens (Mehdi et al. 1984); (5) a reduction in the ability of the lectin, concanavalin A, to induce patching and capping of cell-surface glycoproteins (Stevenson et al. 1981); and (6) an increase in hexose transport into the cell (Warren et al. 1986). Note that most of these changes would result in the cell's becoming refractory to a variety of different stimuli associated with its normal lifestyle. Again, whether these changes are a reflection of perturbations in the structure/function of membrane components involved in these processes (e.g., as a consequence of exposure to potential protein denaturing agents that induce the stress response) is unknown. Alternatively, whether these changes are part of a mechanism by which the stressed cell redirects or shuts down plasma-membrane-mediated events in an effort to protect itself (i.e., become refractory to other stimuli) is difficult to distinguish.

Finally, whether any of these changes are directly responsible for the initiation of the stress response or are more secondary effects remains entirely unclear.

Several studies are in agreement that one early consequence of hyperthermia is a rapid drop in intracellular pH and an increase in cytosolic calcium levels (Hammond et al. 1982; Findly et al. 1983; Weitzel et al. 1985; Drummond et al. 1986; Stevenson et al. 1986). Because changes in pH as well as calcium are now being implicated as mediators by which certain "second messengers" are activated in the cell (e.g., those leading to the activation of gene transcription events), Drummond et al. (1986) examined the possibility that these changes in intracellular pH and Ca^{++} observed after heat shock may be involved in the activation of heat shock gene transcription. No direct evidence was found to support this notion. For example, manipulation of mammalian cells at their normal temperature by external agents that either lower intracellular pH or result in increased intracellular calcium levels did not result in the activation of the heat shock response. Moreover, preventing intracellular pH decreases after heat shock did not inhibit the activation of heat shock gene transcription nor the increase of hsp synthesis. Hence, although changes do occur in intracellular pH and calcium levels following hyperthermia, these changes do not appear sufficient nor required for the induction of the heat shock response. Even more confusing in this regard is a study reporting that artificially increasing intracellular pH will result in the activation of the stress response (Whelan and Hightower 1985).

Finally, one area concerning changes in metabolic pathways arising from physiologic stress that continues to receive considerable attention concerns overall cellular energy metabolism. Most of the earlier attention in this regard arose from the fact that in isolated *Drosophila* salivary glands, a number of agents that either uncouple oxidative phosphorylation or inhibit election transport result in the activation of the stress response (for review, see Ashburner and Bonner 1979). In this regard, it does appear that an early effect of hyperthermia is a rapid reduction in intracellular ATP levels (Leenders et al. 1974a; Ellgaard and Maxwell 1975; Findly et al. 1983). This reduction in ATP is most likely correlated with the observed alterations in the integrity of the mitochondria, most notably their "swollen" appearance and the accompanying changes in the intramitochondrial cristae after heat shock (discussed above). As was also mentioned earlier, these changes in the integrity of the mitochondria after heat shock are similar to those observed in cells exposed to a number of agents that directly inhibit mitochondrial function. The alterations in mitochondrial integrity and function after heat shock may be a consequence of the observed denaturation of various intramembranous mitochondrial proteins (Lepock et al. 1983). Not surprisingly, an increase in anaerobic metabolism accompanies the inhibition of mitochondrial function after heat shock (Christiansen and Kuamma 1969; Mondovi et al. 1969; Dickson and Oswald 1976; Hammond et a 1982; Nickells and Browder 1988). This shift from aerobic to anaerobic energy production may be essential for the cell to survive the hyperthermic treatment. Some support for this idea is the observation that in yeast, those cells growing on dextrose, which is utilized exclusively via the glycolytic pathway, appear more thermoresistant than those growing on acetate, a metabolite that is used exclusively in respiration (Lindquist et al. 1982). Moreover, recent studies in both yeast and *Xenopus* give evidence that the cell may be more dependent on glycolytic energy metabolism after heat shock. In the case of yeast, heat shock treatment results in an increased synthesis of two glycolytic enzymes, enolase and glyceraldehyde 3-phosphate dehydrogenase (GAPDH) (Iida and Yahara 1985). Similarly, *Xenopus* embryos exposed to elevated temperatures exhibit an increased activity and accumulation of a specific isozyme of GAPDH (Nickells and Browder 1988). Thus, in addition to the more classically defined hsps, we probably should add at least two new members of the stress protein family, both of which are components of the glycolytic pathway. In light of these results, it will be interesting to see whether other cellular enzymes involved in energy metabolism also exhibit changes in either their levels or specific activity both during and/or after recovery from stress.

As an aside, it is worth pointing out that these observations regarding changes in cellular energy metabolism may be most relevant for the use of site-directed hyperthermia in the clinical treatment of cancer. For example, most animal tumors exist in a reduced oxygen environment and consequently exhibit a higher dependence on an anaerobic rather than aerobic pathway for energy production. Hence, when considering the situation in yeast where those cells exclusively dependent on anaerobic metabolism appear more thermally resistant, one wonders whether the anaerobic environment of the tumor may similarly render it more resistant to thermal killing. More studies are necessary to delineate the apparent connection between aerobic/anaerobic energy metabolism and cellular thermosensitivity. The results of such studies would likely prove invaluable in the use of site-directed hyperthermia for the treatment of animal tumors.

V. THE MAMMALIAN STRESS PROTEINS

In defining the stress proteins, one may divide the family into two major groups: the heat shock proteins (hsps) and the glucose-regulated proteins

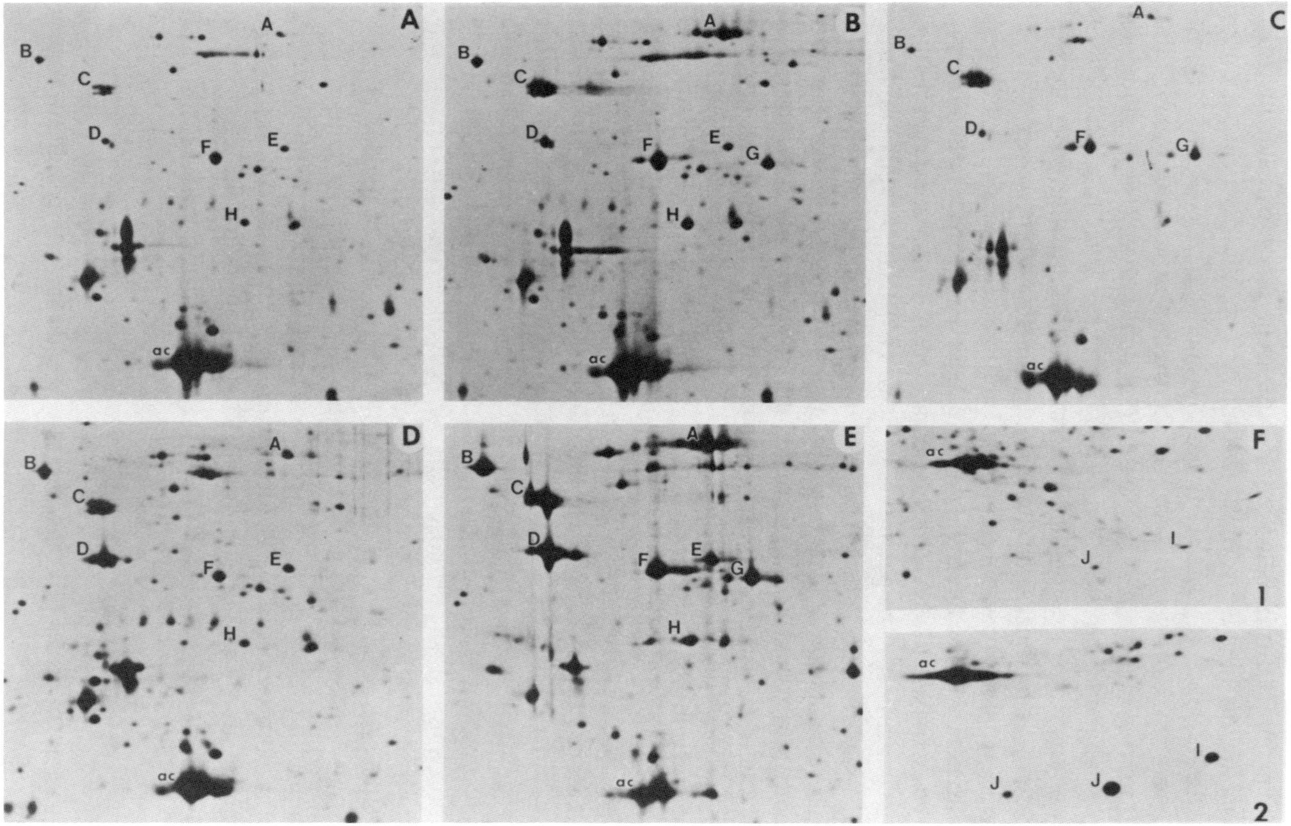

Figure 3 (See facing page for legend.)

(grps). In lieu of a defined function, we refer to the individual stress proteins according to their apparent migration on one- (or two-) dimensional gels. Unfortunately, due to differences between gel systems employed by various investigators in the field, the stress proteins are often ascribed different molecular weights. Whenever possible, I will try to minimize confusion in this regard. Cells exposed to elevated temperatures or to heavy metals commonly exhibit an increased synthesis of proteins with apparent masses of 28, 32, 47, 58, 72, 73, 90, and 110 kD, with a concomitant reduction in the synthesis of most other proteins that were actively being translated prior to the environmental insult (for reviews, see Ashburner and Bonner 1979; Craig 1985; Lindquist 1986). We refer to this group of proteins as the hsps (Fig. 3). The second family of stress proteins, the grps, exhibit increased synthesis in cells (1) deprived of glucose or oxygen, (2) following treatment with agents that perturb calcium homeostasis, or (3) treated with inhibitors of glycosylation (Pouyssegur et al. 1977; Olden et al. 1978; Wu et al. 1981; Welch et al. 1983; Subjeck and Shyy 1986). The major grps exhibit apparent masses of approximately 75, 80, and 100 kD (Fig. 3). In some cell types, and depending on the stress agent used (e.g., amino acid analogs), synthesis of the grps increases along with that of the hsps. As can be seen in the two-dimensional gels shown in Figure 3, (1) most of the major stress-induced proteins are relatively acidic polypeptides with their isoelectric points falling between pH 5 and 7; (2) almost all of the stress proteins are synthesized at modest or even high levels in cells grown under normal conditions; (3) synthesis of most, but not all, normal cellular proteins (i.e., those synthesized in the unstressed cell) is reduced in the cells experiencing stress. Finally, the two groups of stress proteins are often observed to be regulated in an inverse fashion. For example, in cells deprived of glucose or treated with a calcium ionophore, there is an increased synthesis of the grps with a concomitant decreased synthesis of

Figure 3 Stress proteins of mammalian cells. Rat embryo fibroblasts were labeled with [^{35}S]methionine (*A*) at 37°C, (*B*) after a 43°C/90-min heat shock; (*C*) after exposure to 120 μM sodium arsenite; (*D*) after exposure to 10 μM calcium ionophore A23187; (*E*) after exposure to the proline analog AzC (5 mM). After labeling, the cells were harvested and the labeled proteins were analyzed by two-dimensional gel electrophoresis (pH 4–8, acid end to the left). Shown in panels *A–E* are only those regions analyzing the higher-molecular-weight stress proteins. (*F*) A portion of the gels showing the lower-molecular-weight heat shock proteins (1, normal; 2, arsenite-treated cells). Stress proteins indicated are (*A*) 110 kD; (*B*) 100 kD; (*C*) 90 kD; (*D*) 80 kD; (*E*) 75 kD; (*F*) 73 kD; (*G*) 72 kD; (*H*) 58 kD; (*I*) 32 kD; (*J*) 28 kD. Actin is indicated by ac.

some of the hsps (Welch et al. 1983; Subjeck and Shyy 1986). Refeeding the cells glucose or removal of the ionophore results in a reduction of grp synthesis and a corresponding increased production of the hsps (Subjeck and Shyy 1986).

A. The grps and hsps Are Structurally Related

Biochemical, immunological, and cDNA sequence analyses have revealed that the members of the two stress protein families are related. Specifically, both the grp 75- and 80-kD proteins exhibit considerable homology with the hsp 72- and 73-kD proteins (Munro and Pelham 1986; Ting and Lee 1988; Mizzen et al. 1989). Moreover, the 72-, 73-, 75-, and 80-kD proteins all exhibit binding to ATP in vitro (Welch and Feramisco 1985). Similarly, the grp 100-kD protein displays considerable sequence homology with the hsp 90-kD protein (Sargan et al. 1986). Consequently, because of their similar structures and biochemical properties, what we learn about one protein (e.g., grp 80-kD) may be applicable to its related partner (e.g., hsp70). As a result, our task in determining the function of the stress proteins has become simplified to some extent. As discussed below, many of those stress proteins related to one another do appear to function in a similar fashion, but within their own distinct subcellular compartments.

For the sake of brevity, I have summarized much of what we know concerning the structure and properties of the individual mammalian stress proteins. These data, presented in Table 1 and discussed in more detail below, comprise a composite of studies performed in many laboratories and represent a brief but current status of our knowledge concerning the properties, intracellular locale, and phenomenology of the various mammalian stress proteins.

B. The Heat Shock Proteins

1. 28 kD (Also Referred to as the Low-m.w. hsp or hsp27)

In mammalian cells, 28 kD does not label efficiently with [^{35}S]methionine and therefore is most commonly detected using a mixture of [^{3}H]-labeled amino acids (Kim et al. 1984; Welch 1985). In cells grown at 37°C, the steady-state synthesis of 28 kD is relatively low and is often difficult to detect by metabolic labeling using radiolabeled amino acids. In contrast, the protein is easily visualized in the 37°C cell when labeling is done with [^{32}P]H_3PO_4 (Kim et al. 1984; Welch 1985). The protein consists of at least four major isoforms, three of which are phosphorylated. Interestingly, within minutes after exposure of cells to

Table 1 The hsp70 proteins: A family of related ATP-binding proteins present within different intracellular compartments

hsp73:	An abundant and constitutive protein present in the cytoplasm and nucleus. Implicated in the uncoating and/or reformation of clathrin-coated vesicles and in the maintenance of a translocation-competent state of certain proteins that are transferred across intracellular membranes.
hsp72:	Negligible levels in the normal cell but induced to high levels after stress. Present within the nucleus, nucleolus, and cytoplasm. Likely to function, similar to hsp73, in the stressed cell to maintain the solubility of cytosolic and nuclear proteins as well as perhaps facilitate the removal of denatured proteins.
grp 80 kD or BiP:	Abundant constitutive protein present within the lumen of the endoplasmic reticulum. Facilitates the proper assembly of monomeric proteins into larger macromolecular complexes.
grp 75 kD:	Constitutive protein localized within the mitochondria, most likely present in the matrix. Presumed to function, similar to BiP, in the proper assembly of mitochondrial monomeric proteins into larger macromolecular complexes.

defined growth factors or mitogens, certain tumor promoters such as phorbol esters, or to calcium ionophores, the phosphorylation of 28 kD increases significantly (Welch 1985). These agents, unlike heat shock or stress, do not result in an increased synthesis of the protein. Considering that in *Drosophila* the analogous low-molecular-weight heat shock proteins exhibit both differential expression and/or changes in their phosphorylation status during both development and differentiation (Ireland and Burger 1982; Sirotkin and Davidson 1982; Cheney and Shearn 1983; Zimmerman et al. 1983), these observations concerning the phosphorylation of mammalian 28 kD are intriguing. It will be interesting to see if 28 kD is also differentially regulated (either its synthesis or phosphorylation) throughout mammalian development and/or differentiation. Moreover, considering the extremely rapid kinetics by which 28 kD phosphorylation increases in cells exposed to mitogens/growth factors, one wonders whether the protein is somehow involved in the initiation of cellular proliferation events.

Recent data have implicated a connection between 28 kD and steroids. Previous studies had reported a 24-kD protein whose synthesis increased following exposure of human breast tissue culture cells to estrogen (Edwards et al. 1980). In addition, the protein was shown to be estrogen-regulated during the menstrual cycle in normal human endometrium (Ciocca et al. 1983). Recently, a cDNA for this estrogen-regulated 24-kD protein was isolated, and the deduced sequence of the protein revealed it to be identical with the mammalian 28-kD stress protein (Moretti-Rojas et al. 1988; W. McGuire, pers. comm.). These investigators have found that hsp 28 kD is present in high levels in cells that contain either estrogen or progesterone steroid receptors but is absent in receptor negative cell lines. Again these observations sound reminiscent of the situation in *Drosophila*, where the analogous low-molecular-weight hsps appear to be regulated in response to the steroid molting hormone, ecdysone (Ireland and Berger 1982). The significance of these findings and, in particular, the role of hsp28 in the response of cells to steroids is still unclear.

Mammalian 28 kD has been purified from HeLa cells and exhibits a rather high native molecular mass ranging from 200–400 daltons in 37°C cells on up to >2 x 10^6 daltons in cells after heat shock (Arrigo and Welch 1987). Like the low-molecular-weight hsps from other organisms, mammalian 28 kD exhibits some sequence homology to the α-crystallin proteins present in the lens (Ingolia and Craig 1982; Hickey et al. 1986). The α-crystallins similarly form higher-ordered structures, and we suspect that those domains shared between hsp28 and α-crystallin are responsible for the similar self-assembling properties exhibited by the two proteins. Using an antibody raised against the purified protein, a perinuclear distribution of 28 kD in cells maintained at 37°C has been observed. Using double-label staining with a fluorescent lectin specific for some components of the Golgi, we find that much of the perinuclear staining of 28 kD appears coincident with the Golgi complex (Arrigo et al. 1988). Biochemical fractionation studies indicate that 28 kD is apparently not an integral Golgi protein but rather appears to be concentrated around the Golgi cisternae. Following heat shock treatment, much of 28 kD relocalizes within the nucleus and, to a lesser extent, is found dispersed throughout the cytoplasm (Arrigo et al. 1988). Such a dispersal of the protein may be correlated with previous observations that much of the Golgi complex fragments after heat shock and is subsequently found to exist as a collection of vesicles distributed throughout the cytoplasm (Welch and Suhan 1985). As the cells recover from the hyperthermic shock, there is a gradual exit of hsp28 from the nucleus and a corresponding return into the perinuclear region. Biochemical fractionation studies examining the detergent solubility of 28 kD before and after heat shock have provided some interesting results. Although the protein is easily solubilized following nonionic detergent treatment of 37°C cells, most of 28 kD is found in the detergent-insoluble nuclear/cytoskeletal fraction after heat shock. With time of recovery from the heat shock treatment, the protein gradually returns to the detergent-soluble fraction. In cells first made thermotolerant and then exposed to a second heat shock treatment, significantly less of 28 kD partitions into the detergent-insoluble fraction (Arrigo et al. 1988).

2. 32 kD

The induction of the 32-kD stress protein was first observed in cells exposed to various heavy metals, such as arsenite and cadmium (Levinson et al. 1980). Subsequent studies showed that other reagents that also react with vicinal sulfhydryl groups, such as iodoacetamide, similarly increase the synthesis of 32 kD, along with the other classic stress proteins (Caltabiano et al. 1986). Recent work has also demonstrated an induction of the protein in cells exposed to either ultraviolet light or the oxidizing agent, H_2O_2 (Keyse and Tyrell 1987). Although a slight induction of 32 kD is observed in cells after heat shock, the levels appear far lower than that observed in cells following metal exposure. That induction of 32 kD may be associated with overall oxidative stress is supported by the findings that agents that result in elevated intracellular levels of H_2O_2 or superoxides (O_2^-) also elicit an increased synthesis of 32 kD (Keyse and Tyrell 1987). A common target of all these agents that induce 32 kD may be glutathione. For example, if cells are first depleted of intracellular glutathione, significantly less UV light exposure is required for 32 kD induction (Keyse and Tyrell 1987). Although limited (and perhaps conflicting) data exist regarding the role of glutathione in the stress response, some studies have reported that glutathione levels are increased above normal in cells made thermotolerant and that glutathione depletion can often result in cell thermosensitization (Mitchell and Russo 1983; Mitchell et al. 1983; Freeman et al. 1985).

Recently, cDNA clones that encode 32 kD have been isolated from both mouse 3T3 cells and rat myoblasts. Using these probes, it has been shown that an increase in 32 kD expression also occurs in 3T3 cells exposed to various tumor promoters or chemical carcinogens (Kageyma et al. 1988). In the case of rat muscle cells, the levels of 32 kD decline as the rat myoblasts fuse into myotubes. Moreover, once fully differentiated, the myotubes no longer synthesize 32 kD either before or after exposure to arsenite (Shuman and Przybyla 1988). Curiously, the levels of

32 kD also decline when the myoblasts are cultured in the presence of steroid hormones.

The deduced amino acid sequence of 32 kD has revealed the protein to be highly related to heme oxygenase. In addition, the rat heme oxygenase gene itself contains a consensus heat shock promoter element and exhibits some increased transcription after heat shock (Muller et al. 1987; Shibahara et al. 1987). Heme oxygenase is an essential component in the catabolism of heme, catalyzing the first step in the degradation of heme to bilirubin. In addition to its oxygen-binding role in both hemoglobin and myoglobin, heme also is a component of a number of other important enzymes. A notable example is the cytochromes. With the exception of cytochromes b5 and P450, which are components of the ER, most cytochromes reside within the mitochondria where they participate in electron transfer events involved in respiration. Considering that overall mitochondrial function, most notably ATP production, is compromised in cells exposed to various stress agents, one wonders whether alterations in cytochrome(s) structure/function may be a contributing factor. If so, increases in a heme oxygenase like stress protein, 32 kD, may be necessary to facilitate the turnover of some heme-containing proteins, like the cytochromes, that may have become inactivated as a consequence of the stress event.

3. 47 kD

Nagata and colleagues (1986) have described a heat-induced 47-kD protein in chicken fibroblasts. This protein, relative to the other major members of the stress protein family, has an unusually basic pI of approximately 9.0 and consequently is somewhat difficult to resolve in two-dimensional gels. 47 kD is a membrane glycoprotein, contains phosphate, and exhibits an affinity for collagen in a pH-dependent manner (Saga et al. 1987). Curiously, this collagen-binding protein appears to be an intracellular protein present near or within the ER. Following transformation of chick fibroblasts with Rous sarcoma virus, the synthesis of 47 kD decreases approximately two- to threefold, yet its level of phosphorylation appears to increase five- to sevenfold. In contrast, although the synthesis of 47 kD increases after heat shock treatment, the relative levels of 47 kD phosphorylation do not change (Nagata and Yamada 1986). Further studies are needed to ascertain the function of this somewhat novel eukaryotic stress protein.

4. 58 kD (Also Referred to as hsp60 or the GroEL Homolog)

Although noted in some earlier studies in mammalian cells, 58 kD was first documented as a true heat shock protein in Tetrahymena (McMullin and Hallberg 1987). Using an antibody against 58 kD, the protein was shown to localize within the mitochondria. The protein is apparently not exposed on the cytoplasmic side of the mitochondria as ascertained by proteinase protection experiments using isolated mitochondria. Hallberg and colleagues have shown that the Tetrahymena 58-kD protein is immunologically related to the more well characterized GroEL hsp of *Escherichia coli* (McMullin and Hallberg 1988). Tetrahymena 58 kD, like GroEL, exists as an oligomer with both proteins exhibiting a rather large sedimentation value of ~20–25S. In the electron microscope, the two proteins appear morphologically indistinguishable, with each exhibiting a "donut"-like structure.

Genetic studies have shown that the GroEL gene is essential for the morphogenesis of bacteriophages λ, T4, and T5 (Georgopoulos and Hohn 1978). Biochemical studies have shown that GroEL facilitates the correct posttranslational assembly of the bacteriophage head and tail proteins (Coppo et al. 1973; Georgopoulus et al. 1973; Sternberg 1973; Hendrix and Tsui 1978). Recently, a protein present in the stroma of higher plant chloroplasts, termed the RuBisCO-binding protein, has been shown to have a sequence similar to the GroEL protein (Hemmingsen et al. 1988). This protein appears to facilitate the proper posttranslational assembly of RuBisCO, an oligomeric enzyme that catalyzes photosynthetic CO_2 fixation in higher plants (Bochkaneva et al. 1988; Goloubinoff et al. 1989). A picture is emerging that GroEL and its plant equivalent, the RuBisCO-binding protein, serve a more general role in facilitating or chaperoning the proper posttranslational assembly of proteins into larger oligomeric complexes.

In mammalian cells, we have similarly observed the increased synthesis of a 58-kD protein in cells experiencing stress (Mizzen et al. 1989). The protein exhibits an isoelectric point very similar to that reported for the Tetrahymena 58-kD protein, and like the Tetrahymena protein, is present within the mitochondria of all mammalian cells so far examined. Antibodies raised against the Tetrahymena protein cross-react with the mammalian 58-kD protein. Mammalian 58 kD is synthesized in the cytoplasm as a precursor protein of ~60,000 daltons and presumably upon import into the mitochondria undergoes a proteolytic processing event leading to the production of mature hsp58. In cells treated with the potassium ionophore, nonactin, both the import and processing of hsp58 as well as a number of other mitochondrial proteins is inhibited. Consequently, the precursor form of 58 kD accumulates within the cytoplasm in the nonactin-treated cells (Anderson 1981; Mizzen et al. 1989). Because of what is known concerning the homologous GroEL and RuBisCO-binding proteins described above, we suspect that hsp58 may function to facilitate the proper oligomeric assembly of proteins that are first synthesized in the cytoplasm and then translocated into the mitochondria.

5. 72 kD and 73 kD (Also Referred to as hsp70 and hsp72)

There appear to be two major forms of the so-called hsp70 proteins, with each comprising multiple isoelectric forms. The first, 73 kD, is an abundant protein of the normal unstressed cell and is often referred to as the "cognate" or "constitutive" hsp70 protein. The second, 72 kD, is highly related to 73 kD and is synthesized at very high levels after stress. In most cells, synthesis of hsp72 occurs only after induction of the stress response. An exception to this rule occurs in primate cells. So far in all primate cells examined (over 15 human and 3 monkey cell lines), synthesis of both hsp73 and hsp72 is observed in the cells grown under normal conditions (Welch et al. 1983 and unpubl.). Following heat shock treatment, increased synthesis of 72 kD and, to a lesser extent, 73 kD, is observed. A number of interesting observations concerning this constitutive expression of 72 kD in primate cells have emerged: (1) Its expression appears cell-cycle-regulated with its synthesis observed at the Gi/S boundary (Milarski and Morimoto 1986); (2) 72 kD expression, in general, appears to increase in primate cells following transformation (W.J. Welch, unpubl.); and (3) the expression of 72 kD increases in human cells (but not rodent cells) in response to transfection with either of two so-called cooperating oncogenes, E1A or *myc* (Nevins 1982; Kingston et al. 1984; Wu et al. 1986).

This constitutive expression of 72 kD in primate cells deserves further discussion. In addition to its increased expression in cells transfected with either E1A or *myc*, 72 kD is also observed to increase in primate cells following infection with adenovirus, polyomavirus, SV40, and herpes simplex virus type 2 (Nevins 1982; Khandjian and Turler 1983; Kennedy et al. 1985; Lathangue and Latchman 1987; Simmon et al. 1987; White et al. 1988). In all of these cases, the induction of the 72-kD protein appears dependent on one (or more) of the early viral gene products (Notarianni and Preston 1982; Khandjian and Turler 1983; Wu et al. 1986). In addition to its increased expression, a number of studies have shown that 72 kD (and perhaps 73 kD) is apparently interacting with a number of these early viral gene products. For example, relatively stable interactions of the 72-kD protein with the adenovirus E1A protein, SV40 large T antigen, and with polyoma middle T antigen have been observed (Khandjian and Turler 1983; Walter et al. 1987; White et al. 1988; E.T. Sawai and J.S. Butel, in prep.). Similarly, very stable interactions of the

72-kD stress protein with the cellular oncoprotein P53 have been described (Pinashi-Kimhi et al. 1986; Hinds et al. 1987; Sturzbecher et al. 1987). In many of these cases, these interactions of the 72-kD (and perhaps 73-kD) stress protein were observed primarily for only "mutated" forms of these various viral proteins. In some cases, however, the "wild-type" forms of these viral proteins were similarly found to complex with hsp72. At present, the significance of these findings remains unclear. Whether hsp72 (and hsp73) is essential for the proper structure/function of these viral gene products or whether the associations represent a "housekeeping" function of the 70-kD stress proteins remains to be established. Because the *E. coli* equivalent of 72/73 kD, the DnaK protein, appears essential for the replication of bacteriophage λ DNA in the infected bacterium (Itikawa and Ryu 1979; Dodson et al. 1986; Yamamoto et al. 1987), one wonders whether the association of 72/73 kD with SV40 T antigen, for example, facilitates the role of T antigens in SV40 DNA replication. In the case of E1A, one wonders whether its interaction with 72 kD may facilitate the ability of E1A to mediate gene *trans*-activation events and/or other viral morphogenesis events.

Immunological, biochemical, and DNA sequence analyses have demonstrated that whereas the 72-kD and 73-kD proteins are highly related proteins, they are in fact distinct gene products (see reviews by Craig 1985; Lindquist 1986 for a discussion of this complicated gene family). In addition, both proteins exhibit very similar biochemical properties including their stoichiometric copurification during either gel filtration or ion exchange chromatography, and both exhibit similar intracellular locales (Welch and Feramisco 1982, 1984, 1985b). Finally, both proteins exhibit a high affinity for various nucleotides, the highest being for ATP. This property of the proteins has facilitated their relatively simple and rapid purification (Welch and Feramisco 1985b). For example, milligram amounts of the proteins can be isolated from a few liters of suspension-grown HeLa cells by affinity chromatography on Sepharose columns containing covalently linked ATP (see Fig. 4). It should be emphasized that to date, no obvious biochemical or biological differences between the constitutive hsp73 and highly stress-inducible hsp72 have been described. Finally, both hsp72 and 73 have been reported to be methylated polypeptides, and both appear to bind fatty acids (Wang and Lazarides 1984; Guidon and Hightower 1986).

A number of recent observations have implicated a role for the 72/73-kD stress proteins in biochemical pathways essential to the normal, unstressed cell. First, it has been shown in vitro that the 72/73-kD stress proteins facilitate the uncoating and release of clathrin triskelions from clathrin-coated vesicles (Ungewickel 1985; Chappell et al. 1986). The

70-kD stress proteins, in the presence of ATP, appear to bind to the clathrin coat and facilitate removal of the clathrin triskelions. Once the clathrin coat is removed, the endocytotic vesicle can now subsequently fuse with other intracellular organelles such as the lysosome. Uncoating activity appears to require ATP hydrolysis to displace each leg of the clathrin triskelion from the coated vesicle. Curiously, these uncoating events do not appear to be truly catalytic, since the 70-kD proteins appear to remain bound to the released triskelion. Conceivably then, the 70-kD proteins bound to the released triskelion could facilitate the subsequent reformation of a new clathrin-coated vesicle. Yet another role for the hsp70 stress proteins has been suggested by recent studies examining the mechanism of protein translocation across intracellular membranes (Chirico et al. 1988; Deshaies et al. 1988). Specifically, these investigators have observed that the posttranslational translocation of some secretory and/or mitochondrial proteins across the membranes of the ER and mitochondria, respectively, are mediated in part by the hsp 70-kD stress proteins. Again this process requires ATP (and likely ATP hydrolysis) and presumably one or more additional factors in the cell. It is not yet clear whether the 70-kD stress proteins actually participate directly in the translocation event or simply maintain the target protein in a translocation "competent state" prior to its transfer across the mem-

Figure 4 Rapid purification of the 72/73 kD stress proteins by affinity chromatography on ATP agarose. (*A*) HeLa cells, growing in suspension, were heat-shock-treated at 43°C for 2 hr, then returned to 37°C for 8 hr. The cells were collected, incubated in hypotonic buffer, and lysed by Dounce homogenization. A 15,000*g* supernatant was prepared and applied to a DEAE-52 cellulose ion exchange column, the column was washed and the proteins were eluted with a linear gradient of NaCl. A sample of every other fraction was analyzed by SDS-polyacrylamide gel electrophoresis, and the proteins were visualized by Coomassie blue staining of the gel. Lane *A* represents molecular weight markers; lane *B*, the material applied to the column; and lane *C*, the material that flowed directly through the column. The 72/73-kD proteins are present within fractions 30–60 with the peak at approximately fraction 50. (*B*) The peak fractions of 72/73-kD (35–58) eluting of the DEAE column were pooled and directly applied to a Sepharose column containing covalently linked ATP. The column was then developed with 1 mM GTP to elute two proteins of approximately 20 kD and then developed with 5 mM ATP to elute the 72/73-kD stress proteins. A portion of each fraction eluting off the column was analyzed by SDS-PAGE, and the proteins were visualized by Coomassie blue staining of the gel. (Lane *A*) The material applied to the column; (lane *B*) the material that flowed through the column after washing the column with alternating cycles of low- and high-salt buffer.

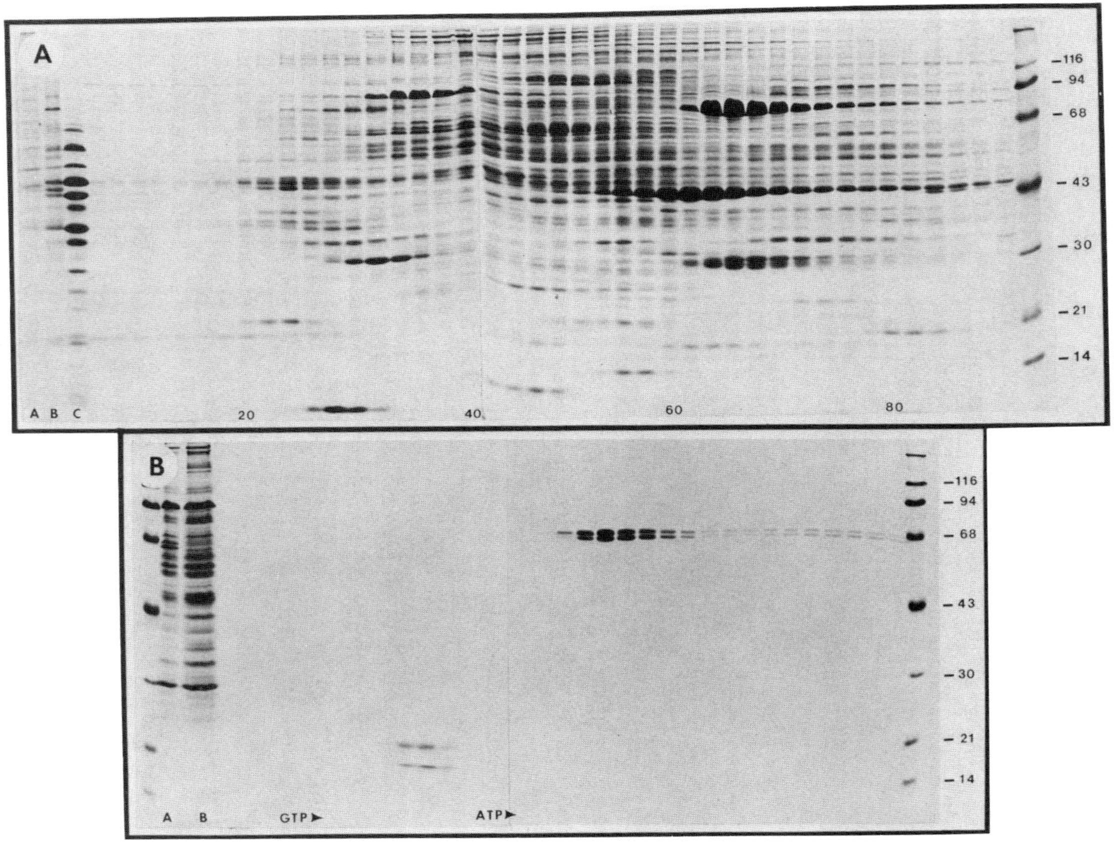

Figure 4 (*See facing page for legend.*)

brane. Whatever the mechanism, these results implicating a role for the 70-kD stress proteins in the translocation of proteins across intracellular membranes are quite exciting and should expedite further work concerning the mechanism(s) by which the 70-kD proteins interact with their target proteins in the cell and the exact role of ATP in such interactions. Moreover, these studies may help us understand the biochemical role of the 72/73-kD stress proteins in the cell experiencing stress (discussed further below).

6. 90 kD (Also Referred to as hsp83 or hsp89)

The mammalian 90-kD stress protein, homologous to the *Drosophila* 83-kD stress protein, is very abundant in cells grown under normal conditions. Synthesis of 90 kD increases approximately three- to fivefold after activation of the stress response. 90 kD is heavily phosphorylated with at least 12 isoforms, half of which appear to contain phosphate (Welch et al. 1983). The protein, when purified to homogeneity, exists as a dimer. Both sucrose gradient and gel filtration analyses indicate the purified protein to have a rod-like structure (Welch and Feramisco 1982). In whole-cell lysates, 90 kD apparently interacts with a number of other cellular components, as evidenced by its extremely heterogenous elution profile during gel filtration (W.J. Welch, unpubl.). Indeed, a few examples of specific interactions of 90 kD with other rather interesting proteins have been described. The first example was the reported association of 90 kD with the viral oncogene protein pp60^{src}. In cells transformed with Rous sarcoma virus, newly synthesized pp60^{src} rapidly associates with both the 90-kD stress protein and another cellular protein of 50 kD (Sefton et al. 1978; Brugge et al. 1981; Opperman et al. 1981). When present in this complex, the Src protein does not exhibit its normal tyrosine kinase activity nor phosphorylated tyrosine residues. As the complex containing pp60^{src} reaches its final destination at the plasma membrane, the complex dissociates, Src is deposited at the inner side of the plasma membrane, and the protein now exhibits both tyrosine kinase activity and autophosphorylated tyrosine. Subsequent studies by a number of laboratories have demonstrated that many retrovirus-encoded oncogene proteins, most of which appear to be tyrosine-specific protein kinases, similarly interact with the cellular proteins p50 and hsp90 (for review, see Brugge 1986). In a perhaps analogous scenario, 90 kD is found associated with a number of steroid hormone receptors, but only with the so-called nontransformed 8S/9S forms of the receptors (Catelli et al. 1985; Sanchez et al. 1985). This form of the receptor, which contains the steroid-binding protein in addition to hsp90 (and perhaps hsp 73), is unable to bind DNA and therefore is incapable of activating gene expression. Following binding of the particular steroid to the receptor, the 90-kD stress protein dissociates from the complex, with the steroid receptor, now in its 4S form, capable of exhibiting its biological activity (i.e., target gene transcription). Consequently, it has been suggested that the presence of 90 kD within the 8S/9S form of the steroid receptor complex prevents the receptor itself from binding its target gene in the absence of hormone, perhaps by interacting directly with that portion of the receptor involved in DNA binding. Finally, very recent data have implicated a role for 90 kD in the regulation of yet another protein kinase. Specifically, 90 kD is observed to copurify with the heme-regulated protein kinase that phosphorylates the α subunit of eukaryotic initiation factor 2 (eif-2α) (Rose et al. 1989). eif-2α appears to be involved in the regulation of translational initiation events in eukaryotic cells. Increased phosphorylation of eif-2α results in a reduction in ribosomal initiation complexes and an inhibition of new translational events (Duncan and Hershey 1984). Although hsp90 does appear to interact with the eif-2α kinase, the exact role of hsp90 in the possible regulation of kinase activity is still under investigation. Interestingly, there are reports that eif-2α exhibits an increased level of phosphorylation in some cells after heat shock, an observation that may explain why overall translation is transiently arrested in some cells after a severe heat shock treatment (DeBenedetti and Baglioni 1986; Scorsone et al. 1987). All of these studies point toward a role for 90 kD in the regulation of a number of rather interesting proteins within the cell. However, it should be noted that the levels of the 90 kD present in the cell are significantly more than what would be needed to interact with only tyrosine kinases, the steroid receptor, and the heme-regulated kinase, proteins that are present at rather low levels in the cell. Hence, we may find that a number of other proteins/enzymes in the cell are similarly regulated by virtue of transient interactions with the 90-kD stress protein.

7. 110 kD

Of all major members of the stress protein family, we know the least regarding the properties of this protein. 110 kD is a constitutively expressed protein whose synthesis increases approximately fivefold after stress. The protein appears present within that region of the nucleolus involved in rRNA transcription and/or processing (Subjeck et al. 1983). Considering that rRNA transcription is inhibited after heat shock, one wonders whether 110 kD is involved in the resumption of normal nucleolar transcription events during recovery from the heat shock insult.

C. The Glucose-regulated Proteins

1. 80 kD (Also Referred to as grp78 or BiP)

80 kD is a relatively abundant protein in cells grown under normal conditions. The protein contains phosphate and/or is ADP-ribosylated and has been shown to be present within the ER (Carlsson and Lazarides 1983; Welch et al. 1983; Munro and Pelham 1986). As was discussed earlier, 80 kD, like hsp72 and 73, binds ATP and exhibits approximately 50% sequence homology with the hsp70 proteins (Munro and Pelham 1986; Ting and Lee 1988). cDNA sequence analysis has shown that grp80 kD is in fact identical to a previously described protein, referred to as BiP (Munro and Pelham 1986). A number of studies have shown that BiP, present within the lumen of the ER, transiently interacts with maturing immunoglobulin IgG heavy and light chains (Haas and Wabl 1983; Bole et al. 1986; Munro and Pelham 1986; Hendershot et al. 1987). It was suggested that BiP mediates the proper assembly of the monomeric heavy (H) and light (L) chains into their mature form, H_2L_2. BiP also binds to a number of mutated proteins that traffic through the ER, but that appear incapable of self-oligomerization (Kassenbrock et al. 1988; Kozutsumi et al. 1988). Thus, it appears that grp80 (or BiP) functions in an ATP-dependent manner to facilitate the stabilization, correct folding, and/or oligomeric assembly of proteins that pass through the ER. In addition, BiP may also serve a role in the recognition and/or removal of some proteins that are secreted into the ER, but that fail to move further along the secretory pathway.

2. 100 kD (Also Referred to as grp94)

The other major member of the grp family, 100 kD, contains both phosphate and carbohydrate and has been localized to the ER, Golgi, and perhaps the plasma membrane (Pouyssegur et al. 1977; Olden et al. 1978; McCormick et al. 1979; Welch et a. 1983). As was mentioned above, grp 100 kD exhibits considerable sequence homology with hsp90, a cytosolic protein (Lee et al. 1984; Kulomao et al. 1986; Mazzarella and Green 1987). The predicted amino acid sequence of grp100 contains a putative transmembrane spanning region near the amino terminus, and the protein may therefore span the ER/Golgi membrane with a considerable portion of the protein present on the cytosolic side of the membrane

(Mazzarella and Green 1987). Although little is known concerning its function, we might expect 100 kD, like its related hsp 90-kD counterpart, to exhibit transient regulatory interactions with other macromolecules, but within its own compartment, the ER and Golgi.

3. 75 kD

Very recent studies in our laboratory have defined a new stress protein, 75 kD, whose synthesis appears affected by many of the same treatments that increase the synthesis of the two major grps, 80 kD and 100 kD. For example, increased synthesis of 75 kD, like 80 kD and 100 kD, is observed in cells depleted of glucose or exposed to calcium ionophores. Our data indicate that 75 kD is yet another member of the 70-kD family of stress proteins (Mizzen et al. 1989). The protein exhibits high affinity for ATP and exhibits immunological cross-reactivity with some (but not all) antibodies raised against the 72/73-kD stress proteins. Using an antibody raised against the protein, (and specific for only 75 kD) as well as biochemical fractionation, we have found that 75 kD is a constitutively expressed protein located in the mitochondria. The protein is synthesized as a precursor and apparently is processed to its mature form as it is translocated into the mitochondria. On the basis of these observations, we suspect that 75 kD is the mammalian equivalent of the SSCI gene product of yeast (Craig et al. 1987). As was discussed by Craig, SSCI is an essential gene in yeast that is synthesized initially as a precursor protein containing a mitochondrial signal leader sequence and is processed into its mature form presumably as it crosses into the mitochondria. Moreover, SSCI exhibits considerable sequence homology with the yeast hsp70 proteins. Thus grp75, like grp80, represents yet another member of the hsp70 family of ATP-binding proteins. We suspect that this compartmentalized 75-kD stress protein, like the grp80 protein present in the ER, serves a general role in the proper assembly and/or disassembly of target proteins, but within its own compartment, the mitochondria.

D. Biochemistry of the hsp70 Proteins in the Cell Experiencing Stress

Having discussed some of the properties and possible functions of the stress proteins in the normal cell, the obvious question arises as to their role in the cell experiencing stress. Although we still do not have definitive answers to this question, progress is being made in this direction. Most of the current data come from studies examining the most highly induced stress protein, 72 kD. As was discussed earlier, synthesis of 72 kD is low or negligible in most cells grown under normal conditions, but after stress, 72 kD represents the major translation product of the cell.

72 kD consists of multiple and related isoforms, the exact number depending on the cell type, the agent used to induce the response, and the severity of the stress treatment (Welch et al. 1983; Welch 1985). Biochemical fractionation and immunological studies have shown that the majority of 72 kD localizes within the nucleus shortly after its synthesis (Velazquez and Lindquist 1984; Welch and Feramisco 1984). Here the protein appears present within the nucleoplasm, the nuclear matrix, and large amounts accumulating within the nucleoli (Fig. 5) (Pelham 1984; Welch and Feramisco 1984). The nucleolar deposition of the protein always appears correlated with marked alterations in the integrity of the nucleoli. As was mentioned earlier, these alterations in the nucleoli appear to coincide with a diminishment of nucleolar function, most notably the shutdown of rRNA synthesis and new ribosomal assembly (Ellgaard and Clever 1971; Rubin and Hogness 1975). Using immunoelectron microscopy, it has been shown that most of the nucleolar-distributed 72 kD is present within the so-called granular portion, that region of the nucleolus involved in the assembly of small ribonucleoproteins and preribosomes (Welch and Suhan 1986). During recovery from the heat shock treatment, and as the nucleoli slowly regain their normal morphology (and presumably their normal function), 72 kD is observed to exit the organelle (Pelham 1984; Velazquez and Lindquist 1984; Welch and Feramisco 1984; Welch and Suhan 1986). Although no definitive data exist, 72 kD within the damaged nucleolus may function to resolubilize denatured preribosomal complexes and/or perhaps facilitate their removal. In addition, the protein could conceivably facilitate the restoration of normal nucleolar function, in particular ribosomal assembly, during recovery from the stress event.

During the later periods of recovery from physiological stress, the majority of 72 kD begins to accumulate within the cytoplasm (Velazquez and Lindquist 1984; Welch and Feramisco 1984). Three distinct cytoplasmic locales are observed: a strong perinuclear distribution; a portion present within unusually large and phase-dense structures; and a portion present directly underneath the plasma membrane, again in association with phase-dense material. Using electron microscopy and double-label indirect immunofluorescence analysis, it was shown that much of the cytoplasmic 72 kD is in very close association with the ribosomes and polysomes (Fig. 6) (Welch and Suhan 1986). Finally, it should be noted that hsp72 is not only interacting with the translational machinery after heat shock. Recent studies, employing gel filtration and immunological methods, indicate ATP-dependent interactions of hsp72/73 with a num-

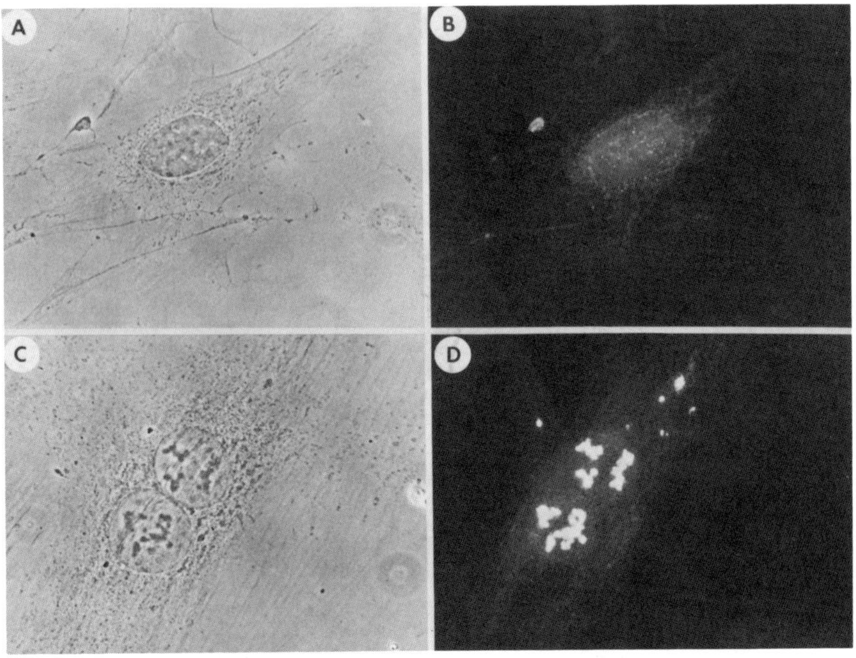

Figure 5 The 72-kD stress protein localizes prominently within the nucleolus in cells immediately after heat shock. Rat fibroblasts, growing on glass coverslips, were incubated at either 37°C or at 43°C for 2 hr. The cells were then fixed and the distribution of the 72 kD stress protein was determined via indirect immunofluorescence using an antibody specific to 72 kD. (*A*, *C*) Phase contrast micrographs; (*B*, *D*) corresponding fluorescent micrographs. Note the relatively low to negligible staining in the calls incubated at 37°C (*A* and *B*) and the intense nucleolar staining following heat shock (*C* and *D*).

ber of other cytoplasmic and nuclear proteins (R. Beckmann et al., in prep.).

E. The Phenomenon of Thermoprotection and Thermotolerance: Are the Stress Proteins Involved?

Probably the most interesting aspect of the stress response as it relates to cell physiology is the phenomenon of thermoprotection and thermotolerance. As was mentioned earlier, cells exposed to a sublethal heat shock treatment, then allowed to recover for a period of time back at their normal growth temperature, exhibit significantly higher survival rates to a second, and what would otherwise be a lethal, heat shock treatment (Gerner and Scheider 1975; Gerner et al. 1976; Henle and Leeper 1976).

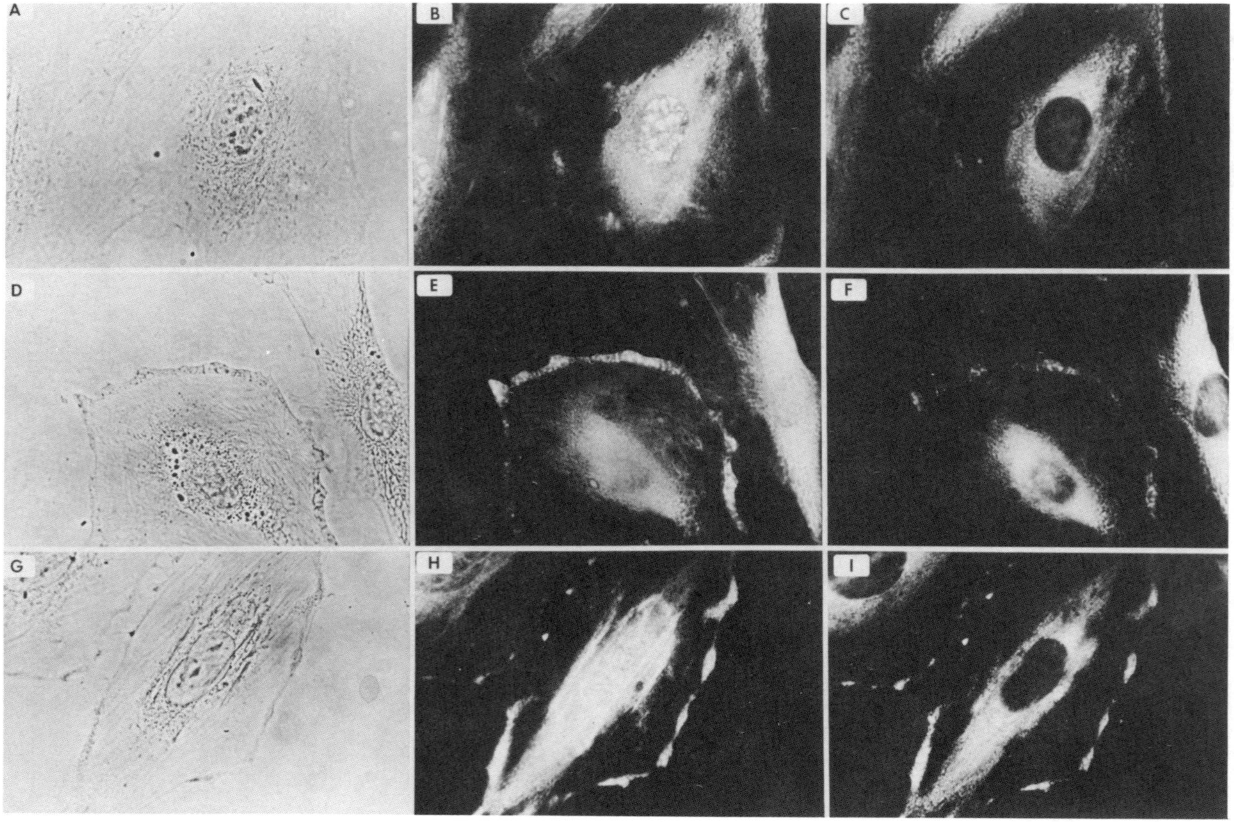

Figure 6 (*See facing page for legend.*)

Although one would suspect that the stress proteins are probably involved in providing this thermotolerant effect, direct experiments to prove this supposition are difficult and, consequently, there is considerable controversy regarding the role of the stress proteins in conferring the thermotolerant state. For example, numerous studies have reported a close correlation between the levels of the stress proteins and the degree of thermotolerance as assayed by cell survival (Henle and Leeper 1982; Li and Werb 1982; Li et al. 1982; Subjeck et al. 1982). In addition, some investigators have noted that thermotolerance appears transient with the decay of tolerance correlated with the half-life of one or more of the hsps (Landry et al. 1982; Li and Werb 1982; Subjeck et al 1982). Other investigators, however, have failed to observe a correlation between de novo stress protein synthesis and the development of thermo-tolerance (Landry and Chretien 1983; Hallberg et al. 1985; for review, see Carper et al. 1987). However, one must remember that many of the stress proteins are constitutive and rather abundant components of the normal unstressed cell. Moreover, these constitutive stress proteins, even in the absence of new stress protein synthesis, exhibit marked changes in their intracellular locale following induction of the stress response (Welch and Mizzen 1988). This is indicative that the constitutive stress proteins may be called upon to perform new tasks that arise due to the new physiological state of the cell (i.e., stress). We still face a formidable task in defining the basis for cellular thermotolerance. In addition, it is probably naive to suggest that only the stress proteins themselves are solely responsible for conferring the thermotolerant state.

Whatever the mechanism involved, recent studies have begun to delineate some of the hallmarks exhibited by the thermotolerant cell. For

Figure 6 Colocalization of the 72-kD stress protein with ribosomes in cells recovering from physiologic stress. REF-52 cells growing on glass coverslips were placed under stress by heat-shock treatment (42.5°C for 3hr), exposure to AzC (5 mM for 10 hr), or exposure to sodium arsenite (80 μM for 1.5 hr). The cells were then returned to normal culture conditions at 37°C for either 4 hr (heat shock) or 8 hr (AzC and arsenite), at which time the cells were fixed and permeabilized. The cells were then incubated with a mixture of a mouse monoclonal anti-72-kD serum and a human autoimmune antiribosomal antibody. The primary antibodies were visualized with a mixture of rhodamine-conjugated goat anti-mouse and fluorescein-conjugated rabbit anti-human antibodies. *A*, *D*, and *G* are phase-contrast micrographs; *B*, *E*, and *H* show the anti-72-kD staining; *C*, *F*, and *I* show the antiribosomal staining. (*A–C*) Cells heat-shock-treated and recovered for 4 hr. (*D–F*) Cells treated with AzC and recovered for 8 hr. (*G–I*) Cells treated with sodium arsenite and recovered for 8 hr.

example, it has been well established that overall translation activities decline as a function of the severity of the hyperthermic shock. This is illustrated in Figure 7, where cells given a mild 43°C 1-hour shock exhibit a slight inhibition of protein synthesis activity for approximately 1 hour (Fig. 7A), whereas cells given a more severe 45°C 30-minute shock show complete inhibition of protein synthesis activities for as long as 5 hours following their return to 37°C (Fig. 7B). If the cells are first given the 43°C/1-hour shock and then allowed to recover at 37°C for varying times, the cells exhibit markedly less inhibition of overall translational activities following the second and more severe 45°C/30-minute challenge (Fig. 7C, D). Both the acquisition and duration of such "translational thermotolerance" appear to correlate with the expression, accumulation, and relative half-lives of functional 70-kD stress proteins (Mizzen and Welch 1988). Other laboratories have similarly observed a correlation between stress protein synthesis, and, in particular, hsp72, in conferring this phenomenon of translational thermotolerance (Petersen and Mitchell 1981; Sciandra and Subjeck 1984). These studies, along with those indicating an apparent colocalization of the 70-kD stress protein with the translation machinery, are suggestive of a role for the 70-kD proteins in the protection and/or restoration of protein synthesis activities during recovery from stress.

A number of other interesting aspects concerning the thermotolerant status of cells have been described. In general, all of these studies indicate that lesions that occur as a consequence of stress are minimized or even prevented in the thermotolerant cell. For example, the heat-induced collapse of the intermediate cytoskeletal network can be prevented or at least repaired with faster kinetics in the thermotolerant cell (Welch and Mizzen 1988). Similarly, nucleolar structure, which is significantly al-

Figure 7 Development of translational thermotolerance in rat fibroblasts. Rat embryo fibroblasts growing on 35-mm dishes were subjected to (*A*) a 43°C/60-min heat shock; (*B*) a 45°C/30-min heat shock; (*C*) a prior 43°C/60-min heat shock, recovery at 37°C for 4 hr, and then a subsequent 45°C/30-min shock; or (*D*) as in *C*, but the cells were allowed to recover at 37°C for 8 hr prior to the second 45°C/30-min heat shock challenge. After the above described treatments, the cells were radiolabeled for 2-hr periods with [^{35}S]methionine at various times during the course of their recovery at 37°C. Following each 2-hr labeling period, the cells were solubilized and an equal number of cells were analyzed by SDS-PAGE. Shown is a fluorograph of the gel. Molecular weight markers are indicated on the left and the relative positions of the major stress proteins on the right. Note that the thermotolerant cells (*C* and *D*) recover protein synthesis activities significantly faster than the nontolerant cells (*B*).

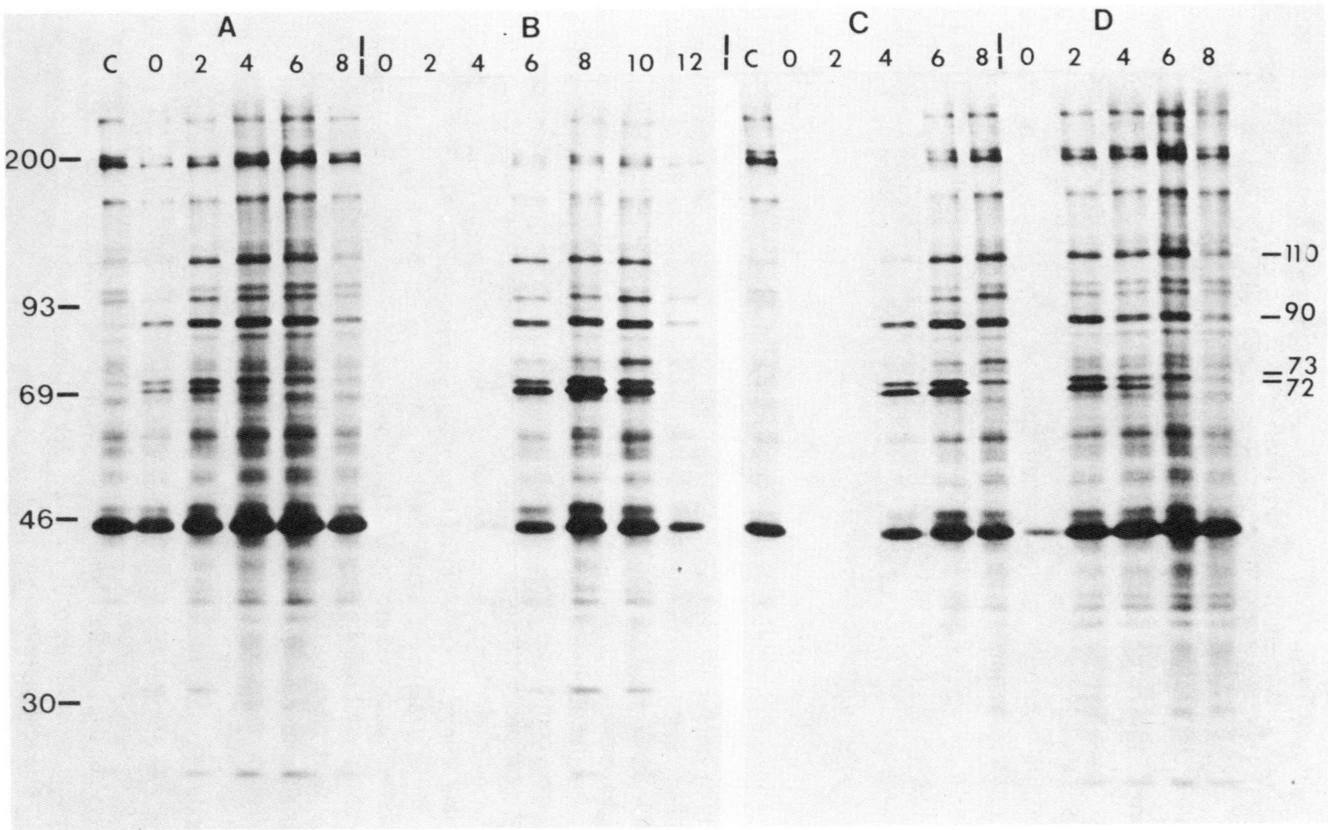

Figure 7 (See facing page for legend.)

tered in the nontolerant cell after hyperthermia, appears to regain its normal morphology markedly faster in the thermotolerant cell (Welch and Mizzen 1988). One presumes that this accelerated recovery of nucleolar morphology will correlate with a faster return of normal nucleolar function. Finally, hnRNA processing, which is inhibited after heat shock treatment in the nontolerant cell, appears markedly less affected in cells first made tolerant (Yost and Lindquist 1986). Consistent with these results are morphological studies indicating a disruption in the architecture of the so-called snRNP components involved in hnRNA processing following heat shock, but significantly less perturbation of snRNP morphology in cells first made tolerant and then heat shock challenged (Welch and Mizzen 1988). Thus, although it does appear that a variety of metabolic pathways can apparently be protected or repaired faster in the thermotolerant cell following stress, these studies still do not delineate the underlying biochemical basis and, in particular, the role of the stress proteins in conferring the thermotolerant state.

Figure 8 Cells microinjected with antibodies to the 72/73-kD stress proteins are unable to survive a heat shock challenge. (*A*, *B*) Rat fibroblasts growing on glass coverslips were injected with anti-72/73-kD stress protein antibodies (injection into the cytoplasm), then heat-shock-treated at 43°C for 2 hr. The cells were fixed and stained with rhodamine-conjugated goat antibody against mouse Ig6 to identify those cells injected with the antibodies (*A*). The distribution of the 72-kD stress protein was then determined by incubation of the cells with a rabbit polyclonal antibody specific for 72 kD followed by incubation with a fluorescein-conjugated goat anti-rabbit-antibody (*B*). Note that in those cells injected with antibodies against hsp 72/73 kD (indicated by arrows in *B*), no nuclear or nucleolar fluorescence was observed, whereas in contrast, the surrounding uninjected cells exhibit a nucleolar locale of the 72-kD protein. (*C–F*) In a similar experiment, rat fibroblasts growing in 35-mm dishes were microinjected with either control goat anti-chicken antibodies or with hsp72/73 mouse monoclonal antibodies. After incubation for 2 hr at 37°C phase contrast micrographs were taken of the cells injected with either the control antibodies (*C*) or the hsp70 antibodies (*E*). The same cells were then heat-shock-treated at 45°C for 30 min, returned to 37°C, and further incubated for 24 hr. The cells were then incubated with horseradish peroxidase conjugated rabbit anti-goat antibodies (to identify those cells injected with the control antibodies, *D*) or with horseradish peroxidase conjugated rabbit anti-mouse antibodies (to identify those cells injected with the anti-hsp70 antibodies, *F*) and subsequent color reaction. Note that in the heat-shock-treated cells injected with the control antibodies, most of the cells survived (*D*), whereas those injected with the anti-hsp70 antibodies appeared dead (*F*).

Although no evidence has been provided to directly ascertain the role of the stress proteins in rendering thermotolerance, there are ample data demonstrating a requirement for the 70-kD stress proteins in the ability of the cell to survive a hyperthermic challenge. In the case of bacteria and yeast, deletion of the genes encoding the 70-kD stress proteins results in an inability of the cells to survive a severe heat shock treatment (Saito and Uchida 1977; Itikawa and Ryu 1979; Craig and Jacobsen 1984). In the case of mammalian cells, two recent studies similarly point toward a requirement for the 70-kD stress proteins in the ability of the cell to survive thermal stress. First, microinjection of rat fibroblasts with monoclonal antibodies specific to the 72/73-kD stress proteins results in a failure of the 70-kD proteins to migrate into the nucleus/nucleolus immediately following thermal shock. Moreover, the microinjected cells are unable to survive the heat shock treatment (Fig. 8) (Riabowol et al.

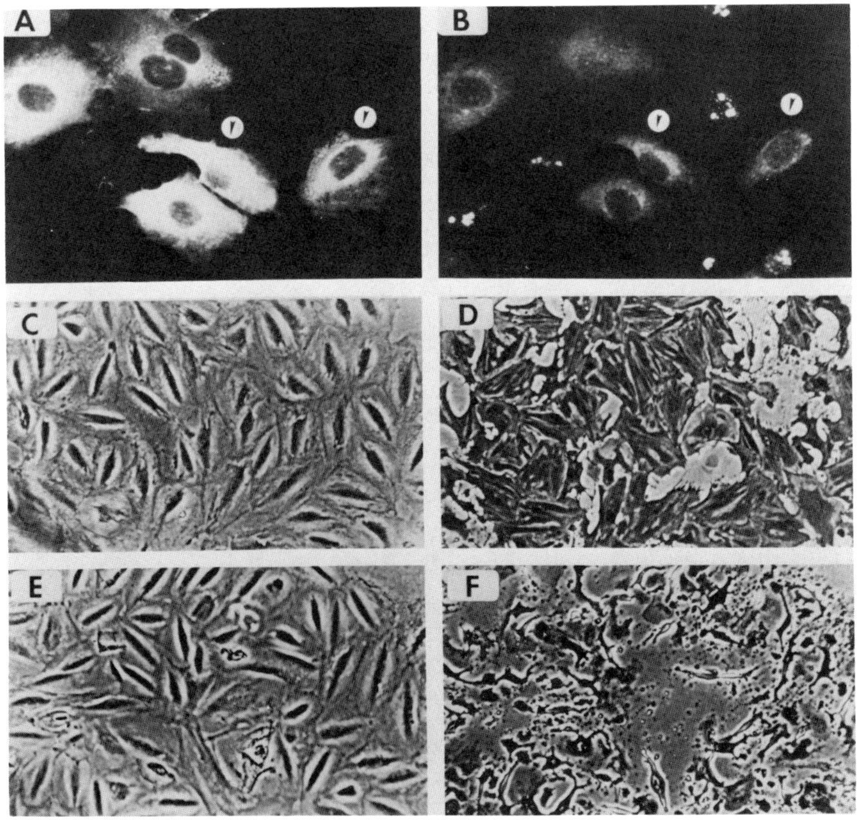

Figure 8 (See facing page for legend.)

1988). In a second study, a rather novel approach was used to obtain a similar result. Chinese hamster ovary cells were transfected with a plasmid containing the upstream control region of the hsp70 gene (e.g., the heat shock transcription faster binding site) along with dihydrofolate reductase. Plasmid copy was increased 20,000-fold by continued selection of the cells in the presence of methotrexate. In these cells containing the amplified plasmid, the heat-inducible expression of hsp70 (e.g., the highly induced 72-kD protein) was reduced by 90%. Moreover, the transfected cells exhibited significant increased thermosensitivity as assayed by cell survival (Johnston and Kacey 1988). Thus, taken together, these two studies support a role for the hsp70 proteins in the protection and recovery of mammalian cells from thermal stress.

F. Possible Functions of the hsp70 Family of Stress Proteins in Both the Normal and Stressed Cell

Although most workers in the field would agree that the stress proteins are involved in thermoprotection and thermotolerance, the obvious question arises as to their biochemical function in conferring such protection. In this regard, most of our information comes from those studies examining the family of related hsp70 proteins that interact with ATP (Table 2). For example, we now know of at least four major members of the 70-kD stress protein family: (1) the constitutive and highly abundant 73 kD, present within the cytoplasm in the 37°C cell, but also found within the nucleus and nucleolus after heat shock; (2) the highly stress-induced 72 kD, very homologous to 73 kD, and also found within the cytoplasm, nucleus, and nucleolus, as well as colocalizing with the translation maachinery after stress; (3) the glucose-regulated 75 kD, present within the mitochondria; and (4) the glucose-regulated 80 kD or BiP protein present within the lumen of the ER. All available evidence indicates that each of these related stress proteins is performing similar functions, but within its own distinct subcellular compartment. Specifically, in all of the processes implicating a role for the hsp70 family of proteins, the protein itself appears to be serving a role in the stabilization and/or solubilization of a particular target protein. For example, the cytosolic hsp 73-kD stress protein has been implicated in the process of posttranslational translocation of various yeast pre-proteins across the membranes of either the ER or mitochondria (Chirico et al. 1988; Deshaies et al. 1988). Because of the relatively high amounts of hsp73 required in this process, one suspects that the protein is serving a stoichiometric rather than catalytic role. The most likely scenario is that hsp73 is simply acting to stabilize or maintain the solubility of the pre-protein in a translocation-competent

Table 2 Properties of mammalian stress proteins

Stress proteins ($x10^{-3}$ dalton)	Modification	Location 37	Location 42	Remarks
hsp 28	phosphorylation	cytoplasm near Golgi	nucleus cytoplasm	increased phosphorylation in response to mitogens/tumor promoters; homology to lens α crystallin; native mass ≥400,000 daltons, forms aggregates
47	glycosylation phosphorylation	ER	ER	basic pI; binds collagen; changes in phosphorylation after transformation
58	phosphorylation?	mitochondria	mitochondria	synthesized as a precursor; homology to *E. coli* GroEL and to RuBisCO-binding protein; catalyzes protein assembly
72	methylation		cytoplasm nucleus nucleoli	most induced: low levels in 37°C cells; affinity for nucleotides, RNA, fatty acids; cell-cycle regulated, role in DNA synthesis?; ↑ expression in cells transfected with "cooperating" oncogenes (E1A, SV40 T, c-*myc*); association with p53, E1A, polyoma mT, SV40 T
73	methylation	cytoplasm	cytoplasm nucleus nucleoli	abundant constitutive protein related to 72: affinity for nucleotides, RNA, and fatty acids; involved in clathrin-coated vesicle assembly/disassembly: protein translocation across membranes
90	phosphorylation methylation	cytoplasm	cytoplasm (nuclear?)	abundant cellular phosphoprotein with multiple isoelectric forms; increased phosphorylation in response to dsDNA; transient association with tyrosine kinase; present in steroid hormone receptor complex; synthesis repressed following glucose/calcium deprivation
110	phosphorylation?	nucleoli	nucleoli cytoplasm	involved in rRNA transcription?
grp 75	—	mitochondria	mitochondria	synthesized as a precursor; binds ATP; homology to hsp72/73 and grp80
80	phosphorylation ADP-ribosylation	ER	ER	increased synthesis following glucose/calcium deprivation; homology to immunoglobulin heavy chain binding protein (BiP); binds ATP; related to hsp72/73
100	phosphorylation glycosylation	Golgi/ER plasma membrane?	Golgi/ER, plasma membrane? nucleus	increased synthesis during glucose/calcium deprivation; portion associated with 90-kD stress protein related to hsp90

state. Why would such a detergent-like role of hsp73 be required? Presumably the pre-protein with its signal sequence externally exposed (to provide information to the translocation machinery) has not reached its final three-dimensional confirmation and therefore might require stabilization until it is transferred across the membrane, where it is then proteolytically processed to remove the signal sequence and now able to fold into its final confirmation.

In yet another example, the cytosolic hsp72/73 proteins have been implicated in the process of clathrin-coated vesicle uncoating and/or coated vesicle assembly (Ungewickel 1985; Chappell et al. 1986). Again considering the high amounts of the hsp70 proteins required and the fact that the hsp70 protein remains bound to the released clathrin triskelion "monomer," one again suspects that hsp70 is acting in a stoichiometric rather than catalytic manner. I would suggest that the clathrin triskelion, when not in the context of a coated vesicle, might require some type of stabilization or solubilization until it is recruited to reassemble back into a coated vesicle. Such stabilization would be provided via its interaction with hsp70. The third example concerns the hsp70-related protein, grp80 or BiP. This protein, present within the lumen of the ER, appears to bind to monomeric proteins that are on their way to a larger oligomeric structure. For example, in the case of IgG synthesis, BiP would maintain the solubility of monomeric immunoglobulin heavy (H) and light (L) chains until their assembly into the mature and freely soluble H_2L_2 final immunoglobulin molecule. Similarly, other secretory (or membrane-bound) proteins that are synthesized as monomers but subsequently assemble into oligomers might require the cooperation of BiP to maintain them in a competent state prior to the oligomerization process. It has already been demonstrated that the influenza virus hemagglutinin monomer protein (HA) transiently interacts with BiP prior to its assembly into its final $(HA)_3$ trimerized configuration (Gething et al. 1986). Presumably, BiP maintains the monomeric HA molecule in an "oligomerization"-competent state until its final assembly into the trimer. Moreover, mutants of HA that are unable to correctly assemble into a trimer remain stably bound to BiP. The fact that BiP remains bound to the various HA mutant molecules likely reflects a failure of the mutant proteins to continue along their proper pathway of trimerization. As a consequence of this now stable association of BiP with the mutant HA monomer, the available free pool of BiP is reduced, and the cell responds by synthesizing more BiP. In this model, then, BiP does *not* serve as a watchdog or specific chelator of abnormally folded proteins. Instead, the observed binding of BiP to abnormally folded proteins only reflects a failure of the mutant proteins to continue along their usual pathway of oligomerization and export out of the ER. Supporting this idea is the fact that simply inhibiting N-linked protein glycosylation (using tunicamycin) results in an inhibition of protein secretion and a corresponding increased synthesis of BiP. Presumably, the inhibition of N-linked glycosylation caused by the drug results in an accumulation of secretory proteins within the ER; many of these remain in a stable complex with BiP, again thereby reducing the available pool of free BiP. Hence, one might suspect that BiP also will interact with other proteins that exist in their final form as a multimer, but which require stabilization prior to some other posttranslational event such as glycosylation. Finally, in the case of the last hsp70 family member, the grp75-kD protein, we suspect the protein will serve a BiP-like function, but within its own compartment, the mitochondria.

What then is the role of ATP in the interaction of the hsp70 family members with their target proteins? Although the available data are limited, one surmises that the hsp70 proteins utilize the energy of ATP hydrolysis to either facilitate the stabilization/solubilization of the target protein or utilize ATP hydrolysis to allow for release of their target protein. Support for the latter alternative is that target proteins that form complexes with the various hsp70 family members appear to dissociate upon addition of ATP (Lewis and Pelham 1985; Clarke et al. 1988). If this is so, how would energy of ATP hydrolysis be coupled to the dissociation event? I suggest that the hsp70 proteins undergo an autophosphorylation event resulting in a change in their *own* confirmation, thereby resulting in their release from the target protein. Following release, the hsp70 protein would quickly dephosphorylate, thereby allowing for its subsequent binding to a new target. Unfortunately, few data exist to support this notion. It is curious, however, that despite the observed high affinity of the 70-kD family of stress proteins for ATP, the purified proteins themselves exhibit incredibly weak ATPase or autokinase activity (Zylicz et al. 1983; Welch and Feramisco 1985b). However, such ATPase or autokinase activity might require the presence of an associated target protein and, therefore, in the absence of an associated target protein the purified hsp70 proteins would be unable to hydrolyze ATP. Finally, the fact that rather low levels of the total amount of cytosolic hsp70 are observed to incorporate $[^{32}P]H_3PO_4$ during steady-state labeling of cells would be consistent with this model (Loomis et al 1982; W.J. Welch, unpubl.). At any time, only a fraction of the hsp70 proteins would release their target via an autophosphorylation event, and then would quickly dephosphorylate to allow for a new round of target interaction.

If this model describing an intracellular ATP-dependent detergent-like role for the hsp70 family of stress proteins is correct, the next question arises as to why increased levels of these proteins would be required in the response of cells to stress. Because so many of the agents that induce the response are capable of perturbing protein structure and therefore assembly, it seems reasonable for the cell to increase the production of a group of proteins present within different intracellular compartments, whose primary function is in the stabilization of overall protein three-dimensional confirmation. For example, within the stressed cell where protein confirmation has been perturbed due to the presence of protein denaturing agents, the increased synthesis of the hsp70 family members could (1) effect solubilization and subsequent removal of proteins that have become denatured; (2) facilitate the refolding of proteins that have become only partly unfolded; or (3) simply facilitate the synthesis and assembly of new proteins to replace those irreversibly denatured as a consequence of the stress event. With the availability of the purified stress proteins, suitable antibodies, and at least some putative target proteins defined, we should now be able to test these various models. However, our task will still not be complete. We have yet to define how the other major stress proteins, 28, 32, 47, 90, 100, and 110 kD, function in both the normal and stressed cell. Obviously, we still face a rather formidable task.

ACKNOWLEDGMENTS

The author is grateful to all those who have been involved in our research efforts trying to understand the physiology and biochemistry of the stress response. Special thanks to J. Feramisco, P. Thomas, and J. Garrels, with whom I began my studies of the heat shock response, and to the subsequent members of our laboratory, including A.P. Arrigo, L.A. Mizzen, J. Suhan, R. Beckmann, and M. Mulcahy, and on occasion, K. Riabowol. I owe thanks for the technical help of P. Renna, M. Szadkowski, N. Halliran, G. Blose, and L. Cipp, S. Gumpert, and T. Kleven. Finally, I thank Robin Anderson (Stanford University) for providing pertinent materials for part of the review. Most of my work has been supported by National Institutes of Health grant GM-33551.

REFERENCES

Amalric, F., R. Simard, and J.P. Zalta. 1969. Effect de la température supraoptimale sur les ribonucleoprotines et le RNA nucleolaire. II. Etude biochimique. *Exp. Cell Res.* **55:** 370–377.

Anderson, L. 1981. Identification of mitochondrial proteins and some of their precursors in two dimensional electrophoretic maps of human cells. *Proc. Natl. Acad. Sci.* **78:** 2407–2411.

Anderson, R. and G.M. Hahn. 1985. Differential effects of hyperthermia on the Na^+, K^+-ATPase of Chinese hamster ovary cells. *Radiat. Res.* **102:** 314–323.

Anderson, R.L. and R. Parker. 1982. Analysis of membrane lipid composition of mammalian cells during the development of thermotolerance. *Int. J. Radiat. Biol.* **42:** 57–69.

Anderson, R.L., K.W. Minton G.C. Li, and G.M. Hahn. 1981. Temperature induced homeoviscous adaptation of Chinese hamster ovary cells. *Biochim. Biophys. Acta* **641:** 334–338.

Ananthan, J., A.L. Goldberg, and R. Voellmy. 1986. Abnormal proteins serve as eukaryotic stress signals and trigger the activation of heat shock genes. *Science* **232:** 252–254.

Arrigo, A.P. and W.J. Welch. 1987. Purification and characterization of the small 28kd mammalian stress protein. *J. Biol. Chem.* **262:** 15359–15369.

Arrigo, A.P., J.P. Suhan, and W.J. Welch. 1988. Dynamic changes in the structure and intracellular locale of the mammalian low-molecular-weight heat shock protein. *Mol. Cell. Biol.* **8:** 5059–5071.

Ashburner, M. and J.J. Bonner. 1979. The induction of gene activity in *Drosophila* by heat shock. *Cell* **17:** 241–254.

Blose, S.H., D.I. Meltzer, and J.R. Feramisco. 1984. 10nm Filaments are induced to collapse in living cells microinjected with monoclonal and polyclonal antibodies against tubulin. *J. Cell. Biol.* **98:** 842–858.

Bochkaneva, E.S., N.M. Lissin, and A.S. Girshovich. 1988. Transient association of newly synthesized unfolded proteins with the heat shock GroEL protein. *Nature* **336:** 254–257.

Bole, D.G., L.M. Hendershot, and J.F. Kearney. 1986. Posttranslational association of immunoglobulin heavy chain binding protein with nascent heavy chains in nonsecreting and secreting hybridomas. *J. Cell Biol.* **102:** 1558–1566.

Bouche, G., F. Amalric, M. Caizergues-Ferrer, and J.P. Zalta. 1979. Effects of heat shock on gene expression and subcellular protein distibution in Chinese hamster ovary cells. *Nucleic Acids Res.* **7:** 1739–1747.

Brugge, J.S. 1986. Interaction of the Rous sarcoma virus protein, $pp60^{src}$, with the cellular proteins pp50 and pp90. *Curr. Topics Microbiol. Immunol.* **123:** 1–23.

Brugge, J., E. Erikson, and R.L. Erikson. 1981. The specific interaction of the Rous sarcoma virus transforming protein, $pp60^{src}$, with cellular proteins. *Cell* **25:** 363–372.

Buffa, P., V. Guarriera-Bobyleva, V. Muscatello, and I. Pasquale-Ronche Hi. 1970. Conformational changes of mitochondria associated with uncoupling of oxidative phosphorylation in vivo and in vitro. *Nature* **226:** 272–274.

Burdon, R.H. and M.M. Cutmore. 1982. Human heat shock gene expression and the modulation of plasma membrane Na^+-K^+ ATPase activity. *FEBS Lett.* **140:** 45–48.

Calderwood, S.K. and G.M. Hahn. 1983. Thermal sensitivity and resistance of insulin receptor binding. *Biochim. Biophys. Acta* **756:** 1–8.

Caltabiano, M.M., T.P. Koestler, G. Poste, and R.G. Greig. 1986. Induction of 32- and 34-kDa stress proteins by sodium arsenite, heavy metals, and thiol-reactive agents. *J. Biol. Chem.* **261:** 13381–13386.

Carlsson, L. and E. Lazarides. 1983. ADP ribosylation of the mr 83,000 stress-inducible and glucose regulated protein in avian and mammalian cells: Modulation by heat shock and glucose starvation. *Proc. Natl. Acad. Sci.* **80:** 4664.

Carper, S.W., J.J. Duffy, and E. Gerner. 1987. Heat shock protein in thermotolerance and other cellular processes. *Cancer Res.* **47:** 5245–5255.

Catelli, M.G., N. Binart, I. Jung-Testas, J.M. Renoir, E.E. Baulieu, J.R. Feramisco, and W.J. Welch. 1985. The common 90KD protein component of nontransformed "8S" steroid receptors is a heat shock protein. *EMBO J.* **4:** 3131–3137.

Cervera, J. 1978. Effects of thermal shock on Hep-2 cells. An ultrastructural and high resolution autoradiographic study. *J. Ultrastruct. Res.* **63:** 51–63.

Cerveza, M., G. Dreyfuss, and S. Penman. 1981. Messenger RNA is translated when associated with the cytoskeletal framework in normal and VSV-infected cells. *Cell* **23:** 11–20.

Chappell, T.G., W.J. Welch, D.M. Schlossman, K.B. Palter, M.J. Schlesinger, and J.E. Rothman. 1986. Uncoating ATPase is a member of the 70 kda family of stress proteins. *Cell* **45:** 3–12.

Cheney, C.M. and A Shearn. 1983. Developmental regulation of *Drosophila* imaginal disc proteins: Synthesis of a heat shock protein under non-heat shock conditions. *Dev. Biol.* **95:** 325–333.

Chirico, W.J., M.G. Waters, and G. Blobel. 1988. 70k Heat shock related proteins stimulate protein translocation into microsomes. *Nature* **333:** 805–810.

Christiansen, E.N. and E. Kuamma. 1969. Effects of thermal treatment on mitochondria, brain, liver, and ascites cells. *Acta Physiol. Scand.* **76:** 472–484.

Ciocca, D.R., R.H. Asch, D.J. Adams, and W.L. McGuire. 1983. Evidence for modulation of a 24k protein in human endometrium during the menstrual cycle. *J. Clin. Endocrinol. Metabol.* **57:** 496–499.

Clarke, C.F., K. Cheng, A.B. Frey, R. Stein, P.W. Hinds, and A.J. Levine. 1988. Purification of complexes of nuclear oncogene p53 with rat and *Escherichia coli* heat shock proteins: In vitro dissociation of hs c70 and dna k from murine p53 by ATP. *Mol. Cell. Biol.* **8:** 1206–1215.

Collier, N.C. and M.J. Schlesinger. 1986. The dynamic state of heat shock proteins in chicken embryo fibroblasts. *J. Cell Biol.* **103:** 1495–1507.

Coppo, A., A. Manzi, J.F. Pulitzer, and H. Takahashi. 1973. Abortive bacteriophage T4 head assembly in mutants of *Escherichia coli*. *J. Mol. Biol.* **76:** 61–87.

Craig, E.A. 1985. The heat shock response. *CRC Crit. Rev. Biochem.* **18:** 239–280.

Craig, E., and K. Jacobsen. 1984. Mutations of the heat inducible 70 kilodalton genes of yeast confer temperature sensitive growth. *Cell* **38:** 841–849.

Craig, E.A., J. Kramer, and J. Kosic-Smithers. 1987. SSCI, a member of the 70 kDa heat shock protein multigene family of *Saccharomyces cerevisiae*, is essential for growth. *Proc. Natl. Acad. Sci.* **84:** 4156–4160.

Cress, A.E. and E.W. Gerner. 1980. Cholesterol levels inversely reflect the thermal sensitivity of mammalian cells in culture. *Nature* **283:** 677–679.

Cress, A.E., P.S. Culver, J. Moon, and E. Gerner. 1982. Correlation between amounts of cellular membrane components and sensitivity to hyperthermia in a variety of mammalian cell lines in culture. *Cancer Res.* **42:** 1716–1721.

DeBenedetti, A. and C. Baglioni. 1986. Activation of hemin-regulated initiation factor-2-kinase in heat shocked Hela cells. *J. Biol. Chem.* **261:** 338–342.

Deshaies, R.J., B.D. Koch, M. Weiner-Washburne, E. Craig, and R. Schekman. 1988. A subfamily of stress proteins facilitates translocation of secretory and mitochondrial precursor polypeptides. *Nature* **332:** 800–805.

Dickson, J.A. and B.E. Oswald. 1976. The sensitivity of a malignant cell line to hyperthermia (42°C) at low intracellular pH. *Br. J. Cancer* **34:** 262–271.

Dodson, M., H. Echols, S. Wickner, C. Alfono, K. Memsa-Wilmot, B. Gomes, J. Le Bowitz, J.D. Roberts, and R. McMacken. 1986. Specialized nucleoprotein structures of the origin of replication of bacteriophage-lambda: Localized unwinding of duplex DNA by a six protein reaction. *Proc. Natl. Acad. Sci.* **83:** 7638–7646.

Drummond, I.A., S.A. McClure, M. Poenie, R.Y. Tsien, and R.A. Steinhardt. 1986. Large changes in intracellular pH and calcium observed during heat shock are not responsible for the induction of heat shock proteins in *Drosophila melanogaster*. *Mol. Cell. Biol.* **6:**

1767–1775.

Duncan, R. and J.W.B. Hershey. 1984. Heat shock-induced translational alterations in Hela cells. *J. Biol. Chem.* **259:** 11882–11889.

Edwards, D.P., D.J. Adams, N. Savage, and W.L. McGuire. 1980. Estrogen induced synthesis of specific proteins in human breast cancer cells. *Biochem. Biophys. Res. Commun.* **93:** 804–812.

Ellgaard, E.G. and U. Clever. 1971. RNA metabolism during puff induction in *Drosophila melanogaster*. *Chromosome* **36:** 60–78.

Ellgaard, E.G. and B.L. Maxwell. 1975. Nucleotide metabolism in *Drosophila melanogaster* salivary glands during temperature and dinitrophenol-induced puffing. *Cell Differ.* **3:** 379–387.

Escribano, J. and E. Rozengurt. 1988. Cyclic AMP increasing agents rapidly stimulate vimentin phosphorylation in Swiss 3T3 cells: Evidence of a dual mechanism of action. *J. Cell. Physiol.* **137:** 223–234.

Falkner, F.G., H. Saumweber, and H. Biessman. 1981. Two *Drosophila melanogaster* proteins related to intermediate filament proteins of vertebrate cells. *J. Cell Biol.* **91:** 175–183.

Findly, R.C., R.J. Gillies, and R.G. Shulman. 1983. *In vivo* phosphorous-31 nuclear magnetic resonance reveals lowered ATP during heat shock of tetrahymena. *Science* **219:** 1223–1225.

Freeman, M.L., A.W. Malcolm, and M.J. Meredith. 1985. Glutathione pool size affects cell survival after hyperthermia treatment. *Cell Biol. Toxicol* **3:** 213–221.

Fukui, Y. and H. Katsumura. 1979. Nuclear actin bundles in amoeba *Dictyostelium* and human Hela cells induced by dimethylsulfoxide. *Exp. Cell. Res.* **120:** 451–455.

Fulton, A.K., A. Wan, and S. Penman. 1980. The spatial distribution of polyribosomes in 3T3 cells and the associated assembly of proteins into the skeletal framework. *Cell* **20:** 849–857.

Georgopoulos, C.P. and B. Hohn. 1978. Identification of a host protein necessary for bacteriophage morphogenesis (the gro E gene product). *Proc. Natl. Acad. Sci.* **75:** 131–135.

Georgopoulos, C.P., R.W. Hendrix, S.R. Casjens, and A.D. Kaiser. 1973. Host participation in bacteriophage lambda head assembly. *J. Mol. Biol.* **76:** 45–60.

Gerner, E.W. and M.J. Scheider. 1975. Induced thermal resistance in Hela cells. *Nature* **256:** 500–502.

Gerner, E.W., R. Boone, W.G. Connor, J.A. Hicks, and M.L.M. Boone. 1976. A transient thermotolerant survival response produced by single thermal doses in Hela cells. *Cancer Res.* **36:** 1035–1040.

Gething, M.-J., K. McCammon, and J. Sambrook. 1986. Expression of wild type and mutant forms of influenza hemagglutinin: The role of folding in intracellular transport. *Cell* **46:** 939–950.

Glass, J.R., R.G. Dewitt, and A.E. Cress. 1985. Rapid loss of stress fibers in Chinese hamster ovary cells after hyperthermia. *Cancer Res.* **45:** 258–262.

Goloubinoff, P., A.D. Gatenby, and G.H. Lorimer. 1989. GroE heat shock proteins promote assembly of foreign prokaryotic ribulose bisphosphate carboxylase oligomers in *E. coli*. *Nature* **337:** 44–47.

Guidon, P.-T. and L.E. Hightower. 1986. Purification and initial characterization of the 71-kilodalton rat heat-shock protein and its cognate as fatty acid binding proteins. *Biochemistry* **25:** 3231–3239.

Haas, I.G. and M. Wabl. 1983. Immunoglobulin heavy chain binding protein. *Nature* **306:** 387–389.

Hallberg, R.-L., K.W. Kraus, and E.M. Hallberg. 1985. Induction of acquired thermotolerance in *Tetrahymena thermophila*: Effects of protein synthesis inhibitors. *Mol. Cell. Biol.* **5:** 2061–2069.

Hammond, G.L., Y.K. Lai, and C.L. Market. 1982. Diverse forms of stress lead to new patterns of gene expression through a common and essential metabolic pathway. *Proc. Natl. Acad. Sci.* **79:** 3485–3488.

Heine, V., L. Sverkak, J. Kondratuck, and R.A. Bonar. 1971. The behavior of Hela S_3 cells under the influence of supranormal temperatures. *J. Ultrastruct. Res.* **34:** 375–396.

Hemmingsen, S.M., C. Woolford, S.M. Vandervies, K. Tilly, D.T. Dennis, C.P. Georgopoulos, R.W. Hendrix, and R.J. Ellis. 1988. Homologous plant and bacterial proteins chaperone oligomeric protein assembly. *Nature* **333:** 330–334.

Hendershot, L., D. Bole, G. Kohler, and J.F. Kearney. 1987. *J. Cell Biol.* **104:** 761–767.

Hendrix, R.W. and L. Tsui. 1978. Role of the host in virus assembly: Cloning of the *Escherichia coli groE* gene and identification of its protein product. *Proc. Natl. Acad. Sci.* **75:** 136–139.

Henle, K.J. and D.B. Leeper. 1976. Interaction of hyperthermia and radiation in CHO cells: Recovery kinetics. *Radiat. Res.* **66:** 505–510.

———. 1982. Modification of the heat response and thermotolerance by cycloheximide, hydroxyurea, and lucanthone in CHO cells. *Radiat. Res.* **90:** 339–347.

Hickey, E.S., E. Brandom, R. Potter, G. Stein, J. Stein, and L.A. Weber. 1986. Sequence and organization of genes encoding the human 27 kDa heat shock protein. *Nucleic Acids Res.* **14:** 4127–4135.

Hinds, P.W., C.A. Finlay, A.B. Frey, and A.J. Levine. 1987. Immunological evidence for the association of p53 with a heat shock protein hsc 70, in p53 plus ras-transformed cells. *Mol. Cell. Biol.* **7:** 2863–2869.

Iida H. and I. Yahara. 1984. A heat shock resistant mutant of *Saccharomyces cerevisiae* shows constitutive synthesis of two heat shock proteins and altered growth. *J. Cell Biol.* **99:** 1441–1450.

———. 1985. Yeast heat shock protein of Mr 48,000 is an isoprotein of enolase. *Nature* **315:** 688–690.

Iida, K., H. Iida, and I. Yahara. 1986. Heat shock induction of intranuclear actin rods in cultured mammalian cells. *Exp. Cell. Res.* **165:** 207–215.

Ingolia, T.D. and E.A. Craig. 1982. *Drosophila* gene related to the major heat shock-induced gene is transcribed at normal temperatures and not induced by heat shock. *Proc. Natl. Acad. Sci.* **79:** 2360–2364.

Itikawa, H. and J.L. Ryu. 1979. Isolation and characterization of a temperature sensitive dna K mutant of *Escherichia coli B. J. Bacteriol.* **138:** 339–344.

Ireland, R. and E. Burger. 1982. Synthesis of low molecular weight heat shock proteins stimulated by ecdysterone in a cultured *Drosophila* cell line. *Proc. Natl. Acad. Sci.* **79:** 855–859.

Johnston, R.N. and B.L. Kacey. 1988. Competitive inhibition of hsp 70 gene expression causes thermosensitivity. *Science* **242:** 1551–1554.

Kageyma, H., T. Hiwasa, K. Tokunaga, and S. Sakiyama. 1988. Isolation and characterization of a complementary DNA clone for a Mr 32,000 protein which is induced with tumor promoters in Balb/c 3T3 cells. *Cancer Res.* **48:** 4795–4798.

Kassenbrock, C.K., P.D. Garcia, P. Waller, and R.B. Kelley. 1988. Heavy-chain binding protein recognizes aberrant polypeptides translocated in vitro. *Nature* **333:** 90–93.

Kennedy, P.G.E., N.B. Lathangue, W.L. Chan, and G.B. Clements. 1985. Cultured human neural cells accumulate a heat shock protein during the acute herpes simplex virus infection. *Neurosci. Lett.* **61:** 321–326.

Keyse, S.M. and R.M. Tyrrell. 1987. Both near ultraviolet radiation and the oxidizing agent hydrogen peroxide induce a 32-kDa stress protein in normal human skin fibroblasts. *J. Biol. Chem.* **262:** 14821–14825.

Khandjian, E.W. and H. Turler. 1983. Simian virus 40 and polyoma virus induce synthesis of heat shock proteins in permissive cells. *Mol. Cell. Biol.* **3:** 1–8.

Kim, Y.J., T. Shuman, M. Sette, and A. Przybla. 1984. Nuclear localization and phosphorylation of three 25-kilodalton rat stress proteins. *Mol. Cell. Biol.* **4:** 468–473.

Kingston, R.E., A.S. Baldwin, and P.A. Sharp. 1984. Regulation of heat shock protein 70 gene expression by c-myc. *Nature* **312:** 280–283.

Kozutsumi, Y., M. Segal, K. Normington, M.J. Gething, and J. Sambrook. 1988. The presence of malfolded proteins in the endoplasmic reticulum signals the induction of glucose regulated proteins. *Nature* **332:** 462–464.

Kulomao, M.S., N.L. Weigel, D.A. Kleinsek, W.G. Beatlie, O.M. Conneely, C. March, T. Zarucki-Schultz, W.T. Schrader, and B.W. O'Malley. 1986. Amino acid sequence of a chicken heat shock protein derived from the complementary DNA nucleotide sequence. *Biochemistry* **25:** 6244–6251.

Lamb, N., A. Fernandez, J.R. Feramisco, and W.J. Welch. 1989. Modulation of vimentin containing intermediate filament distribution in living fibroblasts by the cAMP-dependent protein kinase. *J. Cell Biol.* **108:** 2409–2422.

Landry, J. and P. Chrétien. 1983. Relationship between hyperthermia induced heat shock proteins and thermotolerance in Morris hepatoma cells. *Can. J. Biochem.* **61:** 428–437.

Landry, J., D. Bernier, P. Chrétien, L.M. Nicole, R.M. Tanguay, and N. Marceau. 1982. Synthesis and degradation of heat shock proteins during development and decay of thermotolerance. *Cancer Res.* **42:** 2457–2461.

Lathangue, N.B. and D.S. Latchman. 1987. Nuclear accumulation of a heat shock 70-like protein during herpes simplex virus replication. *Bioscience Rep.* **7:** 475–483.

Lee A.S., J. Bell, and J. Ting. 1984. Biochemical characterization of the 94- and 78-kilodalton glucose-regulated proteins in hamster fibroblasts. *J. Biol. Chem.* **259:** 4616–4621.

Leenders, H.J., A. Kemp, J.F. Koninkx, and J. Rosing. 1974a. Changes in cellular ATP, ADP and AMP levels following treatments affecting cellular respiration and the activity of certain nuclear genes in *Drosophila* salivary glands. *Exp. Cell. Res.* **86:** 25–30.

Leenders, H.J., H.D. Berendes, P.J. Helmsing, J. Derksen, and J.F. Koninkx. 1974b. Nuclear mitochondrial interactions in the control of mitochondrial respiratory metabolism. *Subcell. Biochem.* **3:** 119–147.

Lepock, J.R. 1982. Involvement of membranes in cellular responses to hyperthermia. *Radiat. Res.* **92:** 433–438.

Lepock, J.R., K.H. Cheng, H. Al-Qysi, and J. Kruuv. 1983. Thermotropic lipid and protein transitions in Chinese hamster lung cell membranes: Relationship to hyperthermic cell killing. *Can. J. Biochem. Cell. Biol.* **61:** 421–427.

Lepock, J.R., P. Massicotti-Nolan, G.S. Rule, and J. Kruuv. 1981. Lack of a correlation between hyperthermic cell killing, thermotolerance, and membrane lipid fluidity. *Radiat. Res.* **87:** 300–313.

Levinson, W., H. Opperman, and J. Jackson. 1980. Transition series metals and sulfhydryl reagents induce the synthesis of four proteins in eukaryotic cells. *Biochim. Biophys. Acta* **606:** 170–180.

Lewis, M.J. and H.R.B. Pelham. 1985. Involvement of ATP in the nuclear and nucleolar function of the 70 kd heat shock protein. *EMBO J.* **4:** 3137–3143.

Li, G.C. and Z. Werb. 1982. Correlation between the synthesis of heat shock proteins and the development of thermotolerance in Chinese hamster fibroblasts. *Proc. Natl. Acad. Sci.* **79:** 3918–3922.

Li, G.C., C. Petersen, and H.K. Mitchell. 1982. Induced thermotolerance and heat shock protein synthesis in Chinese hamster ovary cells. *Br. J. Cancer* (suppl.) **5:** 132–136.

Li, G.E. and A. Laszlo. 1985. Amino acid analogs while inducing heat shock proteins sensitize CHO cells to thermal damage. *J. Cell. Physiol.* **122:** 91–97.

Lindquist, S.C. 1986. The heat shock response. *Annu. Rev. Biochem.* **55:** 1151–1191.

Lindquist, S., B. DiDomenico, G. Bugaisky, S. Kurtz, L. Petko, and S. Sonoda. 1982. Regulation of the heat shock response in *Drosophila* and yeast. In *Heat shock from bacteria to man* (ed. M.T. Schlesinger et al.), pp. 167–175. Cold Spring Harbor Laboratory, Cold Spring Harbor, New York.

Littleword, T.D., D.C. Hancock, and G.I. Evan. 1987. Characterization of heat shock induced insoluble complex in the nuclei of cells. *J. Cell Sci.* **88:** 65–72.

Loomis, W.F., S. Wheeler, and A. Schmidt. 1982. Phosphorylation of the major heat-shock protein of *Dictyostelium discoideum. Mol. Cell. Biol.* **2:** 484–489.

Magun, B.E. and C.W. Fennie. 1981. Effects of hypothermia on binding, internalization, and degradation of epidermal growth factor. *Radiat. Res.* **86:** 133–146.

Mayrand, S. and T. Pederson. 1983. Heat shock alters ribonucleoprotein assembly in *Drosophila* cells. *Mol. Cell. Biol.* **3:** 161–171.

Mazzarella, R.A. and M. Green. 1987. ERp 99, an abundant conserved glycoprotein of the endoplasmic reticulum is homologous to the 90-kDa heat shock protein (hsp90) and the glucose regulated protein (grp 94). *J. Biol. Chem.* **262:** 8875–8883.

McCormick, P., B.J. Keys, C. Pucci, and A.J.T. Mills. 1979. Human fibroblast-conditioned medium contains a 100 k dalton glucose-regulated cell surface protein. *Cell* **18:** 173–192.

McMullin, T.W. and R.L. Hallberg. 1987. A normal mitochondrial protein is selectively synthesized and accumulated during heat shock in *Tetrahymena thermophila. Mol. Cell. Biol.* **7:** 4414–4423.

———. 1988. A highly evolutionarily conserved mitochondrial protein is structurally related to the protein encoded by the *Escherichia* gro EL gene. *Mol. Cell. Biol.* **8:** 371–380.

Mehdi, S.Q., D.J. Kecktenwald, L.M. Smith, G.C. Li, E.P. Armour, and G.M. Hahn. 1984. Effect of hyperthermia on murine cell surface histocompatibility antigens. *Cancer Res.* **44:** 3394–3397.

Milarski, K.L. and R. Morimoto. 1986. Expression of human Hsp 70 during the synthetic phase of the cell cycle. *Proc. Natl. Acad. Sci.* **83:** 9517–9521.

Mitchell, J.B. and A. Russo. 1983. Thiols, thiol depletion, and thermosensitivity. *Radiat. Res.* **95:** 481–485.

Mitchell, J.B., A. Russo, J.J. Kinsella, and E. Glutstein. 1983. Glutathione elevation during thermotolerance induction and thermosensitization by glutathione depletion. *Cancer Res.* **43:** 987–991.

Mizzen, L.A. and W.J. Welch. 1988. Characterzation of the thermotolerant cell. I. Effects on protein synthesis activity and the regulation of heat shock protein 70 expression. *J. Cell Biol.* **106:** 1105–1116.

Mizzen, L.A., C. Chang, J.G. Garrels, and W.J. Welch. 1989. Identification, characterization and purification of two mammalian stress proteins present in mitochondria: One related to hsp 70, the other to GroEL. *J. Biol. Chem.* (in press).

Mondovi, B.R., G. Rotilo, A.F. Argo, R. Cavaliere, and A. Rossi Fanelli. 1969. The biochemical mechanism of selective heat sensitivity of cancer cells. I. Studies on cellular respiration. *Eur. J. Cancer* **5:** 129–136.

Moretti-Rojas, I., S.A. Fugua, R.A. Montgomery III, and W.L. McGuire. 1988. A cDNA for estradiol-regulated 24k protein: Control of mRNA levels in MCF-7 cells. *Breast Cancer Res. Treat.* **11:** 155–163.

Muller, R.M., H. Taguchi, and S. Shibahara. 1987. Nucleotide sequence and organization of the rat heme oxygenase gene. *J. Biol. Chem.* **262:** 6795–6802.

Munro, S. and H.R.B. Pelham. 1986. An hsp-70 like protein in the ER: Identity with the 78kd glucose-regulated protein and immunoglobulin heavy chain binding protein. *Cell* **46:** 291–300.

Nagata, K. and K.M. Yamada. 1986. Phosphorylation and transformation sensitivity of a major collagen-binding protein of fibroblasts. *J. Biol. Chem.* **261:** 7531–7536.

Nagata, K., S. Saga, and K.M. Yamada. 1986. A major collagen-binding protein of chick embryo fibroblasts is a novel heat shock protein. *J. Cell Biol.* **103:** 223–229.

Neumann, D., K.D. Scharf, and L. Nover. 1984. Heat shock induced changes of plant cell ultrastructure and autoradiographic localization of heat shock proteins. *Eur. J. Cell Biol.* **34:** 254–264.

Nevins, J.R. 1982. Induction of the synthesis of a 70,000 dalton mammalian heat shock protein by the adenovirus E1A gene product. *Cell* **29:** 913–920.

Nickells, R.W. and L.W. Browder. 1988. A role for glyceraledehyde-3-phosphate dehydrogenase in the development of thermotolerance in *Xenopus laevis* embryos. *J. Cell Biol.* **107:** 1901–1909.

Niewiarowski, A., M.A. Calabiayno, D.S. Bailey, G. Poste, and K. Grieg. 1987. Alterations in lipid-linked oligosaccharide metabolism in human melanoma cells concomitant with induction of stress proteins. *J. Biol. Chem.* **262:** 14815–14820.

Notarianni, E.L. and C.M. Preston. 1982. Activation of cellular heat shock genes by herpes simplex virus temperature sensitive mutants which overproduce immediate early polypeptides. *Virology* **123:** 113–122.

Nover, L. 1984. *Heat shock response of eukaryotic cells*. Springer-Verlag. New York.

Olden, K., R.M. Pratt, and K.M. Yamada. 1978. Role of carbohydrates in protein secretion and turnover: Effects of tunicamycin on the major cell surface glycoprotein of chick embryo fibroblasts. *Cell* **13:** 461–473.

Oppermann, H., W. Levinson, and J.M. Bishop. 1981. A cellular protein that associates with a transforming protein of Rous sarcoma virus is also a heat shock protein. *Proc. Natl. Acad. Sci.* **78:** 1067–1071.

Osborn, M. and K. Weber. 1980. Dimethylsulfoxide and the ionophone A 23187 affect the arrangement of actin and induce nuclear actin paracrystals in PtK_2 cells. *Exp. Cell. Res.* **129:** 103–114.

Pekkala, D., B. Heath, and J.C. Silver. 1984. Changes in chromatin and the phosphorylation of nuclear proteins during heat shock of *Achlya ambisexualis*. *Mol. Cell. Biol.* **4:** 1198–1205.

Pelham, H.R.B. 1984. HSP 70 accelerates the recovery of nucleolar morphology after heat shock. *EMBO J.* **3:** 3095–3100.

Petersen, N.S. and H.K. Mitchell. 1981. Recovery of protein synthesis after heat shock: Prior treatment affects the ability of cells to translate mRNA. *Proc. Natl. Acad. Sci.* **78:** 1708–1711.

Pinhasi-Kimhi, O., D. Michalovitz, A. Ben-Zeev, and M. Oren. 1986. Specific interaction between the p53 cellular tumour antigen and major heat shock proteins. *Nature* **320:** 182–185.

Pouyssegur, J., R.P.C. Shiu, and I. Pastan. 1977. Induction of two transformation sensitive membrane polypeptides in normal fibroblasts by a block in glycoprotein synthesis or glucose deprivation. *Cell* **11:** 941–947.

Rao, P.N. and J. Engelberg. 1965. Hela cells: Effects of temperature on the life cycle. *Science* **148:** 1092–1094.

Riabowol, K.T., L.A. Mizzen, and W.J. Welch. 1988. Heat shock is lethal to fibroblasts microinjected with antibodies against hsp 70. *Science* **242:** 433–436.

Rose, D.W., W.J. Welch, G. Kramer, and B. Hardesty. 1988. Possible involvement of the 90 kDa heat shock protein in the regulation of protein synthesis. *J. Biol. Chem.* **264:** 6239–6244.

Rubin, G.M. and D.S. Hogness. 1975. Effects of heat shock on low molecular weight RNA's in *Drosophila melanogaster:* Accumulation of a novel form of 5S RNA. *Cell* **6:** 207–213.

Saga, S., K. Nagata, W.T. Chen, and K.M. Yamada. 1987. pH-dependent function, purification and intracellular location of a major collagen-binding glycoprotein. *J. Cell Biol.* **105:** 517–527.

Saito, H. and H. Uchida. 1977. Initiation of the DNA replication of bacteriophage lambda in *Escherichia coli* K-12. *J. Mol. Biol.* **113:** 1–15.

Sanchez, E.R., D.O. Toft, M.J. Schlesinger, and W.B. Pratt. 1985. Evidence that the 90 kDa phosphoprotein associated with the untransformed L-cell glucocorticoid receptor is a murine heat shock protein. *J. Biol. Chem.* **260:** 12358–12403.

Sanger, J.W., J.M. Sanger, T.E. Kreis, and B.M. Jockusch. 1980. Reversible translocation of cytoplasmic actin into the nucleus caused by dimethylsulfoxide. *Proc. Natl. Acad. Sci.* **77:** 5268–5272.

Sargan, D.R., M.J. Tsai, and B.W. O'Malley. 1986. hsp 108, a novel heat shock inducible protein of chicken. *Biochemistry* **25:** 6252–6258.

Scheer, V., H. Hinssen, W. Franke, and B. Jockusch. 1984. Microinjection of actin binding proteins and actin antibodies demonstrates involvement of nuclear actin in transcription of lampbrush chromosomes. *Cell* **39:** 111–122.

Sciandra, J.J. and J.R. Subjeck. 1984. Heat shock proteins and protection of proliferation and translation in mammalian cells. *Cancer Res.* **44:** 5188–5194.

Scorsone, K.A., R. Panniers, A.G. Rowlands, and E. Henshaw. 1987. Phosphorylation of eukaryotic initiation factor 2 during physiological stresses which affect protein synthesis. *J. Biol. Chem.* **262:** 14538–14543.

Sefton, B.M., K. Beemon, and T. Hunter. 1978. Comparison of the expression of the *src* gene of Rous sarcoma virus *in vitro* and *in vivo*. *J. Virol.* **28:** 957–971.

Shibahara, S., R. Muller, and H. Taguchi. 1987. Transcriptional control of rat heme oxygenase by heat shock. *J. Biol. Chem.* **262:** 12889–12892.

Shin, S.Y., K. Matsumoto, H. Iida, I. Uno, and T. Ishikawa. 1987. Heat shock response of *Saccharomyces cerevisiae* mutants altered in cyclic-AMP dependent protein phosphorylation. *Mol. Cell. Biol.* **7:** 244–250.

Shuman, J. and A. Przybyla. 1988. Expression of the 31-KD stress protein, in rat myoblasts and hepatocytes. *DNA* **7:** 475–482.

Simard, R. and W. Bernhard. 1967. A heat-sensitive cellular function localized in the nucleolus. *J. Cell Biol.* **34:** 61–76.

Simmon, M.C., K. Kitchener, H.T. Kao, E. Hickey, L. Weber, R. Voellmy, N. Heintz, and J.R. Nevins. 1987. Selective induction of human heat shock gene transcription by the adenovirus E1A gene products including the 12S E1A product. *Mol. Cell. Biol.* **7:** 2884–2890.

Sirotkin, K. and N. Davidson. 1982. Developmentally regulated transcription from *Drosophila melanogaster* chromosomal site 67B. *Dev. Biol.* **89:** 196–205.

Sorger, P.K., M.J. Lewis, and H.R.B. Pelham. 1987. Heat shock factor is regulated differently in yeast and Hela cells. *Nature* **329:** 81–83.

Sternberg, N. 1973. Properties of a mutant of *Escherichia coli* defective in bacteriophage lambda head formation (groEL) II. The propagation of phage lambda. *J. Mol. Biol.* **76:** 25–44.

Stevenson, M.A., S.K. Calderwood, and G.M. Hahn. 1986. Rapid increases in inositol trisphosphate and intracellular Ca^{++} after heat shock. *Biochem. Biophys. Res. Commun.* **137:** 826–833.

Stevenson, M.A., K.W. Minton, and G.M. Hahn. 1981. Survival and concanavalin A-induced capping in CHO fibroblasts after exposure to hyperthermia, ethanol, and irradiation. *Radiat. Res.* **86:** 467–478.

Sturzbecher, H., P. Chumakov, W.J. Welch, and J. Jenkins. 1987. Mutant p53 proteins bind hsp 72/73 cellular heat shock related protein in SV40-transformed monkey cells. *Oncogene* **1:** 201–211.

Subjeck, J.R. and T.T. Shyy. 1986. Stress protein systems of mammalian cells. *Am. J. Physiol.* **250:** C1–C17.

Subjeck, J.R., J. Sciandra, and R.J. Johnson. 1982. Heat shock proteins and thermotolerance: Comparison of induction kinetics. *Br. J. Radiol.* **55:** 579–584.

Subjeck, J.R., T. Shyy, J. Shen, and R.J. Johnson. 1983. Association between mammalian 110,000 dalton heat shock protein and nucleoli. *J. Cell Biol.* **97:** 1389–1398.

Thomas, G.P., W.J. Welch, M.B. Mathews, and J.R. Feramisco. 1982. Molecular and cellular effects of heat shock and related treatments of mammalian tissue culture cells. *Cold Spring Harbor Symp. Quant. Biol.* **46:** 985–996.

Ting, J. and A.-S. Lee. 1988. Human gene encoding the 78,000-dalton glucose-regulated protein and its pseudogene: Structure, conservation and regulation. *DNA* **7:** 275–286.

Ungewickel, E. 1985. The 70 kDa mammalian heat shock proteins are structurally and functionally related to the uncoating protein that releases clathrin triskelia from coated vesicles. *EMBO J.* **4:** 3385–3391.

Velazquez, J.M. and S. Lindquist. 1984. Hsp 70: Nuclear concentration during environmental stress; cytoplasmic storage during recovery. *Cell* **36:** 655–663.

Walter, G., A. Carbone, and W.J. Welch. 1987. Medium tumor antigen of polyomavirus transformation-defective mutant N659 is associated with 73-kilodalton heat shock protein. *J. Virol.* **61:** 405–410.

Wang, C. and E. Lazarides. 1984. Arsenite-induced changes in methylation of the 70,000 dalton heat shock proteins in chicken embryo fibroblasts. *Biochem. Biophys. Res. Commun.* **119:** 735–743.

Warren, A.P., M.H. James, D.E. Menzies, C.C. Widnell, P.A. Whitaker-Dowling, and C.A. Pasternak. 1986. Stress induces an increased hexose uptake in cultured cells. *J. Cell Physiol.* **128:** 383–388.

Weinbach, E.C. and J. Garbus. 1968. Structural changes in mitochondria induced by uncoupling agents. *Biochem. J.* **106:** 711–747.

Weitzel, G., U. Pilatus, and L. Rensing. 1985. Similar dose response of heat shock protein synthesis and intracellular pH change in yeast. *Exp. Cell. Res.* **159:** 252–256.

Welch, W.J. 1985. Phorbol ester, calcium ionophore, or serum added to quiescent rat embryo fibroblast cells all result in the elevated phosphorylation of two 28,000 dalton mammalian stress proteins. *J. Biol. Chem.* **260:** 3058–3065.

Welch, W.J. and J.R. Feramisco. 1982. Purification of the major mammalian heat shock proteins. *J. Biol. Chem.* **257:** 14949–14559.

———. 1984. Nuclear and nucleolar localization of the 72,000 dalton heat shock protein in heat shocked mammalian cells. *J. Biol. Chem.* **259:** 4501–4510.

———. 1985a. Disruption of the three cytoskeletal networks in mammalian cells does not affect translocation, translation or protein translocation changes induced by heat shock. *Mol. Cell. Biol.* **5:** 1571–1581.

———. 1985b. Rapid purification of mammalian 70,000 dalton stress protein: Affinity of the proteins for nucleotides. *Mol. Cell. Biol.* **5:** 1229–1236.

Welch, W.J. and L.A. Mizzen. 1988. Characterization of the thermotolerant cell. II. Effects on the intracellular distribution of heat shock protein 70, intermediate filaments and small ribonucleoprotein complexes. *J. Cell Biol.* **106:** 1117–1130.

Welch, W.J. and J.P. Suhan. 1985. Morphological study of the mammalian stress response: Characterization of changes in cytoplasmic organelles, cytoskeleton, and nucleoli, and appearance of intranuclear actin filaments in rat fibroblasts after heat-shock treatment. *J. Cell Biol.* **101:** 1198–1211.

———. 1986. Cellular and biochemical events in mammalian cells during and after recovery from physiological stress. *J. Cell Biol.* **103:** 2035–2053.

Welch, W.J., J.R. Feramisco, and S.H. Blose. 1985. The mammalian stress response and the cytoskeleton: Alterations in intermediate filaments. *Ann. N.Y. Acad. Sci.* **455:** 57–67.

Welch, W.F., J.G. Garrels, G.P. Thomas, J.J. Lin, and J.R. Feramisco. 1983. Biochemical characterization of the mammalian stress proteins and identification of two stress proteins as glucose and Ca^{+2} ionophore regulated proteins. *J. Biol. Chem.* **258:** 7102–7111.

Westra, A. and W.C. Dewey. 1971. Variation in sensitivity to heat shock during the cell cycle of Chinese hamster ovary cells in vitro. *Int. J. Radiat. Biol.* **19:** 467–477.

Whelan, S.A. and L.E. Hightower. 1985. Differential induction of glucose regulated and heat shock proteins: Effects of pH and sulfhydryl-reducing agents on chicken embryo cells. *J. Cell Physiol.* **125:** 251–258.

White, E., D. Spector, and W.J. Welch. 1988. Differential distribution of the adenovirus E1A proteins and colocalization of E1A with the 70 kilodalton cellular heat shock protein in infected cells. *J. Virol.* **62:** 4153–4166.

Wu, B.J., A.C. Hurst, N.C. Jones, and R.I. Morimoto. 1986. The E1A 13S product of adenovirus 5 activates transcription of the cellular human hsp 70 gene. *Mol. Cell. Biol.* **6:** 2994–2999.

Wu, F.-S., Y.C. Park, D. Roufa, and A. Mantonosi. 1981. Selective stimulation of the synthesis of an 80,000 dalton protein by calcium ionophore. *J. Biol. Chem.* **256:** 5309–5312.

Yamamoto, T., J. McIntyre, S.M. Sell, C. Georgopoulos, D. Skowyra, and M. Zylicz. 1987. Enzymology of the pre-priming steps in lambda dv DNA replication in vitro. *Proc. Natl. Acad. Sci.* **84:** 7956–7999.

Yi, P.N. 1979. Cellular ion content changes during and after hyperthermia. *Biochim. Biophys. Res. Commun.* **91:** 177–182.

Yost, H.J. and S. Lindquist. 1986. RNA splicing is interrupted by heat shock and is rescued by heat shock protein synthesis. *Cell* **45:** 185–193.

Zeuthen, E. 1971. Synchrony in tetrahymena by heat shocks spaced a normal cell generation apart. *Exp. Cell. Res.* **68:** 49–60.

Zimmerman, J.L., W. Petri, and M. Meselson. 1983. Accumulation of a specific subset of *D. melanogaster* heat shock mRNAs in normal development without heat shock. *Cell* **32:** 1161–1168.

Zylicz, M., J.H. Lebowitz, R. McMaken, and C. Georgopoulos. 1983. The dna K protein of *Escherichia coli* possesses an ATPase and autophosphorylating activity and is essential in an *in vitro* DNA replication system. *Proc. Natl. Acad. Sci.* **80:** 6431–6435.

11

Role of hsp70 in Translocation of Proteins across Membranes

Elizabeth A. Craig
Department of Physiological Chemistry
University of Wisconsin-Madison
Madison, Wisconsin 53706

I. INTRODUCTION

A number of functions of heat shock proteins have been postulated, mainly on the basis of results of biochemical experiments. In most cases, however, these proposed roles for hsps have not been supported by physiological or genetic data because strains containing mutations in heat shock genes have not been available. In *Saccharomyces cerevisiae* there is a large family of HSP70 genes that can be divided into at least four subfamilies based on the structure and function of the genes (see Craig, this volume). Recently, strains containing mutations in these genes have been constructed, allowing a genetic analysis of hsp function. In particular, we have analyzed the phenotypes of strains containing mutations in the subfamily. The *SSA* genes, *SSA1*, *SSA2*, *SSA3*, and *SSA4*, compose an essential gene family, in the sense that at least one of the proteins encoded by these genes must be present at relatively high levels for cell growth (Werner-Washburne et al. 1987). An *ssa1ssa2ssa4* strain is inviable; spores of this genotype do not bud. An *ssa1ssa2ssa4* strain can be rescued by a plasmid containing *SSA1* under the control of the galactose-inducible *GAL1* promoter (*GAL1*p:*SSA1*). However, in glucose-based media, the *GAL1* promoter is repressed and *SSA1* production ceases. Be-

Stress Proteins in Biology and Medicine.

cause of dilution due to cell division and protein degradation, cellular levels of Ssa1p decrease, allowing analysis of cells lacking protein.

We chose to analyze the effect of Ssa1p depletion on protein translocation for a number of reasons. The results of a number of biochemical experiments have implicated hsp70 proteins in protein movement within cells. For example, a mammalian hsp70 is able to uncoat clathrin cages (Schlossman and Rothman 1984; Chappel et al. 1986), and grp78 is known to associate with at least some proteins as they pass through the endoplasmic reticulum (ER) (Haas and Wabl 1983; Gething et al. 1986; Munro and Pelham 1986). It has been proposed that hsp70 acts to alter protein conformation (Pelham 1986). This idea and the observation that transport of proteins across the mitochondrial membrane required "unfolding" of the protein being imported indicated a possible role of hsp70 in protein translocation. Therefore, my laboratory, in collaboration with the laboratory of Randy Scheckman, analyzed the ability of cells depleted of *SSA* proteins to translocate and process precursor proteins across the ER and mitochondrial membranes.

II. ANALYSIS OF PRECURSOR ACCUMULATION IN STRAINS LACKING *SSA* PROTEINS

ssa1ssa2ssa4 strains containing a single copy *GAL1*p-*SSA1* fusion plasmid contain six- to sevenfold more *SSA1* protein than an *SSA*$^+$ strain (Deshaies et al. 1988). After shift to glucose-based media, the amount of Ssa1p/cell decreases and by 3 hours reaches the level found in wild-type cells. By 6 hours the levels have dropped substantially below those of wild-type cells. Therefore, one would not expect to see an effect due to the depletion of Ssa1p until more than 3 hours after transfer to glucose. A direct effect of such a depletion ought to be seen by 6 hours.

To test the idea that the *SSA* proteins are involved in the translocation of proteins across membranes, we chose to look at two proteins whose transport has been well characterized: the secretory protein α factor and the β subunit of mitochondrial F_1 ATPase. α factor, a secreted pheromone, is synthesized as a large precursor, preproalpha factor, which contains several repeats of α factor as well as precursor sequences. In wild-type cells, preproalpha factor is rapidly translocated into the ER where it is glycosylated. The prealpha factor is proteolytically processed and the 13-amino-acid α factor is secreted (Julius et al. 1984). The β subunit of F_1 ATPase is synthesized as a precursor containing a 19-amino-acid presequence that is cleaved upon import into mitochondria (Vassarotti et al. 1987). Precursor forms of preproalpha factor and the β subunit of F_1

ATPase ($F_1\beta$) can be easily distinguished from translocated products by polyacrylamide gel electrophoresis.

Extracts were prepared from cells isolated after shift of *ssa1ssa2ssa4* to glucose-based media. The resulting proteins were separated on SDS polyacrylamide gels, blotted to nitrocellulose, and reacted with a number of antibodies. Accumulation of both preproalpha factor and pre$F_1\beta$ was observed, starting at about 6 hours after transfer of *ssa1ssa2ssa4* to glucose-based media (Deshaies et al. 1988). The preproalpha factor that accumulated in the *ssa* mutant cells comigrated with the precursor that accumulates in the translocation mutant, *sec61*. Also, the pre$F_1\beta$ that accumulates in *ssa*$^-$ cells comigrates with that that accumulates in *mas1* cells defective in the cleavage of the leader sequence of proteins imported into mitochondria. A calmodulin deletion mutant containing a *GAL1*-calmodulin plasmid, dying due to depletion of calmodulin caused by a shift from galactose to glucose-based media, was tested for precursor accumulation. Since no accumulation of precursor was observed in the calmodulin mutant, precursor accumulation is not simply due to death caused by the transfer to glucose. Also, no accumulation of precursor was seen in *ssa1* cells shifted from media containing galactose to media containing glucose.

Assaying for pulse-labeled precursor is a more sensitive test for precursor accumulation. Using this method of analysis, precursors of $F_1\beta$ and α factor were detected as well as low levels of precursors of invertase (a secreted protein) and carboxypeptidase Y (a vacuolar protein).

The accumulation of precursors as well as the correspondence of the timing of accumulation with Ssa1p depletion indicates that *SSA* proteins are involved in the import of at least some proteins. The fact that the accumulated preproalpha factor precursor is proteinase sensitive, whereas the glycosylated preproalpha factor is resistant to proteinase digestion, suggests that translocation is blocked at a stage before translocation across the ER membrane (Deshaies et al. 1988). The proteinase-sensitive precursor sediments with the particulate fraction of cell extracts, suggesting that the precursor is specifically associated with the ER membrane. However, the sedimentation with the particulate fraction of cell extracts might be due to nonspecific aggregation.

III. *SSA* PROTEINS FACILITATE TRANSPORT IN VITRO

The results of in vitro assays also indicate involvement of *SSA* proteins in translocation across the ER membrane. Uptake of in-vitro-synthesized preproalpha factor into yeast microsomes is stimulated about fivefold by the addition of cytoplasmic extracts (Waters and Blobel 1986). Fractionation of yeast extracts indicates that stimulation is due to at least two factors. Two groups independently identified hsp70 as one factor. Since hsp70 has an extremely high affinity for ATP, it is possible to partially purify these proteins in one step. Deshaies et al. (1988) showed that the ATP-binding fraction of yeast cytosol isolated from an *ssa1ssa2ssa4* [*GAL1-SSA1*] strain grown on galactose-based media facilitated the posttranslational translocation of preproalpha factor, in the presence of a small amount of postribosomal supernatant. The ATP-binding proteins isolated from the same strain after shift to glucose-based media did not stimulate translocation even in the presence of postribosomal supernatant. These results indicate that translocation is facilitated by Ssa1p in the presence of an additional factor(s).

Blobel and colleagues (Chirico et al. 1988) identified two 70-kD proteins in yeast extracts that have translocation stimulatory activity. These purified proteins comigrated with Ssa1p and Ssa2p on two-dimensional polyacrylamide gels and reacted with a monoclonal antibody directed against mammalian hsp70. The *SSA* proteins stimulate both the rate of posttranslational translocation of preproalpha factor and the amount of precursor translocated. In this study, a second stimulatory activity, which is inhibited by a sulfhydryl alkylating agent (NEM), was also identified. Both these activities are required for efficient translocation. It is likely that this NEM-sensitive factor is the same as the factor present in the postribosomal supernatant described by Deshaies et al. (1988).

IV. POSSIBLE ROLES OF *SSA* PROTEINS IN TRANSLOCATION

The biochemical and genetic results described above indicate a role for *SSA* protein in protein translocation into the ER and mitochondria (Fig. 1). However, the mode of action of *SSA* proteins remains to be elucidated. *SSA* proteins may bind to precursor proteins and catalyze an ATP-dependent alteration in precursor tertiary structure, converting it to a translocation-competent form. There is no direct evidence for this mechanism of action of Ssa1p, but experimental data suggest that unfolding of the precursor is necessary for import. Stabilization of the tertiary structure of a fusion protein has been shown to interfere with import into mitochondria (Eilers and Schatz 1986). That unfolding is NTP-dependent is supported by the fact that nascent polypeptides, which are presumably unfolded, do not require ATP for translocation (Verner and Schatz 1987). Because of the apparent involvement of hsp70 in many types of protein:protein interactions, a role as an unfoldase is indicated. However, even if one assumes a role for hsp70 in translocation that involves alteration of protein conformation, hsp70 need not directly interact with the

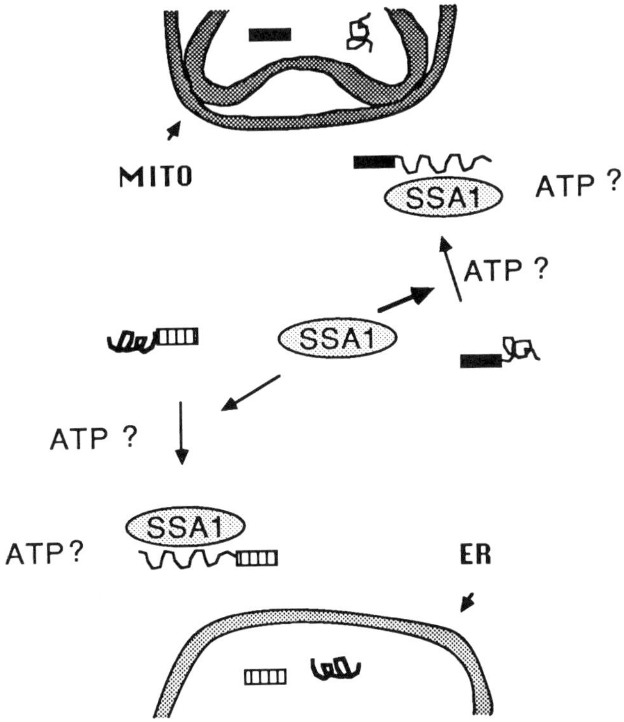

Figure 1 Schematic diagram of possible role of *SSA* proteins in translocation across the endoplasmic reticulum (ER) and mitochondrial (MITO) membrane. The interaction of Ssap with precursor proteins could either be direct as depicted or involve interactions with other components of the translocation machinery.

precursor. It is possible that *SSA* proteins interact with some other component of the translocation system, either on the membrane or in the cytoplasm.

V. IS TRANSLOCATION AN ESSENTIAL FUNCTION OF *SSA* PROTEINS?

The experiments discussed here point to a role for hsp70 in protein translocation. However, these experiments do not indicate whether or not a defect in translocation is responsible for the death of hsp70-deficient cells. There is reason to speculate that *SSA* proteins, as well as other hsp70 proteins, are involved in a number of cellular processes. Therefore, the question arises as to whether *SSA* mutant cells die because of a defect in translocation or for some other reason. Several pieces of evidence suggest that cell death may be due to other effects of hsp70 deficiency.

Although the defect in translocation of preproalpha factor is nearly complete in an *ssa1ssa2ssa4* mutant, translocation of other proteins is affected less dramatically. For example, translocation of carboxypeptidase *Y* and invertase are only partially inhibited, as is the translocation of $F_1\beta$ (Deshaies et al. 1988). It is not clear whether the translocation defect could cause the inhibition of growth observed after the depletion of *SSA* protein, especially since translocation into the ER is thought to occur mainly cotranslationally. Of course, it is possible that the translocation of an essential protein not examined to date is completely blocked in *ssa*⁻ cells.

The *ssa1ssa2* strain is temperature sensitive for growth at 37°C (Craig and Jacobsen 1984). Growth of this strain at lower temperatures is apparently permitted by the aberrant overexpression of Ssa4p (Werner-Washburne et al. 1987). However, Ssa4p appears not to be able to compensate for the absence of Ssa1p and Ssa2p at high temperatures. Unlike an *ssa1ssa2ssa4* strain, *ssa1ssa1* mutants shifted to nonpermissive growth conditions (37°C) do not accumulate untranslocated forms of preproalpha factor, $F_1\beta$ subunit, or carboxypeptidase Y even after long times at the nonpermissive temperature (J. Nelson et al., unpubl.). Therefore, when growth of *ssa1ssa2* cells arrests at 37°C, translocation appears to continue normally, suggesting that the growth arrest is due to factors other than defective translocation. Future genetic and biochemical experiments may well reveal additional functions of *SSA* proteins if they exist.

VI. POSSIBLE FUNCTIONAL RELATIONSHIPS BETWEEN HSP70 PROTEINS OF THE CYTOPLASM, MITOCHONDRIA, AND ER

Recently, it has become apparent that hsp70-related proteins are present in a number of cellular compartments, including the cytoplasm, nucleus, mitochondrium, and ER. A good deal of information is available concerning the function of the ER 70K protein, referred to as grp78 or BiP (Munro and Pelham 1986). This protein is thought to interact with at least some proteins as they pass through the ER membrane. It has been proposed that grp78 acts in an ATP-dependent fashion to alter the conformation of these proteins aiding in the assembly of multimeric structures. Although no information exists as to the role of mitochondrial hsp70, it possibly acts in a similar way, aiding in assembly and/or maintenance of the internal structure of the mitochondria. The role of cytoplasmic hsp70 in protein translocation appears at first glance, perhaps, to be quite different from the proposed role of the ER and mitochondrial proteins. However, there are significant similarities. They are involved in interactions with proteins that have not reached their final destination in the cell, and thus probably not in their final functional conformations, and involve alterations in the conformation of the substrate proteins. In the cytoplasm, the proposed hsp70 activity allows for transport across membranes and, in the ER, allows for productive interactions of proteins to form multimeric complexes. Also, there is evidence to suggest that both types of interactions require ATP. As more is understood about the mechanism of action of different members of the hsp70 family, similarities in function will likely emerge.

VII. CONCLUDING REMARKS

The yeast *SSA* family of proteins has been shown by both biochemical and genetic experiments to function in an early step of posttranslational translocation. These proteins are involved in translocation across both the mitochondrial and ER membranes, suggesting that initial steps in the translocation process are similar for these two organelles. It is likely that they act as unfoldases, altering the structure of precursor proteins into a translocation-competent conformation. However, genetic evidence suggests that there may be other essential functions of *SSA* proteins that remain to be elucidated.

ACKNOWLEDGMENTS

The work carried out in my laboratory was supported by grants from the National Institutes of Health. I thank members of my laboratory for comments on this manuscript, especially John Nelson, who can make sense out of my sentences.

REFERENCES

Chappell, T.G., W.J. Welch, D.M. Schlossman, K.B. Palter, M.J. Schlesinger, and J.E. Rothman. 1986. Uncoating ATPase is a member of the 70-kDa family of stress proteins. *Cell* **45:** 3–13.

Chirico, W., M.G. Waters, and G. Blobel. 1988. 70K heat shock related proteins stimulate protein translocation into microsomes. *Nature* **332:** 805–810.

Craig, E. and K. Jacobsen. 1984. Mutations of the heat-inducible 70 kilodalton genes of yeast confer temperature-sensitive growth. *Cell* **38:** 841–849.

Deshaies, R., B. Koch, M. Werner-Washburne, E. Craig, and R. Schekman. 1988. A subfamily of stress proteins facilitates translocation of secretory and mitochondrial precursor polypeptides. *Nature* **332:** 800–805.

Eilers, M. and G. Schatz. 1986. Binding of a specific ligand inhibits import of a purified precursor protein into mitochondria. *Nature* **322:** 228–232.

Gething, M.-J., K. McCammon, and J. Sambrook. 1986. Expression of wild-type and mutant forms of influenza hemagglutinin: The role of folding in intracellular transport. *Cell* **46:** 939–950.

Haas, I. and M. Wabl. 1983. Immunoglobulin heavy chain binding protein. *Nature* **306:** 387–389.

Julius, D., R. Scheckman, and J. Thorner. 1984. Glycosylation and processing of prepro-alpha factor through the yeast secretory pathway. *Cell* **36:** 309–318.

Munro, S. and H.R.B. Pelham. 1986. An hsp79-like protein in the ER: Identity with the 78kd glucose-regulated protein and immunoglobulin heavy chain binding protein. *Cell* **46:** 291–300.

Pelham, H. 1986. Speculations on the functions of the major heat shock and glucose-regulated proteins. *Cell* **46:** 959–961.

Schlossman, D.M. and J.E. Rothman. 1984. An enzyme that removes clathrin coats: Purification of an uncoating ATPase. *J. Cell Biol.* **99:** 723–733.

Vassarotti, A., W.-J. Chen, C. Smagula, and M. Douglas. 1987. Sequences distal to the mitochondrial targeting sequences are necessary for the maturation of the F_1-ATPase β subunit precursor in mitochondria. *J. Biol. Chem.* **262:** 411–418.

Verner, K. and G. Schatz. 1987. Import of an incompletely folded protein precursor protein into isolated mitochondria requires an energized inner membrane, but no ATP. *EMBO J.* **6:** 2449–2456.

Waters, M.G. and G. Blobel. 1986. Secretory protein translocation in a yeast cell-free system can occur post-translationally and requires ATP hydrolysis. *J. Cell Biol.* **102:** 1543–1550.

Werner-Washburne, M., D. Stone, and E. Craig. 1987. Complex interactions among members of an essential subfamily of HSP70 genes in *Saccharomyces cerevisiae*. *Mol. Cell. Biol.* **7:** 2568–2577.

12

Functions of the hsp70 Protein Family: An Overview

Hugh R.B. Pelham
MRC Laboratory of Molecular Biology
Cambridge CB2 2QH, England

There has been speculation about the function of heat shock proteins ever since their discovery (Tissières et al. 1974). An early assumption was that they somehow serve to restore the status quo in cells that have experienced an acute and potentially damaging stress, and to protect against further injury. In 1980, Hightower noted that most inducers of the stress response lead to the accumulation of denatured or damaged proteins in cells and suggested that hsps somehow help to deal with this material. In recent years, these generalizations have been backed up by more specific studies of hsp function that have led to plausible molecular models for the mechanism of action of at least some of them. It has also become clear that hsps or closely related proteins are present in unstressed cells and perform functions essential for normal growth. In this chapter, I discuss the hsp70 family and argue that the role of these proteins (as well as that of some other hsps) is to promote the refolding and reassembly of proteins that have become denatured as a result of hyperthermia or other abuse. This role is an extension of one of the functions of the hsp70 family in unstressed cells, which is to aid the assembly of newly synthesized proteins.

I. LOCALIZATION OF HSP70 IN STRESSED CELLS

hsp70 is often the most prominent product of protein synthesis in stressed cells, and in heat-shocked *Drosophila* cells, most incorporation

Stress Proteins in Biology and Medicine.

of [^{35}S]methionine is into this protein. It is thus easy to follow the intracellular distribution of hsp70 by autoradiography or subcellular fractionation. Early studies showed that hsp70 was associated with many cellular components, but the most striking observation was that it bound to nuclei and nucleoli in stressed cells, moving to the cytoplasm during recovery (Arrigo et al. 1980; Velazquez et al. 1980). This behavior was observed both in tissue-culture cells and in larval salivary gland cells and was subsequently confirmed by immunofluorescence studies using a monoclonal antibody specific for *Drosophila* hsp70 (Velazquez and Lindquist 1984).

We and other investigators obtained broadly similar results using mammalian cells. We initially studied the behavior of *Drosophila* hsp70 expressed from a cloned gene in monkey COS cells (Munro and Pelham 1984; Pelham 1984). Immunofluorescence microscopy showed that the protein was present in both the nucleus and cytoplasm of unstressed (or recovered) cells but was excluded from nucleoli. The association with the nucleus was weak, the protein being rapidly released upon cell lysis. After a brief heat shock, hsp70 bound tightly to nuclei and then progressively accumulated in nucleoli before reverting to its normal distribution as the cells recovered. Identical behavior was observed when the distribution of the endogenous mammalian hsp70 was examined using antibodies specific for this protein (Welch and Feramisco 1984; Pelham et al. 1985).

The localization data suggested that hsp70 performs some function in the nucleus and, in particular, the nucleolus. It has been known since the 1960s that heat shock causes dramatic but reversible changes in nucleolar morphology. These changes are detectable by light microscopy and correlate well with the movement of hsp70 into and out of the nucleoli (Pelham 1984). Electron microscopy shows that the granular structure of partially assembled ribosomes becomes less distinct; at the same time, processing of rRNA precursors is blocked, indicating that ribosome assembly has ceased (for review, see Simard et al. 1974; Puvion and Moyne 1981). Immunoelectron microscopy suggests that hsp70 associates directly with damaged pre-ribosomes (Welch and Suhan 1986).

These descriptive studies show that hsp70 accumulates in heat-damaged nucleoli, but they do not prove that it performs a useful function there. To test this, we transfected COS cells with a plasmid that expresses *Drosophila* hsp70 at a high constitutive level (Pelham 1984; Pelham et al. 1985). The response of the transfected cells to heat shock was compared with that of untransfected cells, or of cells transfected with a plasmid that did not produce hsp70. After exposure to 43°C for 45 minutes, it took 3.5–4 hours for 50% of the control cells to show normal nucleolar morphology, whereas cells expressing *Drosophila* hsp70 recovered about twice as fast (Pelham et al. 1985). In the presence of actinomycin D, which blocks the transcriptional induction of heat shock genes, the control cells recovered even more slowly, whereas ribosome assembly and export resumed within 4 hours in the majority of the transfected cells (Pelham 1984). Thus, overproduction of hsp70 alone is sufficient to accelerate the reconstruction of a functional nucleolus. This somewhat ill-defined activity of hsp70 remains the most specific known function of the protein in heat-shocked cells.

II. LOCATION OF OTHER HSP70 FAMILY MEMBERS

In eukaryotes, the hsp70 family comprises a number of proteins, many of which are expressed at substantial levels in the absence of stress. Some of these behave much like the strictly heat-inducible *Drosophila* protein, being largely cytoplasmic under normal conditions but moving to the nucleus and/or nucleolus upon heat shock (Lewis and Pelham 1985; Ungewickell 1985). Others have a completely different location. In a search for cDNA clones encoding hsp70-related proteins, we found one that encoded the glucose-regulated protein grp78 and showed that this protein is identical to immunoglobulin heavy-chain binding protein (BiP) (Munro and Pelham 1986). BiP is found in the lumen of the endoplasmic reticulum (ER) and is kept there by a specific retention system (Munro and Pelham 1987). Another member of the hsp70 family, first identified in yeast, is imported into mitochondria and presumably resides in the inner matrix of this organelle (see Craig, this volume). Genetic studies show that hsp70-like proteins must be present in the cytoplasm, ER, and mitochondria for yeast cell viability (Werner-Washburne et al. 1987; Rose et al. 1989; L. Moran, pers. comm.). It is likely that the proteins in all three locations perform biochemically similar functions but have access to distinct sets of substrates.

III. HSP70-LIKE PROTEINS BIND TO ABNORMAL OR UNFOLDED PROTEINS

Although hsp70 associates tightly with heat-shocked nuclei and nucleoli, the precise molecules to which it is bound have not been identified, largely because they are not easily solubilized. However, interaction of hsp70 family members with a variety of substrates in unshocked cells has been detected by coimmunoprecipitation. For example, mutant forms of the transformation-related protein p53 accumulate in a complex with the constitutively expressed cytoplasmic form of hsp70 (Pinhasi-Kimhi et al. 1986; Clarke et al. 1988). In the ER, BiP is known to bind to mutant or misfolded forms of the influenza virus hemagglutinin protein (Gething et al. 1986); to an altered form of SV40 T antigen that contains a signal peptide and thus is delivered to the ER, where it fails to fold correctly (Sharma et al. 1985); and to a large number of secretory and membrane proteins whose normal glycosylation has been prevented by specific inhibitors, which frequently results in misfolding and precipitation of the proteins (Gething et al. 1986; Dorner et al. 1987; Hendershot et al. 1988; Kassenbrock et al. 1988).

In addition to these abnormal substrates, BiP also binds to newly synthesized secretory proteins during the course of their folding and assembly. Thus, it binds to at least some immunoglobulin heavy chains prior to their association with light chains (Bole et al. 1986) and can transiently interact with other immature secretory proteins (Dorner et al. 1987). Binding of BiP to proteins synthesized in an in vitro translation-translocation system can also be demonstrated. In such a system, it is possible to manipulate the redox potential so as to block or promote disulfide bond formation within the microsomes; in this way it was shown that BiP would bind to non-disulfide-bonded prolactin, but not to the mature, oxidized form (Kassenbrock et al. 1988).

All of these studies showed common features of the hsp70– or BiP–substrate interaction. In each case, the complexes could withstand exposure to high salt concentrations (0.5–2 M), and the bound proteins were invariably abnormal or incompletely folded: The mature, wild-type forms of the proteins did not coprecipitate with hsp70 or BiP. These results are consistent with the earlier studies with heat-shocked nuclei: hsp70 resists extraction from these with 2 M salt, DNase, and RNase (Sinibaldi and Morris 1981; Lewis and Pelham 1985) and appears to be part of a hydrophobic aggregate of many nuclear proteins. This aggregation occurs at a critical temperature both in vivo and in vitro and is presumably caused by the partial denaturation or disruption of nuclear structures (Evan and Hancock 1985; Littlewood et al. 1987).

Taken together, the results indicate that hsp70-like proteins preferentially recognize partially unfolded proteins, probably by binding to exposed hydrophobic residues that would be buried in the mature protein. Indeed, biochemical studies show that hsp70 will bind tightly to hydrophobic materials such as phenyl-Sepharose, although whether this property is directly related to its substrate specificity in vivo is unclear.

IV. ROLE OF ATP HYDROLYSIS IN HSP70 FUNCTION

The first evidence that ATP might be involved in hsp70 function was the observation that the *Escherichia coli* hsp70 homolog, the DnaK protein, had weak ATPase activity (Zylicz et al. 1983). It was subsequently shown that mammalian hsp70 binds tightly and specifically to ATP-agarose (Welch and Feramisco 1985). These observations prompted us to test the effects of ATP on the binding of hsp70 to heat-shocked nuclei, and we found that micromolar concentrations of ATP were sufficient to cause the rapid and quantitative release of the protein from nuclei and nucleoli (Lewis and Pelham 1985). Nonhydrolyzable analogs of ATP were ineffective even at millimolar concentrations, suggesting that ATP hydrolysis was required for release. Similar results were obtained when ATP was added to complexes containing the constitutively expressed cytoplasmic form of hsp70 and p53 (Clarke et al. 1988), or BiP and immunoglobulin heavy chains (Munro and Pelham 1986), indicating that ATP-dependent release from their substrates is a common feature of the hsp70 family. However, we and other workers have detected only very low rates of ATP hydrolysis with purified hsp70 or BiP, amounting to about one hydrolysis event per hsp70 molecule per minute, whether or not putative substrates are present.

On the basis of these results, we suggested a possible mechanism of action for the hsp70-like proteins (Lewis and Pelham 1985; Pelham 1986; see Fig. 1). We assume that the normal substrates for these proteins are newly synthesized or recently denatured proteins (depicted by the shaded box in Fig. 1) that are capable of folding into their correct tertiary and quaternary structures. Such proteins have exposed hydrophobic regions that tend to interact to form intractable precipitates. hsp70 binds to such regions; by reducing the effective concentration of hydrophobic surfaces, it inhibits (or reverses) the aggregation process. At intervals, hsp70 hydrolyzes an ATP molecule and undergoes a conformational change that decreases its affinity for the substrate, which is released. The substrate then has a chance to fold before hsp70 reverts to its binding conformation. If hydrophobic regions are still exposed, the cycle is repeated. If no binding sites remain, hsp70 molecules can simply interact with each other to form dimers or higher oligomers (which are indeed observed).

It is useful to compare this simple model with the known effects of detergents on protein folding. A good example is provided by the enzyme rhodanese, which consists of two domains that interact through a relatively hydrophobic interface. During refolding of denatured rhodanese, these domains tend to form intermolecular contacts, leading to precipitation and very little active enzyme. Addition of the detergent lauryl maltoside weakens these incorrect interactions and greatly increases the yield of correctly folded enzyme (Tandon and Horowitz 1986). Thus, the prevention of aggregation can significantly improve the folding process, even without an input of energy.

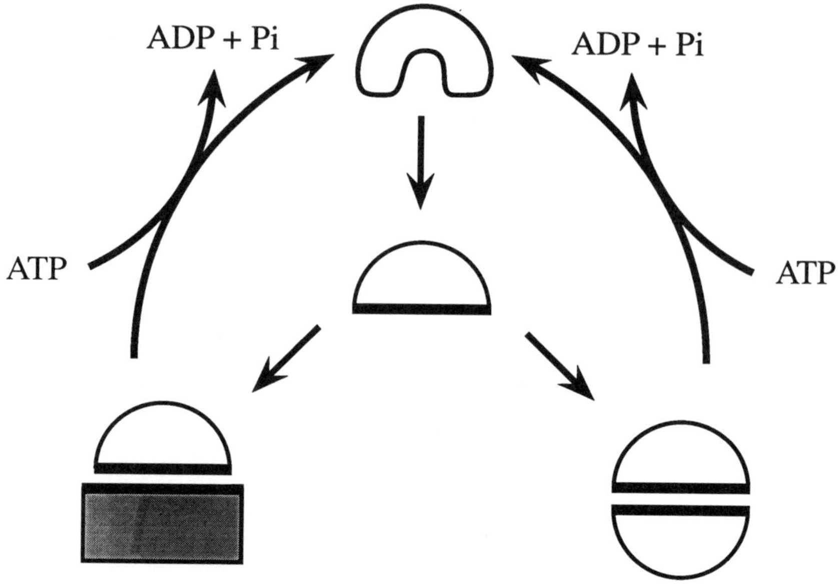

Figure 1 Hypothetical model of hsp70 action. hsp70 (depicted by the open symbol) normally exists as a dimer (lower right) or bound to a hydrophobic substrate (lower left). Binding and/or hydrolysis of ATP results in a conformational change to a nonbinding form (top), which spontaneously reverts to the binding form (center). This form of the protein can interact with a substrate or, if none is available, it can dimerize.

Why does hsp70 hydrolyze ATP when it could achieve much the same result simply by acting as a detergent? The answer may be that the ability of detergents to catalyze folding is limited. To have the desired effect, the detergent must form complexes with an unfolded protein that are more stable than any incorrect aggregates, but not so stable that they hinder the normal folding process. Detergents that bind too tightly (such as dodecyl sulfate) are very effective at dissolving precipitates but block folding completely. hsp70 does not suffer from this limitation—it can form complexes that are much more stable than nonspecific aggregates without these becoming kinetic dead ends, because the energy of ATP hydrolysis can be used to disrupt them. Nevertheless, for some purposes, an energy-independent detergent effect may be sufficient, and there is evidence that hsp70 can also act in this way (see below).

V. STUDIES OF HSP70 AND BIP ACTION IN VITRO

Direct testing of models for hsp70 action requires a biochemical assay for the purified protein. Unfortunately, the recovery of nuclei from heat shock is not an easy phenomenon to study in cell-free systems, and it has proved very difficult even to assemble in vitro the kinds of ATP-disruptible complexes that are found in cells. However, a number of assays for more specialized functions of hsp70 family members have been devised, and these have provided some information about the properties of these proteins.

VI. SOLUBILIZATION OF IMMUNOGLOBULIN CHAINS

We attempted to demonstrate a role for BiP in maintaining the solubility of unfolded immunoglobulin heavy chains. A purified monoclonal antibody was completely denatured and reduced with 5 M guanidine hydrochloride and dithiothreitol, diluted rapidly to 1 M guanidine, and then dialyzed against guanidine-free buffer. Such treatment caused the precipitation of most of the protein. However, if dilution was into a solution containing BiP, most of the immunoglobulin remained soluble; ovalbumin, bovine serum albumin, and the cytoplasmic form of hsp70 did not have this effect (see Fig. 2). Further investigation showed that most of the soluble immunoglobulin had not achieved its correct disulfide-bonded state, but was not bound to BiP.

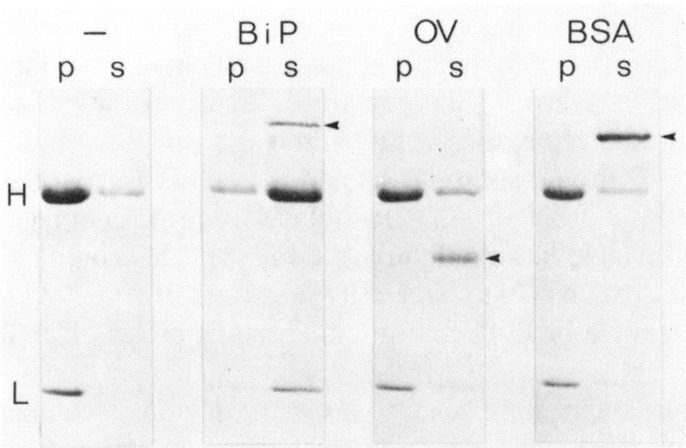

Figure 2 Solubilization of immunoglobulin chains by BiP. Reduced and denatured heavy (H) and light (L) chains of a monoclonal antibody, from a preparation enriched for H chains, were diluted from 5 M guanidine into a solution containing either no protein (–), BiP, ovalbumin (OV), or bovine serum albumin (BSA). The diluted mixture was dialyzed against a physiological salt solution, and the insoluble material was pelleted in a microcentrifuge. Pellet (p) and supernatant (s) were then analyzed by SDS-PAGE and Coomassie blue staining. The non-immunoglobulin proteins are indicated by arrowheads.

Surprisingly, the effect of BiP was quite unaffected by the presence or absence of ATP, and only small amounts of an ATP-disruptible BiP/immunoglobulin complex could be detected. Thus, although BiP had a dramatic and specific effect, this was evidently the result of weak, detergent-like interactions. Such interactions may play an important role in vivo, helping to prevent aggregation and facilitate protein folding in the concentrated environment of the ER.

VII. DISRUPTION OF CLATHRIN CAGES

Rothman and his colleagues described a protein, termed uncoating ATPase, that catalyzed the removal of clathrin from coated vesicles in vitro and was also capable of disrupting cages assembled from purified clathrin (see Rothman and Schmid 1986). This protein was subsequently shown to be a cytoplasmic member of the hsp70 family (Ungewickell 1985; Chappell et al. 1986). Although there is no direct evidence that hsp70 is involved in uncoating in vivo, the in vitro reaction remains one of the best-studied activities of hsp70.

The reaction can be thought of as a special example of the disruption of a preformed protein aggregate. ATP hydrolysis is clearly required for this event, although no ATP-disruptible intermediate has been identified. The released clathrin does form a stoichiometric complex with hsp70, but this is unaffected by ATP and appears to result from quite weak interactions, because clathrin can readily be reassembled from this complex into cages. The role of ATP in the reaction is not obvious: If the mechanism of disruption were simply the capture of transiently exposed portions of the clathrin molecule, then ATP hydrolysis, which normally leads to release of hsp70 from a substrate, should not be necessary. It is possible, however, that a cycle of ATP hydrolysis is required to get hsp70 into a form that is able to bind tightly (see Fig. 1). This could explain why it is so hard to get hsp70 to bind to its substrates in vitro—ATP may be needed to achieve a binding conformation, but if present it will then catalyze release.

The clathrin system is appealing because it clearly demonstrates that hsp70 alone, with a purified substrate, can induce disaggregation in an ATP-dependent manner. However, the precise sequence of events in the reaction remains hard to define.

VIII. INITIATION OF BACTERIOPHAGE λ DNA SYNTHESIS

The DnaK protein (the *E. coli* homolog of hsp70) was originally identified as a host protein required for bacteriophage λ DNA replication, and its function in this process has now been described in some detail (see

Georgopoulos et al., this volume). Initiation of replication involves the binding of λO protein to the origin; λP protein then binds both to O and to the host DnaB helicase, thus bringing the latter to the site where its action is required. The DnaK protein, along with another heat shock protein, DnaJ, is required to separate helicase from the origin complex, so that unwinding of the DNA helix can proceed (Dodson et al. 1986). Interestingly, DnaK is required in massive molar excess to catalyze the reaction, but the amount can be greatly reduced if another protein, GrpE, is present (Zylicz et al. 1989). There may be other situations where the efficient action of hsp70-like proteins requires one or more unidentified cofactors.

The use of DnaK by λ is clearly a special case, as is shown by the fact that many mutations of the *dnaK* gene block this activity without affecting other essential functions of the DnaK protein in *E. coli*. Nevertheless, there is good genetic evidence that the function in λ DNA replication is important in vivo, and good biochemical evidence that this function involves the ATP-dependent disruption of protein/protein interactions (Dodson et al. 1986; Liberek et al. 1988). It seems likely that DnaK plays similar roles in vivo that are mechanistically analogous.

IX. TRANSLOCATION OF PROTEINS ACROSS MEMBRANES

Most membrane and secretory proteins are synthesized on membrane-bound ribosomes and are cotranslationally inserted into the ER. However, some small proteins are capable of posttranslational translocation into the ER, and many cytoplasmically synthesized mitochondrial proteins are also capable of posttranslational import into mitochondria. A number of studies have shown that proteins cross membranes in an unfolded state and that a prime requirement for import is that the precursor protein remains in an unfolded or loosely folded state (for reviews, see Zimmermann and Meyer 1986; Eilers and Schatz 1988). The properties of hsp70 make it an ideal candidate for a protein that can maintain unfolded polypeptides in a soluble form, and it was natural to ask whether an hsp70-like protein is involved in posttranslational import.

In a collaborative study with Zimmermann's group, we examined the effect of the cytoplasmic, constitutively synthesized hsp70 protein in a model assay system, the posttranslational insertion of bacteriophage M13 procoat protein into dog pancreas microsomes (Zimmermann et al. 1988). High concentrations of hsp70 weakly stimulated import of procoat in an ATP-independent manner. This stimulation probably involved direct interaction of hsp70 and procoat, because addition of hsp70 partially protected procoat from digestion by added proteinases. A much more efficient stimulation was seen when an additional cytoplasmic factor (contained in rabbit reticulocyte lysate) was also present, and the effect was then dependent on the addition of ATP. It appears that there is a component of reticulocyte lysate that acts synergistically with hsp70 to confer competence for import on the procoat molecule. It is not clear whether ATP is required for the function of this component, of hsp70, or of both.

Independent proof that hsp70 can stimulate protein transport across membranes was provided by studies of the import of yeast preproalpha factor into microsomes (Chirico et al. 1988; Deshaies et al. 1988) and of the *PUT2* gene product into mitochondria (Murakami et al. 1988). In each case, maximal import required both hsp70 and an additional cytoplasmic component and was dependent on ATP. These requirements could largely be obviated by prior urea denaturation of the preproalpha factor, although the capacity of the denatured precursor to be imported is rapidly lost in the absence of urea. This result supports the idea that the role of hsp70 is to create and maintain an unfolded, nonaggregated state of the precursor.

The most convincing evidence for the physiological relevance of these assays was provided by a genetic experiment in yeast. Deshaies et al. (1988) used a yeast strain in which the only cytoplasmic hsp70 protein was provided by a cloned gene under the control of the GAL promoter. This strain can only grow on galactose, because when glucose is present, expression of hsp70 is inhibited and growth ceases. Shortly after the addition of glucose, the cells begin to accumulate, in the cytoplasm, both preproalpha factor and the precursor to the β subunit of the mitochondrial ATPase. This result demonstrates that hsp70 is required for the successful import of these proteins in vivo.

As translocated proteins emerge into the ER or mitochondria, they are once more exposed to the attentions of hsp70-like proteins, and the role of these proteins can also be tested in yeast. Strikingly, within 15 minutes of a shift of a temperature-sensitive BiP mutant to the nonpermissive temperature, secretory protein precursors have accumulated in the cytoplasm (M. Rose, pers. comm.). This phenomenon suggests that the import machinery becomes clogged by proteins that have failed to complete their import, and thus that BiP plays an important role in maintaining the solubility of unfolded proteins as they enter the ER.

X. CONCLUSIONS

There is now considerable evidence to support the view that hsp70 and its relatives share the ability to prevent the formation of, and to disrupt, certain types of protein/protein interactions. They appear to achieve this by binding to the interacting surfaces and then releasing themselves in an ATP-dependent reaction, although in many cases, a simpler, ATP-independent detergent-like action may be sufficient. Under normal circumstances, the main substrates are probably newly synthesized incompletely folded proteins and protein subunits. Interactions with these substrates are transient: Only mutant or irreversibly altered proteins become permanently associated with hsp70-like proteins. After heat shock, denatured or disrupted proteins and protein-containing structures become substrates, until such time as they have refolded or been degraded. Superimposed on these general roles are several specific functions of members of the hsp70 family that involve disruption of particular protein/protein interactions.

hsp70 is not the only heat shock protein to have such properties. The *E. coli groE* protein and its homologs in mitochondria and chloroplasts can also bind to newly synthesized proteins, release them in the presence of ATP, and catalyze both the assembly of multi-subunit proteins and the translocation of proteins across membranes (for references, see Bochkareva et al. 1988; Hemmingsen et al. 1988). Efficient protein folding and assembly, essential processes in living cells, are probably stimulated by a range of proteins; the heat shock response may simply represent the overexpression of the most useful of these when their services are urgently required.

ACKNOWLEDGMENTS

I thank my colleagues who have survived numerous attempted experiments and discussions about hsp70 function; especially Mike Lewis, who performed the experiment shown in Figure 2, and Sean Munro.

REFERENCES

Arrigo, A.P., S. Fakan, and A. Tissières. 1980. Localization of the heat shock-induced proteins in *Drosophila melanogaster* tissue-culture cells. *Dev. Biol.* **78:** 86–103.

Bochkareva, E.S., N.M. Lissin, and A.S. Girshovich. 1988. Transient association of newly-synthesized unfolded proteins with the heat-shock GroEL protein. *Nature* **336:** 254–257.

Bole, D.G., L.M. Hendershot, and J.F. Kearney. 1986. Posttranslational association of immunoglobulin heavy-chain binding-protein with nascent heavy chains in non-secreting and secreting hybridomas. *J. Cell Biol.* **102:** 1558–1566.

Chappell, T.G., W.J. Welch, D.M. Schlossmann, K.B. Palter, M.J. Schlesinger, and J.E. Rothman. 1986. Uncoating ATPase is a member of the 70 kilodalton family of stress proteins. *Cell* **45:** 3–13.

Chirico, W.J., M.G. Waters, and G. Blobel. 1988. 70K heat shock related proteins stimulate protein translocation into microsomes. *Nature* **332:** 805–810.

Clarke, C.F., K. Cheng, A.B. Frey, R. Stein, P.W. Hinds, and A.J. Levine. 1988. Purification of complexes of nuclear oncogene p53 with rat and *Escherichia coli* heat shock proteins: in vitro dissociation of hsc70 and dnaK from murine p53 by ATP. *Mol. Cell. Biol.* **8:** 1206–1215.

Deshaies, R.J., B.D. Koch, M. Werner-Washburne, E.A. Craig, and R. Schekman. 1988. A subfamily of stress proteins facilitates translocation of secretory and mitochondrial precursor polypeptides. *Nature* **332:** 800–805.

Dodson, M., H. Echols, S. Wickner, C. Alfano, K. Mensa-Wilmot, B. Gomes, J. LeBowitz, J.D. Roberts, and R. McMacken. 1986. Specialized nucleoprotein structures at the origin of replication of bacteriophage lambda: Localized unwinding of duplex DNA by a six-protein reaction. *Proc. Natl. Acad. Sci.* **83:** 7638–7642.

Dorner, A.J., D.G. Bole, and R.J. Kaufman. 1987. The relationship of N-linked glycosylation and heavy chain-binding protein association with the secretion of glycoproteins. *J. Cell Biol.* **105:** 2665–2674.

Eilers, M. and G. Schatz. 1988. Protein folding and the energetics of protein translocation across biological membranes. *Cell* **52:** 481–483.

Evan, G.I. and D.C. Hancock. 1985. Studies on the interaction of the human c-myc protein with cell nuclei: $p62^{c\text{-}myc}$ as a member of a discrete subset of nuclear proteins. *Cell* **43:** 253–261.

Gething, M.-J., K. McCammon, and J. Sambrook. 1986. Expression of wild-type and mutant forms of influenza hemagglutinin: The role of protein folding in intracellular transport. *Cell* **46:** 939–950.

Hemmingsen, S.M., C. Woolford, S.M. van der Vies, K. Tilly, D.T. Dennis, C.P. Georgopoulos, R.W. Hendrix, and R.J. Ellis. 1988. Homologous plant and bacterial proteins chaperone oligomeric protein assembly. *Nature* **333:** 330–334.

Hendershot, L.M., J. Ting, and A.S. Lee. 1988. Identity of the immunoglobulin heavy-chain-binding protein with the 78,000-dalton glucose-regulated protein and the role of posttranslational modifications in its binding function. *Mol. Cell. Biol.* **8:** 4250–4256.

Hightower, L.E. 1980. Cultured animal cells exposed to amino acid analogues or puromycin rapidly synthesize several polypeptides. *J. Cell. Physiol.* **102:** 407–427.

Kassenbrock, C.K., P.D. Garcia, P. Walter, and R.B. Kelly. 1988. Heavy chain binding protein recognizes aberrant polypeptides translated in vitro. *Nature* **333:** 90–93.

Lewis, M.J. and H.R.B. Pelham. 1985. Involvement of ATP in the nuclear and nucleolar functions of the 70 kd heat shock protein. *EMBO J.* **4:** 3137–3143.

Liberek, K., C. Georgopoulos, and M. Zylicz. 1988. Role of the *Escherichia coli* DnaK and DnaJ heat shock proteins in the initiation of bacteriophage lambda DNA replication. *Proc. Natl. Acad. Sci.* **85:** 6632–6636.

Littlewood, T.D., D.C. Hancock, and G.I. Evan. 1987. Characterisation of a heat shock-induced insoluble complex in the nuclei of cells. *J. Cell Sci.* **88:** 65–72.

Munro, S. and H.R.B. Pelham. 1984. Use of peptide tagging to detect proteins expressed from cloned genes: Deletion mapping functional domains of *Drosophila* hsp70. *EMBO J.* **3:** 3087–3093.

———. 1986. An hsp70-like protein in the ER: Identity with the 78kd glucose-regulated protein and immunoglobulin heavy chain binding protein. *Cell* **46:** 291–300.

———. 1987. A C-terminal signal prevents secretion of luminal ER proteins. *Cell* **48:** 899–907.

Murakami, H., D. Pain, and G. Blobel. 1988. 70-kD heat shock-related protein is one of at least two distinct cytosolic factors stimulating protein import into mitochondria. *J. Cell Biol.* **107:** 2051–2057.

Pelham, H.R.B. 1984. Hsp70 accelerates the recovery of nucleolar morphology after heat shock. *EMBO J.* **3:** 3095–3100.

———. 1986. Speculations on the functions of the major heat shock and glucose regulated proteins. *Cell* **46:** 959–961.

Pelham, H., M. Lewis, and S. Munro. 1985. A nucleolar function for mammalian and

Drosophila 70kD heat shock proteins. In *Sequence specificity in transcription and translation* (ed. R. Calendar and L. Gold), pp. 681–688. Alan R. Liss, New York.

Pinhasi-Kimhi, O., D. Michalovitz, A. Ben-Zeev, and M. Oren. 1986. Specific interaction between the p53 cellular tumour antigen and major heat-shock proteins. *Nature* **320:** 182–185.

Puvion, E. and G. Moyne. 1981. In situ localization of RNA structures. In *The cell nucleus* (ed. H. Busch), vol. 8, pp. 59–115. Academic Press, New York.

Rose, M.D., L.D. Misra, and J.P. Vogel. 1989. KAR2, a karyogamy gene, is the yeast homolog of mammalian BiP/GRP78 gene. *Cell* **57:** 1211–1221.

Rothman, J.E. and S.L. Schmid. 1986. Enzymatic recycling of clathrin from coated vesicles. *Cell* **46:** 5–9.

Sharma, S., L. Rodgers, J. Brandsma, M.-J. Gething, and J. Sambrook. 1985. SV40 T antigen and the exocytic pathway. *EMBO J.* **4:** 1479–1489.

Simard, R., Y. Langelier, R. Mandeville, N. Maestracci, and A. Royal. 1974. Inhibitors as tools in elucidating the structure and function of the nucleus. In *The cell nucleus* (ed. H. Busch), vol. 3, pp. 447–487. Academic Press, New York.

Sinibaldi, R.M. and P.W. Morris. 1981. Putative function of *Drosophila melanogaster* heat shock proteins in the nucleoskeleton. *J. Biol. Chem.* **256:** 10735–10738.

Tandon, S. and P. Horowitz. 1986. Detergent-assisted refolding of guanidinium chloride-denatured rhodanese: The effect of lauryl maltoside. *J. Biol. Chem.* **261:** 15615–15681.

Tissières, A., H.K. Mitchell, and U. Tracy. 1974. Protein synthesis in salivary glands of *Drosophila melanogaster*. Relation to chromosome puffs. *J. Mol. Biol.* **84:** 389–398.

Ungewickell, E. 1985. The 70-kd mammalian heat shock proteins are structurally and functionally related to the uncoating protein that releases clathrin triskelia from coated vesicles. *EMBO J.* **4:** 3385–3391.

Velazquez, J.M. and S. Lindquist. 1984. Hsp70: Nuclear concentration during environmental stress and cytoplasmic storage during recovery. *Cell* **36:** 655–662.

Velazquez, J.M., B. DiDomenico, and S. Lindquist. 1980. Intracellular localization of heat shock proteins in *Drosophila. Cell* **20:** 679–689.

Welch, W.J. and J.R. Feramisco. 1984. Nuclear and nucleolar location of the 72,000-dalton heat shock protein in heat-shocked mammalian cells. *J. Biol. Chem.* **259:** 4501–4513.

———. 1985. Rapid purification of mammalian 70,000-dalton stress proteins: Affinity of the proteins for nucleotides. *Mol. Cell. Biol.* **5:** 1229–1237.

Welch, W.J. and J.P. Suhan. 1986. Cellular and biochemical events in mammalian cells during and after recovery from physiological stress. *J. Cell Biol.* **103:** 2035–2052.

Werner-Washburne, M., D.E. Stone, and E.A. Craig. 1987. Complex interactions among members of an essential subfamily of hsp70 genes in *Saccharomyces cerevisiae. Mol. Cell. Biol.* **7:** 2568–2577.

Zimmermann, R. and D.I. Meyer. 1986. 1986: A year of new insights into how proteins cross membranes. *Trends Biochem. Sci.* **11:** 512–515.

Zimmermann, R., M. Sagstetter, M.J. Lewis, and H.R.B. Pelham. 1988. Seventy-kilodalton heat shock proteins and an additional component from reticulocyte lysate stimulate import of M13 procoat protein into microsomes. *EMBO J.* **7:** 2875–2880.

Zylicz, M., D. Ang, K. Liberek, and C. Georgopoulos. 1989. Initiation of lambda DNA replication with purified host- and bacteriophage-encoded proteins: The role of the dnaK, dnaJ and grpE heat shock proteins. *EMBO J.* **8:** 1601–1608.

Zylicz, M., J.H. Lebowitz, R. McMacken, and C. Georgopoulos. 1983. The dnaK protein of *Escherichia coli* possesses an ATPase and autophosphorylating activity and is essential in an in vitro DNA replication system. *Proc. Natl. Acad. Sci.* **80:** 6431–6435.

13

Regulation and Function of the HSP70 Multigene Family of *Saccharomyces cerevisiae*

Elizabeth A. Craig
Department of Physiological Chemistry
University of Wisconsin-Madison
Madison, Wisconsin 53706

I. INTRODUCTION

hsp70 is the most abundant heat shock protein in many organisms. *Escherichia coli* contains a single HSP70 gene, *dnaK*, which is expressed at high levels during normal logarithmic growth and is induced upon a temperature upshift. In most, if not all, eukaryotes, the heat-inducible

Stress Proteins in Biology and Medicine.

HSP70 gene is one of a number of related genes, referred to as the HSP70 multigene family (Ingolia et al. 1982; Watowich and Morimoto 1988; also see Morimoto and Milarski, this volume). The yeast *Saccharomyces cerevisiae* contains a family of at least nine HSP70-related genes. Eight of these genes, originally named YG100–YG107, have been renamed: YG100, YG102, YG106, and YG107 as *SSA1, SSA2, SSA3,* and *SSA4*, respectively; YG101 and YG103 as *SSB1* and *SSB2*, respectively; YC104 as *SSC1;* YG105 as *SSD1*. These genes have been divided into the four subgroups A–D, as indicated by the third letter of the gene designation, on the basis of structural and functional criteria. The first two letters of the gene designation stand for *S*tress *S*eventy. Recently, another member of the yeast HSP70 family, *KAR2*, has been identified (Rose et al. 1989). The structural relationships of the first eight genes identified are shown in Figure 1. This tree is based on DNA sequence similarity and DNA:DNA hybridization data. The structural relationship among the genes is complex, with nucleotide identities ranging from 50% to 96%. In this paper, I discuss the expression of these genes and

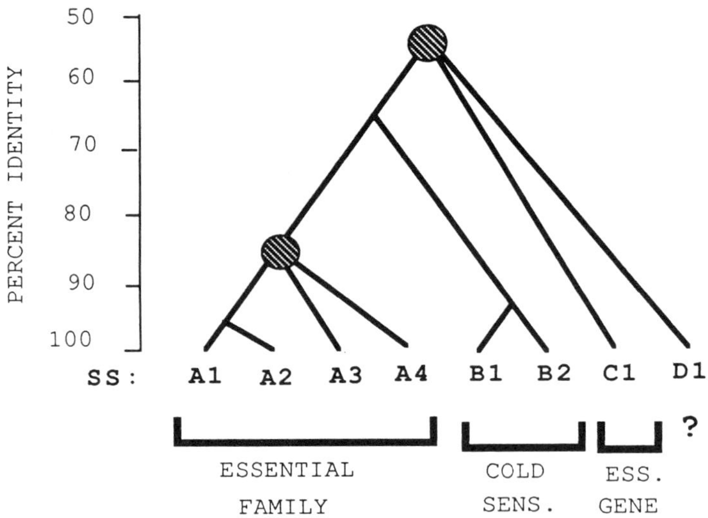

Figure 1 HSP70 multigene family of *S. cerevisiae*. Approximate nucleotide identities are based on complete *(SSA1, SSA2, SSB1)* or partial *(SSA3, SSA4, SSB2, SSC1, SSD1)* DNA sequence data. The phenotypes of strains containing mutations of the *SSA* family are from Craig and Jacobsen (1984) and Werner-Washburne et al. (1987); of the *SSB* family are from Craig and Jacobsen (1985); of *SSC1* and *SSD1* are from Craig et al. (1987).

the genetic experiments carried out to attempt to elucidate the functional relationships among them.

II. EXPRESSION OF THE YEAST HSP70 GENES

A. Response to Heat Shock

1. RNA Levels after Heat Shock

Not all members of the HSP70 family are induced by a heat shock. For example, the levels of mRNA encoded by the two *SSB* genes, *SSB1* and *SSB2*, rapidly decrease after a shift from 23°C to 37 or 39°C (Craig and Jacobsen 1985). Within 5 minutes after a heat shock, *SSB1/2* mRNA levels have decreased 80%; the levels remain low for at least 30 minutes after the upshift, the longest time analyzed. Presumably, this decrease in RNA levels is due both to degradation of the mRNA and to a decrease in *SSB1/2* transcription. *SSA2* expression changes very little, if at all, after a heat shock, whereas the expression of the other members of the family is induced from 10-fold to more than 100-fold. The *SSA3* and *SSA4* genes are what could be referred to as "classic" heat shock genes. Little or no *SSA3* or *SSA4* RNA is present when cells are grown at 23°C, but levels increase dramatically within 5 minutes of a temperature upshift and peak within 10–15 minutes. However, the response is transient (Werner-Washburne et al. 1989). Even when cultures are maintained at the high temperature, *SSA3* and *SSA4* levels return to pre-heat-shock levels within 90 minutes. The other genes, *SSA1*, *SSC1*, and *SSD1* are expressed at moderately high levels during normal growth and are induced about 10-fold upon a heat shock, with kinetics of induction being very similar to those of *SSA3* and *SSA4* (Craig et al. 1987; Werner-Washburne et al. 1989).

2. The Yeast Heat Shock Element

Sequences responsible for the heat shock induction of *SSA1*, *SSA2*, and *SSA3* have been delineated (Slater and Craig 1987; Craig et al. 1988). The 5′ nontranscribed region of each of these genes contains multiple sequence elements that are closely related to the heat shock consensus element (HSE) defined in other eukaryotes (Pelham and Bienz 1982). Each of the genes has three such regions within 350 nucleotides of the transcriptional initiation sites. In each promoter, there are several regions that contain overlapping arrangements of HSEs similar to those found in the 5′ flanking region of the *Drosophila* HSP83 and several other heat-inducible genes. Extensive 5′ flanking regions of *SSA1*, *SSA3*, and *SSA4* have been fused to the *lacZ* gene of *E. coli* to facilitate the study of expression from these promoters. The inducible expression of these fusions, as measured by β-galactosidase activity, parallels the expression of the native genes. Deletion analyses have shown that HSEs are required for the induction of these genes by thermal stress, but the presence of all the HSE-like sequences is not necessary for induction (Slater and Craig 1987; Craig et al. 1988; W. Boorstein and E. Craig, unpubl.).

Synthetic oligonucleotides identical to the HSE sequences of these three genes have been tested for their ability to confer heat inducibility on a heterologous promoter. All of the upstream activating sequences (UASs) of a cytochrome C1 (*CYC1*) *lacZ* fusion vector (Guarente and Ptashne 1981) were replaced by these oligonucleotides. All three sequences were able to activate the *CYC1* downstream promoter elements in a heat-inducible manner and thus constitute functional UAS^{HS}s.

The *SSA1*, *SSA3*, and *SSA4* oligonucleotides that are sufficient to confer heat inducibility all contain sequences that resemble the palindromic heat shock element CNNGAANNTTCNNG (the underlined sequences in Fig. 2). As indicated by the triangles, these sequences contain one or more GAA blocks in addition to those that comprise the "core" of the consensus sequence. These additional HSE-like sequences have been observed adjacent to the HSEs of a number of heat shock genes from widely divergent species. Recently, these nucleotides were shown to be required for HSE activity in a rigorous functional test of the precise requirements for the heat-inducible enhancer in *Drosophila* (Amin et al. 1988; Xiao and Lis 1988). The three functional HSEs shown here conform to the model proposed by Voellmy and colleagues, which postulates the requirements of at least 3 GAA blocks in alternating orientation within rigid space constraints (Amin et al. 1988). These requirements

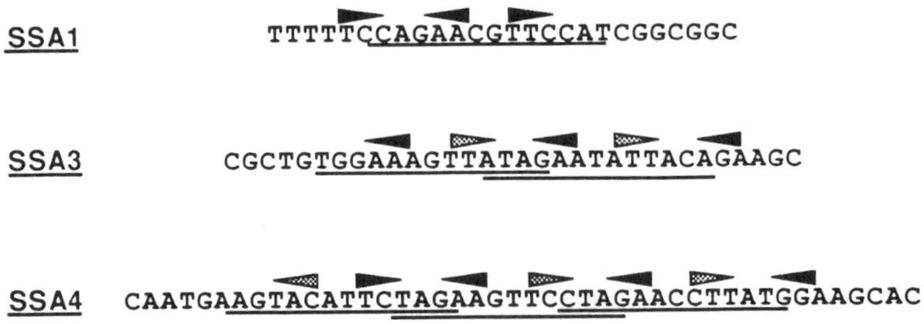

Figure 2 Heat shock elements and flanking sequences from *SSA1*, *SSA3*, and *SSA4*. Bars indicate the positions of the consensus HSEs. The *SSA1* HSE shown here is referred to as "HSE2." Triangles mark the component GAA repeat with 2-nucleotide relative spacing. Black triangles are over exact matches to GAA. Gray triangles are over sequences with a mismatch to the component repeat element. (Reprinted, with permission, from Craig et al. 1988.)

may exist in *S. cerevisiae* as well. The *SSA1* oligonucleotide shown in Figure 2 is a functional UAS^{HS}; however, a smaller oligomer that includes the entire underlined consensus region, but not the 5′TTC sequence, has no UAS^{HS} activity (Slater and Craig 1987).

3. *Role of HSEs and Associated Repressing Sequences in Basal Expression*

A question that has arisen in the study of HSE function is whether HSEs are important in basal expression. Our studies with the constitutively expressed *SSA1* gene indicate that at least in this gene, HSEs are essential for normal basal expression (Slater and Craig 1987; H.O. Park and E. Craig, unpubl.). The basal level of expression of an *SSA1* promoter that has mutations only in a single HSE, denoted HSE2, is reduced 80% as compared to the wild-type promoter. HSE2 has also been shown to be sufficient for both basal and heat-inducible expression, as measured from a *CYC1-lacZ* reporter fusion gene. Our finding that HSE2 is important in basal expression of *SSA1* supports the idea that heat shock transcription factor (HSF) is active and can support transcription, although probably with reduced efficiency, under normal growth conditions. These results are consistent with recent findings (Sorger and Pelham 1988; Wiederreicht et al. 1988) that the HSF is essential for growth under non-heat-shock conditions, indicating a role for HSF in transcription during exponential growth.

HSF has been shown to bind to HSEs and, in *Drosophila*, to activate transcription (Topol et al. 1985; Kingston et al. 1987; Sorger et al. 1987). Further tests for sequences necessary for heat-inducible expression led to the identification of a sequence involved in modulating the basal activity of HSE2. The 26-bp fragment containing HSE2 and its immediately adjoining nucleotides, when inserted in the place of the *CYC1* UAS, conferred a substantial basal expression on the heterologous reporter gene. Unexpectedly, when a larger DNA fragment of 137 nucleotide pairs containing the same HSE was used, the basal, but not the heat-inducible, expression was lowered (Fig. 3). This inhibition of expression suggests that sequences present in the 137-bp DNA fragment negatively regulate the basal expression of *SSA1*. Deletion analysis and subsequent study of in vitro constructed point mutants led to the identification of an upstream repressing sequence (URS). The identified URS, in the inverted orientation, has two mismatches to a consensus sequence, TAGCCGCCGPu4, proposed by T.G. Cooper and colleagues to be a negative control site in yeast (Sumada and Cooper 1987). This URS partially overlaps HSE2. To directly determine if this URS affects the basal expression of *SSA1*, a

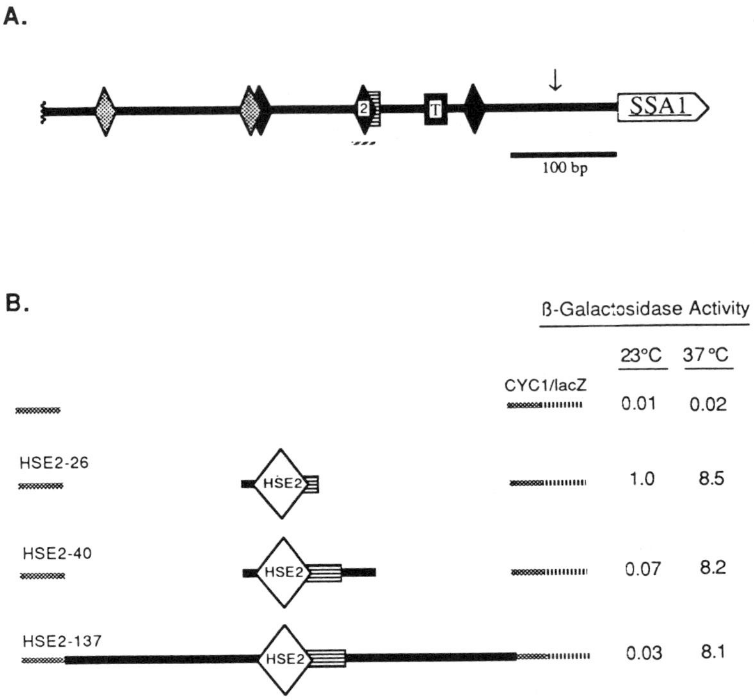

Figure 3 (*A*) Schematic diagram of the 5′ noncoding region of *SSA1*. Open boxes indicate the amino termini of the coding region. Diamonds represent matches to the heat shock element consensus sequence, CNNGAANNTTC NNG. Black diamonds indicate exact matches and single-base matches to this consensus. Stippled diamonds mark sequences that differ from the consensus by 2 nucleotides. The box containing the letter T represents the major transcriptional initiation site. The striped box marks the URS. HSE2 is underlined. (*B*) Activity of *SSA1-CYC1* promoter fusions before and after heat shock. Segments of the *SSA1* promoter were inserted into a *CYC1* promoter-*lacZ* fusion vector (Guarente and Ptashne 1981) in place of the *CYC1* UAS. The enzymatic activity of β-galactosidase in yeast cells transformed with the constructions was determined during logarithmic growth at 23°C and 60 min after shift to 37°C. The β-galactosidase activity of cells grown at 23°C containing the plasmid with the 26-bp segment was arbitrarily given the value of 1. All other β-galactosidase levels are presented as a fraction of the activity present in cells containing the fusion. (Reprinted, with permission, from Craig et al. 1988.)

mutant promoter that contained a single mutation in the URS was constructed. The basal expression of this mutant was about threefold higher than the wild-type promoter, indicating a role of URS in regulating basal expression (Park and Craig 1989).

4. Self-regulation

The results of several studies in *Drosophila melanogaster* and *E. coli* suggest that hsp70 is autoregulatory, that is, hsp70 plays a role in regulating its own synthesis, and may regulate the synthesis of other heat shock proteins as well. We have carried out a set of experiments to test the self-regulation model in *S. cerevisiae* (Craig et al. 1988; D. Stone and E. Craig, unpubl.). Initially, RNA levels in an *ssa1ssa2* strain were analyzed to determine if the levels of any HSP70 RNAs were altered. Higher than normal expression of *SSA3* and *SSA4* mRNAs was observed. Transcripts from the *SSA1* and *SSA2* promoters themselves were also increased severalfold as compared to that found in wild-type cells. These results are consistent with the hypothesis that Ssa1p and Ssa2p are involved in regulating the heat shock response, but other explanations are possible. For example, the absence of Ssa1p and Ssa2p causes a reduced growth rate at all temperatures, and the cell may respond by turning on the heat shock response. Therefore, a complementary experiment was performed. The effect of overproduction of Ssa1p on the activity of its own promoter was determined. The *SSA1* structural gene was fused to the *GAL1* promoter, thus putting Ssa1p synthesis under the control of a galactose-inducible:glucose-repressible promoter. A reporter vector consisting of an *SSA1-lacZ* translational fusion was used to assess *SSA1* promoter activity. A strain containing a single copy of the *GAL1*p-*SSA1* fusion produced approximately threefold the normal heat shock level of Ssa1p when grown on galactose. This overproduction caused a twofold reduction in both the basal and induced levels of β-galactosidase as measured from the *SSA1-lacZ* fusion vector. In a strain containing approximately tenfold more Ssa1p than is found after heat shock, induction of β-galactosidase activity was almost completely blocked.

Since a translational fusion was used to assay *SSA1* expression, the observed effect of Ssa1p overproduction on expression of this promoter could have been exerted at the level of transcription, translation, or both. That the effect was at least in part rendered at the transcriptional level was demonstrated by analyzing the effects of overproduction of Ssa1p on the expression of the *SSA1-CYC1-lacZ* transcriptional fusions described in the previous section. Expression driven by a 137-bp *SSA1* DNA fragment containing HSE2 was almost completely repressed when levels of Ssa1p were high. However, expression of fusions containing a 26-bp fragment containing HSE2, or a 40-bp fragment containing HSE2 and the URS, were not significantly inhibited. Therefore, the 137-bp DNA fragment contains a sequence that is responsive to the amount of Ssa1p in the cell, but the smaller fragments do not. These results imply that this self-regulation does not act solely through the HSF, since inhibition is only seen when sequences in addition to HSE2 are present. Similarly, it can be concluded that repression does not act solely through the URS. These experiments do not rule out the involvement of HSF or the URS binding factor in self-regulation, but do implicate other sequences as important in this form of regulation.

The experiments described above do not address the question of whether Ssa1p is directly involved in its own regulation. It is possible that Ssa1p interacts directly with its promoter. Alternatively, Ssa1p could cause the alteration of another factor, thus affecting *SSA1* expression indirectly. Future experiments will further define the sequences involved in self-regulation, and it is hoped they will lead to the identification of binding factors.

B. Expression of HSP70 Genes under Non-heat-shock Conditions

1. Expression of the HSP70 Genes during Approach to Stationary Phase

We have found that the HSP70 genes of *S. cerevisiae* are regulated differentially during approach to stationary phase (M. Werner-Washburne et al., in prep.). To determine the level of HSP70 transcripts during approach to stationary phase, RNA samples were isolated and analyzed by hybridization to radioactively labeled DNA probes after blotting of the RNA to a solid matrix (Northern blot analysis). In these experiments, *SSA1* and *SSA2* transcripts are indistinguishable, as are *SSB1* and *SSB2* transcripts, since these pairs of genes are 97% and 94% identical, respectively. The strain of *S. cerevisiae* commonly used in our laboratory reaches a final density of about 3×10^8 cells/ml after about 140 hours of growth at 30°C in rich glucose-based media. This corresponds to an optical density of 40 at 600 nm. At a density of about 1.5×10^8 cells/ml, which in a typical experiment is reached after about 15 hours of growth, an abrupt decrease in growth rate is observed. This abrupt change is indicative of the diauxic shift, the time when glucose has been depleted from the media and the cells must begin to utilize nonfermentable carbon sources such as ethanol. The cells continue to grow at a slower rate for a number of days before the cell number ceases to increase and the cells enter stationary phase.

Three different patterns of expression of HSP70 genes were observed (Fig. 4): (1) a gradual decrease in RNA levels as cell growth slows, with the most rapid decline occurring around the time of the diauxic shift; (2) an increase in RNA levels as cells approached the diauxic shift, followed by a decrease; (3) an increase in RNA levels as cells approached stationary phase. Like the expression of many yeast genes, *SSC1*, *SSD1*, and *SSB1/2* follow expression pattern 1. When cells have reached a density of

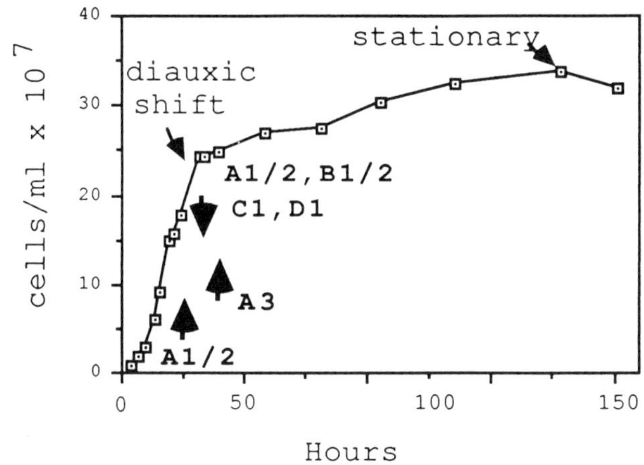

Figure 4 Expression of the HSP70 family of genes during approach to stationary phase. The number of viable cells during growth to stationary phase at 23°C was determined by plating aliquots of cells and determining the number of colonies formed on rich media after incubation at 30°C. The density of viable cells agreed well with total cell number determined using a Coulter counter, except that after the diauxic shift, the number of viable cells was about 20% lower than the total number of cells. Arrows beneath the curve indicate the approximate time at which a dramatic increase (↑) or decrease (↓) in the levels of RNA of the designated genes, as determined by Northern blot analysis, was observed.

1–2 x 10^8 cells/ml, the RNA levels had decreased to less than 1/10 that found during exponential growth. As cells approached stationary phase, the levels of these RNAs approached the limit of detection.

The level of SSA1/2 RNA is moderately high in early exponential phase; as cells approach the diauxic shift (at a cell density of about 1 x 10^8 cells/ml), the levels increase severalfold. Immediately after the diauxic shift, the level of *SSA1/2* RNA begins to decrease, reaching very low levels by the time the cells reach saturation. However, relatively high levels of *SSA1* and *SSA2* proteins persist in stationary phase cells, after the mRNAs encoding them have disappeared. Apparently, these proteins are quite stable, at least under these circumstances. Accumulation of *SSA4* transcripts was variable. When *SSA4* RNA was detectable, it accumulated only to low levels and followed the pattern of *SSA1/2*.

SSA3 exhibits the third pattern of expression. Although *SSA3* RNA was not detected during exponential growth, after the diauxic shift, *SSA3* RNA levels increased more than 100-fold. The RNA levels attained are comparable to those found after a temperature shift from 23°C to 39°C. This pattern of expression is unusual, and to our knowledge, not previously reported. Therefore, we were interested in learning more about the regulation of expression of *SSA3*. Since *SSA3* transcripts appear only as cells approach stationary phase, the expression of *SSA3* in strains defective for stationary phase arrest, *bcy1*⁻ (Matsumoto et al. 1982) and *ard1*⁻ (Whiteway and Szostak 1985) mutants, was analyzed. Although *ard1*⁻ cells had normal accumulation of *SSA3* transcripts, *bcy1*⁻ cells failed to induce the expression of *SSA3* (Werner-Washburne et al. 1989). In both these mutants, expression of the other members of the HSP70 multigene family was the same as in a wild-type strain.

The *BCY1* gene product is a negative regulatory subunit of a cAMP-dependent protein kinase, binding to the catalytic subunit in the absence of cAMP and inhibiting kinase activity. Protein kinase is a positive regulator of cell growth. When cellular levels of cAMP are high, a reflection of nutrient availability, protein kinase activity is high, and cell division continues. When the cellular environment is depleted of nutrients, cAMP levels decrease, causing *BCY1*-mediated repression of protein kinase activity, leading to cell cycle arrest. Since *bcy1*⁻ cells have high protein kinase activity regardless of nutrient availability, they are defective in stationary arrest. These mutants cannot grow on nonfermentable carbon sources and are more sensitive than wild-type strains to both starvation and heat shock (Matsumoto et al. 1983).

The failure of *bcy1*⁻ strains to induce *SSA3* mRNA suggests that a decrease in cAMP-dependent protein kinase activity is involved in regulating the expression of *SSA3*. We wanted to analyze a situation in which cAMP-dependent protein kinase activity is low even when nutrients are readily available. *CYR1* encodes adenyl cyclase (Matsumoto et al. 1982), the enzyme that catalyzes the synthesis of cAMP. When grown in the presence of cAMP, *cyr1*⁻ cells do not accumulate *SSA3* mRNA. However, after removal of cAMP from the media, *SSA3* transcripts accumulate (Werner-Washburne et al. 1989). Therefore, consistent with the failure of *SSA3* RNA to accumulate in *bcy1*⁻ mutants at saturation, *SSA3* is induced in response to low cAMP levels and, therefore, presumably low levels of cAMP-dependent protein kinase activity.

What is the mechanism of *SSA3* expression during approach to stationary phase? We first addressed the question of whether the HSEs that are sufficient for heat inducibility of *SSA3* are also sufficient for stationary phase induction. No induction of a heat-responsive *CYC1-lacZ* fusion driven by the SSA3 HSE occurs as cells reach saturation (W. Boorstein and E. Craig, unpubl.). Therefore, an HSE is not sufficient for stationary phase induction. This result is consistent with the fact that *bcy1*⁻ cells, although they do not induce *SSA3* mRNA at saturation, are capable of inducing *SSA3* upon a heat shock. This stationary phase induction presumably requires factors other than HSF. Experiments are in progress to determine the sequences required for induction.

2. *Expression of HSP70 Genes under Conditions That Induce Sporulation*

During sporulation, a response to nutrient limitation, *S. cerevisiae* undergoes dramatic changes in physiology and gene expression. Therefore, it was of interest to determine if any HSP70 genes undergo changes in expression under conditions that induce meiosis and sporulation in a/α diploids (Werner-Washburne et al. 1989). To distinguish sporu-lation-specific changes in gene expression from changes due to starvation-induced arrest, HSP70 expression in an a/α diploid, which will undergo sporulation in potassium acetate media, was compared to expression in an α/α diploid strain and a haploid strain that will simply arrest in this medium.

SSA1/2 mRNA levels change very little upon transfer to acetate media; however, the levels of the other HSP70 mRNAs change significantly. The levels of *SSB1/SSB2, SSC1,* and *SSD1* transcripts all decrease within a few hours after transfer to sporulation media. The patterns of expression of the transcripts are the same for a/a and a/α diploids, although the magnitude of the changes are different. Interestingly, *SSA3* transcripts increase several hours after transfer. Although the increase is greatest in the a/α diploids, it is not a sporulation-specific effect but a response to starvation, since an increase is also seen in a/a diploids as well as in a haploid strain. However, there may be a sporulation-specific component to the response.

As discussed above, the amount of *SSA3* RNA increases during approach to stationary phase and after transfer to media containing only potassium acetate. One assumes that this accumulation is in response to some signal, likely caused by depletion of an essential nutrient. Addition of glucose to stationary phase cells or cells that had been incubated 5 hours in sporulation medium resulted in a large decrease in *SSA3* mRNA levels within 1 hour after addition. However, when glucose (or nitrogen) alone was removed from cells growing in minimal media, no induction of *SSA3* was observed, indicating that the simple removal of this single essential nutrient was not sufficient to cause induction. In addition, *SSA3* is not expressed constitutively in acetate-based rich media. Therefore, *SSA3* induction is not simply a function of catabolite derepression. It is possible that the addition of glucose back to cells in stationary phase or cells incubated in sporulation medium allows the resynthesis of other essential nutrients, which results in cessation of *SSA3* expression. It is not surprising that the signal for cells to enter stationary phase, the G_0 state, may be more complex than simply the lack of a single nutrient. Since this transition involves major changes in the metabolism of the cell, the shift to the G_0 state may not be initiated until several parameters change, indicating a severe change in circumstances, not merely a transient change in nutritional status.

III. GENETIC AND FUNCTIONAL ANALYSIS OF THE *S. CEREVISIAE* MULTIGENE FAMILY

A. Functional Grouping of the HSP70 Multigene Family Members

1. Strategy of the Genetic Analysis

The structural relationship among the genes of the *S. cerevisiae* HSP70 multigene family is complex. The nucleotide sequence identities range from 50% to 96%. To determine whether the obvious structural similarities between the genes reflect similarities in function as well, we have constructed mutant alleles of these eight genes and subsequently generated strains containing different combinations of these mutations. These alleles were constructed by inserting selectable markers into the protein-coding regions of the genes. In many cases, a portion of the protein-coding region was deleted. The wild-type genes were replaced by these in-vitro-constructed mutant alleles, using the one-step gene replacement method (Rothstein 1983). This procedure relies on the fact that in *S. cerevisiae,* free ends of linear DNA fragments are highly recombinogenic readily undergoing targeted homologous recombination. In such a reaction, the wild-type gene is replaced by the mutant gene that has been transformed into yeast as a linear fragment. Since the mutant gene contains a selectable marker, transformants carrying the mutant allele can be easily selected. Usually diploids were used in these transformations, since haploids would not have been recoverable if the disrupted gene was essential. The phenotype of haploid strains carrying the disrupted gene can be determined by analysis of meiotic segregants. To confirm that a phenotype is due to the absence of functional HSP70 genes, the mutant strain can be transformed by a plasmid carrying the corresponding wild-type gene.

Using this type of genetic analysis, we have placed the members of the HSP70 family into functional groups. As shown in Figure 1, there is a complete correspondence between the grouping based on DNA sequence similarity and that based on mutational analysis. We have divided the genes into subgroups A, B, C, and D. As described below, phenotypes have been found for strains containing mutations in each of the groups, except *SSD*. In addition, Mark Rose has identified the *KAR2* gene as a structurally and functionally unique member of the HSP70 multigene family (Rose et al. 1989).

2. SSA *Family*

The *SSA* genes, comprising *SSA1*, *SSA2*, *SSA3*, and *SSA4*, make up an essential family, in the sense that protein encoded by one of these genes must be present at relatively high levels for growth. The *SSA1* and *SSA2* genes are 96% identical at the nucleotide level, and strains containing a null allele of either one of these genes grow well over a wide temperature range. However, *ssa1ssa2* mutants have a complex phenotype. They are temperature-sensitive for growth, growing well at 23°C but unable to form colonies at 37°C (Craig and Jacobsen 1984). In addition, they are constitutively thermotolerant, able to survive short periods of time at high temperatures (50°C) without a preincubation at moderate temperatures. Under normal growth conditions, *ssa1ssa2* cells synthesize many heat shock proteins, including hsp90, hsp26, and Ssa4p. Presumably, this stressed state is responsible for the constitutive thermoresistance of the strain. However, the reason for the constitutive production of hsps is not clear. Either the cells are responding because they are "sick" due to the absence of the Ssa1p and Ssa2p, or Sssa1p and Ssa2p are part of an important regulatory loop that is broken because of their absence. The second possibility is consistent with the fact, as discussed above, that Ssa1p is involved in regulating expression from its own promoter. However, it is not clear at this point whether it plays a role in regulating the expression of other heat shock genes as well.

Studies of strains containing mutations in either *SSA4* or *SSA3* or in both genes have revealed no differences as compared to wild-type strains (Werner-Washburne et al. 1987). These mutant strains have the same growth and thermotolerance properties as wild-type strains. In an *ssa1ssa2* background, however, an *SSA4* null allele causes inviability, although an *SSA3* allele does not. The inviability of the *ssa1ssa2ssa4* strain indicates that *SSA4*, which is constitutively expressed in the *ssa1ssa2* strain, is allowing growth at low temperatures. *ssa1ssa2ssa4* strains carrying a wild-type *SSA1* gene under the control of the galactose-inducible:glucose-repressible *GAL1* promoter (*GAL1p:SSA1*) are viable when grown on galactose-based media. After shift to glucose-based media, *ssa1ssa2ssa4 (GAL1p:SSA1)* cultures undergo approximately three cell divisions before growth stops. Before shift to glucose-based media, the level of Ssa1p is about sixfold higher than that present in a wild-type strain. Ssa1p levels are equivalent to that in wild-type cells about 3 hours after shift to glucose and are less than 30% of wild-type levels at 6 hours. These results suggest that growth ceases shortly after Ssa1p levels drop below those present in wild-type cells, since the doubling time of *S. cerevisiae* under these conditions is about 2 hours.

The *ssa1ssa2ssa4 (GAL1p-SSA1)* strain allows the analysis of cells that do not contain functional Ssa1p, Ssa2p, or Ssa4p. Studies on the import of proteins across the endoplasmic reticulum (ER) membrane and into mitochondria using this strain indicate that the *SSA* proteins are involved in these processes (Deshaies et al. 1988). As described in more detail in this volume (Craig), several hours after shift to glucose-based media, precursors to α factor (a secreted protein), carboxypeptidase Y (a vacuolar protein), and the β subunit of the F_1ATPase (a mitochondrial protein) begin to accumulate. Cell-free extracts prepared from *ssa1ssa2ssa4 (GAL1*p:*SSA1)* cells after shift to glucose-based media are defective in the import of α factor precursor into microsomal vesicles. Furthermore, two proteins that were purified on the basis of their ability to facilitate import of α factor into microsomes in vitro have been identified as Ssa1p and Ssa2p by Blobel and colleagues (Chirico et al. 1988). Thus, both biochemical and genetic evidence indicates that *SSA* proteins are involved in the transport of proteins across membranes.

3. SSB *Family*

The *SSB* subfamily is composed of the two genes *SSB1* and *SSB2*. Strains containing an insertion mutation in either one of these genes are identical to wild-type. However, strains containing mutations in both genes are relatively cold-sensitive for growth (Craig and Jacobsen 1985). At 37°C, there is only a slight difference between *ssb1ssb2* strains and wild-type, but at 19°C, the double mutant strain grows 2.4-fold more slowly. This cold-sensitive phenotype correlates well with the lack of expression of these genes upon a heat shock, and their higher mRNA levels during steady-state growth at 23°C compared to higher temperatures. These results suggest that *SSB* function may be necessary for optimal growth at lower temperatures, but not at higher temperatures. Perhaps colder temperatures are as stressful as higher temperatures, and thus cells may contain a set of genes necessary for optimal growth under cold temperature conditions.

4. SSC1

SSC1 is an essential gene. *ssc1* spores generated from an *SSC1/ssc1* heterozygote are able to germinate but undergo only about three cell divisions before arrest of growth (Craig et al. 1987). The predicted amino acid sequence of Ssc1p has an additional 27 amino acids at its amino terminus as compared to the Ssa1p. This putative leader has structural features that are very similar to those found in proteins imported into mitochondria. The *SSC1* leader is rich in positively charged amino acids, devoid of acidic amino acids, and rich in the hydroxylated amino acids serine and threonine. Furthermore, Ssc1p synthesized in vitro is imported into mitochondria and converted into a mature form in an in vitro assay. This mature form comigrates with a major yeast mitochondrial protein that reacts with antibody generated against a conserved region of hsp70 proteins (Craig et al. 1989). Therefore, Ssc1p is a mitochondrial hsp70 protein.

5. KAR2

KAR2 mutants were originally identified as strains defective in nuclear fusion (Polaina and Conde 1982). Recently, sequence analysis carried out in Mark Rose's laboratory revealed that *KAR2* is closely related to GRP78, a mammalian member of the hsp70 family found in the ER (Munro and Pelham 1986). As expected of a homolog of GRP78, *KAR2* is induced by the glycosylation inhibitor tunicamycin. As shown by gene disruption experiments, the *KAR2* gene is an essential gene (Rose et al. 1989). The reason for the defect in karyogamy seen in some *kar2*$^-$ strains is not clear. The nuclear envelope may be dependent on proteins or protein complexes that enter and are assembled in the ER. Alternatively, a defective *KAR2* protein may result in a structurally altered ER that prevents the fusion of nuclear membranes during diploid formation.

The genetic analysis in yeast leads to the conclusion that there are at least four separable functions of hsp70s, represented by *SSA, SSB, SSC1*, and *KAR2*. Basic functional differences in the proteins, differences in cellular location of the proteins, or differences in expression under a variety of physiological conditions could be the cause of the observed differences. In the following section, I discuss the genetic results we have obtained in light of possible functional differences among the proteins.

B. Examining the Relationship between Patterns of Expression, Function, and Cellular Location

1. *Are* SSA *and* SSB *Genes Functionally Distinct?*

The results discussed above show that members of the HSP70 multigene family are expressed differentially during exponential growth at different temperatures and during starvation and approach to stationary phase. This differential expression raises the question of whether the functional groups that were defined genetically reflect differences in function or are due simply to inadequate levels of various proteins under particular conditions. For example, are the A and B family differences that we observed due to differences in expression, cellular location, or function? An *ssa1ssa2* mutant is temperature-sensitive for growth, not able to form colonies at 37°C (Craig and Jacobsen 1984). The related *SSB1* and *SSB2* genes are expressed at relatively low levels after a heat shock and at temperatures at which growth of *ssa1ssa2* strains is compromised. Perhaps the proteins encoded by *SSB1* genes have similar functions to *SSA1/2* proteins and could substitute for Ssa1/2p if they were present in greater amounts at the higher temperatures.

ssb1ssb2 strains are cold-sensitive for growth (Craig and Jacobsen 1985). Perhaps Ssa1/2p are not present in sufficient amounts at low temperatures to substitute in the absence of Ssb1/2p. To test this idea, Ssb1p synthesis was placed under the control of the *SSB1* promoter, so that the level of Ssb1p would not decrease at high temperatures. In a complementary experiment, Ssa1p synthesis was placed under the control of the *SSB1* promoter, so its synthesis would not decrease at low temperatures. Even with the aberrant expression of Ssa1p and Ssb1p from the fusion-containing plasmids, the *ssb1ssb2* and *ssa1ssa2* strains remained cold-sensitive and temperature (heat)-sensitive for growth, respectively (Craig and Jacobsen 1985). This result indicates that the phenotypes of the *ssa* and *ssb* mutants cannot be corrected by overexpression of certain other members of the multigene family. Therefore, the phenotypic differences observed are not the result of differential expression of these genes. Since neither the *SSA* nor the *SSB* genes contain sequences that would obviously direct them to a particular cellular compartment, we suspect that the *SSA* and *SSB* proteins are functionally different.

2. *Are All Members of the SSA Family Functionally Identical?*

A similar question relating to the effect of differential expression arises when considering the function of the members of the family of *SSA* genes, *SSA1*, *SSA2*, *SSA3*, and *SSA4*. As mentioned above, *ssa1ssa2* strains are temperature-sensitive for growth; in addition, they constitutively express mRNAs transcribed from most, if not all, heat shock genes, including *SSA3* and *SSA4*. *ssa1ssa2ssa4* strains are inviable. The "abnormal" high expression of *SSA4* in an *ssa1ssa2* strain most likely allows this strain to remain viable at low temperatures. Two questions arise: (1) Why doesn't *SSA3* rescue the *ssa1ssa2ssa4* cells, since *SSA3* mRNA levels are high? (2) Why are *ssa1ssa2* cells that contain wild-type *SSA3*and *SSA4* genes unable to survive at 37°C?

It is possible that Ssa4p and Ssa1/2p are functionally identical, but insufficient quantities of Ssa4p are present in an *ssa1ssa2* strain to allow growth at 37°C. Analysis of proteins on two-dimensional gels indicates that Ssa4p is abundant in *ssa1ssa2* strains and is present in amounts similar to that of Ssa1/2p in wild-type cells. Overexpression of Ssa4p caused by *SSA4* on a multicopy plasmid allows *ssa1ssa2* cells to grow, albeit very slowly, at 37°C (M. Werner-Washburne and E. Craig, unpubl.). These results suggest that *SSA4* (and possibly *SSA3)* are not functionally identical to *SSA1/2*. None of the *SSA* proteins contain sequences that suggest a location other than the cytoplasm. However, since the cellular location of the different *SSA* proteins has not been precisely determined, it is possible that the *SSA* proteins are partially sequestered from one another and therefore unable to completely compensate for the other's absence.

Studies of heat-shocked cells indicate that little Ssa3p is synthesized after a temperature upshift, even though *SSA3* mRNA levels are high, indicating that synthesis of this protein is regulated at the level of protein synthesis (Werner-Washburne et al. 1987). From these results, one might predict that *SSA3* does not rescue *ssa1ssa2ssa4* because little Ssa3p is being synthesized. It is possible that Ssa3p is not functionally similar to Ssa1/2p. To test these possibilities, the protein-coding region of *SSA3* was placed under the control of the *SSA2* promoter (Werner-Washburne et al. 1987). Strains containing this plasmid produce high levels of Ssa3p constitutively. Inviability of the *ssa1ssa2ssa4* strain at 30°C was rescued by the presence of this plasmid. However, the *ssa1ssa2ssa4 (SSA2p:SSA3)* strain is temperature-sensitive for growth. Therefore, the *SSA3* protein is functional and can substitute, at least partially, for the other three proteins in this subfamily.

C. Differences in Cellular Location of *SSC1*-, *KAR2*-, and *SSA*-encoded Proteins

The results described above suggest that differential regulation is not, in most cases, the cause of the inability of members of the HSP70 multigene family to compensate for the absence of other family members. However, the apparent functional differences might be due to differences in cellular location. It is possible that even though the proteins are capable of carrying out the same function, the compartmentalization of the proteins could prevent any functional compensation. It is known that the product of the *SSC1* gene is present in mitochondria (Craig et al. 1989; M. Werner-Washburne et al., in prep.) and that the product of the

KAR2 gene is present in the ER (Rose et al. 1989). Therefore, lack of complementation among *SSC1*, *KAR2*, and the *SSA* subfamily is most likely due to differential compartmentalization. However, currently there are no data to indicate that HSP70 proteins located in different cellular compartments are capable of compensating for the absence of another family member, even if present in the appropriate cellular compartment.

Differential compartmentalization of the hsp70 proteins is also seen in mammalian cells. grp78, the mammalian hsp70 family member most closely related to Kar2p, is located in the ER (Munro and Pelham 1986) and was originally identified as a protein whose rate of synthesis increased when cells were starved for glucose (Shiu et al. 1977; Lee et al. 1984). grp78 is now known to be very similar, if not identical, to BiP, a protein originally identified because of its association with immunoglobulin heavy chains in preB cells that do not make immunoglobulin light chains. In normal B cells and plasma cells, a smaller fraction of the intracellular heavy chain is associated with BiP (Haas and Wabl 1983). BiP also binds transiently to a variety of normal secretory and transmembrane proteins, and permanently to proteins that are unfolded or misfolded. It is thought that grp78/BiP plays a role in facilitating the assembly of multimeric protein complexes in the ER by altering the conformation of proteins. Ssc1p, a yeast mitochondrial protein, may perform a similar role in the assembly of complexes inside the mitochondria.

Ssa1/2p, both predominantly cytoplasmic proteins, have recently been shown to facilitate the translocation of proteins across the membranes of mitochondria and the ER (Chirico et al. 1988; Deshaies et al. 1988). The *SSA* proteins may act to facilitate the unfolding of these proteins prior to translocation. This hypothesized function is similar to that proposed for grp78, in that both involve the altering of protein conformation, in one case to aid in assembly of multimeric complexes, and in the other to aid in disrupting secondary structure. It will be interesting to learn whether these proteins, when present in the same compartment, can compensate for one another.

D. Does the SSA Subfamily Play a Role in Thermotolerance?

The experiments discussed in this paper have focused on the role of the HSP70 family of proteins during normal growth. Their role under heat shock (stress) conditions has not been addressed. Whether the SSA family of proteins plays a role in thermotolerance is difficult to answer since they are essential proteins. Analysis of *ssa3ssa4* and *ssa1ssa3ssa4* strains revealed no differences compared to wild-type strains in thermotolerance assays, but we cannot eliminate the possibility that SSA2 protein performs a role in thermotolerance. Furthermore, thermotolerance is obviously a complex phenomenon, and our assays to measure such tolerance are quite unsophisticated. Perhaps a more sophisticated approach is necessary to elucidate a role for Ssa1p and other heat shock proteins in thermotolerance.

IV. CONCLUDING REMARKS

The HSP70 multigene family of yeast is a structurally and functionally complex multigene family under the control of complex regulatory mechanisms. Four distinct functional groups of HSP70 genes have been identified. The genetic analysis reveals that hsp70 proteins play important and necessary roles in normal cellular growth. Furthermore, hsp70 proteins function in several cellular compartments, including the cytoplasm, mitochondria, and ER.

ACKNOWLEDGMENTS

The work carried out in my laboratory was supported by grants from the National Institutes of Health. I thank members of my laboratory, especially John Nelson, for comments on this manuscript.

REFERENCES

Amin, J., J. Anathan, and R. Voellmy. 1988. Key features of heat shock regulatory elements. *Mol. Cell. Biol.* **8:** 3761–3769.

Chiroco, W.J., M.G. Waters, and G. Blobel. 1988. 70K heat shock-related proteins stimulate protein translocation into microsomes. *Nature* **332:** 805–810.

Craig, E.A. and K. Jacobsen. 1984. Mutations of the heat-inducible 70 kilodalton genes of yeast confer temperature sensitive growth. *Cell* **38:** 841–849.

———. 1985. Mutations in cognate genes of *Saccharomyces cerevisiae* hsp70 result in reduced growth rates at low temperatures. *Mol. Cell. Biol.* **5:** 3517–3524.

Craig, E.A., J. Kramer, and J. Kosic-Smithers. 1987. *SSC1,* a member of the 70-kDa heat shock protein multigene famiy of *Saccharomyces cerevisiae,* is essential for growth. *Proc. Natl. Acad. Sci.* **84:** 4156–4160.

Craig, E.A., W. Boorstein, H.-O. Park, D. Stone, and C. Nicolet. 1988. Complex regulation of three inducible HSP70 genes in *Saccharomyces cerevisiae* in stress-induced proteins. *UCLA Symp. Mol. Cell. Biol. New Ser.* **96:** 51–61.

Craig, E.A., J. Kramer, J. Shilling, M. Werner-Washburne, S. Holmes, J. Kosic-Smithers, and C. Nicolet. 1989. *SSC1,* an essential member of the yeast hsp70 multigene family, encodes a mitochondrial protein. *Mol. Cell. Biol.* **9:** 3000–3008.

Deshaies, R., B. Koch, M. Werner-Washburne, E. Craig, and R. Sheckman. 1988. 70kD stress protein homologues facilitate translocation of secretory and mitochondrial precursor polypeptides. *Nature* **332:** 800–805.

Guarente, L. and M. Ptashne. 1981. Fusion of *Escherichia coli lacZ* to the cytochrome *c* gene of *Saccharomyces cerevisiae. Proc. Natl. Acad. Sci.* **78:** 2199–2203.

Haas, I.G and M. Wabl. 1983. Immunoglobulin heavy chain binding protein. *Nature* **306:** 387–389.

Ingolia, T.D., M. Slater, and E.A. Craig. 1982. *Saccharomyces cerevisiae* contains a complex multigene family related to the major heat shock-inducible gene of *Drosophila. Mol. Cell. Biol.* **2:** 1388–1398.

Kingston, R., T. Schuetz, and Z. Larin 1987. Heat-inducible human factor that binds to a human hsp70 promoter. *Mol. Cell. Biol.* **7:** 1530–1534.

Lee, A.S., J. Bell, and J. Ting. 1984. Biochemical characterization of the 94- and 78-kilodalton glucose-regulated proteins in hamster fibroblasts. *J. Biol. Chem.* **259:** 4616–4621.

Matsumoto, K., I. Uno, and T. Isikawa. 1983. Initiation of meiosis in yeast mutants defective in adenylate cyclase and cAMP-dependent protein kinase. *Cell* **32:** 417–423.

Matsumoto, K., I. Uno, Y. Oshima, and T. Ishikawa. 1982. Isolation and characterization of yeast mutants deficient in adenylate cyclase and cyclic AMP-dependent protein kinase. *Proc. Natl. Acad. Sci.* **79:** 2355–2359.

Munro, S. and H. Pelham. 1986. An hsp70-like protein in the ER: Identity with the 78kd glucose-regulated protein and immunoglobulin heavy chain binding protein. *Cell.* **46:** 291–300.

Park, H.-O. and E.A. Craig. 1989. Positive and negative regulation of basal expression of a yeast HSP70 gene. *Mol. Cell. Biol.* **9:** 2025–2033.

Pelham, H.R.B. and M. Bienz. 1982. A synthetic heat shock promoter element confers heat-inducibility on the herpes simplex virus *tk* gene. *EMBO J.* **1:** 1473–1486.

Polaina, J. and J. Conde. 1982. Genes involved in the control of nuclear fusion during the sexual cycle of *Saccharomyces cerevisiae. Mol. Gen. Genet.* **186:** 253–258.

Rose, M., L.M. Misra, and J.P. Vogel. 1989. *KAR2,* a karyogamy gene, is a yeast homologue of mammalian BiP/GRP78. *Cell* **57:** 1211–1221.

Rothstein, R. 1983. One step gene disruption in yeast. *Methods Enzymol.* **101:** 202–211.

Shiu, R.P.C., J. Pouyssegur, and I. Pastan. 1977. Glucose depletion accounts for the induction of two transformation-sensitive membrane proteins in Rous sarcoma virus-transformed chick embryo fibroblasts. *Proc. Natl. Acad. Sci.* **74:** 3840–3844.

Slater, M. and E.A. Craig. 1987. Transcriptional regulation of an hsp70 heat shock gene in the yeast *Saccharomyces cerevisiae. Mol. Cell. Biol.* **7:** 1906–1916.

Sorger, P. and H.R. Pelham. 1988. Yeast heat shock factor is an essential DNA-binding protein that exhibits temperature-dependent phosphory-lation. *Cell* **54:** 855–864.

Sorger, P., M.J. Lewis, and H.R. Pelham. 1987. Heat shock factor is an essential DNA-binding protein that exhibits temperature-dependent phosphorylation. *Nature* **329:** 81–84.

Sumada, R.A. and T.G. Cooper 1987 Ubiquitous upstream repression sequences control activation of the inducible arginase gene in yeast. *Proc. Natl. Acad. Sci.* **84:** 3997–4001.

Topol, J., D.M. Ruden, and C.S. Parker. 1985. Sequences required for in vitro transcriptional activation of a *Drosophila* HSP70 gene. *Cell* **42:** 527–537.

Watowich, S.S. and R. Morimoto. 1988. Complex regulation of heat shock and glucose-responsive genes in human cells. *Mol. Cell. Biol.* **8:** 393–405.

Werner-Washburne, M., D. Stone, and E.A. Craig. 1987. Complex interactions among members of an essential subfamily of hsp70 genes in *Saccharomyces cerevisiae. Mol. Cell. Biol.* **7:** 2568–2577.

Werner-Washburne, M., J. Becker, J. Kosic-Smithers, and E. Craig. 1989. Yeast Hsp70

RNA levels vary in response to the physiological status of the cell. *J. Bacteriol.* **171:** 2680–2688.
Whiteway, M. and J. Szostak. 1985. The *ARD1* gene of yeast functions in the switch between the mitotic cell cycle and alternative developmental pathways. *Cell* **43:** 483–492.
Wiederrecht, G., D. Sato, and C. Parker. 1988. Isolation of the gene encoding the *S. cerevisiae* heat shock transcription factor. *Cell* **54:** 841–853.
Xiao, H. and J. Lis. 1988. Germline transformation used to define key features of heat-shock response elements. *Science* **238:** 1247–1253.

14

Expression and Function of Vertebrate hsp70 Genes

Richard I. Morimoto and Kim L. Milarski
Department of Biochemistry, Molecular Biology and Cell Biology
Northwestern University
Evanston, Illinois 60208

I. VERTEBRATE HEAT SHOCK GENES

The eukaryotic genome encodes a large multigene family of hsp70-related proteins. In human cells, there are at least five distinct protein members of the hsp70 gene family (Fig. 1) that exhibit constitutive and inducible regulation and share biochemical and antigenic properties. The genes that encode many of these hsp70-related proteins have been

Stress Proteins in Biology and Medicine.

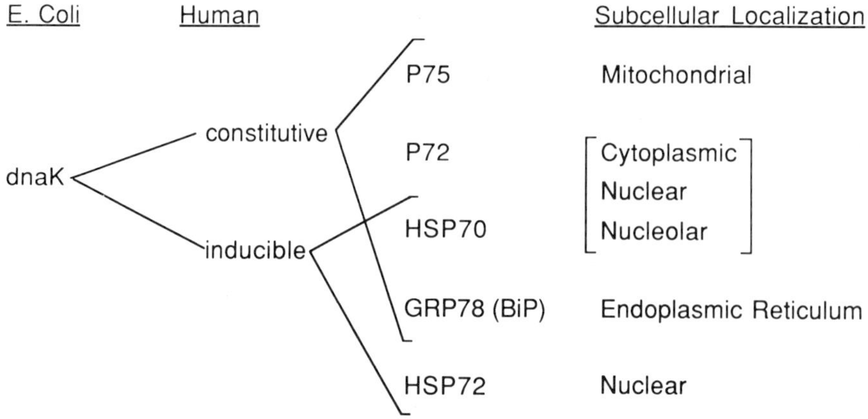

Figure 1 Relationship among members of the hsp70 family from *E. coli* to humans.

cloned, and comparison of the corresponding nucleotide sequences has revealed a high degree of evolutionary conservation among members of the hsp70 family within a single species and between species (Moran et al. 1982; Lowe et al. 1983; Hunt and Morimoto 1985; Mues et al. 1986). For example, *Escherichia coli* DnaK and human hsp70 are 50% identical at the amino acid level (Bardwell and Craig 1984; Hunt and Morimoto 1985) whereas grp78, p72, and hsp70 are approximately 75–85% related. Under normal conditions of cell growth, different members of the hsp70 family are localized to particular cellular compartments and in response to physiological stress are redistributed to different intracellular locales. This feature is best represented by hsp70 and p72 which are typically found in the cytoplasm of cells in the G_1 phase of the cell cycle and, following heat shock or other forms of stress, redistribute from the cytoplasm to the nucleus and nucleolus. Although the function of the hsp70-related proteins has not been firmly established, recent biochemical evidence suggests multiple roles for the hsp70-related proteins in the formation, maintenance, and disaggregation of complex multimeric structures and in protein folding (Chirico et al. 1988; Deshaies et al. 1988; Craig; Pelham; Welch, all this volume).

II. GENOMIC ORGANIZATION OF HUMAN HSP70 GENES

The human genome contains at least ten hsp70 gene-related sequences (Mues et al. 1986). Consistent with this minimal estimate, the results of genomic Southern blot analysis suggest at least two genes for each of the grp78 and p72 genes and a minimum of five hsp70 genes. Representatives of each gene class have been cloned from the mouse, rat, hamster, chicken, and human genomes. Additional members of the hsp70 gene family have been recently cloned, including an hsp70 gene that exhibits high levels of expression during spermatocyte differentiation (Zakeri et al. 1988).

Information on the number of hsp70 genes and their chromosomal location is available from gene cloning studies, in situ hybridization of cloned genes to metaphase chromosomes, and Southern blot analysis of both human DNA and DNA isolated from hamster-human somatic hybrids retaining specific human chromosomes. Another approach was to take advantage of the distinct mobilities on two-dimensional gel electrophoresis of the human and hamster hsp70 proteins. Analysis of the ^{35}S-labeled heat-shock-induced proteins synthesized in a collection of human-hamster somatic-cell hybrids revealed the presence of hsp70-encoding genes on chromosomes 6, 14, and 21. The localization of hsp70 genes to two of these chromosomes was confirmed by in situ hybridization, which identified sites at chromosome 6p21-23 and chromosome 14q22 (Harrison et al. 1987). Further analysis of the hsp70 gene sequences on chromosome 6 by Southern blot analysis of genomic DNA from a human-hamster somatic hybrid retaining chromosome 6 revealed multiple hsp70 gene sequences on a 20-kb *Eco*RI genomic fragment. The cluster of hsp70 genes on chromosome 6p21.3-p23 is in close proximity to the HLA locus. Recent studies on genomic cloning of the HLA class III region have demonstrated that the cluster of hsp70 genes on chromosome 6 is localized to the major histocompatibility complex (MHC) class III region and maps between the genes for complement and tumor necrosis factor α and β (Sargent et al. 1989).

III. REGULATED EXPRESSION OF GRP78, P72, AND HSP70

The genes that encode hsp70-related proteins exhibit distinct patterns of expression. As shown in Figure 2, the expression of hsp70 is induced by a large number of conditions that affect the dynamic process of cell growth. In human cells, hsp70 is induced during cell growth and differentiation, and in response to a wide range of physiological or chemically induced stress conditions (Kelley and Schlessinger 1978; Hightower 1980; Thomas et al. 1981; Hickey and Weber 1982; Wu and Morimoto 1985; Banerji et al. 1986, 1987; Milarski and Morimoto 1986; Mosser et al. 1988; Shaefer et al. 1988; Watowich and Morimoto 1988; Theodorakis et al. 1989; Williams et al. 1989). In many of these examples, gene expression is rapidly induced at the transcriptional level

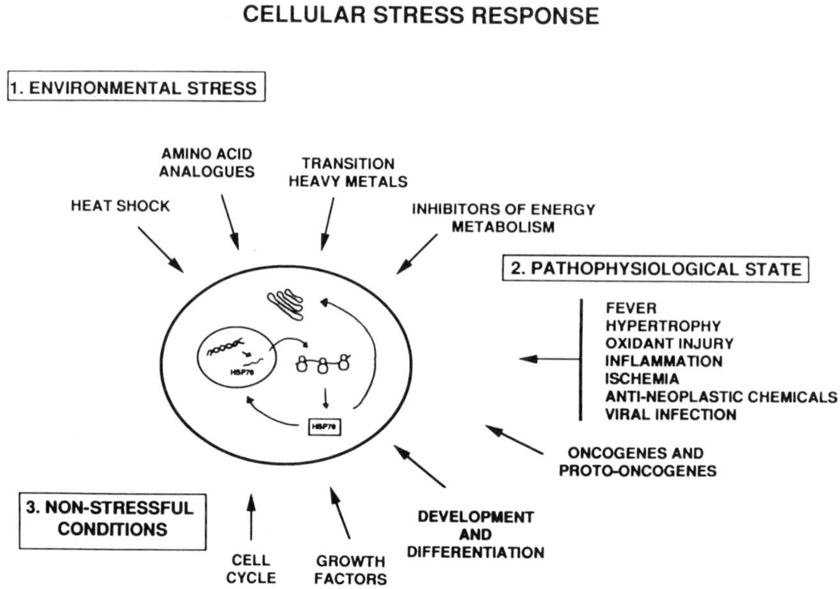

Figure 2 Overview of the conditions that induce hsp70 expression.

through protein factor interactions with distinct arrays of *cis*-acting promoter elements located upstream of the human hsp70 gene (Hunt and Morimoto 1985; B.J. Wu et al. 1986a, 1987; Greene et al. 1987; Morgan et al. 1987; Mosser et al. 1988; Williams et al. 1989). Although transcriptional regulation is the primary form of control for heat shock gene expression, the stability of heat shock mRNAs increases in a physiologically stressed cell (Theodorakis and Morimoto 1987), and the translational efficiency can also dramatically increase (Lindquist 1980, 1981; DiDomenico et al. 1982; Banerji et al. 1984, 1987; Theodorakis and Morimoto 1987).

Studies on hsp70 and grp78 gene expression have suggested a complex interrelationship between different members of the hsp70 family. Although both genes are induced in response to various forms of physiological stress, the conditions that induce grp78 are often distinct from the conditions that activate hsp70 expression. Only grp78 expression is induced by calcium ionophores, inhibitors of posttranslational protein glycosylation including castanospermine, tunicamycin, or glucosamine, in response to anoxia, conditions of low glucose in the growth media, and during infection with negative strand viruses (Fig. 3). In comparison, the p72 gene is constitutively expressed and responds only slightly to many of the regulatory conditions that affect grp78 and hsp70 gene expression.

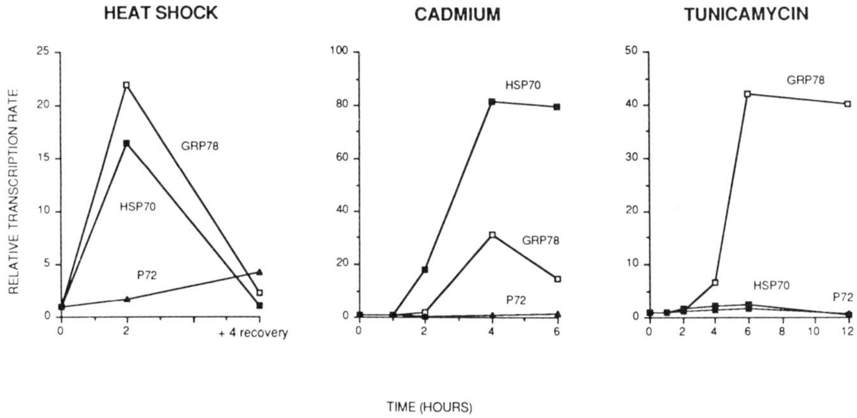

Figure 3 Transcription of hsp70-related genes in response to heat shock, cadmium, and tunicamycin. HeLa cells were treated under conditions of heat shock, cadmium sulfate, or tunicamycin for periods as indicated, and the relative in vitro transcription rates of the hsp70, grp78, and p72 genes were measured by nuclei run-on analysis.

Of particular interest are the effects of transition heavy metals and amino acid analogs that induce the expression of both hsp70 and grp78 (Watowich and Morimoto 1988). The coordinate induction of both genes at the transcriptional level (Fig. 3) suggests a possible common regulatory pathway, although there is no evidence to suggest shared transcription factors. It is unclear how heavy metals and amino acid analogs affect common intracellular events. Both treatments could affect de novo protein synthesis on both free and bound polysomes, thus resulting in the appearance of damaged or altered proteins in the cytosol and the endoplasmic reticulum. Such events induce increased expression of both hsp70 and grp78. In contrast, only grp78 expression is induced by inhibitors of protein glycosylation, which only affect nascent proteins translated on membrane-bound polysomes. Less well understood are certain conditions of grp78 induction, such as treatment with the calcium ionophore A23187, in which the levels of hsp70 mRNA are concomitantly reduced as grp78 mRNA levels increase. This apparent reciprocity in levels of expression suggests a mechanism that regulates the overall levels of hsp70-related gene products in the cell (Subjeck and Shyy 1986; Watowich and Morimoto 1988).

IV. REGULATION OF HSP70 EXPRESSION DURING CELL GROWTH

Growth-regulated expression of the hsp70 gene was demonstrated in growth-arrested HeLa cells following the addition of fresh serum (Wu et al. 1985). Serum deprivation of HeLa or 293 cells resulted in reduced hsp70 transcription and mRNA levels, which upon serum stimulation, were induced approximately 20-fold, attaining maximal levels between 8 and 12 hours poststimulation. More recent experiments have established that increased levels of hsp70 transcription were detected within 2 hours after serum stimulation, whereas increased levels of c-*fos* transcription were detected after 20 minutes. Activation of hsp70 expression also occurs following stimulation of resting human T lymphocytes with the nonspecific mitogen phytohemagglutinin and the polypeptide growth factor interleukin-2 (IL-2) (Ferris et al. 1988). The kinetics for hsp70 expression following IL-2 stimulation of peripheral human T cells supports the growing evidence that hsp70 is a member of the growth-regulated gene family including such genes as c-*myc*, c-*myb*, and c-*fos*. The expression of these genes has been correlated with the transition to the growing state. In support of this, addition of the DNA synthesis inhibitor cytosine arabinoside during serum stimulation of HeLa cells blocks the increase in hsp70 mRNA levels. These results are consistent with the suggestion that DNA synthesis and hsp70 expression are linked (Wu and Morimoto 1985). The phenotype for hsp70 expression during cell growth is complex. Whereas the hsp70 gene transcriptional response to serum factors shares similarities with other growth-regulated immediate-early genes, the coincidence between hsp70 expression and the onset of DNA synthesis reveals that hsp70 gene expression also shares features with the replication-dependent histone genes whose expression is also coupled to DNA synthesis.

In addition to the previously described studies on the serum and growth-factor-activated hsp70 expression, changes in hsp70 mRNA levels and protein synthesis have been observed during the cell cycle. In experiments using mitotically detached populations of synchronized human cells, hsp70 mRNA levels increase at the G_1/S boundary of the cell cycle in unstressed cells and then decline rapidly during mid-to-late S phase (Fig. 4) (Milarski and Morimoto 1986). The rapid increase in hsp70 mRNA levels in S-phase cells and its subsequent disappearance are due to the short half-life of hsp70 mRNA in growing cells (Theodorakis and Morimoto 1987). Coincident with the increase in message levels and hsp70 protein synthesis levels is the nuclear localization of hsp70 in S-phase cells (Milarski and Morimoto 1986).

A role for hsp70 in unstressed cells can be proposed from the following observations: (1) hsp70 transcription is regulated by serum stimulation and is coincident with the onset of DNA replication (Wu and Morimoto 1985), (2) hsp70 expression is cell-cycle-regulated at the G_1/S boundary (Milarski and Morimoto 1986), (3) the bacterial hsp70 homo-

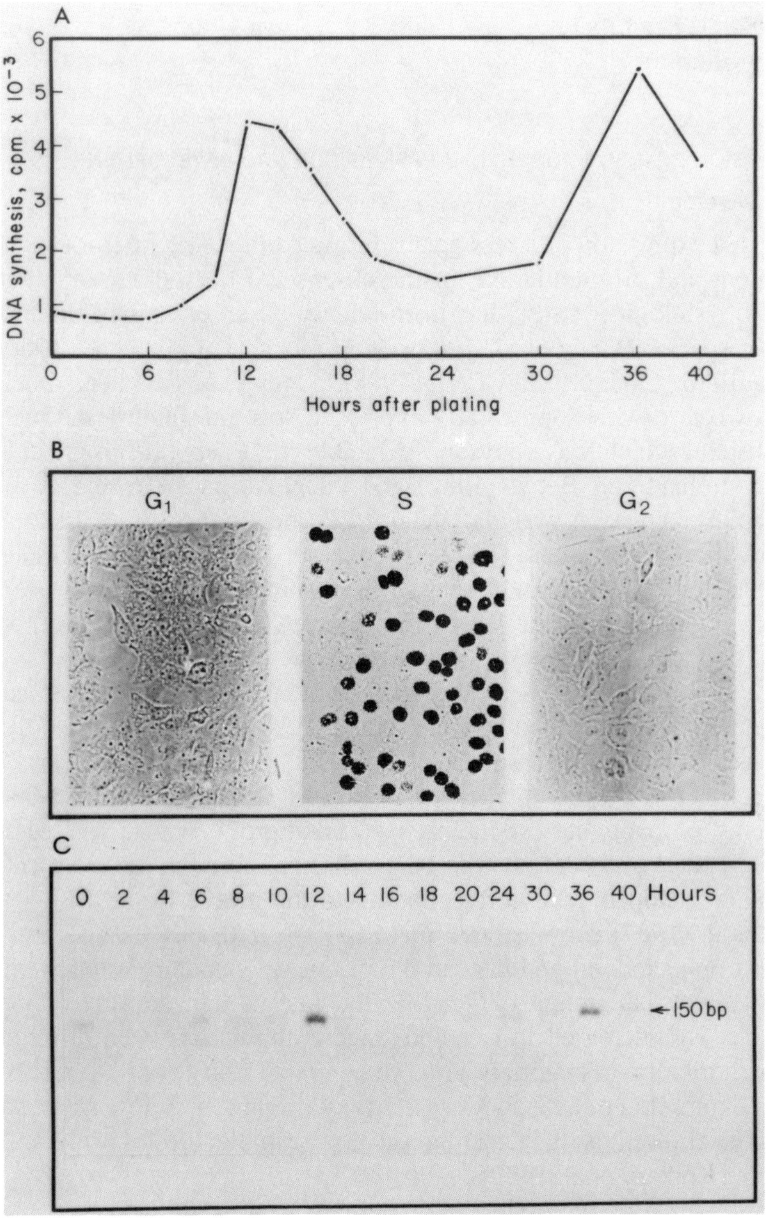

Figure 4 Cell synchrony and levels of hsp70 mRNA. (*A*) DNA synthesis was measured by [^{3}H]thymidine incorporation at the indicated times after plating harvested mitotic HeLa cells. (*B*) Autoradiographic analysis of G_1, S, and G_2 populations of synchronized cells. (*C*) Relative levels of hsp70 mRNA throughout the cell cycle as determined by S1 nuclease analysis.

log, DnaK, which shares 50% amino acid sequence identity with human hsp70, participates in bacteriophage λ and oriC replication (Zylicz et al. 1983, 1984, 1987; Liberek et al. 1988; Sakakibara 1988; Georgopoulos et al., this volume).

V. EXPRESSION OF HSP70 DURING DIFFERENTIATION AND DEVELOPMENT

A. Early Development

The elevated expression of stress genes during early stages of embryonic development and differentiation, in the absence of heat shock or other forms of physiological stress, has been widely observed among diverse eukaryotic species (Craig et al. 1983; Craig 1985; Heikkila et al. 1986; Lindquist 1986; Lindquist and Craig 1988). Among the first heat shock genes shown to be developmentally expressed was the small heat shock protein, hsp23, which was ecdysone-inducible in *Drosophila* tissue culture cells (Ireland and Berger 1982) and imaginal discs (Ireland et al. 1982). hsp23 expression was also detected in the absence of heat shock early in ovarian development (Zimmerman et al. 1983) and in the tissues of imaginal discs (Cheney and Shearn 1983). A more detailed discussion on the developmental expression of the *Drosophila* heat shock genes is presented by Pauli and Tissières (this volume).

The de novo synthesis of a family of 70-kD proteins early in mouse embryogenesis was described by Bensaude et al. (1983) who detected the constitutive expression of a subset of the 70-kD heat shock proteins at the two-cell stage. Further studies by Hahnel et al. (1986) revealed that two proteins, probably corresponding to p72 and grp78, were activated at the two-cell stage and continued to be synthesized through the eight-cell stage. Using a cloned gene corresponding to the mouse hsp68 (hsp70), Kothary et al. (1987) demonstrated that hsp68 (hsp70) mRNA was present in the trophectoderm and later in the primitive endoderm at late fetal stages. A direct assessment of whether the rate of transcription of the hsp70 gene was increased in two-cell-stage embryos has been difficult due to the limitation in materials. A more complete analysis on the developmental expression of heat shock genes may eventually follow from experiments such as those using chimeric hsp promoter vectors in transgenic mice (Kothary et al. 1989).

In studies on the expression of vertebrate heat shock genes, it has often been unclear as to which of the hsp70-related genes are developmentally regulated. In part, this is due to the large number of hsp70-related genes in vertebrate genomes. At the level of gene expression, potential difficulties that arise from the use of heterologous gene probes include an inability to distinguish among members of the hsp70 family.

This problem has been resolved, since many of the yeast, *Drosophila*, rodent, and human hsp70-related genes have now been cloned. Similarly, hsp70-related proteins have similar electrophoretic mobilities and are not easily resolved by one-dimensional gel electrophoresis. The combination of two-dimensional gel electrophoresis and the use of specific antibodies will facilitate future studies.

B. Terminal Differentiation

1. Erythroid Cells

During differentiation of avian erythroid cells, hsp70 is expressed at high levels in primitive day-3 embryonic red cells (Banerji et al. 1987). As the erythroid cells mature, the levels of hsp70 expression decrease; however, high levels of hsp70 are maintained throughout the differentiated state and are accumulated in adult erythroid cells (Morimoto and Fodor 1984). hsp70 expression has also been shown to be induced during erythroid maturation of human K562 cells following hemin stimulation (Singh and Yu 1983). The induced expression of a 70-kD protein antigenically related to the hsp70 family was suggested to represent selective developmental activation of heat shock gene expression independent of heat shock treatment. By using cloned members of the human hsp70 gene family, it has been demonstrated that hemin treatment of K562 cells induces the transcriptional activation of both hsp70 and the grp78 genes (Theodorakis et al. 1989). The effects of hemin on hsp70 transcription were shown to be through activation of the stress-induced form of a heat shock transcription factor. Using a collection of 5′-deletion mutations in the hsp70 promoter, it was demonstrated that the sequences required for hemin responsiveness were coincident with the heat shock element.

The activation of hsp70 during erythroid maturation by hemin suggests either that the hemin-induced stress response may be utilized as a differentiation signal during erythroid maturation or that the biochemical events following hemin stimulation of K562 cells could initiate independent events that trigger the heat shock response and erythroid maturation. Both hsp70 expression and erythroid differentiation are dependent on continued hemin treatment; removal of hemin down-regulates both hemoglobin and hsp70 expression. The suggestion that hemin can mediate its genetic regulatory effects through a stress response is supported by recent studies demonstrating that the heme oxygenase gene is also hemin- and heat-shock-responsive (Shibahara et al. 1987; Yoshida et al. 1988). However, in the case of hsp70 regulation, both stress induction and hemin induction map to a single heat shock promoter element, whereas the hemin-responsive element of the heme oxygenase gene remains to be identified. Additional studies will be necessary to establish both the role of the stress response in differentiation and the role of hsp70 in erythroid maturation.

2. *Spermatogenesis*

During spermatogenesis of mouse and rat, the expression of an hsp70-related gene has been detected in 3-week-old animals, attaining maximal levels by 4 weeks of age and remaining elevated through at least 11 weeks of age (Krawczyk et al. 1987a,b; Allen et al. 1988; Zakeri et al. 1988). The appearance of high levels of hsp70 mRNA corresponds to the period in male gamete differentiation when the first meiotic cells appear and spermatocytes reach the pachytene stage (Zakeri and Wolgemuth 1987; Zakeri et al. 1987).

The constitutive expression and accumulation of high levels of stress proteins in highly differentiated cells such as spermatocytes and erythrocytes may be necessary in cells incapable of activating gene expression. During maturation, both cell types share events leading toward generalized biosynthetic arrest. In addition, the high constitutive levels of hsp70 expression acquired during differentiation support a potential role for hsp70 in the process of differentiation by participating in the systematic disaggregation and degradation of protein complexes. Thus, the accumulation of high levels of hsp70 in red cells may be in response to the increased levels of proteins that accumulate in the cytosol during nuclear condensation and are eventually degraded in the process of terminal differentiation.

The relationship between stress proteins and proteolysis has been suggested from the identification of ubiquitin as a heat shock protein in chicken cells (Bond and Schlesinger 1985) and the *lon* proteinase as an *E. coli* heat shock protein (Goff and Goldberg 1985). Direct evidence to support the model that unfolded proteins signal the induction of the heat shock response in *E. coli* has been provided from experiments using unfolded variants of λ repressor (Parsell and Sauer 1989). In these studies, the synthesis of heat shock proteins appears to be directly modulated by the level of the unfolded proteins. These results are consistent with experiments in which the injection of denatured proteins into eukaryotic cells stimulated the expression of coinjected heat shock genes (Ananthan et al. 1986).

C. Acquisition of the Heat Shock Response during Development

Separate from the stage-specific activation of heat shock gene expression (in the absence of heat shock), the ability to respond to heat shock also appears to be developmentally regulated. An increasing number of observations in eukaryotic cells from a wide range of species have revealed the lack of a heat shock response in cells derived from certain tissues or in certain tissue culture cell lines. In *Drosophila* and sea urchin embryos, the normal repertoire of hsps are not inducible until the 64–128-cell stage (Graziosi et al. 1980; Roccheri et al. 1981). Similarly, in *Drosophila* oocytes, the 83-, 28-, and 26-kD hsps are constitutively expressed, but the 70-, 23-, and 22-kD hsps are neither constitutively expressed nor heat-shock-inducible (Zimmerman et al. 1983). These observations on the acquired inducibility of heat shock genes during *Drosophila* embryogenesis have led to the suggestion that heat shock responsiveness might be acquired in a stage-dependent manner (Graziosi et al. 1980).

The apparent inability of certain teratocarcinoma cell lines to respond to heat shock during differentiation in vitro may be related to the absence of a stress response in two to eight cell-stage mouse embryos (Besaude and Morange 1983; Wittig et al. 1983; Morange et al. 1984). The stress response is, however, present in morula/blastocyst-stage embryos, which are fully capable of inducing hsp70 synthesis. During *Xenopus* embryogenesis, the heat inducibility of the hsp70 gene is not apparent until after the midblastula stage of development (Bienz 1984a; Heikkila et al. 1985, 1987; Krone and Heikkila 1988). Moreover, heat shock responsiveness is not acquired coordinately for the various heat shock genes; for example; the hsp30 gene is not heat-shock-inducible until the tailbud stage of development. These results reveal that additional factors other than the heat shock transcription factor may influence inducible heat shock gene expression.

The expression of the hsp70 genes is developmentally regulated during avian erythroid maturation. Although hsp70 is constitutively expressed in avian cells of both primitive and definitive erythroid lineages, only the definitive cells respond to heat shock by elevated hsp70 synthesis. In primitive cells, hsp70 mRNA levels are induced by day 3 of development; however, these cells are heat-shock-responsive neither at the transcriptional nor at the translational level. (Banerji et al. 1987). However, in definitive erythroid cells that lead to the adult lineage, there is a substantive heat shock response measured by increased synthesis of hsp70. In definitive erythroid cells, heat shock alters the translation rate of preexisting hsp70 mNNA (Morimoto and Fodor 1984; Banerji et al. 1984, 1987; Theodorakis et al. 1988). Prior to heat shock, the preexisting hsp70 mRNA is functionally associated with polysomes but is maintained in a translationally repressed state. Heat shock results in a rapid increase in the rate of elongation of the nascent hsp70 peptide, thus accounting for the increased synthesis of hsp70 without a proportional increase in hsp70 mRNA levels (Theodorakis et al. 1988).

The studies on hsp70 expression during avian erythroid maturation reveal that within a single system, gene expression can be controlled at multiple steps. In early development, the initial increase in hsp70 synthesis is due to the appearance of hsp70 mRNA and its proportional translation. In primitive cells, heat shock neither induces hsp70 gene transcription nor affects hsp70 mRNA translation rates. In contrast, the heat shock response in definitive or adult erythroid cells is mediated entirely at the translational level. The lack of the heat-inducible transcriptional response in differentiating erythroid cells may offer an explanation for similar observations in mouse erythroleukemia and plasmacytoma cell lines that also appear to be deficient in the heat shock response (Aujume 1988).

VI. ONCOGENE ACTIVATION

A. Induction of Stress Protein Expression during Virus Infections

Since the initial observation that adenovirus infection induces the synthesis of a 70-kD heat shock protein (Nevins 1982), a number of other laboratories have reported that the expression of stress genes is induced following infection with a wide range of viruses. The virally induced expression of stress genes may be in response to altered biosynthetic rates or perhaps increased proteolysis concomitant with viral infection and the lytic stage. Alternatively, the transient activation of stress gene expression might reflect specific requirements for stress proteins during the viral life cycle. For example, hsp70 and p72 could participate in viral DNA synthesis analogous to the central role for *dnaK* in λ phage replication. Alternatively, these cellular proteins could facilitate virion assembly analogous to their roles in cellular transport and assembly of proteins into complexes.

Despite the general impression that activation of cellular stress gene expression is a common feature of viral infection, convincing results have been obtained for only a few viruses. Infection of primate cells by adenovirus and herpesvirus induces hsp70 transcription (Nevins 1982; Notarianni and Preston 1982), whereas SV5, Sendai, Newcastle's disease virus, and influenza viruses preferentially induce grp78 synthesis (Collins et al. 1978; Peluso et al. 1978; S. Watowich, unpubl.). Among the DNA viruses, some evidence for selectivity is apparent. For example, adenovirus and herpesvirus infection in human (HeLa) and monkey (CV1) cells induces only hsp70 gene transcription with no significant effects on other members of the hsp70 gene family, whereas infection with

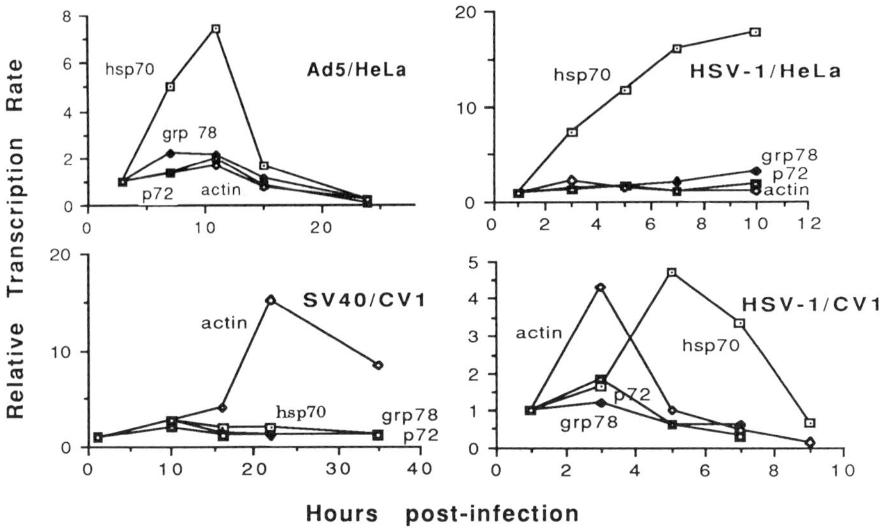

Figure 5 Specificity of hsp70 induction by DNA viruses. HeLa cells were infected with either adenovirus 5, herpes simplex virus 1, or SV40. Over the period of a time course, cell aliquots were taken for in vitro transcription assays in isolated nuclei. Experiments on Ad5 and HSV-1 used HeLa cells; infection of SV40 was performed on CV1 cells. Changes in the rate of transcription were measured following hybridization of ^{32}P-labeled nascent transcripts to nitrocellulose filters containing plasmid DNA for three distinct members of the hsp70 family (hsp70, p72, and grp78).

SV40 has no effect on the expression of any member of the hsp70 gene family (Fig. 5) (B. Phillips, unpubl.). These results are consistent with the suggestion that only viruses that encode a class of permissive transcriptional *trans*-activators induce hsp70 gene transcription. Therefore, stress proteins are not induced during infection with all DNA viruses or in response to all virally encoded transforming proteins. Finally, the induction of hsp70 expression appears to be independent of a productive lytic infection; the level of hsp70 induced in a productive infection in a permissive human cell (HeLa) was similar to that in a restrictive infection in semipermissive monkey (CV1) cells.

B. Adenovirus E1A Inducibility

During adenovirus infection of primate cells, the expression of one cellular gene, that encoding the major heat shock protein hsp70, is strongly induced by 8–12 hours postinfection (Nevins 1982; Wu et al. 1986b). The mechanism of hsp70 transcriptional activation during adenovirus infection requires the E1A region, which is necessary for transcription of the adenovirus early genes. E1A encodes three polypeptides that are products of the 13S, 12S, and 9S mRNAs.

Two experimental approaches have revealed that the 13S product is responsible for hsp70 transcription. Cells were either infected with viruses that encoded only the 12S and 13S products, or they were cotransfected with the cloned 13S or 12S cDNAs and a reporter gene containing the 5′-flanking sequences from the human hsp70 promoter fused to the bacterial CAT gene. The results of both experiments demonstrated that the adenovirus E1A 13S product is sufficient to induce hsp70 gene transcription (Wu et al. 1986b).

VII. TRANSCRIPTION OF THE HUMAN HSP70 GENE

A. Organization of the hsp70 Promoter

Comparisons of the 5′-flanking sequences of different vertebrate hsp70 genes have revealed a common picture of multiple *cis*-acting elements, including one or more heat shock elements (HSEs) (Morimoto et al. 1986; Bienz and Pelham 1987; Greene et al. 1987; Morgan et al. 1987; B.J. Wu et al. 1987; G.T. Williams and R.I. Morimoto, in prep.). Studies on expression of the human hsp70 gene have provided some insight into the complexity of transcriptional controls for vertebrate stress genes. Two distinct human hsp70 genes have been characterized in some detail, and each gene appears to have different transcriptional regulatory properties.

The hsp70 gene isolated in Voellmy's laboratory (Voellmy et al. 1985) exhibits a less complex form of transcriptional regulation, similar to that described for many of the yeast or *Drosophila* heat shock genes. Specifically, basal expression is low, often undetectable, and induction is in response to heat shock. The hsp70 gene cloned by Wu et al. (1985) and located on chromosome 6 exhibits complex transcriptional regulation and is responsive to diverse cellular signals, including serum factors, viral activation, developmental regulation, and stress induction. The *cis*-acting promoter elements of the human hsp70 gene (Fig. 6) confer basal expression, serum and IL-2 inducibility, response to forskolin and TPA, induction following infection with adenovirus and herpes simplex virus, and developmental and tissue-specific expression, as well as induction by heat shock and other forms of stress. In addition to general transcription factors, many promoter-specific factors bind to the upstream elements

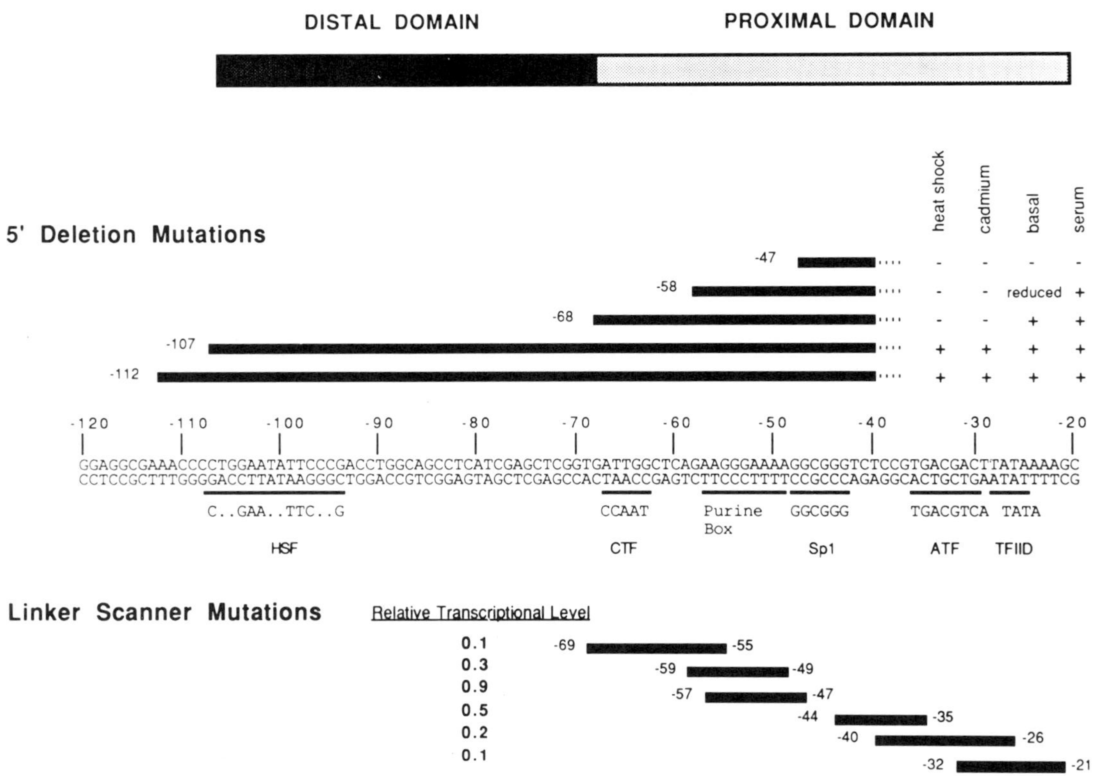

Figure 6 Schematic of human hsp70 promoter regulatory domains. Promoter sequences from −1 to −120 are shown. Locations of transcription factor binding sites in the promoter are indicated with the genetic elements. Above the promoter sequences are the boundaries of regulatory elements identified using 5′-deletion mutations and the phenotypes of each construct. Below are the phenotypes of linker-scanner mutations on basal transcription measured by relative transcription rates.

and may mediate several forms of regulation (Greene et al. 1987; Morgan et al. 1987; B.J. Wu et al. 1987; Williams et al. 1989).

B. Sequences Regulating Basal Expression and Adenovirus E1A Inducibility

The promoter sequences involved in transcriptional regulation of the human hsp70 gene have been examined in rodent and human cell lines. The human hsp70 promoter has two functional domains (Fig. 6). The proximal domain contains the basal promoter, which responds to signals that activate cell growth and responds to transcriptional *trans*-activation by adenovirus E1A. The distal promoter contains the heat shock element, which confers multiple forms of stress induction. Even further upstream of –120 are additional functional basal elements that appear to be redundant relative to sequences in the proximal promoter.

The essential sequences in the proximal and distal hsp70 promoter were identified through four complementary approaches. In experiments using HeLa cells transiently transfected with 5′-deletions and linker-scanner mutations fused to a reporter gene encoding the bacterial chloramphenicol acetyltransferase, multiple promoter elements were identified, each of which contributed to establish the overall level of basal transcription (Fig. 6). Deletion of the CCAAT box at –68 had the most dramatic effect on basal expression, followed by mutations in the TATA element, the ATF, GC(Sp1), and purine boxes. These elements correspond to transcription factor binding sites including those for CTF at –150 and –67, SP1 at –160 and –48, AP-2 at –120 and –20, HSF at –100, ATF (CREB) and AP-1 at –37, and TFIID at –28. The binding sites were identified by DNase I footprinting analysis, gel mobility shift assays, and methylation interference. Less well characterized are factors that bind to a purine-rich element centered at –54 (B.J. Wu et al. 1987) and a GC element at –45 (D.D. Mosser et al., unpubl.). Although sites that interact with protein factors in the basal promoter have been identified, it is unclear how these events regulate transcription in vivo. The human hsp70 promoter reveals a compactness and repetition of transcription elements common to other genes that exhibit complex regulation.

hsp70 transcriptional *trans*-activation by adenovirus E1A has been examined by two approaches (Williams et al. 1989). HeLa cells were cotransfected with a plasmid encoding E1A and hsp70 promoters containing 5′-deletions fused to the CAT gene. In addition, cell lines stably transfected with chimeric hsp70 promoters fused to the herpes simplex thymidine kinase (HSP-TK) gene were infected with adenovirus. The results of cotransfection studies using the 5′-deletion HSP-CAT mutants localized the minimum sequence required for adenovirus responsiveness to the –74 region of the hsp70 promoter. Deletion of the CCAAT box at –68 decreased both basal and E1A-inducible expression by 10–20-fold, whereas mutants extending to –47 were not E1A-inducible. The use of linker-scanner mutations to examine which of the multiple elements within the hsp70 basal promoter were necessary for E1A *trans*-activation revealed that deletion of each element within the proximal promoter reduced the basal level of transcription and, in all cases, the remaining promoter elements were sufficient for E1A responsiveness. Thus, E1A *trans*-activation was not mediated though any single promoter element but appears to activate through the entire transcription complex that assembles on sequences between –74 and the site of transcription initiation.

The role of specific hsp70 promoter elements in transcriptional regulation during adenovirus infection was also examined using a complementary approach in which promoter elements from the hsp70 gene were inserted upstream of the non-E1A-inducible HSP-TK gene. The E1A responsiveness of each promoter construct was tested by infection of stably transfected cell lines with wild-type adenovirus. For example, fusion of a 56-nucleotide fragment, containing the hsp70 CCAAT element and flanking purine-rich sequences, to the TK gene generated a chimeric promoter (HSP-TK) that was induced during adenovirus infection with kinetics that paralleled the increase in hsp70 mRNA. These results extend the conclusions of the mutant promoter studies. Together, the results using transient transfection and infection of stably transfected cell lines reveal that adenovirus induction of the endogenous hsp70 gene is not due to a specific chromosomal location of the hsp70 gene and that induction of the hsp70 promoter is not due to artificially high levels of E1A obtained in cotransfection experiments.

There is less detailed information available on sequences required for growth-regulated expression of the hsp70 gene. Studies using 5′-deletion HSP-CAT mutants and chimeric HSP-TK genes have revealed that sequences within the basal promoter also confer serum responsiveness. There are multiple elements for the growth-regulated response since both AP-2 and AP-1/ATF elements are found in the region of the basal promoter. The molecular basis of serum responsiveness of the hsp70 gene in HeLa cells is likely to be related to IL-2 induction in T cells and signal transduction via the AP-1/ATF element, yet the factors and elements within the basal promoter that mediate this common form of expression may differ. Although the sequences required for growth regulation and E1A *trans*-activation of the hsp70 gene overlap and may correspond to the same elements, there is no information to suggest a common mechanism.

C. *cis*-Acting Elements Involved in Early Embryogenesis

Comparison of the 5′-flanking sequences of the human and *Xenopus* hsp70 (Bienz 1984a) promoter regions reveals considerable similarities in the complexity of *cis*-acting elements (Bienz and Pelham 1987). Both heat shock promoters contain HSEs, CCAAT, GC, and TATA elements in similar locations relative to the transcription initiation site. The collection of 5′-deletion and linker-scanner mutations in the human hsp70 promoter were used to identify the *cis*-acting sequence requirements for constitutive expression of heat shock genes during *Xenopus* embryonic development. HSP-CAT genes microinjected into developing *Xenopus* embryos are transcriptionally activated after the mid-blastula transition (N. Ovsenek et al., in prep.). Maximal expression at mid-to-late blastula requires the presence of multiple elements within the basal promoter such that mutations in a number of the promoter sites within the basal promoter resulted in a complete loss of expression. Each of these promoter elements was shown to bind in a sequence-specific manner with *Xenopus* nuclear proteins, thus suggesting that early developmental activation of heat shock gene transcription was also mediated through the assembly of a multicomponent transcription complex.

The *cis*-acting sequence requirement for high levels of expression at the mid-blastula transition of *Xenopus* embryonic development corresponds to sequences in the basal promoter, including the CCAAT element and additional flanking promoter elements. Although the use of heterologous systems with human promoter constructs limits the conclusions, studies on the sequence requirements for constitutive expression of the *Xenopus* hsp70 gene in *Xenopus* oocytes have shown a requirement for CCAAT and an additional upstream element corresponding to the HSE (Bienz 1984b, 1986).

D. *cis*-Acting Elements Involved in Stress-induced Transcription

Many inducers of the stress response, including heat shock, heavy metals, amino acid analogs, dinitrophenol, and arsenite, activate transcription of the hsp70 gene through sequence-specific interaction of a heat shock factor with the highly conserved HSE (Parker and Topol 1984; Wu 1984; Topol et al. 1985; Shuey and Parker 1986; Kingston et al. 1987; Mosser et al. 1988; Sorger et al. 1987; C. Wu et al. 1987; Zimarino and Wu 1987). The HSE is a conserved sequence motif that often appears as multiple arrays of the consensus sequence CNNGAANNTTCNNG or a minimum of three repeats in either orientation of the pentameric unit NGAAN (see Lis et al., this volume; Holmgren et al. 1981; Pelham 1982; Pelham and Bienz 1982; Amin et al. 1987; Xiao and Lis 1988).

The promoter sequences necessary for heat shock and metal ion induction of the human hsp70 gene are located in the distal domain of the promoter extending from −107 to −68 (B.J. Wu et al. 1987). Within this region is a single 8/8 match to a 14-bp HSE consensus centered at −100 that overlaps a weak HSE (4/8 match to the consensus) centered at −90 and is adjacent to another weak HSE (5/8 match to the consensus) centered at −86 and −110. Using a collection of 5′-deletion and linker-scanner mutations within these boundaries, the minimum sequence requirement for heat shock, heavy metals, and amino acid analog activation of hsp70 transcription was shown to correspond to a single HSE centered at −100 (G.T. Williams and R.I. Morimoto, in prep.).

Within the context of multiple *cis*-acting elements in the hsp70 promoter, mutations in any of the basal elements reduced the level of basal expression and also reduced the absolute level of stress responsiveness (G.T. Williams and R.I. Morimoto, in prep.). Although none of the mutations in the basal promoter resulted in a complete loss of HSE-dependent stress responsiveness, the lack of a TATA element had the most severe phenotype. These results reveal differences in the mechanisms of stress-induced transcription in *Drosophila* and human cells. The *Drosophila* heat shock promoters typically contain multiple HSEs located within close proximity to the TATA element, whereas in the human promoter, the HSE is dependent on interaction with the elements within the basal promoter for the absolute level of inducible transcription. An additional difference between *Drosophila* and human HSE requirements is that the HSE can be located at various positions upstream of the TATA element in the context of the human promoter with little effect on overall inducible transcription. In the *Drosophila* promoter, the spacing between the HSE and TATA elements appears to be strongly dependent on helical rotation (Cohen and Meselson 1988). These differences could be simply explained by the presence of the sequences within the basal promoter that could provide the necessary compensatory interactions.

VIII. HEAT SHOCK AND STRESS-INDUCED TRANSCRIPTION

A. Transcription Rates and HSF Levels

HeLa cells contain two heat-shock-element-specific DNA-binding activities; one found predominantly in extracts from unstressed human cells, the other in extracts from stressed cells. These DNA-binding activities were detected by binding to a radiolabeled synthetic oligonucleotide containing the HSE from the human hsp70 promoter and electrophoresis on low ionic strength polyacrylamide gels (Kingston et al. 1987; Sorger et al. 1987; D.D. Mosser et al., in prep.). The HSE/protein complex found in extracts from unstressed cells has an electrophoretic mobility distinct from that found in extracts made from stressed cells; this latter complex presumably corresponds to the heat shock transcription factor. Although the relationship between these two HSE-binding proteins remains unclear, both HSE-binding activities bind with similar nucleotide specificities based on dimethyl sulfate interference studies. During exposure of cells to elevated temperatures, heavy metal ions, or amino acid analogs, the levels of the constitutive HSE-binding activity diminishes, while the level of the stress-induced activity increases (Fig. 7). When heat-shocked cells are returned to 37°C, the level of the stress-induced HSE-binding complex diminishes, and the level of the constitutive HSE-binding activity increases. This apparent reciprocity between the two HSE-binding activities may correspond to an interconversion of an inactive state of the transcription factor to a transcriptionally active state via a stress-induced posttranslational modification; however, this remains to be rigorously demonstrated.

The HSE/protein complex formed with extracts from heat-shocked human cells has a slower electrophoretic mobility than the complex formed with extracts from unstressed cells. The difference in mobility is due to phosphorylation of the stress-induced form of heat shock transcription factor (HSF) (Larson et al. 1988; Sorger and Pelham 1988). Evidence to support posttranslational modifications of HSF have also been obtained in yeast (Sorger and Pelham 1988). In vitro studies on human HSF (Larson et al. 1988) have led to the suggestion that phosphorylation of HSF is an essential step in transcriptional activation; however, it remains unclear whether this step is necessary for DNA binding or transcriptional activity. Alternatively, the different HSE/protein complexes in human cells might reflect multimeric interactions such as interactions with other proteins in the HSE/HSF complex.

B. Pathways for HSF Activation

Although there is only a limited amount of information on the mechanism by which HSF activity is induced in vivo, it is likely that there are multiple pathways for HSF activation. Two inducers of the stress response, heat shock and cadmium, have different effects on HSF activation based on in vivo studies on stress-induced transcription rates and HSF levels. Cells heat-shocked in the presence of cycloheximide transiently activate HSF and hsp70 gene transcription for only 1 hour. Surprisingly, continued heat shock in cycloheximide-treated cells for longer periods blocks the appearance of HSF. In contrast, cells treated

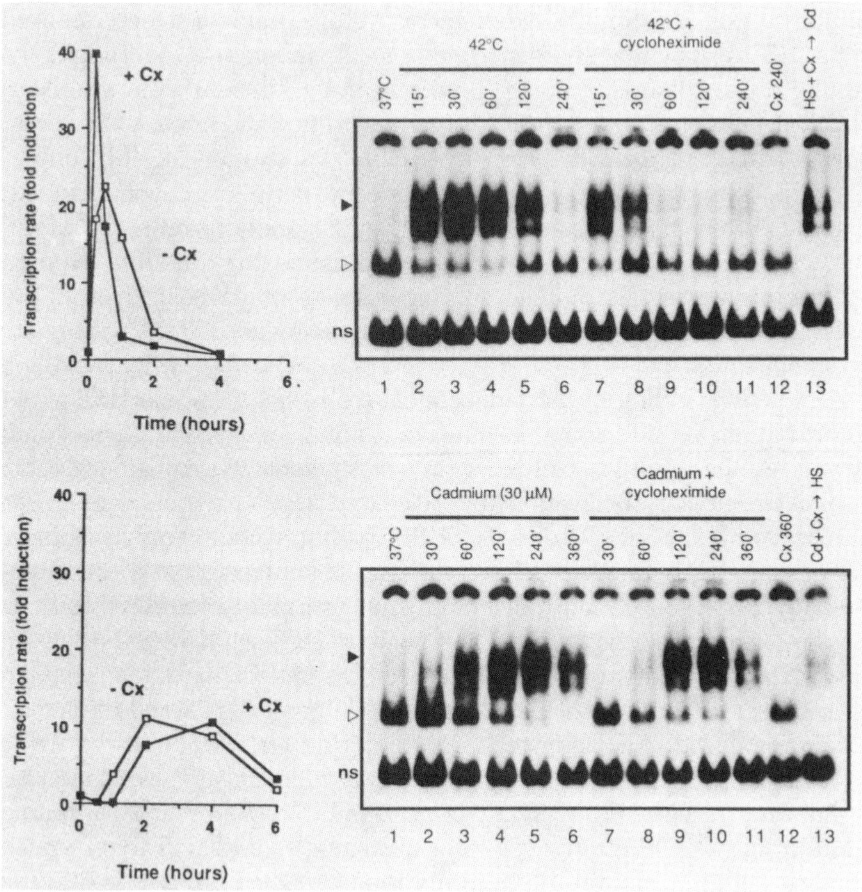

Figure 7 Heat shock transcription rates and HSF levels. hsp70 gene transcription and mRNA levels in heat-shocked and cadmium-treated cells. HeLa cells were incubated at 42°C or treated with 30 μM cadmium sulfate as indicated and in the presence (+Cx) or absence (–Cx) of cycloheximide. In vitro transcription rates were measured by run-on analysis in isolated nuclei from cells treated as described above. The radiolabeled transcripts were isolated and hybridized to filter-bound plasmid DNA encoding the human hsp70 gene. Kinetics of HSF-binding activity following heat shock and cadmium treatment. Whole-cell extracts were prepared from HeLa cells heated at 42°C or treated with 30 μM cadmium sulfate either in the presence or absence of cycloheximide (100 μg/ml) and used for gel shift assays as described. Extracts were also prepared from control (37°C) and cells incubated with cycloheximide (Cx) for 4 or 6 hr. (*Solid arrows*) Stress-induced form of HSF; (*open arrows*) constitutive HSE-binding activity found in control cells; (*ns*) non-specific DNA-binding activity detected with all oligonucleotide probes.

with cadmium in the presence of cycloheximide activate and maintain high levels of HSF for up to 6 hours. The results using cadmium treatment of cycloheximide-treated cells reveal that HSF is not a short-lived protein. Upon addition of cadmium to cycloheximide and heat-shocked cells, HSF reappeared and hsp70 gene transcription was induced. These results suggest that HSF was still present in cycloheximide-treated heat-shocked cells, but in a non-DNA-binding form. Additional evidence to suggest different pathways for stress-induced activation of HSF follows from the slower kinetics of hsp70 gene transcription in metal ion or amino-acid-analog-treated cells relative to heat-shocked cells. For example, incubation of HeLa cells with the proline analog azetidine required 30–60 minutes to accumulate sufficient levels of de novo synthesized azetidine-substituted proteins before the stress-induced HSE-binding activity appeared.

An understanding of the biochemical pathways by which HSF is activated in response to amino acid analogs, metal ions, and heat shock will reveal whether multiple pathways for stress gene activation are available and utilized. The observation that the metal-responsive element in the hsp70 promoter corresponds to the HSE has implications for heavy metal transcriptional regulation. Either the HSE-binding factor is a heavy-metal-responsive protein distinct from metallothionein regulatory proteins that act on the human MTII gene or, perhaps, HSF activity is indirectly regulated by a metal-sensing factor. Comparison of the upstream promoter elements of the glucose-responsive gene grp78 and the metallothionein I and II genes, which are transcriptionally induced following incubation with the metals cadmium, copper, and zinc (Levinson et al. 1980; Karin et al. 1984; Stuart et al. 1984; Watowich and Morimoto 1988), does not suggest a common mechanism of shared transcription factors. For example, the HSE is only found upstream of the hsp70 gene and is not utilized for inducible transcription of either the metallothionein or grp78 genes. Likewise, the metallothionein genes are not heat-shock-responsive. Yet, because transcription of all three genes responds to the same metal ions, it is likely that common metal-ion-sensitive intermediates in the pathway of transcription activation transmit this information to transcription factors such as HSF.

C. Biochemical Studies on HSF

HSF has been purified from yeast, *Drosophila*, and human cells, and corresponds to a 110–130-kD protein that binds specifically to the HSE (see Wu et al., this volume). The *Drosophila* and yeast proteins have been shown to activate transcription in in vitro transcription assays or when coinjected into *Xenopus* oocytes with a heterologous gene containing an HSE in the promoter (Sorger and Pelham 1988; Wiederrecht et al. 1987; C. Wu et al. 1987; Goldenberg et al. 1988). A similar factor, based on DNA-binding studies, methylation interference, and DNA footprint analysis, is found in heat-shocked human HeLa cells (Kingston et al. 1987; Morgan et al. 1987; Mosser et al. 1988).

Activation of HSF in vitro upon incubation of HeLa cell extracts at elevated temperatures (Larson et al. 1988) or upon treatment with low pH buffers (pH 5.8–6.4) or nonionic detergents at physiological temperatures (D.D. Mosser et al., in prep.) has been obtained. A comparison of the HSE-binding activities in extracts from in vivo heat-shocked and in vitro activated extracts is shown in Figure 8. In vitro studies in which extracts from unstressed cells were incubated at elevated temperatures indicate a temperature-dependent heat activation between 38°C and 43°C, with kinetics similar to that observed in vivo. HSF activity can also be activated upon treatment of S100 extracts from unstressed cells with buffers that reduce the pH, or with nonionic detergents. An important feature of the low pH and detergent activation of HSF is that both treatments reduce the temperature requirement for HSF activation to 37°C, whereas cell extracts prepared under normal conditions require extreme temperatures to activate HSF. These studies are consistent with the suggestion that the biochemical events involved in HSF activation require altered protein conformation. Either HSF detects and responds to tran-

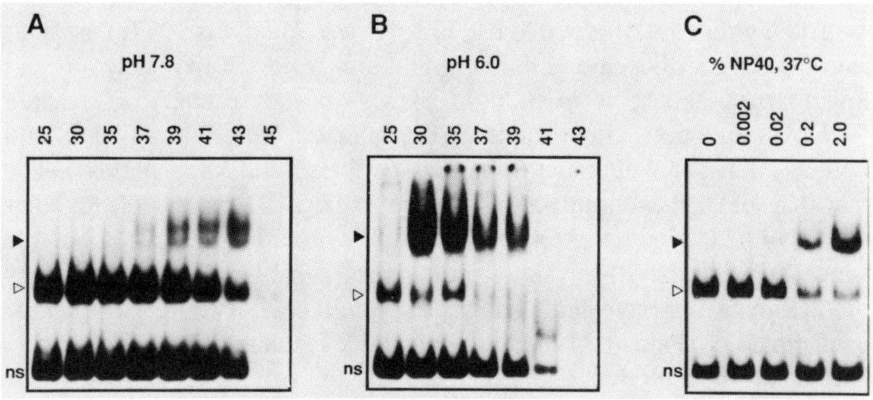

Figure 8 In vitro activation of HSF. S100 extracts prepared from HeLa cells grown at 37°C were maintained either at pH 7.8 (*A*) or at pH 6.0 (*B*) and incubated at various temperatures for 60 min. The HSF/HSE complex was measured using the gel mobility shift assay. S100 extracts incubated with 2% NP-40 (*C*) either on ice at 37°C or at 43°C followed by forming the HSF/HSE complex and gel mobility shift assay. The symbols are as indicated in Fig. 7.

sient changes in the environmental state of the cell through allosteric effects and a conformational change in HSF, or HSF is maintained in an inactive complex by association with other proteins and undergoes a conformational change to acquire DNA binding.

IX. CELLULAR AND BIOCHEMICAL STUDIES ON HSP70

The sequence similarities among the hsp70-related proteins, together with common biochemical properties, have led to the suggestion that these proteins are likely to perform similar tasks, each within its respective intracellular locale (Pelham 1986). For example, grp78 is localized in the endoplasmic reticulum and has been suggested to participate in the assembly within this cellular compartment of multimeric protein complexes that appear on the cell surface or are secreted (Naas and Wabl 1983; Bole et al. 1986; Gething et al. 1986; Munro and Pelham 1986; Watowich and Morimoto 1988). grp75 may similarly aid in the assembly of proteins in the mitochondria (Deshaies et al. 1988).

p72, the constitutive and very abundant cytoplasmic member of the hsp70 family, has been demonstrated to participate in the translocation of polypeptides across membranes of the endoplasmic reticulum and mitochondria (Chirico et al. 1988; Deshaies et al. 1988). In addition, p72 is capable of disassembling clathrin-coated vesicles (Ungewickell 1985; Chappell et al. 1986). hsp70-related proteins have been shown to be associated with other growth-regulated proteins such as the cellular oncogene p53 (Pinhasi-Kimhi et al. 1986; Hinds et al. 1987).

The highly stress-inducible hsp70 protein is found (as is p72) associated with the preribosome-containing granular region of the nucleolus following heat shock; it is thought to aid in the recovery of normal nucleolar function after stress (Velazquez et al. 1980; Welch and Feramisco 1984; Welch and Suhan 1986; Welch and Mizzen 1988). The fact that all of these interactions appear to be ATP-dependent, together with the ATP-binding properties of hsp70 proteins, has prompted the suggestion that the 70-kD stress proteins are capable of using the energy of ATP hydrolysis to effect changes in protein complexes or polypeptide conformation (Pelham 1984; Lewis and Pelham 1985; Welch and Feramisco 1985; Welch; Craig; Pelham; all this volume).

A. Cell-cycle-dependent Variation in hsp70 Associations

Studies in our laboratory have demonstrated that hsp70, like the abundant, constitutively expressed members of the 70-kD stress protein family, is also involved in the normal biochemical processes of the unstressed cell (Milarski et al. 1989). During the cell-cycle-regulated expression of the hsp70 gene, the subcellular localization of hsp70 also varies throughout the cell cycle. These changes in localization may be related to cell-cycle-dependent interactions of hsp70 with specific cellular proteins identified by coprecipitation using nondenaturing immunoprecipitations with hsp70-specific antibodies (Fig. 9). Some of these proteins were associated with hsp70 only at specific points in the cell cycle, whereas others were coprecipitated throughout the entire cell cycle. The cell-cycle-specific interactions were short-lived, indicating that hsp70 is involved in dynamic associations with specific cellular proteins throughout the growth cycle of unstressed cells.

One question on heat shock protein function has been whether the roles of hsp70 during normal cell growth and following exposure to stress are related. In comparing the proteins coprecipitated with hsp70 before and after heat-shock treatment, we found only minor differences. The proteins associated with hsp70 during normal conditions of cell growth remain associated following heat shock, revealing that the function of heat shock proteins is unlikely to be in the formation of specific complexes unique to physiological stress. Although the experimental protocol used may have limited the ability to detect new associations occurring only in stressed cells, the fact that most, if not all, of the proteins associated with hsp70 in unstressed cells were also precipitated with hsp70 following heat shock is of interest. Thus, hsp70 is likely to participate in certain stress-specific functions following heat shock; the protein may also continue to perform some of the same tasks it did in the unstressed cell.

B. Functional Domains of hsp70

There is only a limited amount of information known about the structure of hsp70-related proteins despite the recent progress in understanding the possible biochemical functions. Incubation of hsp70, p72, or DnaK with various proteinases has revealed a common structural feature that is shared among members of the hsp70 family. Digestion with trypsin or chymotrypsin leads to two distinct domains, a stable 45-kD proteinase-resistant fragment derived from the amino terminus of hsp70 and a proteinase-sensitive carboxyl domain. However, following thermal denaturation, the 45-kD amino-terminal fragment becomes completely trypsin-sensitive. Using a collection of hsp70 mutants that retain to various degrees the amino or carboxyl terminus of hsp70, it has been shown that removal of the carboxyl terminus does not affect the folding of the trypsin-resistant amino-terminal domain.

hsp70 can also be divided into two functional domains corresponding to a region necessary for nucleotide binding and a distinct region re-

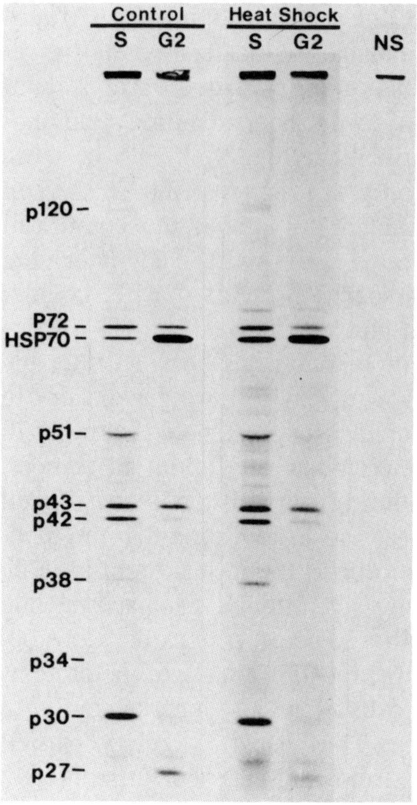

Figure 9 Identification of specific hsp70/cellular protein complexes during the cell cycle. Nondenaturing immunoprecipitations from control and heat-shocked cell lysates of S-phase and G_2 cells. HeLa cells were synchronized by mitotic detachment and labeled with [^{35}S]methionine at 37°C for a total of 6 hr prior to harvesting in S-phase or G_2. In parallel, cells labeled at 37°C for 4.5 hr were incubated, still in the presence of radiolabel, at 43°C for 1.5 hr. The labeled cell lysates were prepared for nondenaturing immunoprecipitation and incubated with monoclonal antibody C92 as described in the Materials and Methods. The immunoprecipitated proteins were analyzed by SDS-PAGE. Shown is a fluorograph of the gel. The estimated molecular weights of precipitated proteins are indicated to the left. (*NS*) Proteins immunoprecipitated with a nonspecific antiserum.

quired for nucleolar localization (Fig. 10) (Milarski and Morimoto 1989). This was demonstrated using the method of peptide tagging to analyze deletion mutations of the human hsp70 protein following transfection of the tagged hsp70 genes into primate cells. The results summarized in Figure 10 indicate that a region in the amino terminus of hsp70 that is highly conserved both between hsp70 proteins of different species and

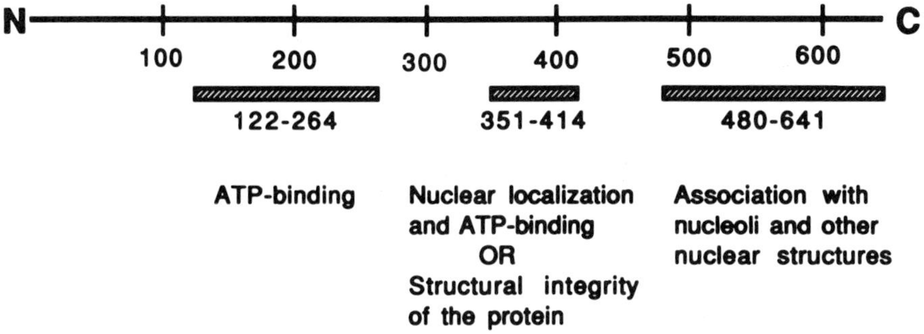

Figure 10 Domains of hsp70 identified by deletion mutation analysis. The ATP-binding domain was abolished by deletion of amino acids 122–264. Removal of amino acids 351–414 affected both intracellular localization and ATP binding and resulted in a trypsin-hypersensitive protein. Sequences involved in nuclear and nucleolar localization correspond to amino acids 480–641.

between the different members of the 70-kD family of a given species is necessary for ATP binding. The carboxy-terminal region of hsp70 contains the antigenic determinants for a wide range of characterized monoclonal and polyclonal antibodies to hsp70. Sequences in the carboxyl terminus of hsp70 are also important for nucleolar localization following heat shock treatment. Nucleolar localization is not a constitutive function of hsp70 and is dependent on biochemical events that occur during physiological stress to translocate hsp70 to the nucleolus. One interpretation of these results is that the carboxyl terminus of hsp70 corresponds to the region that interacts with unfolded domains of proteins. The apparent localization of hsp70 in the nucleolus of heat-shocked cells therefore reflects the increased levels of substrates that interact with hsp70.

The 70-kD proteins are highly conserved in the amino terminus and share the ability to bind ATP, but they differ in their intracellular distribution and specific interactions with associated proteins. The results of the hsp70 deletion mutants are consistent with a role for the amino terminus of the 70-kD heat shock proteins in a common ATP-binding function, whereas the carboxyl terminus may direct intracellular localization and/or substrate interactions. Although grp78 (BiP) interacts in the endoplasmic reticulum with membrane and secreted proteins whereas hsp70 and p72 interactions occur in the cytosol, the ability of these evolutionarily related proteins to transiently bind to proteins with diverse cellular functions is likely to reflect a common mechanism of hsp70/protein interaction.

X. CONCLUSION

Studies on vertebrate heat shock genes have revealed the complexity that underlies the regulation of inducible gene expression. The transcription of human heat shock genes is regulated by a multitude of signals involved in cell growth control, ranging from the effects of specific mitogens and growth factors to both acute and chronic changes in the environmental state of the cell. Regardless of the source of the signal or the mechanism of signal transduction, the end result involves interaction of specific transcription factors with a complex array of regulatory promoter elements. Dissection of the human hsp70 promoter has identified distinct arrays of *cis*-acting elements responsive to growth control signals and various forms of environmental stress. The examination of upstream promoter elements of a wide range of vertebrate heat shock genes, including those from *Xenopus*, mouse, rat, and chicken, have revealed a common pattern of multiple arrays of *cis*-acting elements responsible for the complex forms of transcriptional regulation. The apparent functional redundancy and duplication of promoter elements may be necessary to ensure that hsp70 is regulated in response to related signals in different cell types ranging from early cleavage embryos to blood cells.

Another approach that will reveal new features on the mechanisms of stress gene activation follows from recent data that certain mammalian cell lines appear to be defective in the heat shock response. The available examples include partially differentiated and transformed cell lines such as erythroleukemias and plasmacytomas that appear to be deficient in their ability to induce hsp70. A careful examination of these cell lines should reveal whether they are defective at the transcriptional level. This approach may open up new molecular genetic tools to dissect the pathways of stress gene activation. The possible relationships between different classes of inducible genes may also be examined through a molecular genetic approach, since it seems likely that genes involved in cellular responses to pharmacological agents, carcinogens, UV, and oxidant damage are integrated at some level with the stress response. It is hoped that the results of such molecular genetic approaches will converge on the recent success of cell-free systems in which HSF can be activated in vitro. Important questions remain on the identity of the "sensors" that act on HSF activity and whether these molecule(s) also determine the wide range of temperatures that activate the heat shock response.

Although only a cursory knowledge of the mechanisms of hsp70 gene family regulation is available, there is sufficient information to suggest that members of the hsp70 gene family coordinate the overall level of hsp70 gene expression and thus ensure regulated function. For example, expression of hsp70 and grp78 often shows an inverse relationship in which up-regulation of one member of the hsp70 family has compensatory down-regulatory effects on other members of the family. A clear example of this form of interactive regulation is described for the yeast hsp70 genes by Craig (this volume). What remains unclear is whether autoregulation can occur at both the transcriptional and post-transcriptional levels.

A number of questions remain on the function of hsp70. What are the biochemical and structural features of proteins that associate with stress proteins, and what domain of hsp70 is involved in substrate recognition? Although the role of ATP to solubilize the hsp70-containing complex has been proposed and demonstrated, it is unclear how the energy of ATP hydrolysis is utilized to affect protein structure and whether both the substrate and hsp70 undergo a conformation change.

Finally, an important direction relates to the role of stress proteins in human disease. Does the elevated synthesis of stress proteins indicate cell damage or reflect poor growth conditions that may exist within the constraints of local cellular environments of the pathophysiological state? Under these conditions, the ability to detect the elevated expression or increased levels of stress proteins may provide information on the growth state of the cell. The inappropriate expression of stress proteins via activation of the stress response could have dampening effects on cellular biosynthetic pathways, thus reflecting the compromised cellular state. For example, in metal toxicity diseases including Menkes, Wilson's disease, or hemochromatosis, aberrant and chronic expression of stress proteins could compromise the ability of a cell or tissue to respond to acute physiological and environmental stress. The accumulation of damaged proteins, already implicated as a signal for the stress response, may also be a consequence of oxygen damage during reperfusion with subsequent effects during cellular aging. The "normalcy" of the cellular state may also affect the ability of repair processes that function to reestablish homeostasis; perhaps the efficient repair of cellular damage requires proper regulation, expression, and function of heat shock proteins.

The potential for a role for stress proteins in immunobiology has been suggested from studies in infectious diseases. Although the significance of high titers of anti-hsp70 antibodies in patients with rheumatoid arthritis and lupus is unclear, a significant link has been demonstrated by localization of the hsp70 genes to the MHC class III region. The localization of hsp70 genes within the MHC region, together with the increasing evidence that hsp70 from pathogenic organisms results in a high proportion of T cells directed against heat shock proteins, has led to the speculation that repeated exposure to infections could initiate events leading to autoimmunity.

ACKNOWLEDGMENTS

We thank B. Phillips and K. Abravaya for comments and suggestions, and we acknowledge support from the National Institutes of General Medical Sciences, March of Dimes Foundation, and an American Cancer Society Faculty Research Award (FRA-313). K.L.M. was supported by a National Institutes of General Medical Sciences predoctoral training grant.

REFERENCES

Allen, R.L., D.A. O'Brien, and E.M. Eddy. 1988. A novel hsp70-like protein (P70) is present in mouse spermatogenic cells. *Mol. Cell. Biol.* **8:** 828–832.

Amin, J., R. Mestril, P. Schiller, M., Dreano, and R. Voellmy. 1987. Organization of the *Drosophila melanogaster* hsp70 heat shock regulation unit. *Mol. Cell. Biol.* **7:** 1055–1062.

Ananthan, T., A.L. Goldberg, and R. Voellmy. 1986. Abnormal proteins serve as eukaryotic stress signals and trigger the activation of heat shock genes. *Science* **232:** 522–524.

Aujame, L. 1988. The major heat shock protein hsp68 is not induced by stress in mouse erythroleukemia cell lines. *Biochem. Cell Biol.* **66:** 691–701.

Banerji, S.S., K.L. Berg, and R.I. Morimoto. 1986. Transcriptional and post-transcriptional regulation of avian HSP70 gene expression. *J. Biol. Chem.* **261:** 15740–15745.

Banerji, S.S., N.G. Theodorakis, and R. Morimoto. 1984. Heat shock induced translational control of HSP70 and globin synthesis in chicken reticulocytes. *Mol. Cell. Biol.* **4:** 2437–2448.

Banerji, S.S., K. Laing, and R.I. Morimoto. 1987. Erythroid lineage-specific expression and inducibility of the major heat shock protein HSP70 during avian embryogenesis. *Genes Dev.* **1:** 946–953.

Bardwell, J. and E. Craig. 1984. Major heat shock gene of *Drosophila* and the *Escherichia coli* heat-inducible dna K gene are homologous. *Proc. Natl. Acad. Sci.* **81:** 848–852.

Bensaude, O. and M. Morange. 1983. Spontaneous high expression of heat shock protein in mouse embryonal carcinoma cells and ectoderm from day 8 mouse embryo. *EMBO J.* **2:** 173–177.

Bensaude, O., C. Babinet, M. Morange, and F. Jacob. 1983. Heat shock proteins, first major products of zygotic gene activity in mouse embryos. *Nature* **305:** 331–333.

Bienz, M. 1984a. Developmental control of the heat shock response in *Xenopus. Proc. Natl. Acad. Sci.* **81:** 3138–3142.

———. 1984b. *Xenopus* hsp 70 genes are constitutively expressed in injected oocytes. *EMBO J.* **3:** 2477–2483.

———. 1986. A CCAAT box confers cell-type specific regulation on the *Xenopus* HSP70 gene in oocytes. *Cell* **46:** 1037–1042.

Bienz, M. and H. Pelham. 1987. Mechanisms of heat shock gene activation in higher eukaryotes. *Adv. Genet.* **24:** 31–72.

Bole, D.G., L.M. Hendershot, and J.F. Kearney. 1986. Post-translational association of immunoglobulin heavy chain binding protein with nascent heavy chains in nonsecreting and secreting hybridomas. *J. Cell Biol.* **102:** 1558–1566.

Bond, U. and M.J. Schlesinger. 1985. Ubiquitin is a heat shock protein in chicken embryo fibroblasts. *Mol. Cell. Biol.* **5:** 949–991.

Chappell, T.G., W.J. Welch, D.M. Schlossman, K.B. Palter, M.J. Schlesinger, and J.E. Rothman. 1986. Uncoating ATPase is a member of the 70kDa family of stress proteins. *Cell* **45:** 3–13.

Cheney, C.M. and A. Shearn. 1983. Developmental regulation of *Drosophila* imaginal disc proteins: Synthesis of a heat shock protein under non-heat shock conditions. *Dev. Biol.* **95:** 325–330.

Chirico, W.J., M.G. Waters, and G. Blobel. 1988. 70K heat shock related proteins stimulate protein translocation into microsomes. *Nature* **332:** 805–810.

Cohen, R.S. and M. Meselson. 1988. Periodic interactions of heat shock transcription elements. *Nature* **332:** 856–858.

Collins, P.L., L.E. Hightower, and L.A. Ball. 1978. Transcription and translation of Newcastle disease virus mRNA's in vitro. *J. Virol.* **28:** 324–336.

Craig, E.A. 1985. The heat shock response. *CRC Crit. Rev. Biochem.* **18:** 239–280.

Craig, E.A., T.D. Ingolia, and L.J. Manseau. 1983. Expression of *Drosophila* heat shock cognate genes during heat shock and development. *Dev. Biol.* **99:** 418–426.

Deshaies, R.J., B.D. Koch, M. Werner-Washburne, E.A. Craig, and R. Schekman. 1988. A subfamily of stress proteins facilitates translocation of secretory and mitochondrial precursor polypeptides. *Nature* **332:** 800–805.

DiDomenico, B.J., G.E. Bugaisky, and S. Lindquist. 1982. The heat shock response is self regulated at both the transcriptional and post-trancriptional level. *Cell* **31:** 593–603.

Ferris, D.K., A. Harel-Bellan, R.I. Morimoto, W. Welch, amd W.L. Farrar. 1988. Mitogen and lymphokine stimulation of heat shock proteins in T lymphocytes. *Proc. Natl. Acad. Sci.* **85:** 3850–3854.

Gething, M.-J., K. McCammon, and J. Sambrook. 1986. Expression of wild-type and mutant forms of influenza hemagglutinin: The role of folding in intracellular transport. *Cell* **46:** 939–950.

Goff, S.A. and A.L. Goldberg. 1985. Production of abnormal proteins in *E. coli* stimulate transcription of *lon* and other heat shock genes. *Cell* **41:** 587–595.

Goldenberg, G., Y. Luo, M. Fenna, R. Baler, R. Weinmann, and R. Voellmy. 1988. Purified human factor activates heat-shock promoter in a HeLa cell-free transcription system. *J. Biol. Chem.* **263:** 19734–19739.

Graziosi, G., F. Micali, R. Marzari, F. deCristini, and A. Savoini. 1980. Variability of response of early *Drosophila* embryos to heat shock. *J. Exp. Zool.* **214:** 141–145.

Greene, J.M., Z. Larin, I.C.A. Taylor, H. Prentice, K.A. Gwinn, and R.E. Kingston. 1987. Multiple basal elements of a human hsp70 promoter function differently in human and rodent cell lines. *Mol. Cell. Biol.* **7:** 3646–3655.

Haas, I.G. and M. Wabl. 1983. Immunoglobulin heavy chain binding protein. *Nature* **306:** 387–389.

Hahnel, A.C., D.J. Gifford, J.J. Heikkila, and G.A. Schultz. 1986. Expression of the major heat shock protein (HSP70) family during early mouse embryo development. *Teratog. Carcinog. Mutagen.* **6:** 493–510.

Harrison, G.S., H.A. Drabkin, F.-T. Kao, J. Hartz, I.M. Hart, E.H.Y. Chu, B.J. Wu, and R.I. Morimoto. 1987. Chromosomal location of human genes encoding major heat-shock protein HSP70. *Somatic Cell Mol. Genet.* **13:** 119–130.

Heikkila, J.J., N. Ovsenek, and P. Krone. 1987. Examination of heat shock protein mRNA accumulation on early *Xenopus laevis* embryos. *Biochem. Cell Biol.* **65:** 87–94.

Heikkila, J.J., M. Kloc, J. Bury, G.A. Schultz, and L.W. Browder. 1985. Acquisition of the heat shock response and thermotolerance during early development of *Xenopus laevis. Dev. Biol.* **107:** 483–489.

Heikkila, J.J., L.W. Browder, L. Gedamu, R.W. Nickells, and G.A. Schultz. 1986. Heat shock gene expression in animal embryonic systems. *Can. J. Genet. Cytol.* **28:** 1093–1105.

Hickey, E.D. and L.A. Weber. 1982. Modulation of heat shock polypeptide synthesis in HeLa cells during hyperthermia and recovery. *Biochemistry* **21:** 1513–1521.

Hightower, L.E. 1980. Cultured animal cells exposed to amino acid analogs or puromycin rapidly synthesize several polypeptides. *J. Cell. Physiol.* **102:** 407–427.

Hinds, P.W., C.A. Finlay, A.B. Frey, and A.J. Levine. 1987. Immunological evidence for the association of p53 with a heat shock protein, hsc 70, in p53-plus-ras-transformed cell lines. *Mol. Cell. Biol.* **7:** 2863–2869.

Holmgren, R., V. Corces, R. Morimoto, R. Blackman, and M. Meselson. 1981. Sequence homologies in the 5′ regions of four *Drosophila* heat-shock genes. *Proc. Natl. Acad. Sci.* **78:** 3775–3778.

Hunt, C. and R.I. Morimoto. 1985. Conserved features of eukaryotic HSP70 genes revealed by comparison with the nucleotide sequence of human HSP70. *Proc. Natl. Acad. Sci.* **82:** 6455–6459.

Ireland, R.C. and E.M. Berger. 1982. Synthesis of low molecular weight heat shock peptides stimulated by ecdysterone in a cultured *Drosophila* cell line. *Proc. Natl. Acad. Sci.* **79:** 855–859.

Ireland, R.C., E.M. Berger, K. Sirotkin, M.A. Yund, D. Osterbur, and J. Fristrom. 1982. Ecdysterone induces the transcription of four heat shock genes in *Drosophila* S3 cells and imaginal discs. *Dev. Biol.* **9:** 498–507.

Karin, M., A. Haslinger, H. Holtgreve, R.I. Richards, P. Krauter, H.M. Westphal, and M. Beato. 1984. Characterization of DNA sequences through which cadmium and glucocorticoid hormones induce human metallothionein-IIa gene. *Nature 308:* 513–519.

Kelley, P.M. and M.J. Schlesinger. 1978. The effect of amino acid analogues and heat shock on gene expression in chicken embryo fibroblasts. *Cell* **15:** 1277–1286.

Kingston, R.E., T.J. Schuetz, and Z. Larin. 1987. Heat-inducible human factor that binds to a human hsp70 promoter. *Mol. Cell. Biol.* **7:** 1530–1534.

Kothary, R., M.D. Perry, L.A. Moran, and J. Rossant. 1987. Cell-lineage-specific expression of the mouse HSP68 gene during embryogenesis. *Dev. Biol.* **121:** 342–348.

Kothary, R., S. Clapoff, S. Darling, M.D. Perry, L.A. Moran, and J. Rossant. 1989. Inducible expression of an hsp680lacZ hybrid gene in transgenic mice. *Development* **105:** 707–714.

Krawczyk, Z., N. Szymik, and J. Wisniewski. 1987b. Expression of hsp70-related gene in developing and degenerating rat testis. *Mol. Biol. Rep.* **12:** 35–41.

Krawczyk, Z., J. Wisniewski, and E. Biesiada. 1987b. An hsp70-related gene is constitutively highly expressed in testis of rat and mouse. *Mol. Biol. Rep.* **12:** 27–34.

Krone, P.H. and J.J. Heikkila. 1988. Analysis of hsp30, hsp70 and ubiquitin gene expression in *Xenopus laevis* tadpoles. *Development* **103:** 59–67.

Larson, J.S., T.J. Schuetz, and R.E. Kingston. 1988. Activation in vitro of sequence specific DNA binding by a human regulatory factor. *Nature* **335:** 372–375.

Levinson, W., H. Oppermann, and J. Jackson. 1980. Transition series metals and sulfhydryl reagents induce the synthesis of four proteins in eukaryotic cells. *Biochim. Biophys. Acta* **606:** 170–180.

Lewis, M.J. and H.R.B. Pelham. 1985. Involvement of ATP in the nuclear and nucleolar functions of the 70kd heat shock protein. *EMBO J.* **4:** 3137–3143.

Liberek, K., C. Georgopoulos, and M. Zylicz. 1988. The role of *E. coli* DnaK and DnaJ heat shock proteins in the initiation of bacteriophage λ DNA replication. *Proc. Natl. Acad. Sci.* **85:** 6632–6636.

Lindquist, S. 1980. Varying patterns of protein synthesis in *Drosophila* during heat shock: Implications for regulation. *Dev. Biol.* **77:** 463–479.

———. 1981. Regulation of protein synthesis during heat shock. *Nature* **293:** 311–314.

———. 1986. The heat shock response. *Annu. Rev. Biochem.* **55:** 1151–1191.

Lindquist, S. and E. Craig. 1988. The heat shock proteins. *Annu. Rev. Genet.* **22:** 631–677.

Lowe, D.G., W.D. Fulford, and L.A. Moran. 1983. Mouse and *Drosophila* genes encoding the major heat shock protein (Hsp70) are highly conserved. *Mol. Cell. Biol.* **3:** 1540–1543.

Milarski, K. and R.I. Morimoto. 1986. Expression of human HSP70 during the synthetic phase of the cell cycle. *Proc. Natl. Acad. Sci.* **83:** 9517–9521.

———. 1989. Mutational analysis of the human HSP70 protein: Distinct domains for nucleolar localization and ATP-binding. *J. Cell Biol.* (in press).

Milarski, K., W. Welch, and R.I. Morimoto. 1989. Cell cycle-dependent association of HSP70 with specific cellular proteins. *J. Cell Biol.* **108:** 413–423.

Moran, L.A., M. Chawin, M.E. Kennegy, M. Korri, and D.G. Lowe. 1982. The major heat shock protein (hsp70) gene family: Related sequences in mouse, *Drosophila,* and yeast. *Can. J. Biochem. Cell. Biol.* **51:** 488–499.

Morange, M., A. Diu, O. Bensaude, and C. Babinet. 1984. Altered expression of heat shock proteins in embryonal carcinoma and mouse early embryonic cells. *Mol. Cell. Biol.* **4:** 730–735.

Morgan, W.D., G.T. Williams, R.I. Morimoto, J. Greene, R.E. Kingston, and R. Tjian. 1987. Two transcriptional activators. CCAAT-box binding transcription factor and heat shock transcription factor, interact with a human hsp70 gene promoter. *Mol. Cell. Biol.* **7:** 1129–1138.

Morimoto, R. and E. Fodor. 1984. Cell specific expression of heat shock proteins in chicken reticulocytes and lymphocytes. *J. Cell Biol.* **99:** 1316–1323.

Morimoto, R., C. Hunt, S.-Y. Huang, K.L. Berg, and S.S. Banerji. 1986. Organization, nucleotide sequence and transcription of the chicken HSP70 gene. *J. Biol. Chem.* **261:** 12692–12699.

Mosser, D.D., N.G. Theodorakis, and R.I. Morimoto. 1988. Coordinate changes in heat shock element binding activity and HSP70 gene transcription rates in human cells. *Mol. Cell. Biol.* **8:** 4736–4744.

Mues, G., T.Z. Munn, and J.D. Raese. 1986. A human gene family with sequence homology to *Drosophila melanogaster* HSP70 heat shock genes. *J. Biol. Chem.* **261:** 874–877.

Munro, S. and H.R.B. Pelham. 1986. An HSP70-like protein in the ER: Identity with the 78kD glucose-regulated protein and the immunoglobulin heavy chain binding protein. *Cell* **46:** 291–300.

Nevins, J.R. 1982. Induction of the synthesis of a 70,000 dalton mammalian heat shock protein by the adenovirus E1a gene product. *Cell* **29:** 913–919.

Notarianni, E.L. and C.M. Preston. 1982. Activation of cellular stress protein genes by herpes simplex virus temperature-sensitive mutants which overproduce immediate early polypeptides. *Virology* **123:** 113–122.

Parker, C.S. and J. Topol. 1984. A *Drosophila* RNA polymerase II transcription factor binds to the regulatory site of an HSP70 gene. *Cell* **37:** 273–283.

Parsell, D.A. and R.T. Sauer. 1989. Induction of a heat shock-like response by unfolded protein in *Escherichia coli*: Dependent on protein level not protein degradation. *Genes Dev.* **3:** 1226–1232.

Pelham, H.R.B. 1982. A regulatory upstream promoter element in the *Drosophila* HSP70 heat shock gene. *Cell* **30:** 517–528.

———. 1984. HSP70 accelerates the recovery of nucleolar morphology after heat shock. *EMBO J.* **3:** 3095–3100.

———. 1986. Speculations on the functions of the major heat shock and glucose-regulated proteins. *Cell* **46:** 959–961.

Pelham, H.R.B. and M. Bienz. 1982. A synthetic heat shock promoter element confers heat inducibility on the herpes simplex virus thymidine kinase gene. *EMBO J.* **1:** 1473–1477.

Peluso, R.W., R.A. Lamb, and P.W. Choppin. 1978. Infection with paramyxoviruses stimulates synthesis of cellular polypeptides that are also stimulated in cells transformed by Rous sarcoma virus or deprived of glucose. *Proc. Natl. Acad. Sci.* **75:** 6120–6124.

Pinhasi-Kimhi, O., D. Michalovitz, A. Ben-Zeev, and M. Oren. 1986. Specific interaction between p53 cellular tumor antigen and major heat shock proteins. *Nature* **320:** 182–185.

Roccheri, M.C., M.G. DiBernardo, and G. Giudice. 1981. Synthesis of heat shock proteins in developing sea urchins. *Dev. Biol.* **83:** 173–177.

Sakakibara, Y. 1988. The dnaK gene of *Escherichia coli* functions in initiation of chromosome replication. *J. Bacteriol.* **170:** 972–979.

Sargent, C.A., I. Dunhan, J. Trowsdale, and R.D. Campbell. 1989. Human major histocompatibility complex contains genes for the major heat shock protein HSP70. *Proc. Natl. Acad. Sci.* **86:** 1968–1972.

Shaefer, E.L., R.I. Morimoto, N.G. Theodorakis, and J. Seidenfeld. 1988. Chemical specificity for induction of stress response genes by DNA-damaging drugs in human adenocarcinoma cells. *Carcinogenesis* **9:** 1733–1738.

Shibahara, S., R.M. Muller, and H. Taguchi. 1987. Transcriptional control of rat heme oxygenase by heat shock. *J. Biol. Chem.* **262:** 12889–12892.

Shuey, D.J. and C.S. Parker. 1986. Binding of *Drosophila* heat shock gene transcription factor to the HSP70 promoter. *J. Biol. Chem.* **261:** 7934–7940.

Singh, M.K. and J. Yu. 1983. Accumulation of a heat shock-like protein during differentiation of human erythroid cell line K562. *Nature* **309:** 631–633.

Sorger, P.K. and H.R.B. Pelham. 1988. Purification and characterization of a heat shock element binding protein from yeast. *EMBO J.* **6:** 3035–3041.

Sorger, P.K., M.J. Lewis, and H.R.B. Pelham. 1987. Heat shock factor is regulated differently in yeast and HeLa cells. *Nature* **329:** 81–84.

Stuart, G.W., P.F. Searle, H.Y. Chen, R.L. Brinster, and R.D. Palmiter. 1984. A 12-base-pair DNA motif that is repeated several times in metallothionein gene promoters confers metal regulation to a heterologous gene. *Proc. Natl. Acad. Sci.* **81:** 7318–7322.

Subjeck, J.R. and T.T. Shyy. 1986. Stress protein systems of mammalian cells. *Am. J. Physiol.* **19:** C1–C17.

Theodorakis, N.G. and R.I. Morimoto. 1987. Posttranscriptional regulation of hsp70 expression in human cells: Effects of heat shock, inhibition of protein synthesis, and adenovirus infection on translation and mRNA stability. *Mol. Cell. Biol.* **7:** 4357–4368.

Theodorakis, N.G., S.S. Banerji, and R.I. Morimoto. 1988. HSP70 mRNA translation in chicken reticulocytes is regulated at the level of elongation. *J. Biol. Chem.* **263:** 14579–14585.

Theodorakis, N.G., D.J. Zand, P.T. Kotzbauer, G.T. Williams, and R.I. Morimoto. 1989. Hemin induced transcriptional activation of the HSP70 gene during erythroid maturation in K562 cells is due to a heat shock factor mediated stress response. *Mol. Cell. Biol.* **9:** 3166–3173.

Thomas, G.P., W.J. Welch, M.B. Mathews, and J.R. Feramisco. 1981. Molecular and cellular effects of heat-shock and related treatments of mammalian tissue-culture cells. *Cold Spring Harbor Symp. Mol. Biol.* **45:** 985–996.

Topol, J., D.M. Ruden, and C.S. Parker. 1985. Sequences required for in vitro transcriptional activation of a *Drosophila* HSP70 gene. *Cell* **42:** 527–537.

Ungewickell, E. 1985. The 70-kd mammalian heat shock proteins are structurally and functionally related to the uncoating protein that releases clathrin triskelia from coated vesicles. *EMBO J.* **4:** 3385–3391.

Velazquez, J.M., B.J. DiDomenico, and S. Lindquist. 1980. Intracellular localization of heat shock proteins in *Drosophila. Cell* **20:** 679–689.

Voellmy, R., A. Ahmed, P. Schiller, P. Bromley, and D. Rungger. 1985. Isolation and functional analysis of a human 70,000 dalton heat shock protein gene segment. *Proc. Natl. Acad. Sci.* **82:** 4949–4953.

Watowich, S.S. and R.I. Morimoto. 1988. Complex regulation of heat shock and glucose responsive genes in human cells. *Mol. Cell. Biol.* **8:** 393–405.

Welch, W.J. and J.R. Feramisco. 1984. Nuclear and nucleolar localization of the 72000-dalton heat shock protein in heat-shocked mammalian cells. *J. Biol. Chem.* **259:** 4501–4513.

———. 1985. Rapid purification of mammalian 70,000-dalton stress proteins: Affinity of the proteins for nucleotides. *Mol. Cell. Biol.* **5:** 1229–1237.

Welch, W.J. and L.A. Mizzen. 1988. Characterization of the thermotolerant cell. II. Effects on the intracellular distribution of heat shock protein 70, intermediate filaments and small nuclear ribonucleoprotein complexes. *J. Cell Biol.* **106:** 1117–1130.

Welch, W.J. and J.P. Suhan. 1986. Cellular and biochemical events in mammalian cells during and after recovery from physiological stress. *J. Cell. Biol.* **103:** 2035–2052.

Wiederrecht, G., D.J. Shuey, W.A. Kibbe, and C.S. Parker. 1987. The *Saccharomyces* and *Drosophila* heat shock transcription factors are identical in size and DNA binding properties. *Cell* **48:** 507–515.

Williams, G.T., T.K. McClanahan, and R.I. Morimoto. 1989. E1A-transactivation of the human HSP70 promoter is mediated through the basal transcription complex. *Mol. Cell. Biol.* **9:** 2574–2587.

Wittig, S., S. Hensse, C. Keitel, C. Elsner, and B. Wittig. 1983. Heat shock gene expression is regulated during teratocarcinoma cell differentiation and early embryonic development. *Dev. Biol.* **96:** 507–514.

Wu, B.J. and R.I. Morimoto. 1985. Transcription of the human hsp70 gene is induced by serum stimulation. *Proc. Natl. Acad. Sci.* **82:** 6070–6074.

Wu, B.J., C. Hunt, and R.I. Morimoto. 1985. Structure and expression of the human gene encoding major heat shock protein HSP70. *Mol. Cell. Biol.* **5:** 330–341.

Wu, B.J., R.E. Kingston, and Morimoto, R.I. 1986a. Human HSP70 promoter contains at least two distinct regulatory domains. *Proc. Natl. Acad. Sci.* **83:** 629–633.

Wu, B.J., G.T. Williams, and R.I. Morimoto. 1987. Detection of three protein binding sites in the serum-regulated promoter of the human gene encoding the 70-kDa heat shock protein. *Proc. Natl. Acad. Sci.* **84:** 2203–2207.

Wu, B.J., H.C. Hurst, N.C. Jones, and R.I. Morimoto. 1986b. The E1A 13S product of adenovirus 5 activates transcription of the cellular human HSP70 gene. *Mol. Cell. Biol.* **6:** 2994–2999.

Wu, C. 1984. Activating protein factor binds *in vitro* to upstream control sequences in heat shock gene chromatin. *Nature* **311:** 81–84.

Wu, C., S. Wilson, B. Walker, I. Dawid, T. Paisley, V. Zimarino, and H. Ueda. 1987. Purification and properties of *Drosophila* heat shock activator protein. *Science* **238:** 1247–1253.

Xiao, H. and J.T. Lis. 1988. Germline transformation used to define key features of heat-shock response elements. *Science* **239:** 1139–1142.

Yoshida, T., P. Biro, T. Cohen, R.M. Muller, and S. Shibahara. 1987. Human heme oxygenase cDNA and induction of its mRNA by hemin. *Eur. J. Biochem.* **171:** 457–461.

Zakeri, Z. and D.J. Wolgemuth. 1987. Developmental-stage-specific expression of the HSP70 gene family during differentation of the mammalian male germ line. *Mol. Cell. Biol.* **7:** 1791–1796.

Zakeri, Z., D.J. Wolgemuth, and C.R. Hunt. 1988. Identification and sequence analysis of a new member of the mouse HSP70 gene family and characterization of its unique cellular and developmental pattern of expression in the male germ line. *Mol. Cell. Biol.* **8:** 2925–2932.

Zimarino, V. and C. Wu. 1987. Induction of sequence-specific binding of *Drosophila* heat shock activator protein without protein synthesis. *Nature* **327:** 727–730.

Zimmerman, J.L., W. Petri, and M. Meselson. 1983. Accumulation of a specific subset of *D. melanogaster* heat shock mRNAs in normal development without heat shock. *Cell* **32:** 1161–1170.

Zylicz, M., D. Ang, and C. Georgopoulos. 1987. The grpE protein of *Escherichia coli*. *J. Biol. Chem.* **262:** 17437–17442.

Zylicz, M., J.H. LeBowitz, R. McMacken, and C. Georgopoulos. 1983. The dnaK protein of *Escherichia coli* possesses an ATPase and autophosphorylating activity and is essential in an *in vitro* DNA replication system. *Proc. Natl. Acad. Sci.* **80:** 6431–6435.

———. 1984. The *dnaK* protein of *Escherichia coli* possesses an ATPase and autophosphorylation activity and is essential in an *in vitro* DNA replication system. *Proc. Natl. Acad. Sci.* **80:** 6431–6435.

15

Developmental Expression of the Heat Shock Genes in *Drosophila melanogaster*

Daniel Pauli[1] and Alfred Tissières
Department of Molecular Biology
University of Geneva
1211 Geneva 4, Switzerland

I. INTRODUCTION

Soon after the cloning of the genes encoding the major heat shock proteins (hsps) of *Drosophila melanogaster* in the late 1970s and early 1980s, it was realized that their expression is not restricted only to conditions of stress. This notion arose from at least three different approaches. First, the availability of DNA probes (O'Connor and Lis 1981) and the production of a specific antibody (Chomyn and Mitchell 1982) confirmed what was already suspected from previous studies (e.g., Mirault et al. 1978; Lindquist 1980; Storti et al. 1980): The hsp83 gene is transcribed and translated in appreciable amounts at normal temperature both in tissue culture cells and in several tissues during the fly development. Second, even if the major hsp, hsp70, is not detectable in uninduced cells, at least three highly homologous genes are expressed in flies at normal temperatures (Ingolia and Craig 1982a; Craig et al. 1983). Finally, a study designed to isolate genomic sequences preferentially transcribed in the beginning of the pupal period led to the recloning of the cluster of genes encoding the low-molecular-weight hsps (Sirotkin and Davidson 1982). It is obvious from these instances that most of the heat

[1]Present address: Department of Genetics, School of Medicine, Case Western Reserve University, Cleveland, Ohio 44106.

Stress Proteins in Biology and Medicine.

shock (hs) genes of *D. melanogaster* studied so far are not only stress-inducible,but are also expressed at normal temperatures at least in some tissues and during specific periods of development. This expression of the hs genes in the absence of stress is the subject of this chapter. We first review the structure and the developmental expression of hs genes of *D. melanogaster*. We then discuss the regulation of this expression. We conclude with examples selected from different organisms showing that the fruit fly is not at all an exceptional case.

II. STRUCTURE AND EXPRESSION

A. hsp83

The 83–84-kD hsp is encoded in *D. melanogaster* by a single gene per haploid genome, localized at the chromosomal site 63BC (Holmgren et al. 1979; O'Connor and Lis 1981). The transcription unit is interrupted by one intron, of variable size depending on the fly strain, separating the 5′ untranslated leader from the polypeptide-coding region. The inhibition of the splicing of this intron at high hs temperatures (Yost and Lindquist 1986) accounts for the relatively low optimum (33°C) of expression of hsp83 (Lindquist 1980).

Before DNA probes became available to study the expression of the hsp83 gene, the protein was already thought to be strongly synthesized at normal temperatures (Mirault et al. 1978; Lindquist 1980). This was confirmed by the finding that the hsp83 mRNA is abundant in uninduced tissue culture cells (O'Connor and Lis 1981), essentially as a poly(A)$^-$ RNA (Storti et al. 1980). A hs of 1 hour at 37°C resulted in a 10-fold increase in the expression of the hsp83 gene, measured by in vivo labeling, compared to that at 23°C. However, this resulted in an increase of only 1.2-fold in the total amount of hsp83 mRNA and a 6-fold increase in the abundance of poly(A)$^+$ mRNA (O'Connor and Lis 1981). On the other hand, the rate of transcription measured by nuclear run-on assays was found to increase about 4-fold after a 15-minute hs at 35°C, but only 2-fold at 37°C (J. Vazquez, pers. comm.). A short hs period, therefore, does not strongly modify the amount of hsp83 transcripts but rather shifts them from the poly(A)$^-$ to the poly(A)$^+$ fraction. Such a shift probably increases the efficiency of translation. This phenomenon is seen with several different cell lines (Mirault et al. 1978; Lindquist 1980; Buzin and Petersen 1982). It is probably not valid for whole flies where the basal uninduced mRNA levels appear to be much lower than those in tissue culture cells and where the induction of total hsp83 mRNA might reach 20-fold (cited in O'Connor and Lis 1981).

Mason et al. (1984) have studied the accumulation of the major *D.*

melanogaster hs transcripts at several stages of normal development. They found a relatively high amount of hsp83 mRNA at normal temperatures in all the examined stages. These levels correspond to one half to one third of the abundance of the transcripts encoding the cytoplasmic actin, except in adults where the actin messages are much less abundant. Unfortunately, this analysis concerned only poly(A)$^+$ mRNA, and the absolute abundance of the other hs transcripts was not determined. In any case, these experiments indicate that the levels of hsp83 mRNA do not vary by more than 3-fold during fly development. In a qualitative study, the hsp83 gene was found to be transcribed in the adult ovaries, specifically in the nurse cells, derived from the germ line, but not in the follicle cells of somatic origin (Zimmerman et al. 1983). The transcription starts early in the egg chamber development well before the vitellogenic stages during which most of the growth of the oocyte takes place. The transcripts synthesized by the nurse cells are transported and stored in the maturing oocytes. They are still present during the first 2–3 hours of embryonic development. These observations corroborate an earlier finding indicating that hsp83 is made in embryos at normal temperatures (Graziosi et al. 1980).

Even if the hsp83 mRNA is abundant and translated in uninduced cells, this does not guarantee that the protein is also abundant, because it could be very unstable at normal temperatures. Therefore, the production of specific antibodies was important in assessing the steady-state levels of the polypeptide. One step in this direction was achieved by Chomyn and Mitchell (1982): By immunoprecipitation of proteins labeled at normal temperature in excised tissues, they found hsp83 in the salivary glands and brains of larvae and prepupae, in the thoracic epithelium of young pupae, and possibly in the ovaries of adults. Unfortunately, this study was not pursued further to investigate the steady-state level of the hsp83 polypeptide throughout normal fly development and in tissue culture cells or to determine precisely the localization of the protein by immunohistological methods.

A reasonable conclusion of these different analyses is that the hsp83 gene is strongly expressed in unstressed *D. melanogaster* tissue culture cells in which hsp83 is an abundant constitutive protein. The situation seems more complex if we consider the whole organism and its development. In some tissues, the hsp83 protein is probably quite abundant (differentiated female germ cells, salivary glands, and brains of third-instar larvae and prepupae), whereas the gene is not active in others (for instance, in the follicle cells). Finally, the results of Mason et al. (1984) suggest that many other tissues during other stages of development also express the hsp83 gene.

B. The hsp70 Family

The major hsp, hsp70, is encoded in *D. melanogaster* by several very homologous genes that have been found by genetic analysis at two chromosomal sites (Ish-Horowicz et al. 1979b). Two hsp70 genes are localized in opposite orientation at locus 87A7 (Mirault et al. 1979; Goldschmidt-Clermont 1980; Ish-Horowicz and Pinchin 1980). The 87C1 locus contains three to five copies of the hsp70 genes, depending on the fly stock or tissue culture cell (Craig et al. 1979; Ish-Horowicz et al. 1979a; Mirault et al. 1979). Restriction sites and partial sequence analyses indicate that the transcribed regions are almost or totally identical between copies in the same locus, suggesting a high rate of intralocus gene conversion (Brown and Ish-Horowicz 1981; Mason et al. 1982). Sequence comparison between copies from each locus showed a 4.2% nucleotide divergence within the coding region (Karch et al. 1981). Even so, most of these substitutions are silent, resulting in only 16 amino acid changes (2.5%) and 2 single amino acid residue insertions in the 87A7 copy. Another hsp, hsp68, encoded by a single gene at chromosomal site 95D, turned out to be extremely homologous to hsp70, the melting of heteroduplexes indicating about 15% divergence at the DNA sequence level (Holmgren et al. 1979). Since the differences might be concentrated in the transcribed but nontranslated 5′ and 3′ regions or be silent substitutions, the actual similarity between hsp68 and hsp70 may be greater than 90%. (Surprisingly, Craig et al. [1982] reported a DNA sequence homology of only 75%, which might be due to the small part of the gene so far sequenced.)

Early studies have given contradictory results concerning the expression of the hsp70 genes at normal temperatures (see, e.g., McKenzie et al. 1975; Mirault et al. 1978). Nevertheless, it seems that their uninduced expression is indeed negligible: The amount of hsp70 mRNA in nonstressed tissue culture cells was found to be less than 1/1000th the level achieved after an induction of 1 hour at 37°C, and the difference may be even larger for the protein level (Velasquez et al. 1983). The basal level is strongly increased by some culture conditions (for instance, large volume of media leading to partial anoxia) or particular experimental manipulations such as vigorous pipetting or too-hard centrifugation (cited in Velasquez et al. 1983). The question of the expression of hsp70 and hsp68 genes during fly development is more complex. Mason et al. (1984) detected only very low levels of their mRNAs during development. These low levels may be due to inadvertant stresses during sample collection. A small peak that we estimate to be roughly 10% of that of the hsp83 transcript was also found during the pupal period. In this case, the explanation might be that in pupae, extensive organ reorganization occurs, including the lysis of many tissues that may induce a stress response.

Besides these genes that seem strictly expressed in stress conditions, the hsp70 family of *D. melanogaster* contains at least three other genes called heat shock cognates, hsc1, 2, and 4 (Ingolia and Craig 1982a; Craig et al. 1983). They are localized at chromosomal sites 70C, 87D (near the rosy or xanthine dehydrogenase gene), and 88E, respectively, which are not loci of heat shock puffs. The sequences of the 5′ end of these three genes have been reported (Craig et al. 1983), allowing a comparison with hsp70 along about 15% of the proteins at their amino terminus. No amino acid insertion or deletion has been found in that region. The three hscs, like several mammalian 70K proteins, possess three additional residues at their amino terminus compared to the *D. melanogaster* hsp70. Table 1 shows that the identity between any two hsc or any hsc and hsp70 averages 80%. The best conservation is between hsc4 and either hsc1 or the 87A7 hsp70, the lowest being between hsc1 and hsc2. At least for hsc4, the homology seems even slightly better with mammalian 70K proteins either heat-inducible or constitutive. Nevertheless, the confirmation of this observation would require the complete sequencing of the *D. melanogaster* cognate genes.

In the portion of the gene analyzed so far, both hsc1 and hsc2 have been found to contain an intron (Ingolia and Craig 1982a; Craig et al.

Table 1 Percentage of homology between the hsp70 proteins and the heat shock cognates along their 110 amino-terminal amino acids

	hsp70 (87C1)	hsc1	hsc2	hsc4	human hsp70	hsc71–73	testis hsc70
hsp70 (87A7)	97	77	79	83	80	79	82
hsp70 (87C1)		75	78	81	79	78	80
hsc1 (70C)			74	83	80	83	83
hsc2 (87D)				78	77	76	74
hsc4 (88E)					83	85	85
human hsp70						86	85
hsc71–73							87

D. melanogaster proteins: The sequence of the hsp70 encoded at locus 87A7 is from Karch et al. (1981); those of hsp70 at locus 87C1, hscl, hsc2, and hsc4 are from Craig et al. (1982).

Mammalian proteins: The sequence of the human heat-inducible hsp70 is from Hunt and Morimoto (1985); hsc71–73 is a very abundant constitutive and noninducible polypeptide of the 70K family that is identical (except for one amino acid outside the region considered here) in rat (O'Malley et al. 1985; Sorger and Pelham 1987), mouse (Giebel et al. 1988), and human (Dwornic-zak and Mirault 1987); the testis hsc70 is constitutively expressed in the mouse testis (Zakeri et al. 1988).

1983). The intervening sequence of hsc1 deserves two remarks. First, its position is identical in *D. melanogaster* and *D. simulans* (Ingolia and Craig 1982a), in the rat hsc73 gene (Sorger and Pelham 1987), and in the human hsc71 (Dworniczak and Mirault 1987). Second, the *D. melanogaster* and *D. simulans* introns are rather well conserved. Indeed, hsc1 has been isolated in two steps. The first was the cloning of the *D. simulans* gene by homology with a *D. melanogaster* hsp70, and the second was the isolation of the *D. melanogaster* cognate gene by hybridization with the *D. simulans* intron (Ingolia and Craig 1982a).

The term cognate has been used to emphasize two properties of these genes: their homology to the 70K hsps and the fact that they are expressed at normal temperatures and are not further induced by stress. However, Palter et al. (1986) have reported that a hs of 1 hour at 37°C augments about twofold the amount of hsc4 mRNA in adult males. The significance of this increase is not understood, since the amount of the protein encoded by hsc4 (hsc70) does not change even after 24 hours at 37°C. One possibility is that the promoter of the hsc4 gene possesses heat shock regulatory elements (HSE) and is therefore a true hs gene. In view of its extremely weak induction, we would suggest an alternative explanation. In contrast with the promoters of many genes (for instance that of the cytoplasmic actin), the hsc4 promoter may not be repressed at high temperatures. If this is the case, and since the rate of transcription at 37°C is certainly at least twice that seen at 25°C, the small increase in the amount of transcript would not be due to a positive activation of the transcription of the gene. This explanation would apply to many mRNAs whose amounts are only slightly increased by hs. It should be easily testable by the analysis of the effect of other stresses that do not involve an increase in temperature. As shown by nuclear run-on assays, the transcription of hsc4 is almost unaffected by hs, except for a small increase at 35°C (J. Vazquez, pers. comm.).

The three identified hsp70 cognates are expressed at very different levels during the stages of *D. melanogaster* development that have been analyzed (Craig et al. 1983). The hsc4 message has been found at a rather constant level in embryos, larvae, and adults. hsc1 and hsc2 transcripts have been found only in adults and are 30–70 times less abundant than hsc4 mRNA, with the hsc1 mRNA being two- to threefold more abundant than that of hsc2. Consistent with the RNA analysis, the hsc4 protein is extremely abundant during development, especially during embryogenesis (Palter et al. 1986). Even with maximal stress induction, hsp70 never becomes as abundant as this hsc polypeptide. Further work is needed to identify all the hsp70 cognates of *D. melanogaster* and to determine their temporal and spatial pattern of expression.

C. Locus 67B and the Low-molecular-weight hsps

Four hsps of low molecular weights (22, 23, 26, and 27 kD) have been identified in *D. melanogaster* (McKenzie et al. 1975; Mirault et al. 1978). Each is encoded by a single gene, and the four genes are clustered within 12 kb of DNA at locus 67B (Petersen et al. 1979; Corces et al. 1980; Craig and McCarthy 1980; Wadsworth et al. 1980; Voellmy et al. 1981). The same segment of DNA contains three other genes that were first identified by their expression at some specific stages of development (Sirotkin and Davidson 1982). Since their protein products have not been identified, they have been named gene 1, 2, and 3 (Southgate et al. 1985). Subsequently, it was shown that these three genes are also hs genes: The abundance of their transcripts is indeed increased by a hs (Ayme and Tissières 1985). We note that these developmentally regulated genes are weakly expressed at high temperatures: For instance, the induction of gene 2 is at least an order of magnitude lower than that of the hsp22 transcription unit (Pauli et al. 1988). Figure 1 summarizes the organization of the seven hs genes of locus 67B.

Structural and sequence analyses have shown that the seven genes can be separated into two families (Ingolia and Craig 1982b; Southgate et al. 1983; Ayme and Tissières 1985; Pauli and Tonka 1987; Pauli et al. 1988a). One family includes the four small hs genes, gene 1, and gene 3. Gene 2 is the only member of the second group. Each gene of the first family possesses a single site of initiation and produces a single major message, both during development and hs (gene 1 is an exception with two mRNAs differing by the length of their 3′ untranslated region; J. Vazquez, pers. comm.). Following hs, some minor transcripts that represent downstream extension of the major mRNA have been found for hsp23, 26, 27 (Berger et al. 1985) and gene 3 (Pauli and Tonka 1987). None of the six genes is interrupted by an intron. An additional major characteristic of the six genes is the sequence homology present in two domains of their coding region (Fig. 2). The first domain, comprising the 15 amino-terminal amino acids, is very hydrophobic and might allow

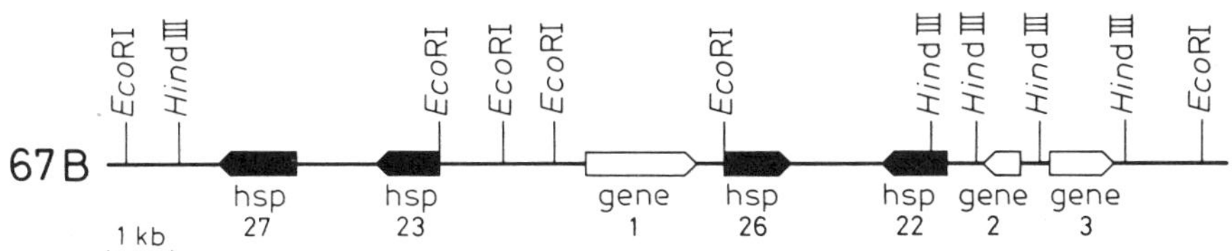

Figure 1 Organization of the locus 67B. Filled arrows represent the direction of transcription of the low-molecular-weight hsp genes. Open arrows show genes 1, 2, and 3.

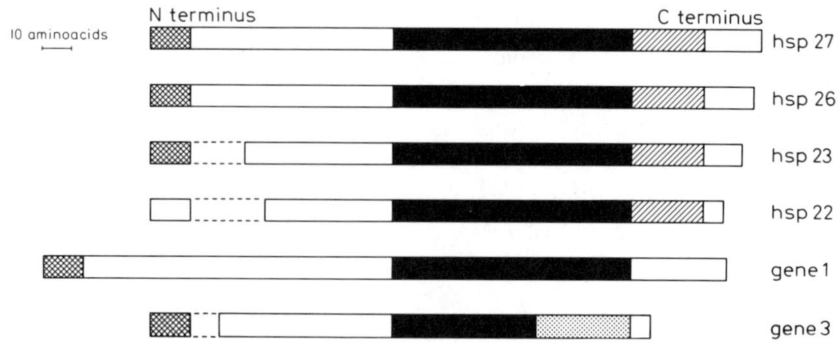

Figure 2 Homologies between six hsp polypeptides encoded at locus 67B. Hatched boxes represent the amino-terminal hydrophobic domain, black boxes represent the α-crystallin homology, and dashed boxes represent the similarity found only in hsp22, 23, 26, and 27. The "crystallin domain" is shorter in gene 3, as indicated by the gray box.

some interactions with membranes (Arrigo and Pauli 1988). The second conserved region in the second half of the polypeptides contains about 80 residues. Ingolia and Craig (1982b) have shown that this domain is homologous to a carboxy-terminal region of a mammalian family of polypeptides, the α-crystallins, which are the major structural proteins of the lens. For gene 3, the homology to α-crystallin is restricted to the first 50 amino acids. In that subregion, the conservation ranges from 40% between hsp22 and gene 3 to more than 70% between hsp23, 26, and 27 (Pauli and Tonka 1987). Immediately following the crystallin domain, the four low-molecular-weight hsps have an additional conserved region of 25 residues that is absent from the gene 1 and 3 putative polypeptides. These overall amino acid homologies strongly suggest that these six genes derive from a common ancestral gene.

The seventh transcription unit of locus 67B, gene 2, exhibits completely different characteristics (Pauli et al. 1988). It produces three classes of mRNAs starting at two different initiation sites. The upstream promoter seems to be used only in the male germ line and directs the synthesis of a 780-nucleotide transcript. The other initiation site localized 200 bp downstream is common to two mRNAs. One, totaling 560 nucleotides, is expressed at several stages of development and is induced by a mild heat shock, in the range of 30–32°C (D. Pauli, unpubl.). This transcript terminates at two close polyadenylation sites that are also used by the 780-nucleotide mRNA. These termination sites are not used at high temperatures (35–36°C) and, therefore, the 560-nucleotide transcript is extended to a 2000-nucleotide mRNA. The latter is a bicistronic message that contains, in addition, the transcription unit downstream of gene 2, i.e., that of the hsp22 gene. Two different possibilities can explain this transcriptional readthrough. First, it could be due to the expression of hsp22. Indeed, the two genes are very close to each other (about 500 bp separate the "normal" end of gene 2 and the initiation site of hsp22) and, thus, the promoter sequences of hsp22 may overlap with the termination sequence of gene 2. When hsp22 is quite active, and this is the case only during a strong hs, the binding of some transcription factors to its promoter may mask the termination site of gene 2, thereby leading to the readthrough. A similar impairment of the transcription termination by a highly expressed close downstream gene has already been reported for another *D. melanogaster* locus (Vincent et al. 1986). Second, the readthrough could be the result of a direct effect of the stress on the termination per se. In support of this possibility, it seems that the transcription termination of the heat shock genes is quite sensitive to thermal stress (J. Vazquez, in prep.). Besides the complexity of its transcription pattern, gene 2 shows two additional important differences compared to the six other genes of locus 67B. First, it is interrupted by several very small introns. Second, its putative polypeptide, which would be about half the size of hsp27, does not show any similarity with the low-molecular-weight hsps or any known protein.

The first indication that the *D. melanogaster* small hs genes are active in the absence of stress did not come from studies on the hs response but rather from experiments designed to isolate genes expressed at some specific stages of the fly development (Sirotkin and Davidson 1982). Even if this analysis did not define precisely the timing of the expression and the abundance of the transcripts, it suggested that the mRNAs transcribed from the genes of locus 67B are very abundant at the beginning of the pupal period. This observation has been confirmed and extended in several reports, either at the mRNA level or by the detection of the individual polypeptides. Zimmerman et al. (1983) have found that the hsp26 and hsp27 genes are transcribed in the ovarian nurse cells (germ-line origin) and then transported and stored in the maturing oocyte. The expression starts at stage 8 or 9 of oogenesis, later than the hsp83 gene transcription. The hsp83, 27, and 26 mRNAs can be detected in preblastoderm embryos (2–3 hr of development). The very strong transcriptional activity of the hsp26 and 27 genes in the female germ line has been confirmed by the analysis of the polyadenylated mRNA extracted from whole organisms: Although the hsp26 and 27 transcripts are almost undetectable in male adults, they are very abundant in mature females (Mason et al. 1984). This study also confirmed the presence of these transcripts in young embryos. We have recently shown that the hsp23 mRNA is also present during the first part of the embryogenesis (Pauli et al. 1989). Two other genes of locus 67B are expressed in the middle of the embryogenesis, i.e., gene 2 (Pauli et al. 1988) and gene 3 (Pauli and Tonka 1987).

The beginning of the pupal period is the only stage of the *D. melanogaster* development during which the seven hs genes of locus 67B are expressed (Mason et al. 1984: Ayme and Tissières 1985; Pauli and Tonka 1987; Pauli et al. 1988). Nevertheless, the amount of their transcripts differs widely, ranging from the high levels of the hsp23 message, which is as abundant as the mRNA encoding the cytoplasmic actin, to the low levels of the hsp22 transcript, which is about 100-fold less abundant (Mason et al. 1984; Pauli et al. 1989).

One important characteristic of the expression of the small hs genes during normal fly development is that it is restricted to particular tissues. This is in contrast to the stress-induced expression that occurs in almost any cell except differentiated gametes and preblastoderm embryos (Dura 1981; Zimmerman et al. 1983; Bonner et al. 1984). The hsp26 gene has been the most extensively characterized, and its transcript has been found at different developmental stages in the male and the female germ lines, the epithelium, the proventriculus, the imaginal discs, and the central nervous system (Glaser et al. 1986). We have shown that gene 2 is also active in the male germ line, where it produces an mRNA larger than that found during the embryogenesis and at the beginning of the pupal period (Pauli et al. 1988). The expression of both genes appears rather early in the male germ-line differentiation, since their transcripts are already detectable in larvae.

The synthesis of the *D. melanogaster* low-molecular-weight hsps in the absence of stress has been analyzed in two ways. First, in vivo labeling experiments showed that imaginal discs isolated from late third instar larvae synthesize hsp23, whereas those isolated from mid-third instar larvae do not, indicating a regulated expression throughout development (Cheney and Shearn 1983). Second, antibodies specific for hsp23 (Arrigo 1987) or hsp27 (Arrigo and Pauli 1988) allowed the determination of the steady-state levels of the small hsps. The major conclusion from these studies is that the low-molecular-weight hsps are more stable than their mRNA and are present during the whole pupal period for hsp23 and during the whole embryogenesis and the whole pupal period for hsp27 (Pauli et al. 1989). We have used antibodies against hsp27 to determine the localization of this polypeptide during fly development. Although hsp27 is uniformly distributed in the embryos, it is restricted to specific larval and adult tissues: the male and the female germ lines, some neurocytes of the central nervous system, and the imaginal discs of late third instar larvae and early pupae (D. Pauli et al., in prep.).

In summary, the seven hs genes of *D. melanogaster* locus 67B constitute two families of homologous genes. They show a complex pattern of temporal and spatial expression during normal fly development. Except perhaps for the pupal expression (see below), they appear not to be coordinately regulated in the absence of stress.

III. REGULATION

Almost at the same time as Sirotkin and Davidson (1982) showed that the small hs genes of locus 67B were expressed during normal *D. melanogaster* development, Ireland and Berger (1982) proposed a possible regulatory mechanism for this non-stress-induced expression. They observed that treatment of *Drosophila* tissue culture cells with the steroid molting hormone β-ecdysone induced the synthesis of the low-molecular-weight hsps. A similar induction was found by hormone treatment of imaginal discs cultivated in vitro (Ireland et al. 1982). These results led to the hypothesis that the developmental expression of the small hs genes is regulated by ecdysteroids. The idea has been strengthened by two studies. First, Cheney and Shearn (1983) showed that imaginal discs isolated from early third instar larvae do not synthesize the low-molecular-weight hsps, but discs isolated from older larvae, in which the concentration of β-ecdysone is higher, do synthesize hsp23. Second, Thomas and Lengyel (1986) showed that the peaks of hsp83, 27, and 26 mRNA levels, as determined by Northern blots, seem to correlate with the peaks of accumulation of β-ecdysone throughout most of the fly development. Moreover, the use of a thermosensitive mutant for the synthesis of the ecdysteroid revealed that expression of the small hs genes in the late larval/early pupal stages depends on the presence of the molting hormone.

Transfections of *D. melanogaster* tissue culture cells have been used to define the DNA sequences upstream of the small hs genes necessary for ecdysone induction. Surprisingly, the sequences found are different for hsp23 (Mestril et al. 1986) and hsp27 (Riddihough and Pelham 1986, 1987). The problem is further complicated by the fact that the results obtained by reintroduction of the hsp27 gene into the *D. melanogaster* genome, by P-element transformation, are different from those obtained by transfection (Hoffman et al. 1987). Two phenomena may combine to create these discrepancies. First, the steroid hormone most likely acts not only at the transcriptional level, but also on the stability of the mRNAS (Brock and Shapiro 1983; Vitek and Berger 1984). Second, it was assumed in the above studies that the expression of the small hs genes is either uniform in the whole fly (like the thermal induction) or that the same regulatory elements are used in the few active tissues. Indeed, neither of these two premises is fulfilled. The expression of the small hs genes in the absence of stress is restricted to some tissues. For instance, in late third instar larvae, at the time of the highest concentration of β-ecdysone, the hsp26 (Glaser et al. 1986) and the hsp27 (D. Pauli et al., in prep.) genes are active almost exclusively in the male germ line, in parts of the central nervous system, and in the imaginal discs. Furthermore, the regulatory sequences appear to differ for various tissues: For hsp26, expression in the female germ line depends on sequences 350–500 bp upstream of the cap site of the gene (Cohen and Meselson 1985), whereas male germ-line activity requires the 200 bp in front of the gene (Glaser et al. 1986). There are at least two independent elements capable of specifying the spermatocyte expression, the one located between –352 and –52 bp having the properties of an enhancer (R.L. Glaser and J.T. Lis, pers. comm.). These regulatory elements are different from those involved in stress induction (Cohen and Meselson 1985; Glaser et al. 1986; Pauli et al. 1986), which is also true for the hsp83 gene (Xiao and Lis 1989). In one case, the hsp22 gene, the sequence necessary for early pupal expression seems to coincide with the consensus sequences found for the HSEs (Klemenz and Gehring 1986). Since the hsp22 gene is only very slightly active during development, its expression might be due to mild stresses, for instance, the lysis affecting some tissues during the extensive remodeling occurring during the pupal stage.

The above presentation suggests that the only means to precisely define the DNA sequences involved in the tissue-specific expression of the hs genes during normal development is the reintroduction of these genes into the *D. melanogaster* genome by P-element-mediated transformation. This should provide a detailed picture of the elements that control the activity of the hs genes in the absence of stress.

IV. CONCLUSION

We have reviewed the expression of *D. melanogaster* hs genes in the absence of stress. Our major conclusion is that most of the fruit fly hs genes are indeed active during several developmental stages without any external stress. In the case of the hsp70 genes that are expressed only during stress, there are at least three highly homologous genes active at some stages of *D. melanogaster* development. This suggests that all the hs polypeptides (or some closely related proteins) have a function both during normal conditions and during stress.

A great deal of evidence suggests that this is also true for the mammalian hs proteins. Several mRNAs homologous to the hsp70 genes have been detected in mammals in the absence of hs. The number of these hsp70 cognate genes has not yet been determined, but they are clearly expressed in some specific tissues, in particular, in the male germ line (Krawczyk et al. 1987a,b; Zakeri and Wolgemuth 1987; Allen et al. 1988a,b; Giebel et al. 1988; Zakeri et al. 1988). This is also true for the hsp68 genes (Kothary et al. 1987). Furthermore, several proteins of the hsp70 family appear to be the first major polypeptides derived from the zygotic gene transcription at the two-cell stage of mouse embryogenesis (Bensaude et al. 1983). Some members of the different families of hsps, therefore, seem to be involved in insect as well as mammalian development.

Considering that many of the hs genes or their homologs are active during particular stages of development in some specific tissues, we must ask what roles their products might play. Recent advances on the functions of the hsp70 gene family and also on the hsp83 gene are reviewed in this volume (see Craig; Pelham; Georgopoulos et al.; Gross et al.; Morimoto and Milarski, all this volume). Although we know very little about the roles of the low-molecular-weight hsps, several observations are probably relevant. First, six small hsps have homologies between themselves and with the α-crystallins. They probably possess related functions that have diverged enough to explain their differential expression in different tissues. Second, they share some common interesting biochemical properties. During a hs, they form huge aggregates of more than 2×10^6 daltons inside the cell nuclei, a phenomenon observed from yeast to man (Arrigo et al. 1980; Arrigo and Ahmad-Zadeh 1981; Collier and Schlessinger 1986; Arrigo 1987; Arrigo and Welch 1987; Rossi and Lindquist 1989). The insolubilization of the low-molecular-weight hsps inside the nucleus is proportional to the severity of the hs and inversely related to the degree of thermotolerance (Arrigo 1987; Arrigo et al. 1988), an observation that could be used to measure the thermotolerant state of a cell. At normal temperatures, the small hsps are localized in the cytoplasm or at the nuclear periphery as aggregates of similar size (5×10^5–8×10^5 daltons) to those formed by α-crystallins. The biochemistry of proteins able to form aggregates of different sizes is likely to be very complex: Their aggregation state could modulate their putative enzymatic activities. Considering the homology of the low-molecular-weight hsps and the α-crystallins, it is interesting to note that some important metabolic enzymes might have been recruited throughout evolution as structural proteins of the lens (Wistow and Piatigorsky 1987; Dubin et al. 1989).

ACKNOWLEDGMENTS

We thank J. Vazquez, R.L. Glaser, and J.T. Lis for communicating results prior to publication; A.-P. Arrigo, C. Georgopoulos, and J. Vazquez for critical reading of the manuscript; and O. Jenni and F. Ebener for the preparation of the figures.

REFERENCES

Allen, R.L., D.A. O'Brien, and E.M. Eddy. 1988a. A novel hsp70-like protein (P70) is present in mouse spermatogenic cells. *Mol. Cell. Biol.* **8:** 828–832.

Allen, R.L., D.A. O'Brien, C.C. Jones, D.L. Rockett, and E.M. Eddy. 1988b. Expression of heat shock proteins by isolated mouse spermatogenic cells. *Mol. Cell. Biol.* **8:** 3260–3266.

Arrigo, A.-P. 1987. Cellular localization of hsp23 during *Drosophila* development and following subsequent heat shock. *Dev. Biol.* **122:** 39–48.

Arrigo, A.-P. and C. Ahmad-Zadeh. 1981. Immunofluorescence localization of a small heat shock protein (hsp23) in salivary gland cells of *Drosophila melanogaster*. *Mol. Gen. Genet.* **184:** 73–79.

Arrigo, A.-P. and D. Pauli. 1988. Characterization of hsp27 and three immunologically related polypeptides during *Drosophila* development. *Exp. Cell Res.* **175:** 169–183.

Arrigo, A.-P. and W.J. Welch. 1987. Characterization and purification of the small 28,000-daltons mammalian heat shock protein. *J. Biol. Chem.* **262:** 15359–15369.

Arrigo, A.-P., S. Fakan, and A. Tissières. 1980. Localization of the heat shock induced proteins in *Drosophila melanogaster* tissue culture cells. *Dev. Biol.* **78:** 86–103.

Arrigo, A.-P., J. Suhan, and W.J. Welch. 1988. Dynamic changes in the structure and locale of the mammalian low molecular weight heat shock protein. *Mol. Cell. Biol.* **8:** 5059–5071.

Ayme, A. and A. Tissières. 1985. Locus 67B of *Drosophila melanogaster* contains seven, not four, closely related heat shock genes. *EMBO J.* **4:** 2949–2954.

Bensaude, O., C. Babinet, M. Morange, and F. Jacob. 1983. Heat shock proteins, first major products of zygotic gene activity in mouse embryos. *Nature* **305:** 331–333.

Berger, E.M., M.P. Vitek, and C.M. Morganelli. 1985. Transcript length heterogeneity at the small heat shock genes of *Drosophila*. *J. Mol. Biol.* **186:** 137–148.

Bonner, J.J., C. Parks, J. Parker-Thornburg, M.A. Mortin, and H.R.B. Pelham. 1984. The use of promoter fusions in *Drosophila* genetics: Isolation of mutations affecting the heat shock response. *Cell* **37:** 979–991.

Brock, M.L. and D.J. Shapiro. 1983. Estrogen stabilizes vitellogenin mRNA against cytoplasmic degradation. *Cell* **34:** 207–214.

Brown, A.J.L. and D. Ish-Horowicz. 1981. Evolution of the 87A and 87C heat-shock loci in *Drosophila*. *Nature* **290:** 677–682.

Buzin, C.H. and N.S. Petersen. 1982. A comparison of the multiple *Drosophila* heat shock proteins in cell lines and larval salivary glands by two-dimensional gel electrophoresis. *J. Mol. Biol.* **158:** 181–201.

Cheney, C.M. and A. Shearn. 1983. Developmental regulation of *Drosophila* imaginal disc proteins: Synthesis of a heat shock protein under non-heat-shock conditions. *Dev. Biol.* **95:** 325–330.

Chomyn, A. and H.K. Mitchell. 1982. Synthesis of the 84,000 dalton protein in normal and heat shocked *Drosophila melanogaster* cells as detected by specific antibody. *Insect Biochem.* **12:** 105–114.

Cohen, R.S. and M. Meselson. 1985. Separate regulatory elements for the heat-inducible and ovarian expression of the *Drosophila* hsp26 gene. *Cell* **43:** 737–746.

Collier, N.C. and M. Schlesinger. 1986. The dynamic state of heat shock protein in chicken embryo fibroblasts. *J. Cell. Biol.* **103:** 1495–1507.

Corces, V., R. Holmgren, R. Freund, R. Morimoto, and M. Meselson. 1980. Four heat shock proteins of *Drosophila melanogaster* coded within a 12-kilobase region in chromosome subdivision 67B. *Proc. Natl. Acad. Sci.* **77:** 5390–5393.

Craig, E.A. and B.J. McCarthy. 1980. Four *Drosophila* heat shock genes at 67B: Characterization of recombinant plasmids. *Nucleic Acids Res.* **8:** 4441–4457.

Craig, E.A., T.D. Ingolia, and L.J. Manseau. 1983. Expression of *Drosophila* heat-shock cognate genes during heat shock and development. *Dev. Biol.* **99:** 418–426.

Craig, E.A., B.J. McCarthy, and S.C. Wadsworth. 1979. Sequence organization of two recombinant plasmids containing genes for the major heat shock-induced protein in *D. melanogaster*. *Cell* **16:** 575–588.

Craig, E., T. Ingolia, M. Slater, L. Manseau, and J. Bardwell. 1982. *Drosophila,* yeast, and *E. coli* genes related to the *Drosophila* heat shock genes. In *Heat shock: From bacteria to man* (ed. M. Schlesinger et al.), pp. 11–18. Cold Spring Harbor Laboratory, Cold Spring Harbor, New York.

Dubin, R.A., E.F. Wawrousek, and J. Piatigorsky. 1989. Expression of the murine αB-crystallin gene is not restricted to the lens. *Mol. Cell. Biol.* **9:** 1083–1091.

Dura, J.-M. 1981. Stage dependent synthesis of heat shock induced proteins in early embryos of *Drosophila melanogaster*. *Mol. Gen. Genet.* **184:** 381–385.

Dworniczak, B. and M.-E. Mirault. 1987. Structure and expression of a human gene coding for a 71 kd heat shock "cognate" protein. *Nucleic Acids Res.* **15:** 5181–5197.

Giebel, L.B., B. Dworniczak, and E.K.F. Bautz. 1988. Developmental regulation of a constitutively expressed mouse mRNA encoding a 72-kDa heat shock-like protein. *Dev. Biol.* **125:** 200–207.

Glaser, R.L., M.F. Wolfner, and J.T. Lis. 1986. Spatial and temporal pattern of hsp26 expression during normal development. *EMBO J.* **5:** 747–754.

Goldschmidt-Clermont, M. 1980. Two genes for the major heat-shock protein of *Drosophila melanogaster* arranged as an inverted repeat. *Nucleic Acids Res.* **8:** 235–252.

Graziosi, G., F. Micali, R. Marzari, F. de Cristini, and A. Savoini. 1980. Variability of response of early *Drosophila* embryos to heat shock. *J. Exp. Zool.* **214:** 141–145.

Hoffman, E.P., S.L. Gerring, and V.G. Corces. 1987. The ovarian, ecdysterone, and heat-shock-responsive promoters of the *Drosophila melanogaster* hsp27 gene react very differently to perturbations of DNA sequence. *Mol. Cell. Biol.* **7:** 973–981.

Holmgren, R., K. Livak, R. Morimoto, R. Freund, and M. Meselson. 1979. Studies of cloned sequences from four *Drosophila* heat shock loci. *Cell* **18:** 1359–1370.

Hunt, C. and R. Morimoto. 1985. Conserved features of eukaryotic hsp70 genes revealed by comparison with the nucleotide sequence of human hsp70. *Proc. Natl. Acad. Sci.* **82:** 6455–6459.

Ingolia, T.D. and E.A. Craig. 1982a. *Drosophila* gene related to the major heat shock-induced gene is transcribed at normal temperatures and not induced by heat shock. *Proc. Natl. Acad. Sci.* **79:** 525–529.

———. 1982b. Four small *Drosophila* heat shock proteins are related to each other and to mammalian α-crystallin. *Proc. Natl. Acad. Sci.* **79:** 2360–2364.

Ireland, R.C. and E.M. Berger. 1982. Synthesis of low molecular weight heat shock peptides stimulated by ecdysterone in a cultured *Drosophila* cell line. *Proc. Natl. Acad. Sci.* **79:** 855–859.

Ireland, R.C., E.M. Berger, K. Sirotkin, M.A. Yund, D. Osterbur, and J. Fristrom. 1982. Ecdysterone induces the transcription of four heat-shock genes in *Drosophila* S3 cells and imaginal discs. *Dev. Biol.* **93:** 498–507.

Ish-Horowicz, D. and S.M. Pinchin. 1980. Genomic organization of the 87A7 and 87C1 heat-induced loci of *Drosophila melanogaster. J. Mol. Biol.* **142:** 231–245.

Ish-Horowicz, D., S.M. Pinchin, P. Schedl, S. Artavanis-Tsakonas, and M.-E. Mirault. 1979a. Genetic and molecular analysis of the 87A7 and 87C1 heat-inducible loci of *D. melanogaster. Cell* **18:** 1351–1358.

Ish-Horowicz, D., S.M. Pinchin, J. Gausz, H. Gyurkovics, G. Bencze, M. Goldschmidt-Clermont, and J.J. Holden. 1979b. Deletion mapping of two *D. melanogaster* loci that code for the 70,000 dalton heat-induced protein. *Cell* **17:** 565–571.

Karch, F., I. Török, and A. Tissières. 1981. Extensive regions of homology in front of the two hsp70 heat shock variant genes in *Drosophila melanogaster. J. Mol. Biol.* **148:** 219–230.

Klemenz, R. and W.J. Gehring. 1986. Sequence requirement for expression of the *Drosophila melanogaster* heat shock protein hsp22 gene during heat shock and normal development. *Mol. Cell. Biol.* **6:** 2011–2019.

Kothary, R., M.D. Perry, L.A. Moran, and J. Rossant. 1987. Cell-lineage-specific expression of the mouse hsp68 gene during embryogenesis. *Dev. Biol.* **121:** 342–348.

Krawczyk, Z., N. Szymik, and J. Wisniewski. 1987a. Expression of hsp70-related gene in developing and degenerating rat testis. *Mol. Biol. Rep.* **12:** 35–41.

Krawczyk, Z., J. Wisniewski, and E. Biesiada. 1987b. A hsp70-related gene is constitutively highly expressed in testis of rat and mouse. *Mol. Biol. Rep.* **12:** 27–34.

Lindquist, S. 1980. Varying patterns of protein synthesis in *Drosophila* during heat shock: Implications for regulation. *Dev. Biol.* **77:** 463–479.

Mason, P.J., L.M.C. Hall, and J. Gausz. 1984. The expression of heat shock genes during normal development in *Drosophila melanogaster* (heat shock/abundant transcripts/ developmental regulation). *Mol. Gen. Genet.* **194:** 73–78.

Mason, P.J., I. Török, I. Kiss, F. Karch, and A. Udvardy. 1982. Evolutionary implications of a complex pattern of DNA sequence homology extending far upstream of the hsp70 genes at loci 87A7 and 87C1 in *Drosophila melanogaster. J. Mol. Biol.* **156:** 21–35.

McKenzie, S.L., S. Henikoff, and M. Meselson. 1975. Localization of RNA from heat-induced polysomes at puff sites in *Drosophila melanogaster. Proc. Natl. Acad. Sci.* **72:** 1117–1121.

Mestril, R., P. Schiller, J. Amin, H. Klapper, A. Jayakumar, and R. Voellmy. 1986. Heat shock and ecdysterone activation of *Drosophila melanogaster* hsp23 gene; a sequence element implied in developmental regulation. *EMBO J.* **5:** 1667–1673.

Mirault, M.-E., M. Goldschmidt-Clermont, S. Artavanis-Tsakonas, and P. Schedl. 1979. Organization of the multiple genes for the 70,000-dalton heat shock protein in *Drosophila melanogaster. Proc. Natl. Acad. Sci.* **76:** 5254–5258.

Mirault, M.-E., M. Goldschmidt-Clermont, L. Moran, A.-P. Arrigo, and A. Tissières. 1978. The effect of heat shock on gene expression in *Drosophila melanogaster. Cold Spring Harbor Symp. Quant. Biol.* **42:** 819–827.

O'Connor, D. and J.T. Lis. 1981. Two closely linked transcription units within the 63B heat shock puff of *D. melanogaster* display strikingly different regulation. *Nucleic Acids Res.* **9:** 5075–5092.

O'Malley, K., A. Mauron, J.D. Barchas, and L. Kedes. 1985. Constitutively expressed rat mRNA encoding a 70-kilodalton heat-shock-like protein. *Mol. Cell. Biol.* **5:** 3476–3483.

Palter, K.B., M. Watanabe, L. Stinson, A.P. Mahowald, and E.A. Craig. 1986. Expression and localization of *Drosophila melanogaster* hsp70 cognate proteins. *Mol. Cell. Biol.* **6:** 1187–1203.

Pauli, D. and C.-H. Tonka. 1987. A *Drosophila* heat shock gene from locus 67B is expressed during embryogenesis and pupation. *J. Mol. Biol.* **198:** 235–240.

Pauli, D., A. Spierer, and A. Tissières. 1986. Several hundred base pairs upstream of *Drosophila* hsp23 and 26 genes are required for their heat induction in transformed flies. *EMBO J.* **5:** 755–761.

Pauli, D., C.-H. Tonka, and A. Ayme-Southgate. 1988. An unusual split *Drosophila* heat shock gene expressed during embryogenesis, pupation and in testis. *J. Mol. Biol.* **200:** 47–53.

Pauli, D., A.-P. Arrigo, J. Vazquez, C.-H. Tonka, and A. Tissières. 1989. Expression of the small heat shock genes during *Drosophila* development: Comparison of the accumulation of hsp23 and hsp27 mRNAs and polypeptides. *Genome* **31:** (in press).

Petersen, N., G. Moller, and H.K. Mitchell. 1979. Genetic mapping of the coding regions for three heat-shock proteins in *Drosophila melanogaster*. *Genetics* **92:** 891–902.

Riddihough, G. and H.R.B. Pelham. 1986. Activation of the *Drosophila* hsp27 promoter by heat shock and by ecdysone involves independent and remote regulatory sequences. *EMBO J.* **5:** 1653–1658.

———. 1987. An ecdysone response element in the *Drosophila* hsp27 promoter. *EMBO J.* **6:** 3729–3734.

Rossi, J.M. and S. Lindquist. 1989. The intracellular location of yeast heat shock protein 26 varies with metabolism. *J. Cell Biol.* **108:** 425–439.

Sirotkin, K. and N. Davidson. 1982. Developmentally regulated transcription from *Drosophila melanogaster* chromosomal site 67B. *Dev. Biol.* **89:** 196–210.

Sorger, P.K. and H.R.B. Pelham. 1987. Cloning and expression of a gene encoding hsc73, the major hsp70-like protein in unstressed rat cells. *EMBO J.* **6:** 993–998.

Southgate, R., A. Ayme, and R. Voellmy. 1983. Nucleotide sequence analysis of the *Drosophila* small heat shock gene cluster at locus 67B. *J. Mol. Biol.* **165:** 35–57.

Southgate, R., M.-E. Mirault, A. Ayme, and A. Tissières. 1985. Organization, sequences, and induction of heat shock genes. In *Changes in gene expression in response to environmental stress* (ed. B.G. Atkinson and D.B. Walden), pp. 1–30. Academic Press, New York.

Storti, R.V., M.P. Scott, A. Rich, and M.L. Pardue. 1980. Translational control of protein synthesis in response to heat shock in *D. melanogaster* cells. *Cell* **22:** 825–834.

Thomas, S.R. and J.A. Lengyel. 1986. Ecdysteroid-regulated heat-shock gene expression during *Drosophila melanogaster* development. *Dev. Biol.* **115:** 434–438.

Velazquez, J.M., S. Sonoda, G. Bugaisky, and S. Lindquist. 1983. Is the major *Drosophila* heat shock protein present in cells that have not been heat shocked. *J. Cell Biol.* **96:** 286–290.

Vincent, A., H.V. Colot, and M. Rosbash. 1986. Blastoderm-specific and read-through transcription of the sry alpha gene transformed into the *Drosophila* genome. *Dev. Biol.* **118:** 480–487.

Vitek, M.P. and E.M. Berger. 1984. Steroid and high-temperature induction of the small heat-shock protein genes in *Drosophila*. *J. Mol. Biol.* **178:** 173–189.

Voellmy, R., M. Goldschmidt-Clermont, R. Southgate, A. Tissières, R. Levis, and W. Gehring. 1981. A DNA segment isolated from chromosomal site 67B in *D. melanogaster* contains four closely linked heat-shock genes. *Cell* **23:** 261–270.

Wadsworth, S.C., E.A. Craig, and B.J. McCarthy. 1980. Genes for three *Drosophila* heat-shock-induced proteins at a single locus. *Proc. Natl. Acad. Sci.* **77:** 2134–2137.

Wistow, G. and J. Piatigorsky. 1987. Recruitment of enzymes as lens structural proteins.

Science **236:** 1554–1556.

Xiao, H. and J.T. Lis. 1989. Heat shock and developmental regulation of the *Drosophila melanogaster* hsp83 gene. *Mol. Cell. Biol.* **9:** 1746–1753.

Yost, H.J. and S. Lindquist. 1986. RNA splicing is interrupted by heat shock and is rescued by heat shock protein synthesis. *Cell* **45:** 185–193.

Zakeri, Z.F. and D.J. Wolgemuth. 1987. Developmental-stage-specific expression of hsp70 gene family during differentiation of the mammalian male germ line. *Mol. Cell. Biol.* **7:** 1791–1796.

Zakeri, Z.F., D.J. Wolgemuth, and C.R. Hunt. 1988. Identification and sequence analysis of a new member of the mouse hsp70 gene family and characterization of its unique cellular and developmental pattern of expression in the male germ line. *Mol. Cell. Biol.* **8:** 2925–2932.

Zimmerman, J.L., W. Petri, and M. Meselson. 1983. Accumulation of a specific subset of *D. melanogaster* heat shock mRNAs in normal development without heat shock. *Cell* **32:** 1161–1170.

16

Posttranscriptional Regulation of Heat Shock Protein Synthesis in *Drosophila*

H. Joseph Yost
Department of Molecular and Cell Biology
University of California
Berkeley, California 94720

Robert B. Petersen and Susan Lindquist
Howard Hughes Medical Institute
Department of Molecular Genetics and Cell Biology
University of Chicago
Chicago, Illinois 60637

I. INTRODUCTION

All cells respond to temperatures above their normal growth temperatures by inducing the synthesis of a small group of evolutionarily conserved polypeptides known as the heat shock proteins (hsps). Some of the hsps are required to help cells grow at the upper end of their normal temperature range, others to help cells withstand the toxic effects of more extreme temperatures. Many of the proteins are also produced at normal temperatures or have close relatives that are produced at normal temperatures (the "heat shock cognate" proteins). The hsps and their relatives

Stress Proteins in Biology and Medicine.

participate in an extraordinary variety of cellular processes, including DNA synthesis, protein secretion, the response to steroid hormones, and the "chaperoning" of newly synthesized proteins. The variety and complexity of these interactions are only beginning to be unraveled. What defines the hsps as a group, however, is the immediate and very dramatic increase in their synthesis when cells are exposed to higher temperatures (Lindquist and Craig 1988).

Considering the extremely ancient nature of the heat shock response and the extraordinary variety of organisms and cell types in which it is found, it is not surprising that several different mechanisms have evolved to ensure that the hsps will be produced as rapidly as possible after a shift to high temperature. In all organisms investigated, transcriptional regulation plays an important role in the induction of the hsps. In eukaryotes, a heat-shock-activated transcription factor binds to a promoter/enhancer element that provides heat-inducible transcription (for review, see Chapters 17 and 18). In prokaryotes, a heat-inducible σ factor redirects core polymerase to transcribe genes that carry heat shock promoter elements (for review, see Chapter 8).

Regulatory mechanisms acting at the levels of RNA processing, translation, and mRNA degradation are less well understood but have an equal, if not greater, effect on gene expression. The interaction of these mechanisms at normal temperatures, during heat shock, and during recovery from heat shock form the focus of this review. This chapter is not intended to be a fully comprehensive survey of the literature. We have concentrated on the work in *Drosophila* because more is known about posttranscriptional regulation in the heat shock response of this organism than in any other and because our own research on this subject has concentrated on this organism. We cite studies on other organisms where they provide further information or a different perspective.

II. EFFECT OF HEAT SHOCK ON RNA PROCESSING

A. Splicing of Intervening Sequences from mRNA Precursors

In comparison with the classic heat shock proteins in *Drosophila*, hsp82 is very unusual in its pattern of expression. It is produced at normal temperatures and, although strongly induced by moderate heat shocks, is poorly induced by severe heat shocks. It has long been established that this defect in expression at high temperatures is not due to a defect in translation of the hsp82 message. Messages induced by a mild heat shock continue to be translated at high levels when cells are shifted to higher temperatures (Lindquist 1980b). The lack of hsp82 expression after a sudden, high-temperature heat shock is due to a failure of the hsp82 message to accumulate under these conditions.

It was initially assumed that this defect derived from a failure in the activity of the hsp82 promoter. However, sequence analysis of the *HSP82* gene later suggested its promoter contained all of the elements required for full activity at high temperatures (Hackett and Lis 1983). It also revealed that the *HSP82* gene contains a single intervening sequence located immediately upstream of the translation-initiating AUG codon. The genes of higher eukaryotes generally contain intervening sequences that must be spliced out of their initial transcripts in order to produce a functional message. Most genes in *Drosophila*, including genes that are closely related to the heat shock genes but are expressed at normal temperatures, conform to this general rule. Most genes encoding hsps do not. They are generally free of intervening sequences (Holmgren et al. 1979; Ingolia et al. 1980). Thus, hsp82 is a marked exception.

These findings led to a reexamination of the defect in hsp82 expression that occurs with a severe heat shock. When cells are heat-shocked at more severe temperatures, the splicing of hsp82 mRNA precursors is disrupted. The hsp82 promoter remains very active at these temperatures, however, and large quantities of full-length, intron-containing hsp82 RNAs accumulate (Fig. 1). Thus, it is the block in splicing that is responsible for the low level of hsp82 expression during high-temperature heat shocks. Depending on the severity of the heat shock, this block in splicing may persist for several hours after cells are returned to normal temperatures for recovery. The time during recovery when splicing is restored coincides with the time when synthesis of hsp82 resumes (Yost and Lindquist 1986).

This disruption of precursor splicing is not sequence-specific. Intron-containing hsp82 RNAs accumulate during heat shock in four *Drosophila* species in which the intron sequences greatly differ (Blackman and Meselson 1986). The splicing of transcripts from the *D. melanogaster* heat-inducible chromosomal locus 93D is partially inhibited at high temperatures (Garbe and Pardue 1986). These transcripts have little or no sequence homology with hsp82 transcripts.

Because the transcription of most genes is reduced during heat shock (Berendes 1968; Spradling et al. 1975; Jamrich et al. 1977; Findly and Pederson 1981) and preexisting mRNAs are stable (Storti et al. 1980; Lindquist 1981; Kelley and Schlesinger 1982), it was not immediately clear whether the transcripts of other, non-heat-inducible genes would also be blocked from splicing at high temperatures. To circumvent this problem, the intron-containing alcohol dehydrogenase gene of *Drosophila* was placed under the control of an hsp70 promoter. At intermedi-

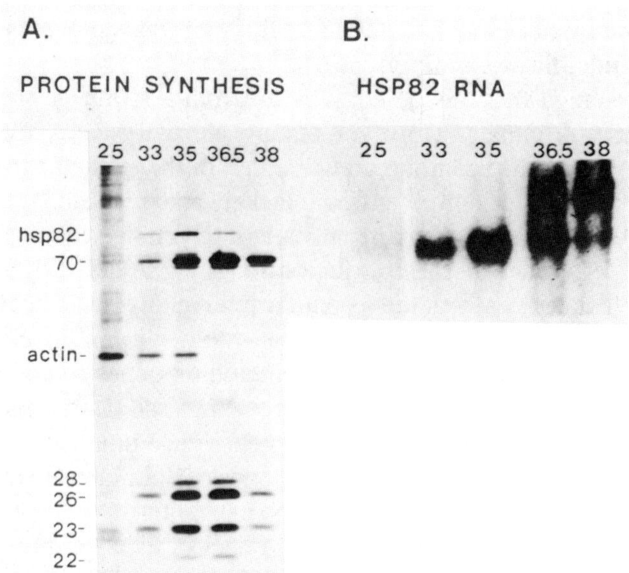

Figure 1 The effect of heat shock on *HSP82* transcripts in *Drosophila*. (*A*) Protein synthesis at various temperatures. Tissue culture cells, grown at 25°C, were shifted to the indicated temperature for 60 min with [^{3}H]leucine present for the last 45 min. (*B*) Accumulation of hsp82 transcripts at various temperatures. At the same time that protein samples were extracted, RNAs were extracted from duplicate aliquots of cells and analyzed by Northern analysis. (Reprinted, with permission, from Yost and Lindquist 1986.)

ate heat shock temperatures, the transcripts of this gene were accurately spliced into mRNAs. At more severe heat shock temperatures, the transcripts were not spliced, and intron-containing precursors accumulate (Yost and Lindquist 1986, 1988). Thus, heat shock produces a very general block in the splicing of mRNA precursors.

The basic phenomenon, that high temperatures disrupt the splicing of intervening sequences from mRNA precursors, has been demonstrated recently in yeast (Fig. 2) (H.J. Yost and S. Lindquist, in prep.), chicken (Bond and Schlesinger 1986), mammals (Kay et al. 1987; Bond 1988), and *Dictyostelium* (Maniak and Nellen 1988). The block in splicing results in the accumulation of substantial quantities of full-length precursor RNAs. Free 5′ exons were not detected, either in *Drosophila* cells or in yeast cells, suggesting that the block in RNA splicing occurs before cleavage of the 5′ splice site (Yost and Lindquist 1988; H.J. Yost and S. Lindquist, in prep.). Some of the transcripts studied in these citations are from non-heat-shock genes, again demonstrating that the block in pre-mRNA splicing is not specific to heat-induced transcripts, but can be viewed as a very general toxic effect of heat.

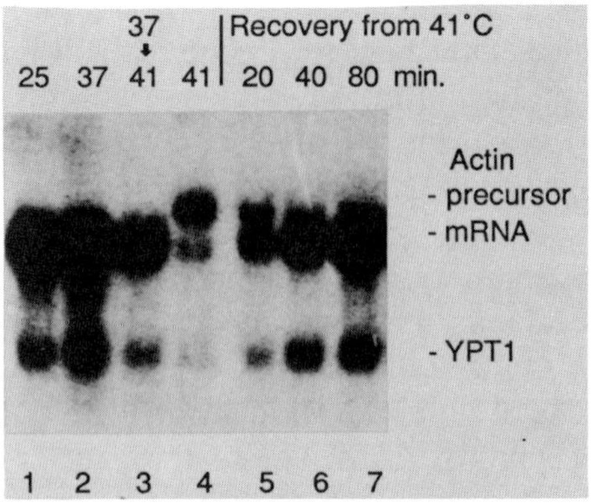

Figure 2 Yeast mRNA splicing is blocked by heat shock and protected by a mild pretreatment. Yeast cells were incubated at 25°C (lane *1*), 37°C (lane *2*), or 41°C (lane *4*), for 60 min, and then RNAs were extracted. The sample in lane *3* was incubated at 37°C for 45 min, and then shifted to 41°C for 60 min before RNA was extracted. The samples on the right were prepared at the times indicated after return to 25°C following a 60-min 41°C heat shock. RNAs were analyzed on a Northern blot with a ^{32}P-labeled actin probe.

A particularly interesting example of the effects of heat on splicing occurs in trypanosomes. All known mature mRNAs in this organism share a 39-nucleotide 5′ leader sequence, a miniexon, that is encoded in a separate genetic locus from that of the protein-coding sequences. The miniexon sequence is derived from a longer, capped RNA, called the miniexon-derived RNA. The 39-nucleotide miniexons are joined to protein-coding exons via a bimolecular (*trans*) splicing mechanism. Muhich and Boothroyd (1988) have recently shown that when trypomastigotes are exposed to a short, severe heat shock, the levels of both α- and β-tubulin mRNAs decrease while larger polycistronic tubulin transcripts accumulate. Furthermore, during heat shock, the concentrations of full-length miniexon-derived RNAs increase, and the levels of splicing branch-intermediates decrease. These results suggest that mature mRNAs in trypanosomes are formed by a *trans*-splicing pathway and that heat shock disrupts this pathway before the initial catalytic step. Thus, the effects of heat shock on *trans*-splicing are analogous to those described above for *cis*-splicing in *Drosophila*, in which heat shock results in the accumulation of full-length, uncut hsp82 transcripts. In both organisms, RNAs that were excluded from the splicing pathway during heat shock do not contribute to mRNA formation when the cells have recovered from heat shock.

B. Mild Heat Pretreatments Protect Splicing

In virtually all organisms investigated, mild heat treatments, which induce the synthesis of hsps, provide protection from the toxic effects of a second more severe heat shock. In organisms as diverse as soybeans, *Drosophila*, mammals, and bacteria, such pretreatments can increase survival by as much as 2 or 3 orders of magnitude. Similarly, mild heat treatments provide protection from a variety of developmental defects elicited by severe sublethal heat treatments. Pretreatments can also provide protection from the disruption of RNA splicing by high temperatures (Fig. 3). When *Drosophila* cells are given a short incubation at a temperature that allows them to accumulate hsps, before they are exposed to a temperature that would otherwise block splicing, splicing is not disrupted. The degree of protection, measured as the percentage of new transcripts that are properly spliced, varies in a dose-dependent fashion with the duration of the pretreatment and with the accumulation of

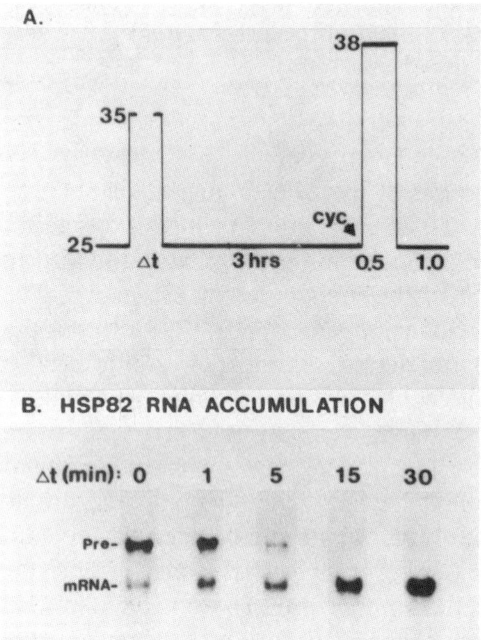

Figure 3 Pretreatment protects RNA splicing from a high-temperature block. (*A*) Experimental regimen indicating the time of pretreatment at 35°C (t), the time of addition of cycloheximide (cyc), the length of the severe heat shock (0.5 hr), and the recovery period after heat shock before RNA extraction. (*B*) Accumulation of hsp82 RNAs. After 1 hr of recovery, RNAs were extracted and analyzed by Northern blotting. The length of the pretreatment is indicated above each sample. (Reprinted, with permission, from Yost and Lindquist 1986.)

hsps. Furthermore, the protective effect of the pretreatment can be eliminated by the addition of cycloheximide, an inhibitor of cytoplasmic protein synthesis, before, but not after, hsps are produced. These results suggest that hsps may be involved in protecting, or quickly repairing, RNA splicing (Yost and Lindquist 1986). The protection of RNA splicing by mild heat pretreatments has also been demonstrated in yeast (H.J. Yost and S. Lindquist, in prep.) and mammalian cells (Bond 1988).

C. Nature of the Block in Splicing and the Role of hsps

Several different experimental approaches have provided further insights on the nature of the heat-induced block in mRNA precursor splicing. By gradient centrifugation analysis, ribonucleoprotein particles (RNPs) from heat-shocked cells display a marked decrease in the protein to RNA ratio compared to RNPs from non-heat-shocked cells (Mayrand and Pederson 1983). Electron microscopy of active chromatin from *Drosophila* embryos suggests that heat shock causes a change in the structure of nascent transcripts. Beyer and Osheim (1988) have shown that nascent RNA polymerase II transcripts contain splicing-associated structures and display cotranscriptional splicing. After heat shock, the nascent transcripts tend to be more rigid and thicker than in control conditions (average diameter 10–15 nm instead of 5–10 nm), suggesting a change in RNP packaging that is apparently not sequence-specific (A.L. Beyer and S.A. Amero, pers. comm.).

Immunological analysis of *Drosophila* embryos with antibodies to RNP particles also suggests that heat shock produces a change in the structure or conformation of the RNP complexes involved in splicing. Sections from both heat-shocked embryos and non-heat-shocked embryos react identically with anti-nRNP antibodies, which recognize determinants specific to U1 snRNPs. Sections from non-heat-shocked embryos react with anti-Sm antibodies, which recognize determinants common to the U1, U2, U4, U5, and U6 particles, but sections from heat-shocked embryos do not. Similar heat-induced alterations in Sm antigenicity have been observed in mammalian cells (Welch and Mizzen 1988). The reactivity of the heat-shocked *Drosophila* embryos can be restored by a brief salt wash. The relationship between this change in antigenicity and the change in splicing activity is strengthened by the fact that mild pre-heat treatments prevent both the heat-induced block in splicing, as discussed above, and the heat-induced loss of Sm antigenicity (Wright-Sandor et al. 1989). Finally, the elegant work of Bond (1988) with in vitro extracts prepared from heat-shocked mammalian cells shows that certain snRNPs are altered at high temperatures and form aberrant splicing complexes.

Although the precise nature of the heat-induced block in processing is unknown, it must occur early in the processing pathway, since full-length precursor RNAs accumulate with heat shock in *Drosophila* cells (Yost and Lindquist 1988), in *Trypanosoma* (Muhich and Boothroyd 1988), and in cell-free extracts of mammalian cells (Bond 1988). These results, together with the results of the biochemical and microscopic analyses discussed above, suggest that if hsps are directly involved in the protection of RNA splicing, they may play a role in the formation, or the reformation, of functional spliceosome complexes. It is not yet clear which of the hsps have a critical role in protecting RNA splicing. However, hsp70 is the only hsp that has been unequivocably shown to concentrate in nuclei (Velazquez and Lindquist 1984). It has been proposed that the hsp70 family of proteins is involved in guiding and correcting protein folding and protein complex formation (Pelham 1986, 1988). Thus, a role in spliceosome assembly at high temperatures is commensurate with other proposed functions of this protein family. It should be emphasized, however, that other models are compatible with the data. For example, the hsps may be involved in protecting some other nuclear structure that is indirectly required for splicing.

D. Effects of Heat on Other RNA Processing Reactions

Heat shock also disrupts other types of RNA processing. Among the first reported effects of high temperature on *Drosophila* cells was the accumulation of large rRNA precursors (Ellgaard and Clever 1971). Whether this heat-induced block in rRNA processing is biologically related to the block in mRNA precursor splicing is doubtful. The proper processing of rRNA precursors requires the continued synthesis of ribosomal proteins. Since heat shock suppresses the production of these proteins (see below), it is not surprising that it also suppresses processing of rRNA (Bell et al. 1988). In fact, the disruption of rRNA processing can be induced at normal temperatures by incubating cells in the presence of cycloheximide (Bugaisky 1981), an inhibitor of cytoplasmic protein synthesis. This treatment does not produce a block in the splicing of mRNA precursors (Yost and Lindquist 1986). Heat shock also produces a dramatic decline in the synthesis of mature 5S RNAs with a concomitant increase in the accumulation of a precursor transcript with an additional 15 nucleotides at its 3′ end (Rubin and Hogness 1975; Jacq et al. 1977). The nature of the defect has not been determined.

In *Xenopus* and human cells, heat shock results in the appearance of improperly terminated rRNAs (Labhart and Reeder 1987; Parker and Bond 1989), due either to improper transcriptional termination or to improper 3′ end processing. A defect in the termination of pol II transcripts may also be a common effect of heat. In *Drosophila*, the small hsp genes show a complex pattern of induction during development and during heat shock. Some of the transcripts produced during heat shock are longer than those produced during development and carry extensions of more than 1 kb at their 3′ ends (Berger et al. 1985; Pauli et al. 1988). The *HSP70* and *HSP82* genes produce longer transcripts after an extreme heat shock than they do after a mild heat shock (R.P. Dellavalle and S. Lindquist, unpubl.). It is not yet clear whether the longer transcripts are a result of improper transcriptional termination, 3′ end processing, or both. Significantly, mild heat pretreatments, which induce the synthesis of hsps, potentiate the production of the shorter transcripts during a subsequent severe heat shock. Since RNP complexes are believed to be involved in the formation of 3′ termini (Proudfoot and Whitelaw 1988), it is tempting to think that the hsps play a general role in protecting the function of such complexes. (The effects of heat shock on mRNA degradation, which may also be viewed as an RNA processing event, are described below.)

E. Perspectives on Heat Shock Genes and Introns

The fact that heat shock blocks the splicing of intervening sequences from mRNA precursors provides the key to understanding the strong selective pressures that have operated in the evolution of intronless hsp genes in *Drosophila*. The organization of these genes is in striking contrast to the organization of most other genes. Among members of the hsp70 protein family, hsp70 itself has no intervening sequences (Holmgren et al. 1979; Ingolia et al. 1980), whereas other members of the family do (Ingolia and Craig 1982; E. Craig and K. Palter, pers. comm.). Of course, an intron would prevent the synthesis of a functional hsp when it is most needed, during a high-temperature heat shock. However, this alone does not explain why every copy of the heat-induced hsp70 genes (there are at least five copies per haploid genome) is intron-free in *Drosophila*. Additional pressure may be provided by the fact that precursor RNAs, blocked from splicing by heat shock, can enter the cytoplasm and be translated into aberrant polypeptides (Yost and Lindquist 1988). In separate experiments, it was found that truncated hsp70 proteins, produced by mutagenesis, can have deleterious effects on the cell (J. Rossi et al., in prep.). Thus, intervening sequences in the coding regions of hsp genes would not only reduce their levels of expression at high temperatures, but might also result in the expression of abnormal, deleterious proteins. That intron-containing RNAs can escape to the cytoplasm and be translated might also have provided a driving force in evolution for decreasing the transcription of non-heat-shock genes at high temperatures.

Why has the hsp82 gene retained an intron? First, hsp82 is abundant at normal temperatures, which would presumably mitigate the effects of decreased production at high temperatures. Second, the single hsp82 intron directly precedes the coding region and contains multiple stop codons that would prevent the expression of all but a 7-amino-acid polypeptide. Thus, translation of the hsp82 precursor RNA would not result in production of a truncated, deleterious hsp. Correspondingly, it is the position, and not the sequence, of the hsp82 intron that is conserved in various *Drosophila* species (Blackman and Meselson 1986) as well as in humans (Hickey et al. 1989).

Given that heat shock disrupts RNA splicing in organisms as diverse as fruit flies, trypanosomes, mammals, and yeast, one might expect the heat shock genes in all organisms to be free of intervening sequences. Indeed, the human and mouse hsp70 genes do not contain intervening sequences, whereas closely related, but constitutively expressed, members of the hsp70 gene family in these organisms do (Dworniczk and Mirault 1987). This general rule, however, cannot be universally applied. A few genes in other organisms now fall into the same category as the *Drosophila HSP82* gene: They produce heat-inducible transcripts that contain intervening sequences, the splicing of which is at least partially inhibited by heat shock (Bond and Schlesinger 1986; Kay et al. 1987; Bond 1988). It may be that in these cases a substantial level of function is already provided at normal temperatures, making an immediate burst of new synthesis less critical than it is for hsp70. It is also possible that in some organisms the splicing of essential heat shock transcripts is more resistant to the toxic effects of heat than is the splicing of other transcripts. In *Trypanosoma brucei, trans*-splicing of tubulin transcripts is blocked by heat treatments that do not block the *trans*-splicing of hsp70 and hsp82 transcripts (Muhich and Boothroyd 1989; Muhich et al. 1989). The hsp70 transcript does not have a consensus branch site within 50 nucleotides of the 3′ splice site (Py-Py-Pu-A*-Py) (Borst 1986; Glass et al. 1986). It will be interesting to learn by what mechanism hsp70 transcripts circumvent the highly conserved heat-induced block in splicing in trypanosomes.

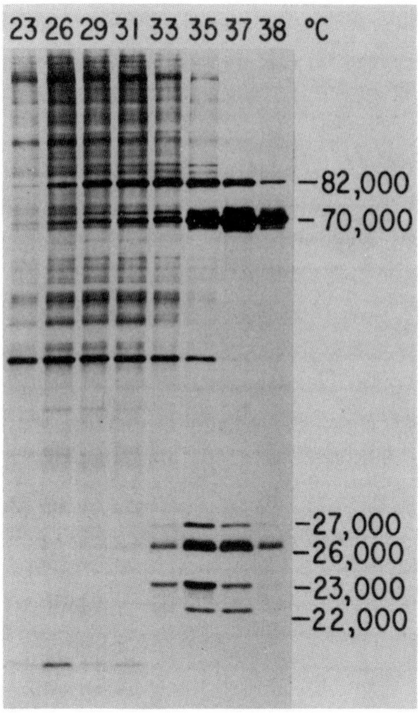

Figure 4 The effect of temperature on the patterns of protein synthesis in *Drosophila* cells. Tissue culture cells were grown at 23°C and incubated at the indicated temperatures for 2 hr with [^{3}H]leucine added during the final hour of incubation. The apparent molecular weights of the heat shock proteins are indicated at right. (Reprinted, with permission, from Lindquist 1980b.)

III. SELECTIVE TRANSLATION OF HEAT SHOCK MESSAGES DURING HEAT SHOCK

A. General Description of the Change in Translational Specificity

Of the various effects that heat shock has on gene expression at the post-transcriptional level, the most dramatic, and that which has received the most attention, is the change in protein synthesis. After a temperature elevation from 25°C to 37°C, the translation of preexisting messages is blocked, and preexisting polysomes disappear. As newly transcribed hsp mRNAs begin to appear in the cytoplasm, polysomes are reformed on the hsp mRNAs, and hsps quickly become the major products of protein synthesis in the cell (Lindquist 1980b).

As dramatic as this response may be, it is finely modulated according to the severity of the heat treatment. A significant increase in hsp synthesis can be detected with a temperature increase of as little as 4°C (see Fig. 4). At 33°C, the translation of preexisting mRNAs is partially repressed. At 36°C and above, virtually all translational activity is devoted to hsps. The rate, as well as the extent, to which 25°C protein synthesis is repressed increases with temperature, requiring about 30 minutes at 35°C but less than 10 minutes at 37°C (Lindquist and DiDomenico 1985).

Although heat shocks at 36°C and above effect a complete change in the specificity of translation, they do not have a deleterious effect on the efficiency of translation. The rate of ribosome initiation on hsp70 mRNAs, 9–14 initiations per minute per message, compares favorably with rates of initiation in rabbit reticulocytes (Lindquist 1980a). In fact, if cells are raised gradually to high temperatures, instead of rapidly, they continue to translate hsp mRNAs efficiently at temperatures up to 42°C. Thus, the change in translation does not reflect any general damage to the translational machinery, but rather a vigorous commitment to a new pattern of protein synthesis.

B. State of Preexisting Messages in Heat-shocked Cells

In an early, carefully conducted study, preexisting poly(A)-containing RNAs were rapidly lost from cytoplasmic fractions during heat shock (Spradling et al. 1977), suggesting that the repression of preexisting protein synthesis is brought about by the degradation of 25°C messages. It is now clear, however, that these messages are quantitatively retained in heat-shocked cells (McKenzie 1976; Storti et al. 1980; Findly and Pederson 1981; Lindquist 1981; Petersen and Mitchell 1981; DiDomenico et al. 1982b). An explanation for this apparent contradiction follows from the intriguing observation that intermediate filaments collapse around the nucleus during heat shock and fractionate with nuclei in routine cell lysis procedures (Biessmann et al. 1982). In other systems, message attachment to the cytoskeleton appears to play an important role in determining the translational activity of the message (Cervera et al. 1981). If preexisting poly(A)-containing mRNAs were pulled into the collapsing intermediate filament network during heat shock, they would fractionate with nuclei, making it appear that there was a loss of poly(A)-containing RNA from the cytoplasm. The subcellular location of preexisting mRNAs deserves examination.

Since 25°C messages extracted from heat-shocked cells are fully capable of translation in cell-free systems, it is unlikely that they undergo any major modifications that would account for their lack of translation in heat-shocked cells. It remains a formal possibility that the messages are altered in a subtle manner that would specifically block their translation in heat-shocked cells. The cap structures of 25°C messages at 25°C are similar to those on heat-induced RNAs (Levis and Penman 1978), but no one has demonstrated that the caps on preexisting messages remain unaltered during heat shock, nor, to our knowledge, have other possible modifications been examined.

Considering the extremely diverse nature of 25°C messages, they behave in a remarkably uniform way with regard to the rate and extent of their repression under different heat shock conditions (Lindquist 1980b; Lindquist and DiDomenico 1985; see Figs. 4 and 5). Repression would appear, therefore, to act through a very general mechanism. There is evidence, however, that different cellular messages are blocked at different points in the translation process. Some appear to be associated with polysomes and are presumably blocked in elongation, others are found in smaller complexes, presumably blocked at the level of initiation (Ballinger and Pardue 1983). (These findings, however, are subject to alternative interpretations [Sirkin and Lindquist 1985].)

Whatever the state of preexisting mRNAs may be, it is clear that they are not actively competing with hsp mRNAs for ribosomes. For one thing, the repression of 25°C protein synthesis is extremely rapid and occurs before hsp mRNAs have appeared in appreciable numbers (McKenzie et al. 1975; Lindquist 1981). Furthermore, if the appearance of hsp messages is blocked altogether (by adding actinomycin D to cells prior to the increase in temperature), 25°C protein synthesis is still repressed (Lindquist 1981).

Several lines of evidence argue that the change in translation is not a direct effect of temperature on the secondary structure of the message. (That is, if a particular structure were required for translation, it might be melted by high temperatures on 25°C mRNAs but not on hsp mRNAs.) First, cell-free lysates of heat-shocked *Drosophila* cells preferentially translate hsp mRNAs over 25°C mRNAs. Lysates of 25°C cells, incubated at the same temperature as those from heat-shocked cells, show no discrimination. Thus, once the change in specificity is established it remains a property of the lysate (Scott et al. 1979; Storti et al. 1980; Kruger and Benecke 1981; Scott and Pardue 1981; Sanders et al. 1986). Second, when 25°C mRNAs and hsp mRNAs are mixed and translated at different temperatures in heterologous cell-free lysates, both classes of messages are affected by temperature in the same way (Kruger and Benecke 1981; Lindquist 1981; Scott and Pardue 1981). Finally, when cells are returned to normal temperatures for recovery, 25°C protein synthesis is not restored immediately. Depending on the severity of the preceding heat shock, translation may be exclusively devoted to hsp mRNAs for up to several hours (Fig. 5). Thus, once the change in translation is induced, its maintenance depends on the physiological state of the cell rather than the temperature itself (DiDomenico et al. 1982a).

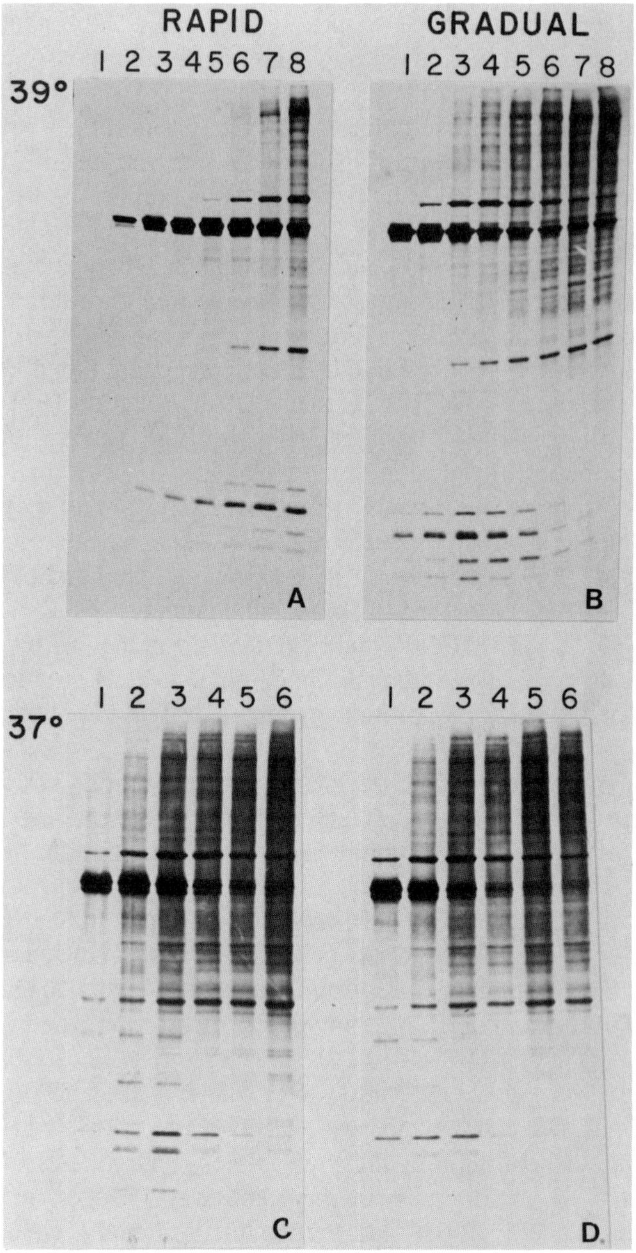

Figure 5 Recovery of normal protein synthesis following heat treatments of varying severity. Cells were heat-treated at either 39°C or 37°C for 30 min, then returned to 25°C. Aliquots were pulse-labeled with [^{3}H]leucine in 30-min intervals. (*A*, *C*) Cells shifted directly to high temperature (rapid heat treatment). (*B*, *D*) Incubation temperature increased gradually (2°C every 15 min). (Reprinted, with permission, from DiDomenico et al. 1982b.)

Temperature does play an important role in translation, nevertheless. The transcription of heat shock genes can be induced by a remarkably broad array of other stress treatments. Many of these stress treatments have adverse effects on protein synthesis, but their effects are nonspecific (Lewis et al. 1975; Vincent and Tanguay 1982; S. Lindquist, unpubl.). Thus, although high temperatures are not required to maintain the change in translational specificity, they appear to be important in triggering it. Temperature is also a factor in the restoration of 25°C protein synthesis (see below for further details). As long as cells are maintained at high temperatures, they continue to produce hsps almost exclusively. Two conditions are generally required for recovery: A certain quantity of hsp must be made, and the temperature must be returned to normal (DiDomenico et al. 1982a). (The synthesis of hsps may not be required for recovery from very short or mild heat shocks.)

C. Recognition of hsp mRNAs

Since hsp mRNAs are translated and coexisting 25°C mRNAs are not, they must be different in some way. Two alternatives have been suggested. (1) hsp mRNAs are different simply because they are new. For example, all preexisting messages might be sequestered in a particular subcellular compartment during heat shock, with new messages being excluded from that compartment; or heat shock might activate or inactivate a particular RNA modification system. Whatever the specifics, in this case, the change in translational activity would be a simple and direct consequence of the change in transcriptional specificity. (2) Alternatively, hsp mRNAs might be distinguished by some feature encoded within their nucleic acid sequences.

An early experiment placed constraints on the first model. Cells were treated with actinomycin D immediately after heat shock and were then put through several additional heat shock/recovery cycles. Both classes of messages were translated during recovery, but only hsp mRNAs were translated at high temperatures (Lindquist 1980b). Therefore, discrimination could not be operating according to whether a message was old or new per se, but only on the basis of whether a message had been transcribed during heat shock or at normal temperatures.

Subsequently, mutational analysis of heat shock genes confirmed the second model. First, when the 5′ end of the *hsp70* gene, including the untranslated message leader and first few codons, was fused to other coding sequences, the chimeric transcripts of these genes were translated at high temperatures (DiNocera and Dawid 1983; Bonner et al. 1984). The results suggest that recognition sequences, if they exist, are confined to the 5′ end of the message. Several unusual structural features shared by the untranslated leader sequences of hsp mRNAs provided candidates for recognition. All hsp message leaders in *Drosophila* are unusually long (on the order of 151–250 nucleotides), are very rich in adenine (on the order of 45–49%), and share two regions of sequence homology, one at their extreme 5′ ends and another located approximately in the middle of the 5′ untranslated region (for review, see Mathews 1986).

In one series of experiments, a wild-type *hsp70* gene was marked by a deletion in its protein-coding sequence and stably transformed into *Drosophila* tissue culture cells (McGarry and Lindquist 1985). Transcripts of the mutant gene were translated as efficiently as wild-type hsp70 messages. *Bam* linker-insertion mutations were then introduced into the 5′ untranslated leader region of the gene and recombined to create a variety of other mutations. The results obtained with one of these mutations, a large deletion removing most of the untranslated leader from nucleotide +2 to +205 (out of a total of 242), are shown in Figure 6. Transcripts of this gene are produced at a high rate during heat shock, but they are not translated. The effect of the mutation is highly specific. When cells are returned to 25°C for recovery, the message is translated, but when cells are returned to 37°C, translation is again inhibited.

A complementary result was obtained with gene fusion experiments (Klemenz et al. 1985). When the *hsp70* promoter was used to drive transcription of the wild-type *adh* gene during heat shock, the message was not translated during heat shock, but only after return to normal temperatures. When the first 95 nucleotides of the hsp70 message were added to the construct, the message was translated at high temperatures.

Transcription at high temperatures, therefore, is not sufficient to guarantee translation. Some feature encoded in the nucleic acid sequence at the 5′ end of the message is required. Determining the precise nature of this feature has proven more difficult. As previously mentioned, hsp mRNA leaders are unusually long, but length per se is clearly not the determining factor. First, messages carrying rather large deletions (nucleotides +37 to +205 of the 249-nucleotide hsp70 leader and nucleotides +27 to +242 of the 250-nucleotide hsp22 leader) are translated well at high temperatures. Second, an inversion mutation, in which nucleotides +2 to +205 of the *hsp70* gene are present in inverse orientation, is not translated at high temperatures (McGarry and Lindquist 1986). Third, two related insertion mutations are translated at very different rates. Both insertions occur at nucleotide position +2 in the hsp70 leader. One, which carries an additional 39 nucleotides normally found upstream of the transcription start site, is not translated at high temperatures (McGarry and Lindquist 1985). The other, which carries a duplica-

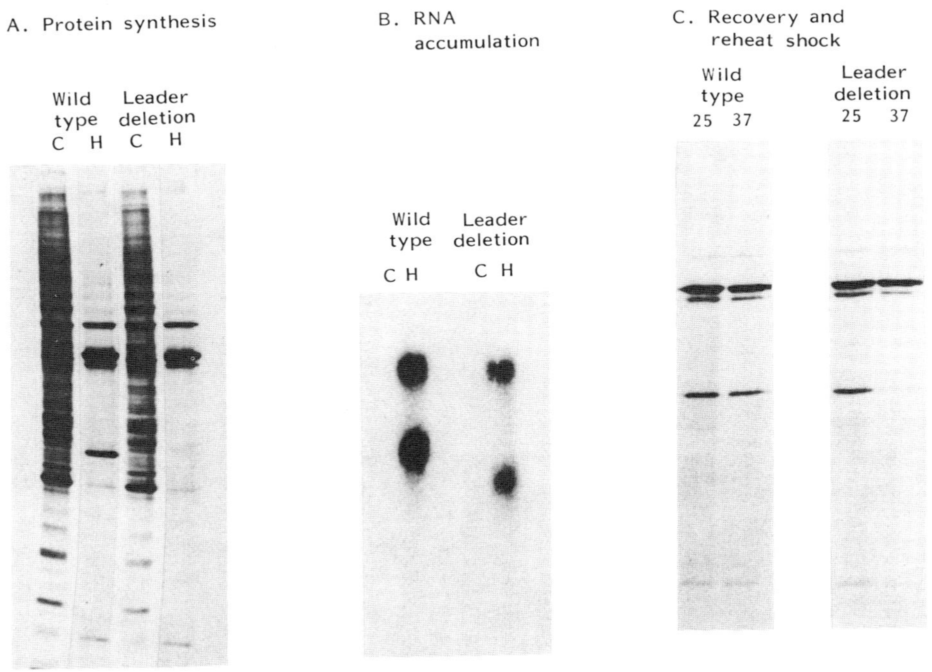

Figure 6 hsp mRNAs carrying a large deletion in their 5′ -untranslated leaders are not translated during heat shock. Cells were transformed with modified hsp70 genes marked by a deletion in the protein-coding sequence that produces a 44-kD protein. When such genes contain wild-type 5′ leader sequences, their messages are translated well at high temperatures. When the leader region is deleted from nucleotides +2 to +205, they are transcribed at high temperatures, but translated only at low temperatures. (*A*) Proteins labeled with [^{3}H]isoleucine during heat shock. (*B*) Electrophoretically separated mRNAs detected by hybridization with ^{32}P-labeled hsp70 DNA. (*C*) Proteins labeled during recovery from heat shock at 25°C and during a second heat shock at 37°C. (Composite produced, with permission, from data in McGarry and Lindquist 1985.)

tion of the first 37 nucleotides of the normal message leader, is translated (Mathews 1986).

The results of these experiments suggest that the crucial element for heat-shock translation is the conserved sequence at the 5′ ends of hsp mRNAs. However, this conserved sequence cannot be the sole determinant, since deletion of the first 27 nucleotides of the hsp70 message does not block translation. Messages carrying deletions in the conserved element in the middle of the hsp70 leader or deletions of both conserved sequences are also translated during heat shock (McGarry and Lindquist 1985; Mathews 1986; for review, see Lindquist 1987).

One possibility is that more than one element in the leader contributes to translation during heat shock. Since large deletions in both the hsp22 and hsp70 message leaders are compatible with heat shock translation as long as they do not extend into the conserved sequence at the 5′ end, this sequence is likely to play an important role (McGarry and Lindquist 1985; Hultmark et al. 1986). The mere presence of this sequence on the message, however, is not sufficient, since messages containing an insertion that moves it 39 nucleotides relative to the 5′ end are not translated. Furthermore, in the complete absence of this sequence, another feature of the leader must be able to provide heat shock translatability. An obvious possibility is that secondary structures in the leader also play an important role.

D. What Cellular Component Discriminates between Messages?

Although we do not yet know precisely what constitutes a proper heat shock leader sequence, it is clear that messages will be translated during heat shock if, and only if, they possess this sequence. Fractionation and supplementation experiments with *Drosophila* cell-free translation lysates hold the greatest promise for determining how it works. As shown by Pardue and her colleagues, when lysates from 25°C cells and heat-shocked cells are mixed, translational activity is an average of that expected from either lysate alone (Scott and Pardue 1981). Thus, heat shock lysates do not contain a soluble, dominant inhibitor of 25°C mRNAs, nor do 25°C lysates contain a soluble, dominant inhibitor of heat shock translation.

In supplementation experiments with crude ribosomal pellets and supernatant fractions, the supernatants did not affect the specificity of translation in either lysate. However, when the crude ribosomal pellet of control cells was added to heat-shocked lysates, it dramatically stimulated the translation of 25°C mRNAs. The crude ribosomal pellet from heat-shocked cells did not affect translation in control lysates. These results suggest that some component of the ribosome, required for the translation of 25°C messages, is inactivated in heat-shocked cells. Experiments performed by Sanders and her colleagues (1986) are in general agreement with those of the Pardue laboratory, except that the stimulatory activity was found in supernatant fractions rather than in the ribosomal pellets. Since a higher salt concentration was used in these experiments, this difference may simply reflect the different fractionation properties of the same factor under different conditions.

At any rate, the change in translational specificity during heat shock apparently results from the inactivation or modification of a factor that is required for the translation of 25°C mRNAs. hsp mRNAs, then, either do not require this factor or are able to utilize it in its modified form. The S6-like ribosomal protein of *Drosophila* cells was an early candidate for such a factor, since it is quantitatively dephosphorylated immediately after temperature elevation (Glover 1982). However, under other conditions, the phosphorylation state of this protein does not correlate with heat shock translation (Olsen et al. 1986). Modifications of other ribosomal proteins in *Drosophila* cells have been observed, but none appears to play a major role in translational specificity.

Two recent studies from the laboratory of Jose Sierra suggest that the selective translation of hs mRNAs may be the result of altered cap binding activity. First, in lysates prepared from heat-shocked *Drosophila* embryos, translation of hs mRNAs is virtually unaffected by addition of cap analogs, whereas residual synthesis from 25°C mRNAs is further repressed (Maroto and Sierra 1988). Second, an antibody was prepared against the *Drosophila* cap binding protein (Maroto and Sierra 1989). Using this antibody to block cap binding activity in cell-free lysates, translation of hsp70 was unaffected, whereas translation of the bulk of 25°C mRNAs was diminished (J. Sierra, pers. comm.). Taken together, these results suggest that hsp70 lacks a requirement for cap binding complex. Although no one has directly demonstrated that hsp70 mRNA is capped, it almost certainly is. The bulk of heat-inducible mRNAs have normal cap structures, and attempts to 5′ end-label hybrid selected hsp70 mRNAs have been unsuccessful (L. Ghibelli and S. Lindquist, unpubl.). Whether hsp70 mRNA is capped or not, heat shock translation may be cap independent. Since the 5′ untranslated region of hsp70 contains 46% adenine residues, it probably contains very little secondary structure. As a consequence, translation of hsp70 mRNA may have a reduced requirement for the ATP-dependent RNA unwinding activity associated with cap binding complex.

These experiments complement the results of the mutational analyses described above, permitting a tentative model for regulation to be drawn:

Preexisting, 25°C mRNAs require cap binding factor for efficient translation, possibly to unwind secondary structure in the message leader. Heat shock inactivates this factor. hsp mRNAs, by virtue of their long, unstructured leaders, are able to escape the requirement for cap binding factor and are therefore translated. In its simplest form, however, this model is not sufficient to explain all the data. For example, we have replaced leader sequences in the hsp70 message with randomly generated A-rich or U-rich sequences. Although all should be similarly unstructured, they are translated at very different rates (R. Petersen and S. Lindquist, unpubl.). Together with the mutational deletion analyses presented above, it appears that some sequence component, presumably something contained within the first 25 nucleotides of the message, is also required for translation.

At a recent Cold Spring Harbor meeting on translational control, it was reported that a 25-kD protein specifically recognizes an element within the first 25 nucleotides of hsp70 mRNA (L. Sheldon and E. Berger, pers. comm.). In contrast, this factor does not recognize the 5′ end of actin mRNA. The binding of the factor to the hsp70 5′ untranslated region was not competed by the cap analog $m^7G(5')ppp(5')N$, indicating that the 25-kD protein is probably not a cap binding factor. This suggests that translation of hs mRNAs during heat shock may be facilitated by a specific protein factor.

IV. REGULATION OF RECOVERY

A. Return to Normal Patterns of Synthesis

The intense and dramatic change in protein synthesis induced by heat shock in *Drosophila* cells is fully reversible. When cells are returned to normal temperatures, normal patterns of synthesis are restored (DiDomenico et al. 1982a,b). Unlike the sudden induction of hsp synthesis, the restoration of 25°C protein synthesis is a gradual process with two distinct phases. Immediately after the return to 25°C, cells continue to produce hsps exclusively. This phase may last for up to several hours, depending on the severity of the preceding heat treatment (Fig. 5). Then, a gradual increase in the synthesis of 25°C proteins and a gradual decline in the synthesis of hsps is initiated. This phase, too, may require from one to several hours for completion.

As is the case with their repression during heat shock, 25°C messages are reactivated as a group during recovery. Considering the diverse nature of these messages, the similarity in their behavior is striking (Figs. 4 and 5). In sharp contrast, hsp mRNAs do not behave as a group and are repressed very asynchronously. Nevertheless, there is a distinct and reproducible pattern to their repression (Fig. 5) (DiDomenico et al. 1982b; Lindquist and DiDomenico 1985). hsp70 is always the first protein to be repressed and hsp82 is always the last. In some cases, the repression of these two messages is completely asynchronous. The repression of hsp70 may be nearly complete at a time when expression of hsp82 is still increasing.

Another major difference in the behavior of 25°C messages and hsp messages concerns their fate during repression. During heat shock, translationally repressed 25°C mRNAs are quantitatively retained in the cell. During recovery, hsp mRNAs are degraded as they are repressed (Fig. 7). In fact, their repression is so tightly coupled to their degradation that the two processes cannot be distinguished (DiDomenico et al. 1982b; Petersen and Lindquist 1988).

B. Regulation of hsp70 Message Degradation

The repression and degradation of hsp mRNAs is a highly regulated process. When cells are maintained at high temperatures, they continue to produce hsps, and virtually nothing but hsps, until after 24–48 hours, they begin to die (S. Lindquist, unpubl.). The same phenomenon is observed even when cells are treated with actinomycin D 1 hour after

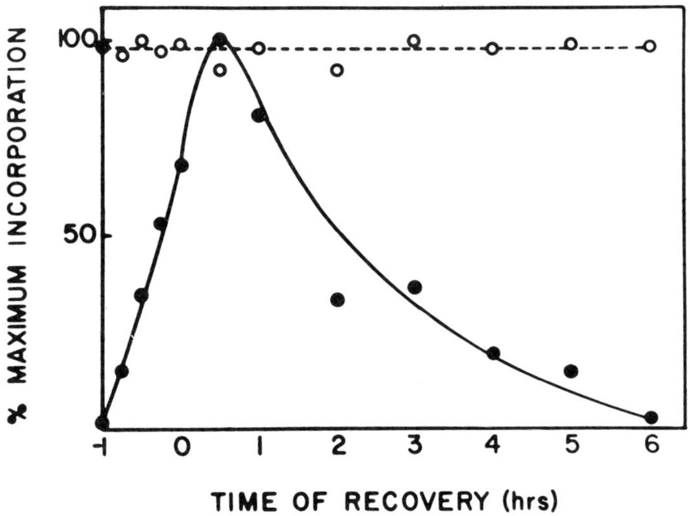

Figure 7 Concentration of hsp70 (filled circle) and actin (open circle) mRNAs during heat shock and recovery. Cells were heat-shocked at 36.5°C for 60 min and then returned to 25°C for recovery. RNA was extracted at the indicated times and analyzed by Northern blotting. (Reprinted, with permission, from DiDomenico et al. 1982b.)

temperature elevation. Thus, the continued synthesis of hsps at high temperatures is not dependent on continued transcription of hsp genes, but on continued translation of stable messages.

These same messages are degraded during recovery at 25°C with a timing that depends on the severity of the preceding heat shock. After a 30-minute heat shock at 37°C, synthesis of hsp70 is repressed within 5 hours of the return to 25°C. After a 30-minute heat shock at 39°C, synthesis of the protein continues unabated for at least 8 hours before it is repressed. Again, the same phenomena are observed if actinomycin D is added shortly after the heat shock. Degradation of hsp mRNAs is highly regulated and not simply a fixed consequence of the change in temperature. With a more severe heat shock, presumably, more hsp is required and degradation is regulated accordingly.

Since the hsp70 message is the most abundant of the heat shock messages and since it shows the greatest degree of regulation, it has received the most study. Is this an inherently unstable message that is stabilized by heat shock, or is it an inherently stable message that is destabilized during recovery? To answer this question, modified hsp70 coding sequences were placed under the control of the metallothionein promoter, so that its message could be induced at normal temperatures with very low concentrations of copper. Under these conditions, the message is extremely short-lived, with a half-life of less than 15 minutes. When these cells are heat shocked, however, the message is stabilized and has a half-life of greater than 4 hours. Apparently, a mechanism for the degradation of hsp70 message preexists in *Drosophila* cells at normal temperatures (Petersen and Lindquist 1988).

The sequences that target this message for regulated degradation have been partially characterized. First, an X-ray-induced *hsp70* deletion mutation, which replaces the 3′ half of the *hsp70* gene with an unknown sequence, was found to produce a message that is more stable than the wild-type message during recovery (Simcox et al. 1985). More recently, a variety of mutations were deliberately introduced into the *hsp70* gene, in the message leader, protein-coding, and 3′ untranslated regions (UTR), and examined for their effects on mRNA degradation. None of the protein-coding mutations, which covered, in different constructs, every codon of the message, affected the stability of the message in *cis*. (Because they produced aberrant proteins, certain of these mutations affected the recovery process as a whole, but none specifically affected the mutated message [see below; Rossi 1988; J. Rossi et al., in prep.].)

Mutations that replaced the 3′ UTR of the hsp70 message with sequences from an actin message, altering the stability of the message, suggested that these sequences were important for regulation of hsp70 mRNA during recovery (T. McGarry, unpubl.). To determine if the 3′ UTR was sufficient to target a message for regulated degradation, two *hsp70-alcohol dehydrogenase* gene fusions were examined (Petersen and Lindquist 1989). The first contained the *hsp70* promoter and message leader region fused to *adh* coding and 3′ untranslated sequences. This gene produced a message that was translated well at high temperatures and continued to be expressed at a high rate during recovery, whereas endogenous hsp70 messages were degraded. The second *adh* construct was identical to the first, except that the 3′ UTR was replaced with the 3′ UTR of the hsp70 message. This gene produced a message that was degraded with the same kinetics as the hsp70 message during recovery from a mild heat shock. Moreover, the degradation of the message was delayed, in the same way that degradation of the hsp70 message was delayed, after a more severe heat shock (Petersen and Lindquist 1989). Finally, when constitutive levels of synthesis were compared, the gene that carried the hsp70 3′ UTR was found to have a much lower level of adh expression (Fig. 8). Thus, the 3′ UTR of the hsp70 message plays a critical role in regulating its degradation, both during recovery from heat shock and in the absence of heat shock at normal temperatures.

The 3′ UTR of the hsp70 message is AU-rich and contains sequence elements that resemble those implicated in the turnover of unstable messages in mammalian cells, such as c-myc, c-fos, and various lymphokine mRNAs. Specifically, an AUUUA-like motif is present 54 nucleotides downstream from the translation stop codon and 104 nucleotides upstream of the polyadenylation site. Although the nucleotides specifically responsible for destabilizing the hsp70 message have not been identified, it seems more than a coincidence that both the c-myc and c-fos messages are stabilized by heat shock in mammalian cells (Sadis et al. 1988). Moreover, Theodorakis and Morimoto (1987) have demonstrated that the hsp70 message of human cells is much more stable after heat shock than before heat shock.

We propose that a highly conserved mechanism for the degradation of unstable messages is employed by hsp mRNAs to reduce their constitutive levels of expression at normal temperatures. The inactivation of this mechanism by heat shock, together with the transcriptional activation of the heat shock genes and the selective translation of hsp mRNAs, facilitates the extraordinarily rapid accumulation of heat shock proteins. It is the reactivation of this degradation mechanism, together with the transcriptional repression of heat shock genes (Ashburner 1970; DiDomenico et al. 1982a), that mediates the repression of hsp synthesis during recovery.

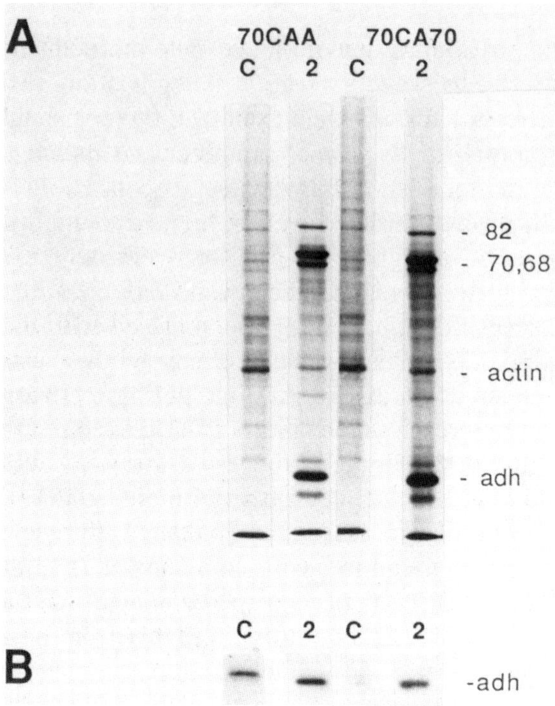

Figure 8 Constitutive expression is higher for adh mRNAs with the natural adh 3′ end than for those with the hsp70 3′ end. (*A*) Cells transformed with the plasmid p70CAA, with the adh 3′ end, or p70CA70, with the hsp70 3′ end, were labeled with [^{3}H]leucine at 25°C (*C*), or 2 hr after a 37°C heat shock (*2*). The positions of the heat shock proteins and adh are indicated on the right. (*B*) Protein samples from panel *A* were analyzed by Western blotting using an antibody against *Drosophila* adh and [^{125}I]-labeled Protein A. (Reprinted, with persmission, from Petersen and Lindquist 1989.)

C. Role of hsps in the Recovery Process

The repression of hsp mRNAs and the reactivation of 25°C messages during recovery is a remarkably precise and highly reproducible process (DiDomenico et al. 1982a). Several lines of evidence suggest that hsps themselves play an important role in it. First, when cells are given a mild heat treatment, which induces the synthesis of hsps, and are then given a second, severe heat treatment, they recover 25°C protein synthesis much more rapidly than do cells that are exposed to the severe heat treatment alone (Petersen and Mitchell 1981). Similarly, when cells are gradually raised to high temperatures, they recover from that exposure much more rapidly than cells that are suddenly raised to the same temperature. Certainly, this regimen of heat treatment more closely mimics conditions organisms encounter in their natural environment and presumably reflects the protective effects of hsp synthesis.

Second, for a given heat treatment, a specific quantity of heat shock protein is always made before 25°C protein synthesis is restored (DiDomenico et al. 1982a; Lindquist and DiDomenico 1985). A variety of treatments that interfere with the synthesis of hsps also interfere with the recovery process. For example, in one series of experiments, cells were treated with actinomycin D at various times during heat shock to limit the quantity of hsp message they produced. When these cells were returned to 25°C for recovery, those that produced hsps at a slower rate (because they had a reduced concentration of hsp message) took proportionately longer to return to normal patterns of synthesis (DiDomenico et al. 1982a). In another series of experiments, cells were treated with cycloheximide, to block hsp synthesis, immediately prior to a mild heat shock. When returned to 25°C, hsp70 messages were stable in these cells for at least 10 hours. When the addition of cycloheximide was delayed until after the heat shock, allowing the cells to produce a substantial quantity of hsps, the message was rapidly degraded (DiDomenico et al. 1982a).

Although it is not yet clear which hsps are required for recovery, the evidence suggests that hsp70, at least, plays an important role. First, hsp70 is the only protein whose accumulation is quantitatively related to the recovery of 25°C protein synthesis (DiDomenico et al. 1982a). That is, for a given heat shock, a specific quantity of hsp70 is always produced before recovery begins. The synthesis of other hsps is considerably more variable. Second, when the synthesis of hsp70 is selectively reduced, by transforming cells with genes constructed to produce antisense hsp70 RNA, the rate at which they return to normal patterns of synthesis is reduced in proportion (Melton 1988). Reducing the rate of hsp26 synthesis with antisense RNA has no such effect (McGarry and Lindquist 1986). Third, cells transformed with certain hsp70 protein-coding mutations show a marked delay in the recovery of normal protein synthesis after heat shock (Rossi 1988; J. Rossi et al., in prep.).

Superficially, there are two ways to view the involvement of hsps in restoring the normal pattern of protein synthesis: First, the block in the translation of 25°C messages and the block in the degradation of hsp messages might simply be regulatory mechanisms employed to maximize the synthesis of hsps. They would be released only when a sufficient quantity of hsp has been produced. Alternatively, high temperatures might damage the translation and message-degradation machinery of the cell. hsps, then, would be required to repair the damage. The difference between these two views, however, is largely artificial. Heat damages a wide variety of cellular processes. The heat shock genes have evolved within this context to bypass the damage. Not surprisingly, then, they effectively take advantage of the inability of the 25°C genes and mRNAs to function to maximize their own expression. Since the general role of the heat shock proteins is to repair normal cellular processes damaged by heat, certain of their functions are, by their very nature, autoregulatory. That is, they remove those conditions that they had evolved to exploit for their own synthesis. The challenge ahead lies in defining the nature of these regulatory mechanisms and their interactions with hsps in precise molecular terms.

V. SUMMARY

It has long been appreciated that transcriptional regulation plays an important role in the heat shock response. In certain organisms, the preferential translation of hsp mRNAs at high temperatures is also a well-established phenomenon. Most recently, mechanisms acting at the levels of RNA processing and selective message degradation have been shown to exert a profound effect on gene expression both during heat shock and during recovery from heat shock. These mechanisms have been most carefully studied in *Drosophila*, but they may be nearly universal in their application. The concerted action of all these mechanisms is responsible for the remarkably rapid and intense induction of hsps at high temperatures. That so many different mechanisms are coordinated to accomplish this induction is in keeping with the crucial roles these proteins play in providing protection from the toxic effects of heat. Given the extraordinary variety of cellular processes in which hsps participate, it is not surprising that they also affect those processes that influence their own synthesis.

ACKNOWLEDGMENTS

The authors thank the following individuals for sharing unpublished data: S.A. Amero, E. Berger, A.L. Beyer, E. Craig, R.P. Dellavalle, K. Golic, T. McGarry, K. Palter, J. Rossi, L. Sheldon, J.M. Sierra, and J. Solomon.

REFERENCES

Ashburner, M. 1970. The genetic analysis of puffing in polytene chromosomes of *Drosophila. Proc. R. Soc. Lond. B* **176:** 319–327.

Ballinger, D.G. and M.L. Pardue 1983. The control of protein synthesis during heat shock in *Drosophila* cells involves altered polypeptide elongation rates. *Cell* **33:** 103–114.

Bell, J., L. Neilson, and M. Pellegrini. 1988. Effect of heat shock on ribosome synthesis in *Drosophila melanogaster. Mol. Cell. Biol.* **8:** 91–95.

Berendes, H.D. 1968. Factors involved in the expression of gene activity in polytene chromosomes. *Chromosoma* **24:** 418–437.

Berger, E.M., M.P. Vitek, and C.M. Morganelli. 1985. Transcript length heterogeneity at the small heat shock protein genes of *Drosophila. J. Mol. Biol.* **186:** 137–148.

Beyer, A.L. and Y.N. Osheim. 1988. Splice site selection, rate of splicing, and alternative splicing on nascent transcripts. *Genes Dev.* **2:** 754–765.

Biessmann, H., F.-G. Falkner, H. Saumweber, and M.F. Walter. 1982. Disruption of the vimentin cytoskeleton may play a role in heat-shock response. In *Heat shock from bacteria to man* (ed. M.J. Schlesinger et al.), pp. 275–281. Cold Spring Harbor Laboratory, Cold Spring Harbor, New York.

Blackman, R.K. and M. Meselson. 1986. Interspecific nucleotide sequence comparisons used to identify regulatory and structural features of the *Drosophila* hsp82 gene. *J. Mol. Biol.* **188:** 499–515.

Bond, U. 1988. Heat shock but not other stress inducers leads to the disruption of a subset of snRNPs and inhibition of in vitro splicing in HeLa cells. *EMBO J.* **7:** 3509–3518.

Bond, U. and M.J. Schlesinger. 1986. The chicken ubiquitin gene contains a heat shock promoter and expresses an unstable mRNA in heat-shocked cells. *Mol. Cell. Biol.* **12:** 4602–4610.

Bonner, J.J., C. Parks, J. Parker-Thornburg, M.M. Mortin, and H.R.B. Pelham. 1984. The use of promoter fusions in *Drosophila* genetics: Isolation of mutations affecting the heat shock response. *Cell* **37:** 979–991.

Borst, P. 1986. Discontinuous transcription and antigenic variation in trypanosomes. *Annu. Rev. Biochem.* **55:** 701–732.

Bugaisky, G. 1981. "RNA metabolism during heat shock and recovery in *Drosophila*." Ph.D. thesis, University of Chicago.

Cervera, M., G. Dreyfuss, and S. Penman. 1981. Messenger RNA is translated when associated with the cytoskeletal framework in normal and vesicular stomatitis virus infected HeLa cells. *Cell* **23:** 113–120.

DiDomenico, B.J., G.E. Bugaisky, and S. Lindquist. 1982a. The heat shock response is self-regulated at both the transcriptional and posttranscriptional levels. *Cell* **31:** 593–603.

———. 1982b. Heat shock and recovery are mediated by different translational mechanisms. *Proc. Natl. Acad. Sci.* **79:** 6181–6185.

DiNocera, P. P. and I.B. Dawid. 1983. Transient expression of genes introduced into cultured cells of *Drosophila. Proc. Natl. Acad. Sci.* **80:** 7095–7098.

Dworniczk, B. and M.E. Mirault. 1987. Structure and expression of a human gene coding for a 70kd heat shock cognate protein. *Nucleic Acids Res.* **15:** 5181–5197.

Ellgaard, E.G. and U. Clever. 1971. RNA metabolism during puff induction in *Drosophila melanogaster. Chromosoma* **36:** 60–78.

Findly, R.C. and T. Pederson. 1981. Regulated transcription of the genes for actin and heat-shock proteins in cultured *Drosophila* cells. *J. Cell Biol.* **88:** 323–328.

Garbe, J.C. and M.L. Pardue. 1986. Heat shock locus 93D of *Drosophila melanogaster:* A spliced RNA most strongly conserved in the intron sequence. *Proc. Natl. Acad. Sci.* **83:** 1812–1816.

Glass, D.J., R.I. Polvere, and L.H.T. Van der Ploeg. 1986. Conserved sequences and transcription of the hsp70 gene family in *Trypanosoma brucei. Mol. Cell. Biol.* **6:** 4655–4666.

Glover, C.V.C. 1982. Heat shock induces rapid dephosphorylation of a ribosomal protein in *Drosophila*. *Proc. Natl. Acad. Sci.* **79:** 1781–1785.

Hackett, R.W. and J.T. Lis. 1983. Localization of the hsp82 transcript within a 3292 nucleotide sequence from the 63B heat shock locus of *D. melanogaster*. *Nucleic Acids Res.* **11:** 7011–7030.

Hickey, E., S.E. Brandon, G. Smale, D. Lloyd, and L.A. Weber. 1989. Sequence and regulation of a gene encoding a human 89-kilodalton heat shock protein. *Mol. Cell. Biol.* **9:** 2615–2626.

Holmgren, R., K. Livak, R. Morimoto, R. Freund, and M. Meselson. 1979. Studies of cloned sequences from four *Drosophila* heat shock loci. *Cell* **18:** 1359–1370.

Hultmark, D., R. Klemenz, and W. Gehring 1986. Translational and transcriptional control elements in the untranslated leader of the heat-shock gene hsp22. *Cell* **44:** 429–438.

Ingolia, T.D. and E.A. Craig. 1982. *Drosophila* gene related to the major heat shock induced gene is transcribed at normal temperature and not induced by heat shock. *Proc. Natl. Acad. Sci.* **79:** 525–539.

Ingolia, T.D., E.A. Craig, and B.J. McCarthy. 1980. Sequence of three copies of the gene for the major *Drosophila* heat shock induced protein and their flanking regions. *Cell* **21:** 669–679.

Jacq, B., R. Jourdan, and B.R. Jordan. 1977. Structure and processing of precursor 5S RNA in *Drosophila melanogaster*. *J. Mol. Biol.* **117:** 785–795.

Jamrich, M., A.L. Greenleaf, and E.K.F. Bautz. 1977. Localization of RNA polymerase in polytene chromosomes of *Drosophila melanogaster*. *Proc. Natl. Acad. Sci.* **74:** 2079–2083.

Kay, R.J., R.H. Russnak, D. Jones, C. Mathias, and E.P. Candido. 1987. Expression of intron-containing *C. elegans* heat shock genes in mouse cells demonstrates divergence of 3′ splice site recognition sequences between nematodes and vertebrates, and an inhibitory effect of heat shock on the mammalian splicing apparatus. *Nucleic Acids Res.* **15:** 3723–3741.

Kelley, P. and M. Schlesinger. 1982. Antibodies to two major chicken heat shock proteins cross-react with similar proteins in widely divergent species. *Mol. Cell. Biol.* **2:** 267–274.

Klemenz, R., D. Hultmark, and W. Gehring. 1985. Selective translation of heat shock mRNA in *Drosophila melanogaster* depends on sequence information in the leader. *EMBO J.* **4:** 2053–2060.

Kruger, C. and B.-J. Benecke. 1981. In vitro translation of *Drosophila* heat shock and non-heat shock mRNAs in heterologous and homologous cell-free systems. *Cell* **23:** 595–603.

Labhart, P. and R.H. Reeder. 1987. Heat shock stabilizes highly unstable transcripts of the *Xenopus* ribosomal gene spacer. *Proc. Natl. Acad. Sci.* **84:** 56–60.

Lewis, M., P.J. Helmsing, and M. Ashburner. 1975. Parallel changes in puffing activity and patterns of protein synthesis in salivary glands of *Drosophila*. *Proc. Natl. Acad. Sci.* **72:** 3604–3608.

Levis, E. and S. Penman. 1978. 5′ terminal structures of polyadenylate positive cytoplasmic messenger RNA and of polyadenylate positive and polyadenylate negative heterogeneous nuclear RNA of cells of dipteran *Drosophila melanogaster*. *J. Mol. Biol.* **120:** 487–516.

Lindquist, S. 1980a. Translational efficiency of heat induced messages in *Drosophila melanogaster* cells. *J. Mol. Biol.* **137:** 151–158.

———. 1980b. Varying patterns of protein synthesis in *Drosophila* during heat shock:

Implications for regulation. *Dev. Biol.* **77:** 463–469.
———. 1981. Regulation of protein synthesis during heat shock. *Nature* **294:** 311–314.
———. 1987. Translational regulation in the heat-shock response of *Drosophila* cells. In *Translational regulation of gene expression* (ed. J. Ilan), pp. 187–207. Plenum Publishing, New York.

Lindquist, S. and Craig, E.A. 1988. The heat-shock proteins. *Annu. Rev. Genet.* **22:** 631–677.

Lindquist, S. and B. DiDomenico. 1985. *Coordinate and noncoordinate gene expression during heat shock: A model for regulation.* Academic Press, New York.

Maniak, M. and W. Nellen. 1988. A developmentally regulated membrane protein gene in *Dictyostelium discoideum* is also induced by heat shock and cold shock. *Mol. Cell. Biol.* **8:** 153–159.

Maroto, F.G. and J.M. Sierra. 1988. Translational control in heat-shocked *Drosophila* embryos. *J. Biol. Chem.* **263:** 15720–15725.

———. 1989. Purification and characterization of mRNA cap-binding protein from *Drosophila melanogaster* embryos. *Mol. Cell Biol.* **9:** 2181–2190.

Mathews, M., ed. 1986. *Current communications in molecular biology: Translational control*, pp. 86–90. Cold Spring Harbor Laboratory, Cold Spring Harbor, New York.

Mayrand, S. and T. Pederson. 1983. Heat shock alters nuclear ribonucleoprotein assembly in *Drosophila* cells. *Mol. Cell. Biol.* **3:** 161–171.

McGarry, T.J. and S.L. Lindquist. 1985. The preferential translation of *Drosophila* hsp70 mRNA requires sequences in the untranslated leader. *Cell* **42:** 903–911.

———. 1986. Inhibition of heat shock protein synthesis by heat-inducible antisense RNA. *Proc. Natl. Acad. Sci.* **83:** 399–403.

McKenzie, S.L. 1976. "Protein and RNA synthesis induced by heat in *Drosophila.*" Ph.D. thesis, Harvard University, Cambridge.

McKenzie, S.L., S. Henikoff, and M. Meselson. 1975. Localization of RNA from heat-induced polysomes at puff sites in *Drosophila melanogaster*. *Proc. Natl. Acad. Sci.* **72:** 1117–1121.

Melton, D.A., ed. 1988. *Current communications in molecular biology: Antisense RNA and DNA*, pp. 71-78. Cold Spring Harbor Laboratory, Cold Spring Harbor, New York.

Muhich, M.L. and J.C. Boothroyd. 1988. Polycistronic transcripts in trypanosomes and their accumulation during heat shock: Evidence for a precursor role in mRNA synthesis. *Mol. Cell. Biol.* **8:** 3837–3846.

———. 1989. Synthesis of trypanosome hsp70 mRNA is resistant to disruption of trans-splicing by heat shock. *J. Biol. Chem.* **264:** 7107–7110.

Muhich, M.L., M.P. Hsu, and J.C. Boothroyd. 1989. Heat shock disruption of trans-splicing in trypanosomes: Effect on Hsp70, Hsp85 and tubulin mRNA synthesis. *Gene* **82:** 169–175.

Olsen, A.S., D.F. Triemer, and M.M. Sanders. 1986. Dephosphorylation of S6 and expression of the heat shock response in *Drosophila melanogaster*. *Mol. Cell. Biol.* **3:** 2017–2027.

Parker, K.A. and U. Bond. 1989. Analysis of pre-rRNAs in heat-shocked HeLa cells allows identification of the upstream termination site of human polymerase I transcription. *Mol. Cell. Biol.* **9:** 2500–2512.

Pauli, D., C.-W. Tonka, and A. Ayme-Southgate. 1988. An unusual split *Drosophila* heat shock gene expressed during embryogenesis, pupation and in testis. *J. Mol. Biol.* **200:** 47–53.

Pelham, H.R.B. 1986. Speculations on the functions of the major heat shock and glucose-regulated proteins. *Cell* **46:** 959–961.

———. 1988. Heat-shock proteins: Coming in from the cold. *Nature* **332:** 776–777.

Petersen, N. and H.K. Mitchell. 1981. Recovery of protein synthesis following heat shock; pre-heat treatment affects mRNA translation. *Proc. Natl. Acad. Sci.* **78:** 1708–1711.

Petersen, R. and S. Lindquist. 1988. The *Drosophila* hsp70 message is rapidly degraded at normal temperatures and stabilized by heat shock. *Gene* **72:** 161–168.

———. 1989. Regulation of hsp70 synthesis by messenger RNA degradation. *Cell Reg.* **1:** 135–149.

Proudfoot, N.J. and E. Whitelaw. 1988. Termination and 3′ end processing of eukaryotic RNA. In *Frontiers of molecular biology*, pp. 97–129. IRL Press, Oxford.

Rossi, J. 1988. "Investigations into the function of heat shock proteins." Ph.D. thesis, University of Chicago.

Rubin, G.M. and D.S. Hogness. 1975. Effect of heat shock on the synthesis of low molecular weight RNAs in *Drosophila:* Accumulation of a novel form of 5S RNA. *Cell* **6:** 207–213.

Sadis, S., E. Hickey, and L.A. Weber. 1988. Effect of heat shock on RNA metabolism in HeLa cells. *J. Cell Physiol.* **135:** 377–386.

Sanders, M.M., D.F. Triemer, and A.S. Olsen. 1986. Regulation of protein synthesis in heat-shocked *Drosophila* cells: Soluble factors control translation in vitro. *J. Biol. Chem.* **261:** 2189–2196.

Scott, M.P. and M.L. Pardue. 1981. Translational control in lysates of *Drosophila melanogaster* cells. *Proc. Natl. Acad. Sci.* **78:** 3353–3357.

Scott, M.P., R.V. Storti, M.L. Pardue, and A. Rich. 1979. Cell-free protein synthesis in lysates of *Drosophila melanogaster* cells. *J. Biol. Chem.* **18:** 1588–1593.

Simcox, A.A., C.M. Cheney, E.P. Hoffman, and A. Shearn. 1985. A deletion of the 3′ end of the *Drosophila melanogaster HSP70* gene increases stability of mutant RNA during recovery from heat shock. *Mol. Cell Biol.* **5:** 3397–3402.

Sirkin, E. and S. Lindquist. 1985. Translational regulation in the *Drosophila* heat shock response. In *Sequence specificity in transcription and translation* (ed. R. Calender and L. Gold), pp. 669–679. Alan R. Liss, New York.

Spradling, A., M.L. Pardue, and S. Penman. 1977. Messenger RNA in heat-shocked *Drosophila* cells. *J. Mol. Biol.* **109:** 559–587.

Spradling, A., S. Penman, and M.L. Pardue. 1975. Analysis of *Drosophila* mRNA by in situ hybridization: Sequences transcribed in normal and heat shocked cultured cells. *Cell* **4:** 395–404.

Storti, R.V., M.P. Scott, A. Rich, and M.L. Pardue. 1980. Translational control of protein synthesis in response to heat shock in *D. melanogaster* cells. *Cell* **22:** 825–834.

Theodorakis, N.G. and R.I. Morimoto. 1987. Posttranscriptional regulation of hsp70 expression in human cells: Effects of heat shock, inhibition of protein synthesis, and adenovirus infection on translation and mRNA stability. *Mol. Cell Biol.* **7:** 4357–4368.

Velazquez, J.M. and S. Lindquist. 1984. Hsp70: Nuclear concentration during environmental stress and cytoplasmic storage during recovery. *Cell* **36:** 655–662.

Vincent, M. and R.M. Tanguay. 1982. Different intracellular distributions of heat-shock and arsenite-induced proteins in *Drosophila* Kc cells: Possible relation with the phosphorylation and translation of a major cytoskeletal protein. *J. Mol. Biol.* **162:** 365–378.

Welch, W.J. and L.A. Mizzen. 1988. Characterization of the thermotolerant cell. II. Effects on the intracellular distribution of heat-shock protein 70, intermediate filaments, and small nuclear ribonucleoprotein complexes. *J. Cell Biol.* **106:** 1117–1130.

Wright-Sandor, L.G., M. Reichlin, and S.L. Tobin. 1989. Alteration by heat shock and immunological characterization of *Drosophila* small nuclear ribonucleoproteins. *J. Cell Biol.* **108:** 2007–2016.

Yost, H.J. and S. Lindquist. 1986. RNA splicing is interrupted by heat shock and is rescued by heat shock protein synthesis. *Cell* **45:** 185–193.

———. 1988. Translation of unspliced transcripts after heat shock. *Science* **242:** 1544–1548.

17

Modular Units of Heat Shock Regulatory Regions: Structure and Function

John T. Lis, Hua Xiao, and Olga Perisic
Section of Biochemistry, Molecular and Cell Biology
Cornell University
Ithaca, New York 14853

I. INTRODUCTION

Heat shock genes possess multiple copies of a sequence called the heat shock element (HSE) that is required for the stress-induced transcription of these genes. This element binds tightly to a regulatory protein called heat shock transcription factor (HSF). In this chapter, we summarize genetic and biochemical data that have led to the definition of both the DNA sequences required to create a heat-shock-inducible regulatory region and the interaction of these sequences with HSF. Also, we speculate on how the large complex of HSE/HSF may influence the frequency of transcription of heat shock genes.

Pelham (1982) and Mirault et al. (1982) reported the first functional analyses of the regulatory region of the *Drosophila* hsp70 gene by examining the expression of a series of deletion mutations that were transfected into monkey COS cells. Pelham showed that sequences between nucleotides –66 and –47 upstream of the hsp70 gene are necessary for heat-induced expression, and he identified a consensus CTGGAAT-

Stress Proteins in Biology and Medicine.

TTCTAGA sequence (later simplified to the 14 bp sequence C--GAA--TTC--G) that is found upstream of all *Drosophila* heat shock genes. Regions upstream of a variety of heat shock genes in organisms as diverse as yeast and man possess sequences that match part or all of the HSE consensus (Nover 1987). The HSF protein, which binds to these sequences, has been purified from *Drosophila*, yeast, and human cells (Sorger and Pelham 1987; Wiederrecht et al. 1987; Wu et al. 1987; Goldenberg et al. 1988), and it stimulates transcription of heat shock genes in vitro (Wiederrecht et al. 1987; Goldenberg et al. 1988) and in *Xenopus* oocytes (Wu et al. 1987). The gene coding for HSF has been cloned from yeast (Sorger and Pelham 1988; Wiederrecht et al. 1988).

Pelham and Bienz (1982) demonstrated that the fusion of the HSE consensus sequence to the promoter region of the viral thymidine kinase gene caused this gene to be heat-shock-inducible when transfected into COS cells (Pelham and Bienz 1982). However, analyses of 5′ deletion mutants of the *Drosophila* hsp70 gene in *Drosophila* cells (Amin et al. 1985) or germ-line transformants (Dudler and Travers 1984; Simon et al. 1985) revealed that the promoter-proximal HSE alone, which contained one perfect match to the consensus sequence, conferred only approximately 1% of the wild-type level of heat-induced expression. The minimal additional upstream sequences that were required to specify the wild-type level of heat-induced hsp70 expression contain an additional partial match to the 14-bp HSE consensus sequence as first noted by Dudler and Travers (1984). Thus, it was hypothesized that two copies of the 14-bp HSE consensus sequence are required for wild-type levels of heat-shock-induced expression of the *Drosophila* hsp70 gene in *Drosophila*. As summarized here, the tests of this hypothesis by assays of expression of regulatory region variants in transgenic flies and by in vitro protein/DNA binding assays of purified HSF to variant regulatory regions have led to a new view of both the structural organization of the heat shock regulatory elements and the interaction of these elements with HSF (Xiao and Lis 1988; Perisic et al. 1989).

II. DISCUSSION

A. Heat-shock-induced Expression of Genes Possessing Variant Regulatory Regions

The 14-bp consensus sequence for heat shock elements has been used to identify potential HSEs in heat shock genes from a wide variety of organisms; however, the importance of specific sequences in and around known functional HSEs have only recently been examined in detail (Amin et al. 1988; Xiao and Lis 1988). An hsp70 gene deletion to –89, which contains two separate matches to the heat shock consensus, sites I and II (Fig. 1a), is expressed in transgenic animals at a level similar to that of the wild-type endogenous gene. Surprisingly, we found that the replacement of the native heat shock regulatory region of this –89 deletion mutant with a sequence containing two perfect copies of the 14-bp heat shock consensus reduces the level of induced hsp70 gene expression in transgenic flies to 7% of the wild type (Fig. 1b) (Xiao and Lis 1988). The two synthetic consensus sequences in this construct are in the same positions as the two corresponding native copies; that is, the number of base pairs between the two consensus sequences as well as between the proximal consensus sequence and the TATA box are identical to those of the native hsp70 regulatory region. However, the sequence of nucleotides immediately flanking the synthetic copies of the 14-bp consensus differs from those flanking the corresponding wild-type copies.

The restoration of the flanking sequence immediately downstream from site I to the native hsp70 sequence (see underlined nucleotides of Fig. 1b) increases the level of heat-induced expression from 7% to 27% (Fig. 1c) (Xiao and Lis 1988). Recently, Voellmy and his colleagues have shown independently that the sequences flanking the proximal heat

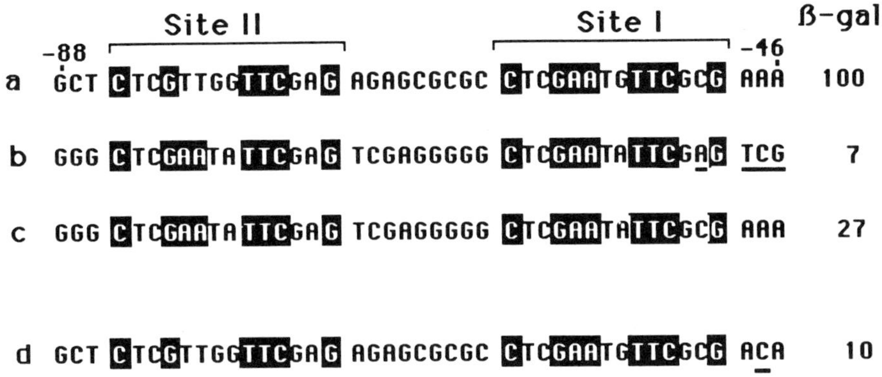

Figure 1 Sequences upstream of hsp70-lacZ genes and the relative level of heat-induced β-galactosidase activity produced by these genes. Constructs *a–c* were assayed as transgenic *Drosophila* lines (Xiao and Lis 1988), and construct *d* was assayed by transfection of *Drosophila* S3 cells (Amin et al. 1988). The pairs of 14-bp consensus sequences in all constructs have the same spacing relative to each other and to the TATA box as the wild-type gene. Transformants in *a–c* contained single inserts. The average β-galactosidase level (β-gal) of several transformants of each variant is listed to the right of each sequence as the percentage of wild-type activity of the –89 deletion shown in *a* (Xiao and Lis 1988). Construct *d* was also reported as the percent activity relative to a wild-type construct, in this case a deletion to –88 (Amin et al. 1988).

shock consensus sequences of site I have a role in the induction of the hsp70 gene (Amin et al. 1988). The substitution of a single base at a position two nucleotides downstream from the native site I reduces the heat-induced expression of hsp70 gene to 10% that of wild type as assayed in transfected *Drosophila* cells (Fig. 1d). The results obtained with these and with additional regulatory region variants indicate that sequences immediately flanking the 14-bp consensus sequences at both site I and site II have a role in the heat shock response (Amin et al. 1988; Xiao and Lis 1988).

B. A 5-bp Sequence -GAA- Is the Basic Repeating Unit of Heat Shock Regulatory Regions of *Drosophila*

Inspection of the genetically defined regulatory regions of a variety of *Drosophila* heat shock genes reveals a simple repeating sequence that includes the nucelotides flanking the consensus (Amin et al. 1988; Xiao and Lis 1988). The last nucleotide of the 14-bp consensus at site I of hsp70, together with the two adjacent flanking nucleotides, generates the trinucleotide GAA that is also present twice as an inverted repeat (-GAA--TTC-) at the center of the 14-bp consensus (Fig. 2). Likewise, the first base of the site II consensus of the hsp70 gene at 87A and the two adjacent flanking nucleotides generate the trinucleotide TTC. Nucleotides GAA, or their complement TTC, are the conserved core of a 5-bp repeating unit of native heat shock regulatory regions (Fig. 3A) (Amin et al. 1988; Xiao and Lis 1988). For example, sites I and II of the hsp70 gene can each be regarded as an imperfect match to an array of four 5-bp units, where each unit is inverted relative to adjacent ones (Xiao and Lis 1988).

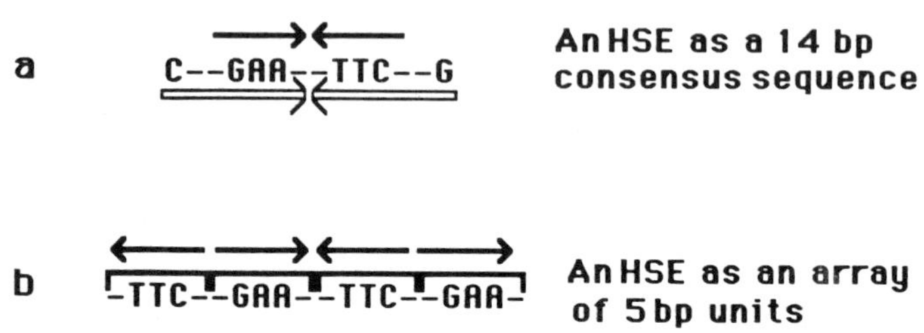

Figure 2 Relationship of the original 14-bp consensus to the repeating arrays of the 5-bp unit. Solid arrows are drawn above each 5-bp unit, and open arrows are drawn below each half of the 14-bp consensus sequence. A pair of 5-bp units in opposite orientation overlap the center of the 14-bp sequence. An array of four 5-bp repeats shown at the bottom of the figure encompasses the original 14-bp consensus sequence.

(A)

```
                           cgcCtcGAAtgTTCgcGAAa  -46       hsp70
                           aTTCtcGttgcTTCgaGAga  -69(H8)
                           gcTCtcGttggTTCgaGAga  -69(E3)
                      aGAAaacTCgaGAAatTTCt      -182
                                cGAAtaTTCtaGAAt -247(H8)
                                aGAAtaTTCtaGAAt -247(E3)
caTCcaGAAgccTCtaGAAgtTTCtaGAgacTTCcaGttc         -49       hsp83
          gTgCcaGAAagagCcaGAAgaTgCgaGAga        -270       hsp27
                aGAAatgTCaaGAAgtTTCtgGttc       -328
                           aGAAacTcCcaGAAa      -353
                     tTTCcgGActcTTCtaGAAa        -45       hsp26
                     tcTCtaGAAacTTCg            -338
      aGAAaaTTCgaGAgagTgCcgGtAttTTCtaGAtt        -59       hsp22
                aGAgagccCcaGAAacTTCc            -185
          gccCgaGAAgtTTCgtGtcccTTCtcGtgt        -232       hsp23
          tgTCtcGAAgtTTCgcGAAttTaCt             -391
```

(B)

	Base Usage at Each Position of the 5 bp Unit				
	1 a	2 G	3 A	4 A	5 a
A	49 (0.00)	0 (5.55)	68 (0.00)	58 (0.00)	36 (0.00)
T	2 (2.99)	0 (5.55)	8 (2.09)	7 (2.05)	18 (0.68)
G	12 (0.97)	85 (0.00)	5 (2.12)	14 (0.99)	19 (0.22)
C	22 (0.38)	0 (5.14)	4 (2.32)	6 (1.79)	12 (0.67)

Figure 3 (*A*) Nucleotide sequences containing the conserved 5-bp repeating unit in the upstream regions of *D. melanogaster* heat shock genes. Bases that match the three conserved nucleotides TTC or GAA of the 5-bp unit are in capital letters. The position of the last nucleotide is indicated to the right of each sequence, and the numbering used is from the original publications. H8 indicates sequences of the hsp70 variant from cytological locus 87A, and E3 indicates sequences of the hsp70 variant from 87C. Only those elements that have been shown to be functional are given: hsp70 (Dudler and Travers 1984; Amin et al. 1985, 1987; Topol et al. 1985; Simon et al. 1985; Simon and Lis 1987), hsp83 (Zimarino and Wu 1987; Xiao and Lis 1989), hsp27 (Hoffman and Corces 1986; Riddihough and Pelham 1986), hsp26 (Cohen and Meselson 1985; Pauli et al. 1986; Simon and Lis 1987; Thomas and Elgin 1988), hsp22 (Ayme et al. 1985; Klemenz and Gehring 1986), and hsp23 (Mestril et al. 1985, 1986). (*B*) Base frequency at each position of the 5-bp unit and the reductions in free energy of binding by particular nucleotide changes from the consensus as calculated using the theory of Berg and von Hippel (1988). The calculation is based on a total of 85 5-bp units listed in *A*, which are compiled as a -GAA- motif. The calculation takes into account the frequency of base occurrence at each position and the base composition of the genome. The most frequent base at a position is assigned as the consensus base. The numbers that are not in parentheses indicate the actual number of occurrences of each base at each of the five positions. The calculated estimates of the reduction (in units of kT, where k is the Boltzmann constant and T is absolute temperature) in free energy of binding when a base substitutes for a consensus base are in parentheses. The summation of these contributions at each position within a binding site provides an approximation of relative free energy of binding of that sequence.

We constructed a synthetic regulatory region containing two arrays of four contiguous inverted 5-bp repeats and tested its activity by the germ-line transformation assay (Xiao and Lis 1988). In this variant, the two arrays are separated by 11 bp. The hsp70-lacZ gene, to which this regulatory region is fused, is expressed at a sixfold higher level than the hsp70 gene containing the pair of perfect 14-bp consensus sequences (Fig. 1b), and at a fivefold higher level than an hsp70 gene that has two native HSE elements separated by 11 bp (Xiao and Lis 1988).

The number of the 5-bp units in a functional heat shock element can vary. For example, although sites I, II, and III of the hsp70 gene each contain an imperfect match to an array of four 5-bp units (Figs. 1 and 2), site IV in the hsp70 gene, which binds HSF tightly (Topol et al. 1985), contains only three 5-bp units. In addition, the arrangement (phase) of the 5-bp unit can also vary. Sites I and II terminate with -GAA-, and site III terminates with -TTC-, although all three contain the same number of the 5-bp units. Even larger arrays of the 5-bp units are found in regulatory regions of heat shock genes. For example, the hsp83 heat shock regulatory region can be viewed as eight contiguous 5-bp units, although as with hsp70, not all are a perfect match to the consensus (Fig. 3A) (Xiao and Lis 1988). Thus, we propose that the HSE can be viewed as a contiguous array of the 5-bp module -GAA- (or its complement -TTC-), where each module is inverted relative to the immediately adjacent modules.

The hsp83 heat shock regulatory region was identified originally as having three overlapping 14-bp consensus sequences, and it was speculated that this overlap may cause strong cooperative binding of the transcription regulatory protein, HSF, and thereby account for high basal level of expression of this gene (Wu 1984; Xiao and Lis 1986; Bienz and Pelham 1987). However, analyses of the expression of 5′ deletion mutants of the hsp83 gene in transgenic flies reveal that the high basal activity is dependent on sequences further upstream of the HSE, and the deletion of these basal level sequences causes the hsp83 gene regulation to resemble that of hsp70 (Xiao and Lis 1989). In addition, in vivo polymer competition assays showed that the regulatory regions of hsp83 and hsp70 have similar affinities for a heat shock regulatory factor (Xiao and Lis 1989). Thus, the heat shock regulatory regions of the hsp70 and hsp83 genes are functionally equivalent, which is consistent with the view that both are composed of contiguous nonoverlapping arrays of the same 5-bp unit.

The significance of nucleotide identity at all five positions of the repeat is not known. Although a complete mutagenic analysis of all five sites of the repeat changed to all possible alternative bases has not yet been performed, analyses of a few substitution mutants and the comparison of HSEs of many genes have revealed important nucleotides. Within the five nucleotides in this basic -GAA- unit, the G is absolutely conserved. A single-base substitution at this position as in derivatives of construct c of Figure 1 can reduce the heat-induced expression to background levels (Xiao and Lis 1988). The two As are highly conserved, although substitution of G for the second A may be permissible (Fig. 3). Also, sequence comparisons of nucleotides in the less conserved positions 1 and 5 reveal that the base composition of these two positions may be nonrandom (Fig. 3B). In particular, an A appears frequently at position 1, whereas T at this position is found rarely among the 5-bp units of genetically defined heat shock regulatory regions. Substitution of T for A at this position in the hsp70 heat shock regulatory region has recently been shown to cause a dramatic reduction in heat shock expression of the hsp70 gene (Amin et al. 1988).

C. The 5-bp Unit Has Been Conserved through Evolution

The heat shock response is highly conserved in organisms ranging from bacterium to man. Even the regulatory regions of yeast and human heat shock genes have at their core a sequence that matches well the *Drosophila* HSE (for review, see Nover 1987). Moreover, heat shock regulatory regions that are functional in one organism usually function efficiently when transferred to another organism. Although one variant HSE was found that specifies binding of HSF and heat-inducible activity in human but not in yeast cells (Sorger et al. 1987), the conservation through evolution of the structure and function of the HSEs is extraordinary.

The generality of the view that the HSE is a contiguous array of alternately oriented 5-bp units has support from studies in yeast. Slater and Craig (1987) examined the minimal sequence required to specify heat shock inducibility of a yeast hsp70 gene. They found that an HSE that has only the 14-bp consensus and one 3′ flanking nucleotide was not active in yeast cells as a heat shock element when placed upstream of the CYC1 gene. However, the inclusion of 4 bases upstream and 4 bases downstream from the consensus rendered the gene heat-inducible. The inclusion of the extra pair of nucleotides upstream of the 14-bp region in this construct preserved a third 5-bp unit, -TTC-.

Heat shock regulatory regions have been identified in a variety of organisms. They all possess sequences that match the HSE consensus (Nover 1987). The new information on the key features of the heat shock consensus should help in searching for heat shock regulatory elements and in discerning which of the putative heat shock elements that have already been identified by sequence searches are likely to be functional.

D. The Interaction of the 5-bp Units with HSF

If each 5-bp unit represents a target of interaction with HSF, then the length of sequence covered by HSF should be proportional to the number of 5-bp units. We have generated a series of cloned DNAs that contain an array consisting of three to nine alternately oriented 5-bp units. Purified HSF binds to these arrays and protects the repeats from DNase I cleavage. The size of the resulting footprints increases by 5-bp increments (on average) as the length of an array is expanded from 15 bp (3 repeats) to 45 bp (9 repeats) (Perisic et al. 1989). The actual increase in footprint size with addition of each 5-bp unit to the array appears to cycle between 4 and 6 bp (Fig. 4). This is not surprising, since each bound HSF protein is arranged in an orientation opposite that of the adjacent HSF. For every addition of 10 bp, a dyad of two 5-bp repeats, the length of the footprint increases by precisely 10 bp, as predicted in our model.

The smallest array that shows stable binding to HSF is the 10-bp dyad symmetry element (-TTC--GAA-), which contains two 5-bp units (Perisic et al. 1989). Surprisingly, the 10-bp sequence, -GAA--TTC-, which has the 5-bp units in a permuted configuration, also binds HSF. We have found that both of these 5-bp unit arrangements, which we refer to as tail-to-tail and head-to-head, respectively, bind to HSF with similar affinity. High-resolution footprinting assays using the chemical cleavage agent methidiumpropyl-EDTA·Fe(II) (MPE) show that the binding of HSF covers and is centered on each of these 10-bp dyad symmetry elements (Fig. 5). These results suggest that the 5-bp unit is a binding site for HSF and that stable binding is provided by two adjacent binding sites in opposing orientation.

Further support of a model in which the 5-bp sequence represents a unit of interaction with HSF is provided by detailed DNA interference analyses. We examined the pattern of interference on the two 10-bp sequences modified by methylation with dimethylsulfate or ethylation with ethylnitrosourea, or by nicking the phosphate backbone with DNase I (Perisic et al. 1989). The sites of modification that interfere with binding of HSF show perfect symmetry about the central axes in both head-to-head and tail-to-tail arrangements of two inverted 5-bp units. This was true of all three modifying agents. This provides strong support for each 5-bp unit's being a target of interaction with HSF protein. Since the HSF monomer possesses a rather discrete DNA-binding domain that is not ob-

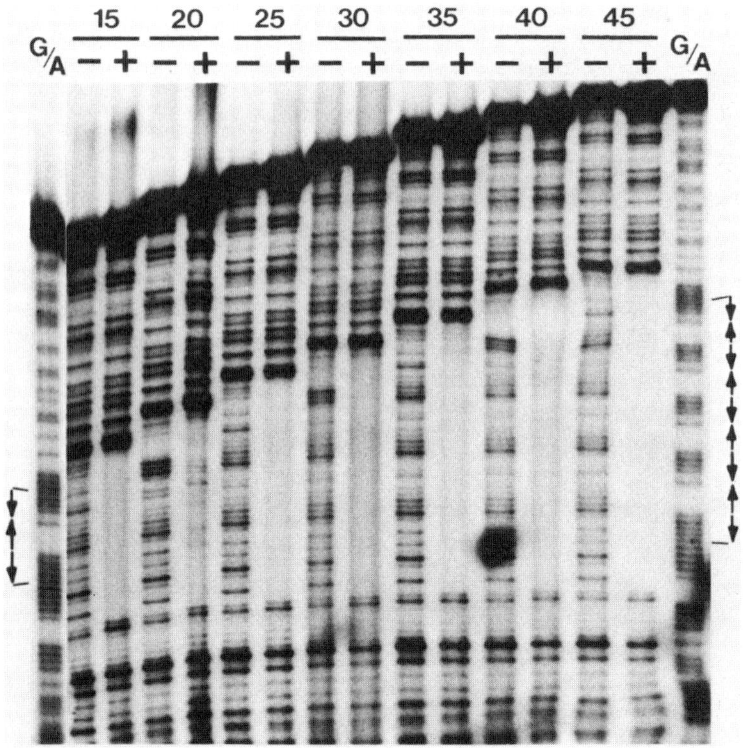

Figure 4 DNase I footprints of HSF bound to a series of arrays containing increasing numbers of inverted 5-bp units. DNA fragments containing different lengths of arrays of inverted 5-bp units (from 15 bp to 45 bp) were incubated with HSF purified from heat-shocked *Drosophila* Kc cells. Complexes (+) or protein-free DNA (–) were cleaved with DNase I, and DNA fragments were analyzed on a sequencing gel. Data obtained with one strand are presented. Maxam and Gilbert sequencing markers (G+A) were used to locate the footprinted regions. Arrows indicate the orientation and location of 5-bp units in the longest array (45 bp) and the shortest array (15 bp).

viously repeated in the molecule (Wiederrecht et al. 1988), we propose that an HSF molecule is bound to each 5-bp unit.

How does HSF bind to the variants of HSEs that contain different numbers and arrangements of 5-bp units? This flexibility in binding seems to be provided by the multimeric nature of HSF. In both *Drosophila* (Perisic et al. 1989) and yeast (Sorger and Nelson 1989), HSF appears to exist as a protein trimer when free or bound to DNA. The trimeric nature of *Drosophila* HSF was first suggested by comparing the electrophoretic mobility of HSF/DNA complexes generated with dif-

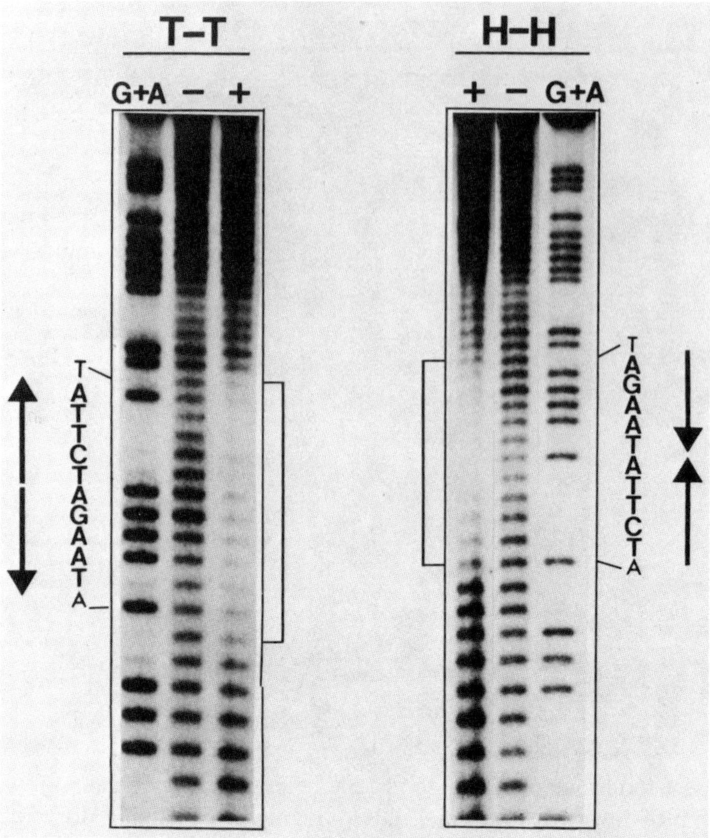

Figure 5 MPE footprints of tail-to-tail (T-T) and head-to-head (H-H) 10-bp dyad symmetry sequences. End-labeled DNA fragments containing 10-bp dyad symmetry sequences were incubated with HSF and then cleaved with MPE. The cleavage pattern of DNA isolated from the complexes (+) was compared with the pattern obtained with protein-free DNA (–). Brackets indicate protected regions. Arrows indicate the orientations of 5-bp units.

ferent arrays of 5-bp units (Perisic et al. 1989). The sizes of the complexes increase most distinctly with every three 5-bp units, whereas footprinting assays showed that the contacts of HSF with DNA grow with the addition of every 5-bp unit. This apparent discrepancy can be explained if HSF binds to DNA as a trimer in which contacts of two subunits with DNA are sufficient to generate a stable protein/DNA complex. HSEs with two or three 5-bp units produce complexes with the same apparent size on native gels. We predict that each of these complexes has a single HSF trimer bound, but in the case of an HSE containing two 5-bp units only two HSF subunits interact with DNA (Fig. 6). The addition of the third 5-bp unit would allow the third subunit of the HSF trimer to interact with DNA without a concomitant increase in the size of the HSF/DNA complex (Fig. 6). An array with four 5-bp units could potentially bind two trimers with two subunits of each trimer directly interacting with DNA. Arrays with five and six subunits would also have two trimers bound, but the number of HSF subunits in contact with DNA would increase with each 5-bp unit. As illustrated in Figure 6, the trimer model also explains the equivalent affinity of HSF for the head-to-head and tail-to-tail dimers of 5-bp units.

Additional support for the trimeric structure of *Drosophila* HSF comes from chemical cross-linking studies of purified HSF. The major cross-linked form has an apparent molecular weight of 350,000, approximately the size expected for an HSF trimer (Perisic et al. 1989).

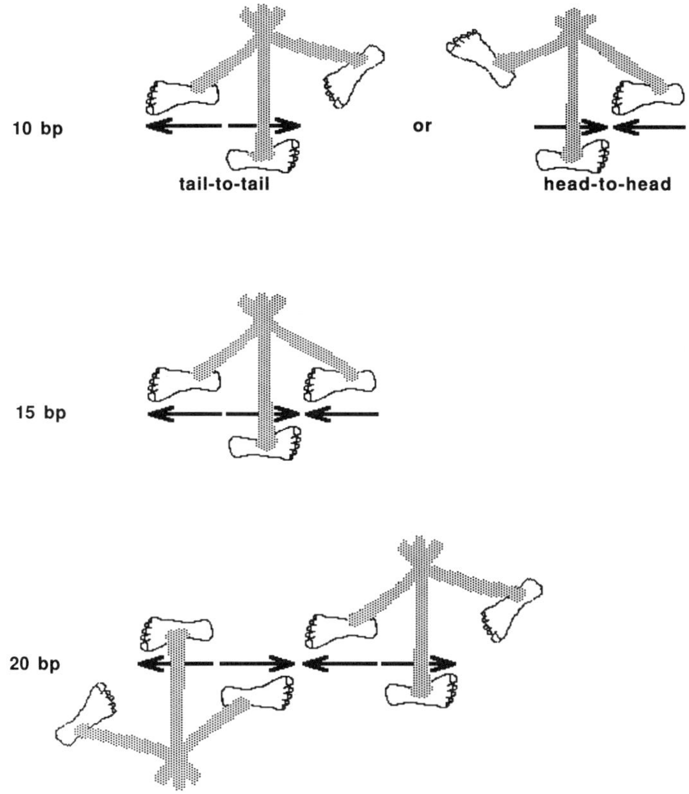

Figure 6 Model for interaction of HSF trimers with different arrays of 5-bp units. Arrows designate the 5-bp units and their relative orientation. The foot represents the DNA-binding domain of HSF, and the legs join the subunits of the trimer. In yeast, the HSF trimerization domain is localized to a region postulated to form three-stranded coiled-coil (Sorger and Nelson 1989).

E. Significance of the Revised View of Heat Shock Elements

The revised view of an HSE affects several facets of heat shock regulatory regions and their interaction with HSF. First, it predicts a different stoichiometry of HSF binding than does the 14-bp consensus. The 14-bp consensus contains two inverted 7-bp sequences, and sites I and II of the hsp70 gene would each be expected in the simplest model to bind a dimer of HSF. If these sites are regarded instead as two imperfect arrays of four inverted 5-bp units, they could each bind two trimers of HSF during heat shock. At an apparent molecular weight of 110,000 per monomer (Wu et al. 1987), the protein component in this complex would exceed 1300 kD. Together with the more upstream sites III and IV there could be over 2300 kD of HSF upstream of the hsp70 promoter. Although HSF molecules at sites I and II appear to interact (Topol et al. 1985; Shuey and Parker 1986), it has not been demonstrated whether HSF molecules bound to the further upstream sites III and IV of the hsp70 gene come together by DNA looping to form a single giant complex.

Second, overlapping HSEs have been noted for hsp83 and other heat shock genes. The manner in which two molecules of HSF recognize the same (overlapping) bases has been a matter of concern (Bienz and Pelham 1987). Now, the overlapping and nonoverlapping 14-bp HSEs can both be considered simply as arrays of contiguous nonoverlapping 5-bp units. Indeed, this structural similarity of overlapping and nonoverlapping 14-bp HSEs is consistent with their recently demonstrated functional similarity (Xiao and Lis 1989).

Third, genetic tests have shown that a few bases flanking the 14-bp consensus in natural and synthetic heat shock regulatory regions are required to achieve high levels of heat-shock-induced expression in yeast (Slater and Craig 1987; H. Xiao and J.T. Lis, unpubl.) and *Drosophila* (Amin et al. 1988; Xiao and Lis 1988). These results can be understood in terms of the 5-bp unit. Heat-shock-induced transcription requires multiple copies of DNA-bound HSF, and the flanking bases provide up to twice the number of binding sites for HSF as does the 14-bp sequence by itself.

Fourth, a large number of putative heat shock elements have been identified on the basis of their similarity to the 14-bp sequence. Often matches at five or six of the eight conserved positions of the consensus are considered as candidates for active HSEs. In some cases, these putative elements do not bind HSF as measured in intact nuclei (Thomas and Elgin 1988), and thus they may not be functional. We suggest that stronger predictions as to which sequences are regulatory can be made by searching for HSEs based on the repeating 5-bp unit sequence. We discuss the details of the strategy for such a search in Figure 3 and below.

Figure 3A shows sequences within the genetically defined heat shock regulatory regions of the *Drosophila* heat shock genes that match the 5-bp repeating unit. Many of these sequences have also been shown to be binding sites for the transcription factor HSF (Topol et al. 1985; Wu et al. 1987). Recently, Berg and von Hippel (1988) have derived a selection theory and have applied their theory to the analysis of binding affinity of several *Escherichia coli* proteins to their known binding sites. They found a good correlation between the experimentally measured binding affinity of a regulatory protein to a particular sequence and a statistical measure expressing the difference of the sequence from the consensus. A statistical analysis of the genetically defined heat shock regulatory regions (Fig. 3A) suggests that sites containing arrays of the 5-bp sequence AGAAA may have the highest affinity for the transcription factor HSF (Fig. 3B). Although two 5-bp units provide detectable binding, three 5-bp units are required to generate a strong binding site (Perisic et al. 1989). We suggest using this trimeric, 15-bp sequence (in both orientations) in the search for heat shock regulatory elements. For searches of *Drosophila* DNA, the weights assigned to each base at each position can be taken from Figure 3B (numbers in parentheses). These weights are based on the frequency of usage of bases at these positions in the defined 5-bp repeats in Figure 3A. The lowest scores identify the best matches, and computer programs designed to select strings of sequence with scores below a certain prescribed value would obviously facilitate analysis of long stretches of DNA sequence. Comparable tables of weights could be constructed for other organisms as more data on defined heat shock elements in these organisms are accumulated.

It is likely that larger arrays of contiguous 5-bp units can compensate for substitutions at certain conserved positions of the 5-bp units (Fig. 3A). In some cases, sites that contain only two contiguous 5-bp units or weaker matches may function when present in conjunction with a strong site (Topol et al. 1985; Xiao and Lis 1988; see also Fig. 3). Thus, the criteria for identifying heat shock elements will continue to evolve as more functionally active sites are identified and additional in-vitro-constructed variants are tested for function.

F. Comparing the Number of 5-bp Units Required for Heat-induced Gene Expression and for Stable Binding of HSF

The number of 5-bp units needed for high levels of heat-induced expression of a reporter gene has been found to depend on the system used for analysis. Expression of the hsp70 gene in transfected *Drosophila* cells or in transgenic fly lines showed that the hsp70 gene deletion mutants with only one HSE (site I) are heat-inducible, but they are expressed at a level approximately 1% of the wild-type gene (Dudler and Travers 1984; Amin et al. 1985, 1987; Simon et al. 1985). A second copy of an HSE (site II of hsp70) was also required for full heat shock expression in *Drosophila*. In contrast, a reporter gene with one HSE shows high levels of heat-induced expression when introduced into COS cells or *Xenopus* oocytes (Pelham and Bienz 1982). This may be a consequence of the high template number per cell in these assays, approximately 10^5 and 10^7 per cell, respectively. At high cellular concentrations of DNA template, we suggest that HSF is likely to be limiting; therefore, the additional binding sites provided by upstream HSEs may provide little contribution to expression. The template genes at this high concentration could still encode the copious amounts of hsp70 RNA observed in these experiments, even if each gene is expressed at only a small fraction of its maximal efficiency. More recently, the requirement of additional HSEs for maximal expression of the human hsp70 gene has been shown to be much more stringent when the gene is present in human cells at a low rather than at a high copy number (Schiller et al. 1988).

The heat-induced transcriptional activities of hsp 70 variants in transgenic flies increase dramatically as the number of 5-bp units in the repeating arrays increases from three to eight (Dudler and Travers 1984; Xiao 1989). The affinity with which HSF binds to DNA also increases with the number of 5-bp units (Topol et al. 1985; H. Xiao et al., unpubl.). Thus, the 5-bp units can act synergistically on the binding of HSF and on the activation of transcription. It remains to be determined if there is a strict correlation between the cooperativity of binding and the level of transcriptional activation.

G. Speculations on the Catalytic Target of Action of HSF

HSF binds to the arrays of the 5-bp unit sequences that reside upstream of heat shock genes. However, the mechanism by which this binding stimulates transcription of these genes remains obscure. Any of several steps required for RNA synthesis could potentially be a rate-limiting step, and presumably, the active HSF/HSE complex acts catalytically to accelerate the rate of that step.

In general, it has been assumed that transcription factors accelerate the rate of transcription initiation by facilitating recruitment of RNA polymerase to the promoter. At odds with this is the observation that a molecule of RNA polymerase II is already present near the transcription start of the hsp70 gene in cells that have not been heat-shocked (Gilmour and Lis 1986). Moreover, this molecule of RNA polymerase has initiated transcription and has synthesized an approximately 25-nucleotide RNA (Rougvie and Lis 1988). In the absence of heat shock induction, the polymerase is blocked from further elongation. These observations indicate that in uninduced cells, RNA polymerase II not only has access to the hsp70 promoter, but it can initiate transcription and synthesize a short RNA molecule. Thus, the rate-limiting step under these conditions is the movement of polymerase out of this early elongation arrest and into the body of the hsp70 gene. We have proposed that HSF may act by accelerating the rate of RNA polymerase release from this arrested configuration (Rougvie and Lis 1988). HSF may also have a catalytic role in the recruitment and initiation of additional RNA polymerase molecules.

The design of models for the mechanism by which HSF accelerates transcription must take into account the requirement for multiple copies of binding sites for HSF. HSF can bind in vitro to as few as two 5-bp units, and it binds very tightly to three contiguous 5-bp units. Yet, it is clear that an array of three perfect 5-bp units has little activity in stimulating heat shock expression of the hsp70 gene in *Drosophila* (Dudler and Travers 1984; Amin et al. 1985; Simon et al. 1985). The regulatory regions of major heat shock genes of *Drosophila* possess several arrays of three or four 5-bp units or single longer arrays (Fig. 3). The strong synergistic activity of the arrays of 5-bp units on transcription suggests that multiple HSF molecules act in a concerted fashion to release arrested RNA polymerase II, or stimulate transcriptional initiation, or both.

III. SUMMARY

The HSEs are short regions of DNA sequence that act in *cis* to specify the heat-inducible activation of heat shock genes. We propose that an HSE is composed of a contiguous array of the alternately oriented, 5-bp modular unit, -GAA-. The number of repeating units in an HSE can vary. HSF can bind in vitro with dyad symmetry to as few as two inverted 5-bp units in either head-to-head (-GAA--TTC-) or tail-to-tail (-TTC--GAA-) arrangement. HSF binds to longer arrays of 5-bp units and has the flexibility to make contact with all of the 5-bp units in such arrays. This ability of HSF to interact with a variety of HSEs of different arrangement and length can be understood in terms of the unusual multimeric nature of this DNA-binding protein. HSF isolated from heat shock cells appears to be a trimer in which the number of subunits in contact with a particular HSE is equal to the number of correctly positioned 5-bp units.

ACKNOWLEDGMENTS

We thank the Cell Culture Center of Massachusetts Institute of Technology for providing us with heat-shocked *Drosophila* Kc cells that were used for purification of HSF. The study was supported by a U.S. Public Health Service grant GM-25232 from the National Institutes of Health.

REFERENCES

Amin, J., J. Ananthan, and R. Voellmy. 1988. Key features of heat shock regulatory elements. *Mol. Cell. Biol.* **8:** 3761–3769.

Amin, J., R. Mestril, P. Schiller, M. Dreano, and R. Voellmy. 1987. Organization of the *Drosophila melanogaster* hsp70 heat-shock regulation unit. *Mol. Cell. Biol.* **7:** 1055–1062.

Amin, J., R. Mestril, R. Lawson, H. Klapper, and R. Voellmy. 1985. The heat shock consensus sequence is not sufficient for hsp70 gene expression in *Drosophila melanogaster*. *Mol. Cell. Biol.* **5:** 197–203.

Ayme, A., R. Southgate, and A. Tissières. 1985. Nucleotide sequences responsible for the thermal inducibility of the *Drosophila* small heat-shock protein genes in monkey COS cells. *J. Mol. Biol.* **182:** 469–475.

Berg, O.G. and P.H. von Hippel. 1988. Selection of DNA binding sites by regulatory proteins. *Trends Biochem. Sci.* **13:** 207–211.

Bienz, M. and H.R.B. Pelham. 1987. Mechanisms of heat-shock gene activation in higher eukaryotes. *Adv. Genet.* **24:** 31–72.

Cohen, R.S. and M. Meselson. 1985. Separate regulatory elements for the heat-inducible and ovarian expression of the *Drosophila* hsp26 gene. *Cell* **43:** 737–746.

Dudler, R. and A.A. Travers. 1984. Upstream elements necessary for optimal function of the hsp70 promoter in transformed flies. *Cell* **38:** 391–398.

Gilmour, D.S. and J.T. Lis. 1986. RNA polymerase II interacts with the promoter region of the noninduced hsp70 gene in *Drosophila melanogaster* cells. *Mol. Cell. Biol.* **6:** 3984–3989.

Goldenberg, C.J., Y. Luo, M. Fenna, R. Baler, R. Weinmann, and R. Voellmy. 1988. Purified human factor activates heat shock promoter in a HeLa cell-free transcription system. *J. Biol. Chem.* **263:** 19734–19739.

Hoffman, E. and V. Corces. 1986. Sequences involved in temperature and ecdysterone-induced transcription are located in separate regions of a *Drosophila melanogaster* heat shock gene. *Mol. Cell. Biol.* **6:** 663–673.

Klemenz, R. and W.J. Gehring. 1986. Sequence requirement for expression of the *Drosophila melanogaster* heat shock protein hsp22 gene during heat shock and normal development. *Mol. Cell. Biol.* **6:** 2011–2019.

Mestril, R., D. Rungger, P. Schiller, and R. Voellmy. 1985. Identification of a sequence element in the promoter of the *Drosophila melanogaster* hsp23 gene that is required for its heat activation. *EMBO J.* **4:** 2971–2976.

Mestril, R., P. Schiller, J. Amin, H. Klapper, J. Ananthan, and R. Voellmy. 1986. Heat shock and ecdysterone activation of the *Drosophila melanogaster* hsp23 gene; a sequence element implied in developmental regulation. *EMBO J.* **5:** 1667–1673.

Mirault, M.E., D. Southgate, and E. Delwart. 1982. Regulation of heat-shock genes: A DNA sequence upstream of *Drosophila* hsp70 genes is essential for their induction in monkey cells. *EMBO J.* **1:** 1279–1285.

Nover, L. 1987. Expression of heat shock genes in homologous and heterologous systems. *Enzyme Microb. Technol.* **9:** 130–144.

Pauli, D., A. Spierer, and A. Tissières. 1986. Several hundred base pairs upstream of *Drosophila* hsp23 and 26 genes are required for their heat induction in transformed flies. *EMBO J.* **5:** 755–761.

Pelham, H.R.B. 1982. A regulatory upstream promoter element in the *Drosophila* hsp70 heat-shock gene. *Cell* **30:** 517–528.

Pelham, H.R.B. and M. Bienz. 1982. A synthetic heat-shock promoter element confers heat inducibility on the herpes simplex virus thymidine kinase gene. *EMBO J.* **1:** 1473–1477.

Perisic, O., H. Xiao, and J. Lis. 1989. Stable binding of *Drosophila* heat shock factor to head-to-tail and tail-to-tail repeats of a conserved 5 bp recognition unit. *Cell* **59:** (in press).

Riddihough, G. and H.R.B. Pelham. 1986. Activation of the *Drosophila* hsp27 promoter by heat shock and by ecdysone involves independent and remote regulatory sequences. *EMBO J.* **5:** 1653–1658.

Rougvie, A.E. and J.T. Lis. 1988. The RNA polymerase II molecule at the 5′ end of the uninduced hsp70 gene of *D. melanogaster* is transcriptionally engaged. *Cell* **54:** 795–804.

Schiller, P., J. Amin, J. Ananthan, M.E. Brown, W.A. Scott, and R. Voellmy. 1988. *Cis*-acting elements involved in the regulated expression of a human hsp70 gene. *J. Mol. Biol.* **203:** 97–105.

Shuey, D.J. and C.S. Parker. 1986. Binding of *Drosophila* heat-shock gene transcription factor to the hsp70 promoter: Evidence for symmetric and dynamic interactions. *J. Biol. Chem.* **261:** 7934–7940.

Simon, J.A. and J.T. Lis. 1987. A germline transformation analysis reveals flexibility in the organization of heat shock consensus elements. *Nucleic Acids Res.* **15:** 2971–2988.

Simon, J.A, C.A. Sutton, R.B. Lobell, R.L. Glaser, and J.T. Lis. 1985. Determinants of heat shock-induced chromosome puffing. *Cell* **40:** 805–817.

Slater, R.M. and E.A. Craig. 1987. Transcriptional regulation of an hsp70 heat shock gene in the yeast *Saccharomyces cerevisiae*. *Mol. Cell. Biol.* **7:** 1906–1916.

Sorger, P.K. and H.R.B. Pelham. 1987. Purification and characterization of heat-shock element binding protein from yeast. *EMBO J.* **6:** 3035–3041.

———. 1988. Yeast heat shock factor is an essential DNA-binding protein that exhibits temperature-dependent phosphorylation. *Cell* **54:** 855–864.

Sorger, P.K. and H.C.M. Nelson. 1989. Trimerization of a yeast transcriptional activator via a coiled-coil motif. *Cell* **59:** (in press).

Sorger, P.K., M.J. Lewis, and H.R.B. Pelham. 1987. Heat shock factor is regulated differently in yeast and HeLa cells. *Nature* **329:** 81–84.

Thomas, G.H. and S.C.R. Elgin. 1988. Protein/DNA architecture of the DNase I hypersensitive region of the *Drosophila* hsp26 promoter. *EMBO J.* **7:** 2191–2201.

Topol, J., D.M. Ruden, and C.S. Parker. 1985. Sequences required for in vitro transcriptional activation of a *Drosophila* hsp70 gene. *Cell* **42:** 527–537.

Wiederrecht, G., D. Seto, and C.S. Parker. 1988. Isolation of the gene encoding the *S. cerevisiae* heat shock transcription factor. *Cell* **54:** 841–853.

Wiederrecht, G., D.J. Shuey, W.A. Kibbe, and C.S. Parker. 1987. The *Saccharomyces* and *Drosophila* heat shock transcription factors are identical in size and DNA binding properties. *Cell* **48:** 507–515.

Wu, C. 1984. Activating protein factor binds in vitro to upstream control sequences in heat shock gene chromatin. *Nature* **311:** 81–84.

Wu, C., S. Wilson, B. Walker, I. Dawid, T. Paisley, V. Zimarino, and H. Ueda. 1987. Purification and properties of *Drosophila* heat shock activator protein. *Science* **238:** 1247–1253.

Xiao, H. 1989. "The nature of heat shock response elements." Ph. D. thesis, Cornell University, Ithaca, New York.

Xiao, H. and J.T. Lis. 1986. A consensus sequence polymer inhibits in vivo expression of heat shock genes. *Mol. Cell. Biol.* **6:** 3200–3206.

———. 1988. Germline transformation used to define key features of heat-shock response elements. *Science* **239:** 1139–1142.

———. 1989. Heat shock and developmental regulation of the *Drosophila melanogaster* hsp83 gene. *Mol. Cell. Biol.* **9:** 1746–1753.

Zimarino, V. and C. Wu. 1987. Induction of sequence-specific binding of *Drosophila* heat shock activator protein without protein synthesis. *Nature* **327:** 727–730.

18

Transcriptional Regulation of Heat Shock Genes

Carl Wu, Vincenzo Zimarino, Charles Tsai, Barbara Walker,[1] and Susan Wilson[1]
Laboratory of Biochemistry
National Cancer Institute
National Institutes of Health
Bethesda, Maryland 20892

I. INTRODUCTION

In all eukaryotes, transcriptional regulation of heat shock genes in response to elevated ambient temperatures is mediated by a common, positive, *cis*-acting sequence, the heat shock element (HSE), that is present in multiple copies upstream of the transcriptional start site. Originally defined as a 14-nucleotide sequence CT-GAA--TTC-AG (Pelham 1982), the HSE has recently been revised by further analysis to two or three modules consisting of alternating GAA or TTC blocks, arranged in alternating orientations and at 2-nucleotide intervals (Amin et al. 1988; Xiao and Lis 1988; see Lis et al., this volume). In this chapter, we focus on our current understanding of the transcriptional activator protein that stimulates the synthesis of heat shock mRNA through interaction with the HSE. Transcription factors that regulate some heat shock genes through *cis* elements that are unrelated to the HSE are not covered in this discussion.

Initial evidence for a factor that interacts with the HSE came from studies of protein/DNA interactions in *Drosophila* cell nuclei, in which a

[1]Present address: Genex Corporation, Gaithersberg, Maryland 20877.

Stress Proteins in Biology and Medicine.

factor, heat shock activator protein, was found to bind specifically to the HSE only upon heat shock stimulation (Wu 1984a,b); parallel studies using DNA-binding and in vitro transcription assays also identified a heat shock transcription factor that bound specifically to the HSE (Parker and Topol 1984). At the 1987 Cold Spring Harbor Meeting entitled "The Role of Heat Shock and Stress Response in Biology and Human Disease," HSE-binding factors were given a general designation: heat shock factor (HSF). In the following discussion, we describe the properties of HSF isolated from a variety of eukaryotic species.

II. INDUCIBLE HSE-BINDING ACTIVITY OF HSF IN HIGHER EUKARYOTES

Figure 1A shows a survey of heat-shock-inducible HSE-binding activities that can be extracted from tissue culture cells derived from a range of higher eukaryotes. The HSE-binding activities were measured by the gel mobility shift technique, using saturating amounts of a 38-bp synthetic ^{32}P-labeled heat shock consensus HSE sequence that was adapted from a *Drosophila* hsp70 gene HSE (Wu et al. 1987). This consensus sequence completely fulfills the requirements of both original and expanded definitions of the HSE and was used as a standard to screen for specific HSE-binding activities in the indicated species. In all of the cell lines, a low level of HSE-binding activity at the normal growth temperature is observed by the appearance of a specific gel mobility shift complex (arrowhead) that is competed by an excess of unlabeled HSE. The low HSE-binding activity present in nonshocked cells probably reflects a physiologically relevant basal activity, although this activity can vary depending on the handling of the cells during the preparation of cell extracts. At elevated temperatures, however, a clear increase in HSE-binding activity is observed for all the cell lines tested. The extent of the increase is about 10–20-fold, depending on the cell type, and the rate of increase is rapid, reaching maximal levels after 5 minutes of heat shock (Zimarino and Wu 1987; C. Tsai and V. Zimarino, unpubl.). Hence, in higher eukaryotes, HSF is induced to bind to the HSE in response to heat shock. To a first approximation, the extent of HSE-binding activity that is induced directly correlates with the severity of heat shock, suggesting that the temperature-dependent modulation of HSE-binding activity is a critical regulatory switch in the activation of heat shock gene transcription. Inducible HSE-binding activities have also been reported by other workers for human and mouse cells (Kingston et al. 1987; Sorger et al. 1987; Goldenberg et al. 1988; Mossier et al. 1988; Mezger et al. 1989).

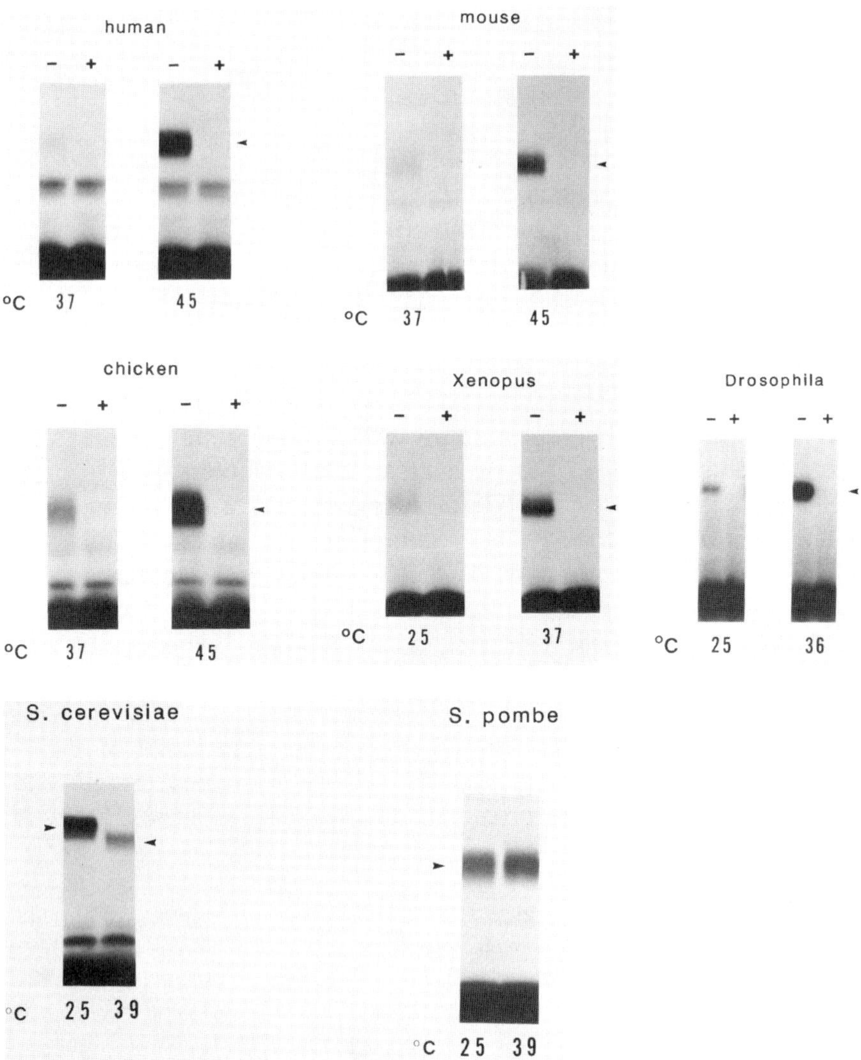

Figure 1 Inducible and constitutive DNA-binding activity of HSF from different species. Whole-cell extracts (Zimarino and Wu 1987) were prepared from human (HeLa), mouse (L), chicken (MSB), *Xenopus* (XTC), and *Drosophila* (SL2) cells, grown or shocked for 20 min at the indicated temperatures. Extracts were assayed by the gel mobility shift technique, using a ^{32}P-labeled consensus HSE (Wu et al. 1987). Yeast cell extracts were prepared according to the method of Zimarino et al. (1990) and assayed similarly, except that the labeled DNA was a 44-bp double-stranded oligonucleotide sequence (HSE 82; Wu et al. 1987) derived from the hsp82 gene. (+) and (–) refer to the presence or absence of a 40-fold molar excess of unlabeled, specific binding sequence.

III. CONSTITUTIVE HSE-BINDING ACTIVITY IN FUNGI

Unlike the inducible HSE-binding activity in higher eukaryotes, a significant level of HSE-binding activity is already present at normal temperature (25°C) in the fungi *Saccharomyces cerevisiae* and *Schizosaccharomyces pombe* (Fig. 1B). Heat shock at 39°C does not increase the level of HSE-binding activity. In *S. cerevisiae*, but not in *S. pombe*, the activity actually seems to decrease upon heat shock, and in addition, the heat-shocked *S. cerevisiae* HSF/HSE complex migrates with a faster electrophoretic mobility on the native agarose gel, suggesting that the factor may be posttranslationally modified in response to heat shock. Sorger and Pelham (1987) have observed a similar constitutive level of HSE-binding activity in *S. cerevisiae*, although they showed, using native polyacrylamide gels, that heat-shock-activated HSF had a slower, instead of faster, electrophoretic mobility; this difference may be due to the different compositions of the gel matrix. These workers also reported that phosphatase treatment of heat-shock-activated *S. cerevisiae* HSF results in a progressive return to the normal gel mobility, suggesting that multiple phosphorylation may be the heat-shock-induced posttranslational modification of HSF in *S. cerevisiae*.

The specificity of HSF binding to heat shock promoters can be analyzed further by DNase I footprinting analysis. Figure 2A shows the DNase I footprints produced by *S. cerevisiae* HSF on the *Drosophila* hsp82 gene promoter, which has seven tandem repetitions of the 5-nucleotide modular unit of the heat shock regulatory region. Among natural *Drosophila* heat shock promoters, this sequence seems to bind with

Figure 2 DNase I footprinting analysis of (*A*) *S. cerevisiae* HSF and (*B*) human (HeLa) HSF on the *Drosophila* hsp82 promoter. Extracts of nonshocked (25°C) and heat-shocked (40°C) yeast, and of heat-shocked HeLa cells, were incubated with 5′ ^{32}P end-labeled hsp82 promoter DNA (Wu et al. 1987) and treated with DNase I before separation of the bound and the free DNA by the gel mobility shift technique. The bound (B), and free (F) DNAs were extracted, electrophoresed on a sequencing gel, and autoradiographed. Solid lines indicate the span of the footprint, dashed lines indicate uncertainties in the footprint boundaries due to poor cleavage of the DNA alone at those regions. Arrowheads denote the sites of hypersensitive cleavage on the promoter bound to the yeast factor. (*C*) Summary of sequences on the hsp82 promoter that are protected from DNase I cleavage by bound human, yeast, and *Drosophila* HSF. Filled circles indicate matches and open circles indicate mismatches between the hsp82 heat shock element and the GAA or TTC blocks within the 5-nucleotide heat shock regulatory module (Amin et al. 1987).

the highest relative affinity to HSF isolated from various eukaryotes. The footprints show that the binding specificities of HSF isolated from non-shocked and heat-shocked *S. cerevisiae* are identical. The DNase I footprint of heat-shock-activated human HSF on the same promoter is shown in Figure 2B. Comparison of the footprints produced by human, *S. cerevisiae,* and *Drosophila* HSF reveals generally similar sequence specificity for the hsp82 promoter, although the *S. cerevisiae* HSF footprint is distinguished by multiple DNase I hypersensitive cleavages on both sides of the protected region that are not observed in the other footprints. In addition, subtle differences in the boundary positions of the

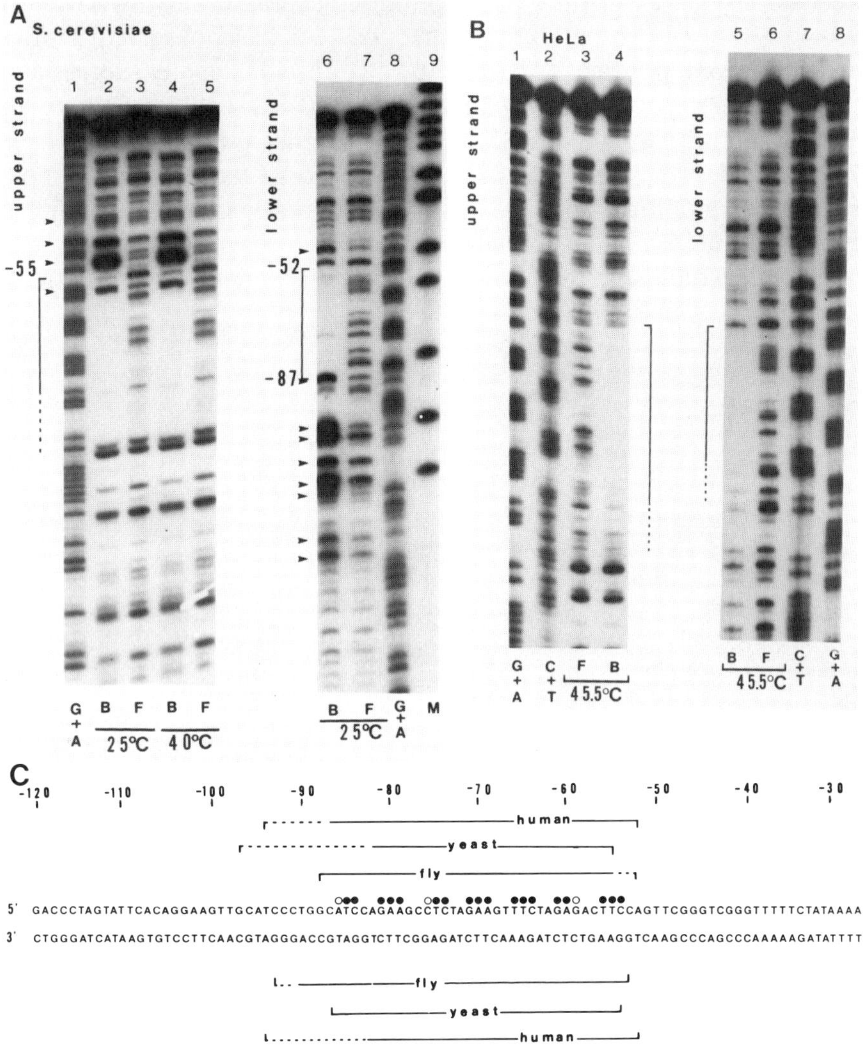

Figure 2 (See facing page for legend.)

DNase I protected regions can be observed in the footprints generated by the different HSFs (summarized in Fig. 2C). The differences could reflect subtle differences in the DNA-binding domain among HSFs from these species, despite overall conservation of the ability to recognize the same HSE sequence.

IV. DIFFERENT MOLECULAR SIZES OF HSF FROM DIFFERENT SPECIES

Further differences among HSFs from different species are revealed by molecular size analysis on SDS-PAGE (Fig. 3). Photoaffinity-labeled HSF from human (HeLa) cells, *Drosophila,* and *S. cerevisiae* separate on a denaturing gel as distinct polypeptides of approximately 90 kD, 110 kD, and 130 kD, respectively. Other workers have observed similar molecular sizes: purified *S. cerevisiae* HSF, 150 kD doublet (Sorger and Pelham 1987); purified *Drosophila* HSF, 110 kD doublet (Wu et al. 1987); and purified human HSF, 83 kD (Goldenberg et al. 1988). These size differences suggest that some divergence in the size and/or post translational modification of HSF may have occurred in evolution. In this regard, it is interesting that polyclonal antibodies raised against *Drosophila* HSF react only with the *Drosophila* factor and not with HSF from

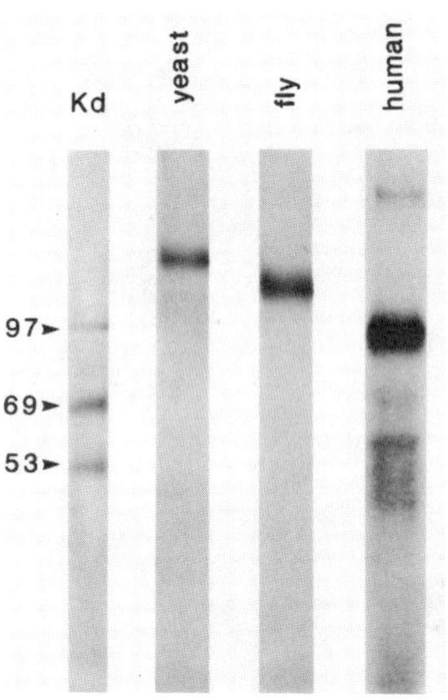

Figure 3 SDS-PAGE analysis of photoaffinity-labeled HSF. The indicated cell extracts were incubated with ^{32}P-labeled, BrdU-substituted HSE 82. Bound DNA was separated from free DNA by the gel mobility shift technique and the bound complex was UV cross-linked in situ and analyzed by SDS-PAGE (Wu et al. 1987). The indicated marker proteins were electrophoresed on the same gel. (Reprinted, with permission, from Zimarino et al. 1990.)

other species (S. Wilson and V. Zimarino, unpubl.). Whether the differences reflect a fundamental divergence of HSF protein structure or merely reflect accumulated evolutionary changes compatible with a common underlying structure and function awaits a comparative analysis of the HSF gene and protein structure in these species, when the cloning and sequence analysis of the HSF genes have been completed.

V. PURIFICATION OF *DROSOPHILA* HSF

An important prerequisite for the analysis of the mechanism(s) underlying activation of HSF is the ability to study the properties of the factor purified to homogeneity. Figure 4 shows the proteins analyzed by SDS-PAGE at each chromatographic stage during purification of HSF from nuclear extracts of *Drosophila* tissue culture cells (Wu et al. 1987, 1988). The purified factor migrates as a 110-kD doublet band on SDS-PAGE and binds to the consensus HSE with high affinity (K_d = 10^{-11} M to 10^{-12} M), and specificity (K_d for *E. coli* DNA = 10^{-6} M). Figure 5 shows that purified *Drosophila* HSF can stimulate transcription of a *Drosophila* hsp70 gene when microinjected along into *Xenopus* oocytes. A 70-kD *Drosophila* HSF, probably a proteolytic product of the 110-kD polypeptide, has been purified by Wiederrecht et al. (1987). *S. cerevisiae* HSF

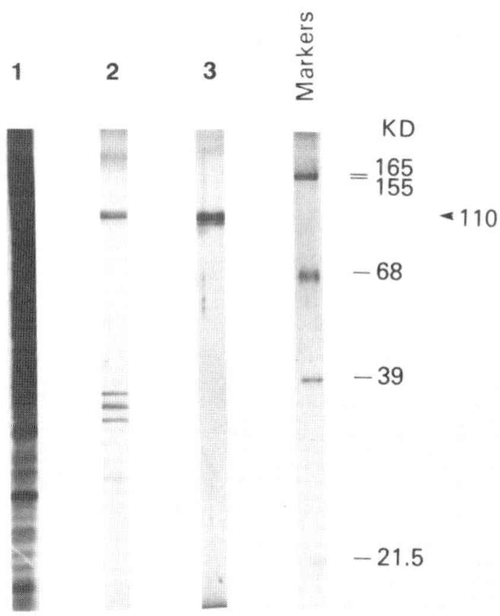

Figure 4 Purification of *Drosophila* HSF. Silver stain of proteins separated by SDS-PAGE. Active (HSE-binding) fractions are displayed after (1) heparin-Sepharose; (2) sequence-specific DNA affinity chromatography, second pass; and (3) Mono S (Pharmacia) ion exchange chromatography. For experimental details, see Wu et al. (1987). (Reprinted, with permission, from Wu et al. 1988.)

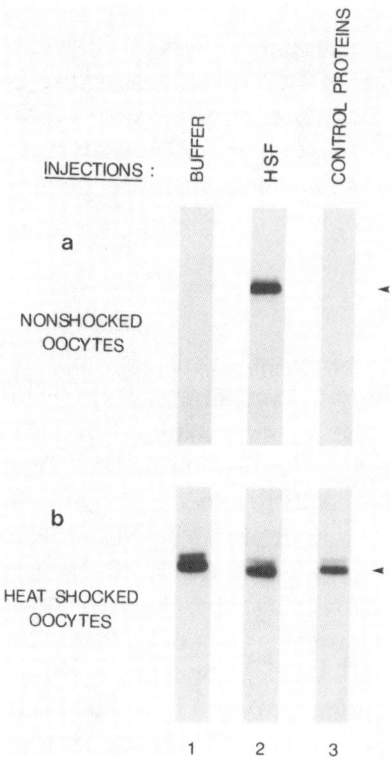

Figure 5 Transcriptional activation in *Xenopus* oocytes. A *Drosophila* hsp70 gene was injected into oocytes, followed by a second injection of buffer, HSF, or control proteins lacking HSF. The oocytes were kept at 18°C, or heat shocked, prior to isolation of RNA. The gel lanes show primer extension analysis of *Drosophila* hsp70 RNA synthesized in the oocytes treated as indicated (Wu et al. 1987).

(150-kD polypeptide: Sorger and Pelham 1987; 70-kD polypeptide [a proteolytic product]: Wiederrecht et al. 1987), and human HSF (Goldenberg et al. 1988) have also been successfully purified.

These purifications of HSF were aided by the use of sequence-specific DNA affinity chromatography methods, which exploited the high affinity of HSF for the HSE, coupled to a chromatographic matrix, to achieve several-hundredfold enrichment in one step. Nonetheless, the purifications have been a daunting biochemical enterprise, due to the very low intracellular abundance of HSF (about 4000 molecules per *Drosophila* cell). Only several micrograms of protein purified to 95% homogeneity could be obtained from 30 liters of suspension cell culture (Wu et al. 1987) Limitations in increasing the scale of cell culture have constrained further biochemical studies and an expeditious cloning of the

Drosophila HSF gene. Similar constraints apply to human HSF, but not to *S. cerevisiae* HSF, which has recently been cloned and shown to be encoded by a single-copy, essential gene (Sorger and Pelham 1988; Wiederrecht et al. 1988). The cloning of *S. cerevisiae* HSF overcomes a major obstacle to progress in elucidating the heat shock activation pathway in yeast.

VI. PROTEIN SYNTHESIS IS NOT REQUIRED FOR INDUCTION OF HSF BINDING AT HIGH TEMPERATURE

Early experiments on heat shock activation of *Drosophila* polytene chromosomes provided the first evidence that protein synthesis was not required for the induction of heat shock puffs (Ritossa and Pulitzer 1963; Ashburner 1970). We used the inducibility of HSE-binding activity in higher eukaryotes as a measure of HSF activation when protein synthesis was inhibited by cycloheximide treatment. Figure 6 shows that treatment of *Drosophila* tissue culture cells with cycloheximide (98% inhibition of protein synthesis) has no effect on the induction of HSF binding when the cells were stimulated at a high heat shock temperature of 37.5°C; there was also no effect on subsequent reversal and reinduction of HSF binding through five sequential cycles of heat shock and recovery (Zimarino and Wu 1987). Similar results were obtained in an experiment involving two cycles of heat shock in the presence of cycloheximide, using human (HeLa) cells heat-shocked at a high temperature of 45°C, or *Xenopus* tissue culture cells at a high temperature of 37°C (Zimarino et al. 1990). The effect is also observed when protein synthesis was inhibited in *Drosophila* cells by other inhibitors, puromycin and emetine (Zimarino and Wu 1987; J.T. Westwood and C. Wu, unpubl.). These findings concur with the results of similar experiments performed on HeLa cells (Kingston et al. 1987; Mossier et al. 1988) and suggest that a posttranslational mechanism is likely to be the general pathway by which higher eukaryotes activate HSF when stimulated by a high-temperature heat shock.

Unlike the lack of requirement for protein synthesis in the induction of HSF binding at high heat shock temperatures, protein synthesis is required for activation of *Drosophila,* chicken, and human HSF at intermediate heat shock temperatures (30–34°C for *Drosophila* cells, 40–42°C or 40–43°C for HeLa and chicken cells, respectively; Zimarino et al. 1990). The requirement for new protein synthesis in inducing HSF binding at intermediate heat shock temperatures is not due to bulk degradation of HSF, since heat shock at intermediate temperature in the presence of cycloheximide, followed by a second shock at high temperature, in-

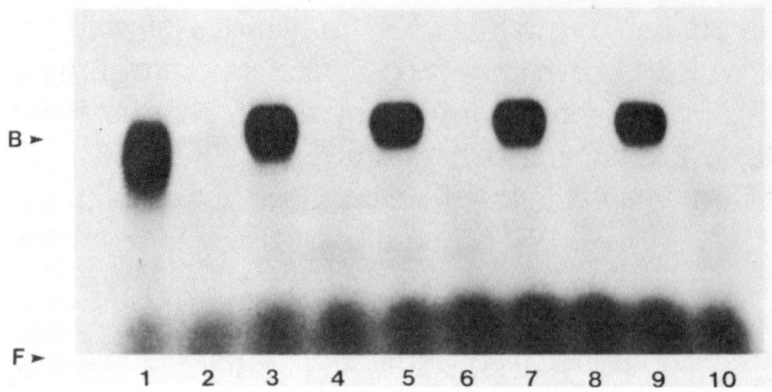

Figure 6 Sequential induction, reversal, and reinduction of HSE-binding activity in absence of protein synthesis. *Drosophila* SL2 cells were treated with 118 μM cycloheximide, heat shocked at 37.5°C for 10 min (lane *1*), and allowed to recover at 25°C for 30 min (lane *2*). Whole-cell extracts were assayed by the gel mobility shift technique (Zimarino and Wu 1987). The heat shock and relaxation was repeated successively four times, and HSE-binding activity was assayed at the end of each heat shock and relaxation (lanes *2–10*). (F) Free ^{32}P-labeled HSE; (B) bound HSE. (Reprinted, with permission, from Zimarino and Wu 1987.)

duces the HSE-binding activity to the same high level as the activity measured without cycloheximide treatment. The inability to induce HSF binding at intermediate temperatures in the presence of cycloheximide might be due to a requirement for newly synthesized HSF itself or to a component of the HSF activation pathway that is labile at those temperatures.

VII. DISCUSSION

A survey of HSE-binding activities in eukaryotic cell extracts has revealed a specific HSE-binding factor in all species examined. In this and other laboratories, HSE-binding activities have been purified to homogeneity or near-homogeneity from *S. cerevisiae, Drosophila,* and human. In all three cases, the factor HSF has been shown to activate transcription of heat shock genes in vivo or in vitro. It is likely that the HSE-binding activities observed in other species also correspond to HSF. Although all HSFs are able to bind to the same DNA sequence, there appears to be considerable variation in the protein structure in different species, and this variation may presage new complexities that may emerge in future studies of the heat shock activation pathway.

In higher eukaryotes, heat shock induces HSF binding to the HSE. Moreover, the lack of a requirement for protein synthesis in the induction, reversal, and reinduction of HSF binding at high heat shock temperatures suggests that a posttranslational mechanism, general to higher eukaryotes, may regulate HSF binding under these conditions. The posttranslational mechanism may be a direct temperature-dependent induction of HSF binding, as was recently suggested from the ability to induce HSF binding in vitro using HeLa cytosol (Larson et al. 1988) and confirmed using *Drosophila* cytosol (V. Zimarino, unpubl.). However, in contrast with the induction in vivo, the induction in vitro is irreversible, and there is as yet no demonstration of the reversibility of HSF binding in vitro. An alternative regulatory mechanism may involve additional factors that act on HSF stoichiometrically or catalytically to regulate its binding activity. To distinguish between these mechanisms, which are not mutually exclusive, further analysis of HSF purified from non-shocked cells will be necessary.

The requirement for new protein synthesis in the induction of HSF binding at intermediate heat shock temperatures in vertebrate and invertebrate cells is intriguing. The full induction of HSF binding in *Drosophila* cells that were first subjected to mild heat shock in the presence of cycloheximide and then to a severe heat shock demonstrates clearly that the bulk of intracellular HSF is not degraded under mild heat shock conditions. The inability to induce HSF binding under such conditions might instead be due to a component of the HSF activation pathway that is labile at intermediate heat shock temperatures, but not at high heat shock temperatures. An alternative explanation might be that the newly synthesized and preexisting pools of HSF have distinct properties: the preexisting pool of HSF being stable and responsive only to a severe heat shock, and the newly synthesized HSF pool having a lower temperature threshold at which HSF binding is induced.

The inducibility of HSF binding in higher eukaryotes is in concordance with studies of in situ protein/DNA interactions on *Drosophila* heat shock promoters, which show that the HSE is freely accessible to nuclease cleavage in nuclear chromatin under normal conditions but is protected from cleavage after heat shock (Wu 1984; Thomas and Elgin 1988). In contrast, HSF binding in fungal cell extracts is constitutive. Indeed, in unstressed *S. cerevisiae,* the HSE is apparently bound to HSF in vivo (Szent-Gyorgyi et al. 1987; Gross and Garrard 1988; Jakobsen and Pelham 1988). Hence, the regulation of HSF in yeast must occur subsequent to the binding to the HSE. Our results concur with those of Sorger et al. (1987), who reported further that phosphatase treatment of heat-shock-activated *S. cerevisiae* HSF results in a progressive return to the normal gel mobility, suggesting that multiple phosphorylation may be the heat-shock-induced modification of HSF in *S. cerevisiae* (Sorger and Pelham 1988). These workers have proposed that the addition of negatively charged phosphates to the HSE-bound factor enhances its transcriptional activity during heat shock. Whether a similar heat-induced phosphorylation modifies *S. pombe* HSF is presently unknown. There is apparently no significant change in the mobility of the *S. pombe* HSF/HSE complex, although modifications not revealed by gel mobility changes could possibly have occurred. In Figure 7, the apparently different paths by which HSF is activated in fungi and higher eukaryotes are summarized.

With regard to the properties of HSF in *S. cerevisiae,* it is conceivable that HSF in higher eukaryotes is regulated at two levels. At the first level, specific binding to the HSE could be mediated by a direct temperature sensitivity of HSF, and/or by additional regulatory factors. At the second level, the transcriptional ability of HSF could be activated in a manner similar to the proposed phosphorylation of *S. cerevisiae* HSF. Indeed, Larson et al. (1988) have suggested such a two-step model for the activation of human HSF. Further analysis of the observed complexities in the regulation of HSF activity should provide new insights into this key regulatory protein.

ACKNOWLEDGMENTS

We thank Rick Morimoto and Tom Sargent for providing chicken MSB cells and *Xenopus* XTC cells, respectively. We thank Amar Klar, Dennis

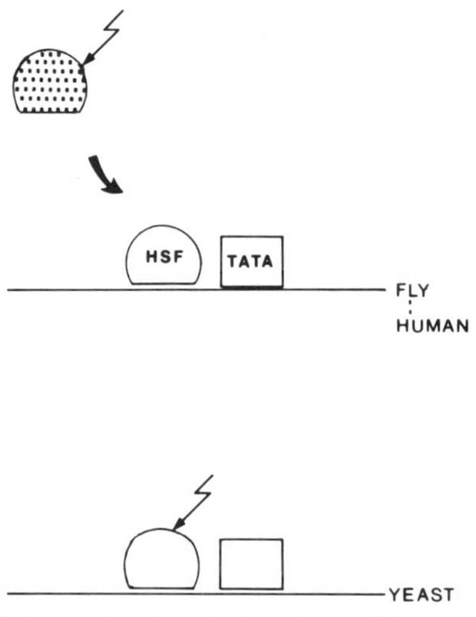

Figure 7 Model showing different pathways of HSF activation. The TATA box sequence at –30 is bound to a factor (square), presumably TFIID, irrespective of heat shock (Wu 1984).

Thiele, Kelly Tatchell, and Enrico Cabib for yeast stains. C.T. was supported by a Howard Hughes Medical Institute/National Institutes of Health Research Scholarship.

REFERENCES

Amin, J., J. Ananthan, and R. Voellmy. 1988. Key features of the heat shock regulatory elements. *Mol. Cell. Biol.* **8:** 3761–3769.

Ashburner, M. 1970. Patterns of puffing activity in the salivary gland chromosomes of *Drosophila. Chromosoma* **31:** 356–376.

Goldenberg, C.J., Y. Luo, M. Fenna, R. Baler, R. Weinmann, and R. Voellmy. 1988. Purified human factor activates heat shock promoter in a HeLa cell-free transcription system. *J. Biol. Chem.* **263:** 19734–19739.

Gross, D.S. and W.T. Garrard. 1988. Nuclease hypersensitive sites in chromatin. *Annu. Rev. Biochem.* **57:** 159–197.

Jakobsen, B.K. and H.R.B. Pelham. 1988. Constitutive binding of yeast heat shock factor to DNA in vivo. *Mol. Cell. Biol.* **8:** 5040–5042.

Kingston, R.E., T.J. Schuetz, and Z. Larin. 1987. Heat inducible human factor that binds to a human hsp70 promoter. *Mol. Cell. Biol.* **7:** 1530–1534.

Larson, J.S., T.J. Schuetz, and R.E. Kingston. 1988. Activation in vitro of sequence specific DNA binding by a human regulatory factor. *Nature* **335:** 372–375.

Mezger, V., O. Bensaude, and M. Morange. 1989. Unusual levels of HSE-binding activity in embryonal carcinoma cells. *Mol. Cell Biol.* (in press).

Mossier, O.D., N.G. Theodorakis, and R. Morimoto. 1988. Coordinate changes in heat shock element binding activity and hsp70 gene transcription rates in human cells. *Mol. Cell. Biol.* **8:** 4736–4744.

Parker, C.S. and J. Topol. 1984. A *Drosophila* RNA polymerase II transcription factor contains a promoter-region-specific DNA binding activity. *Cell* **36:** 357–369.

Pelham, H.R.B. 1982. A regulatory upstream promoter element in the *Drosophila* hsp70 heat shock gene. *Cell* **30:** 517–528.

Ritossa, F. and J.F. Pulitzer. 1963. Aspects of structure of polytene chromosome puffs of *Drosophila busckii* derived from experiments with antibiotics. *J. Cell Biol.* **19:** 60A.

Sorger, P.K. and H.R.B. Pelham. 1987. Purification and characterization of a heat-shock element binding protein from yeast. *EMBO J.* **6:** 3035–3041.

———. 1988. Yeast heat shock factor is an essential DNA-binding protein that exhibits temperature dependent phosphorylation. *Cell* **54:** 855–864.

Sorger, P.K., M.J. Lewis, and H.R.B. Pelham. 1987. Heat shock factor is regulated differently in yeast and HeLa cells. *Nature* **329:** 81–84.

Szent-Gyorgyi, C., D.B. Finkelstein, and W.T. Garrard. 1987. Sharp boundaries demarcate the chromatin structure of a yeast heat shock gene. *J. Mol. Biol.* **193:** 71–80.

Thomas, G.H. and S.C.R. Elgin. 1988. Protein/DNA architecture of the DNase I hypersensitive region of the *Drosophila* hsp26 promoter. *EMBO J.* **7:** 2191–2201.

Wiederrecht, G., D. Seto, and C.S. Parker. 1988. Isolation of the gene encoding the *S. cerevisiae* heat shock transcription factor. *Cell* **54:** 841–853.

Wiederrecht, G., D.J. Shuey, W.A. Kibbe, and C.S. Parker. 1987. The *Saccharomyces* and *Drosophila* heat shock transcription factors are identical in size and DNA binding properties. *Cell* **48:** 507–515.

Wu, C. 1984a. Two protein-binding sites in chromatin implicated in the activation of heat shock genes. *Nature* **309:** 229–234.

———. 1984b. Activating protein factor binds in vitro to upstream control sequences in heat shock gene chromatin. *Nature* **311:** 81–84.

Wu, C., C. Tsai, and S. Wilson. 1988. Affinity chromatography of sequence-specific DNA-binding proteins. *Genet. Eng.* **10:** 67–74.

Wu, C., S. Wilson, B. Walker, I. Dawid, T. Paisley, V. Zimarino, and H. Ueda. 1987. Purification and properties of *Drosophila* heat shock activator protein. *Science* **238:** 1247–1253.

Xiao, H. and J.T. Lis. 1988. Germline transformation used to define key features of heat-shock response elements. *Science* **239:** 1139–1142.

Zimarino, V. and C. Wu. 1987. Induction of sequence specific binding of *Drosophila* heat shock activator protein without protein synthesis. *Nature* **327:** 727–730.

Zimarino, V., C. Tsai, and C. Wu. 1990. Multiple mechanisms of activation of heat shock factor. *Mol. Cell. Biol.* (in press).

Subject Index